NMR SPECTRA OF SIMPLE HETEROCYCLES

NMR SPECTRA OF SIMPLE HETEROCYCLES

T. J. BATTERHAM

Department of Medical Chemistry
John Curtin School of Medical Research
The Australian National University, Canberra.

ROBERT E. KRIEGER PUBLISHING COMPANY
MALABAR, FLORIDA
1982

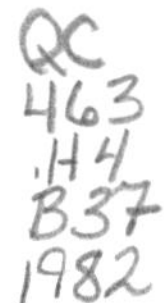

Original Edition 1973
Reprint Edition 1982

Printed and Published by
ROBERT E. KRIEGER PUBLISHING COMPANY, INC.
KRIEGER DRIVE
MALABAR, FLORIDA 32950

Printed in the United States of America

Library of Congress Cataloging in Publication Data

Batterham, T J 1933-1972.
NMR spectra of simple heterocycles.

"A Wiley-Interscience publication."
Reprint of the ed. published by Wiley, New York, in series: General heterocyclic chemistry series.
Includes bibliographical references and index.
1. Heterocyclic compounds—Spectra. 2. Nuclear magnetic resonance spectroscopy. I. Title. [DNLM: 1. Spectrum analysis. 2. Nuclear magnetic resonance. 3. Heterocyclic compounds. QC463.H4 B335n 1973a]
[QC463.H4B37 1982] 547'.59'046 80-11724
ISBN 0-89874-140-8

THOMAS JAMES BATTERHAM was born in Cessnock, N.S.W., on October 29, 1933, and died in Canberra on November 5, 1972. His last activity was the correction of galley proofs for this book, work that he could never have completed in his weakened state without the devoted help of his wife Elaine.

Apart from his family, Tom had two main interests in life. The first was nuclear magnetic resonance on which he published nearly 50 fine research papers and to which this book will serve as a fitting memorial. The second was helping those less fortunate than himself: much of his leisure time was given to Outreach, a voluntary organization for the reintegration of wayward young people into society. He is sadly missed by his many friends and colleagues at the John Curtin School of Medical Research and elsewhere.

It has been a willingly accepted task and an honor for me to complete the page proofs and index on behalf of the author.

DES BROWN

Canberra,
December 1972

INTRODUCTION TO THE SERIES

General Heterocyclic Chemistry

The series, "The Chemistry of Heterocyclic Compounds," published since 1950 by Wiley-Interscience, is organized according to compound classes. Each volume deals with syntheses, reactions, properties, structure, physical chemistry, etc., of compounds belonging to a specific class, such as pyridines, thiophenes, pyrimidines, three-membered ring systems. This series has become the basic reference collection for information on heterocyclic compounds.

Many aspects of heterocyclic chemistry have been established as disciplines of *general* significance and application. Furthermore, many reactions, transformations, and uses of heterocyclic compounds have specific significance. We plan, therefore, to publish monographs that will treat such topics as nuclear magnetic resonance of heterocyclic compounds, mass spectra of heterocyclic compounds, photochemistry of heterocyclic compounds, X-ray structure determination of heterocyclic compounds, UV and IR spectroscopy of heterocyclic compounds, and the utility of heterocyclic compounds in organic synthesis. These treatises should be of interest to *all* organic chemists as well as to those whose particular concern is heterocyclic chemistry. The new series, organized as described above, will survey under each title *the whole field of heterocyclic chemistry* and is entitled "General Heterocyclic Chemistry." The editors express their profound gratitude to Dr. D. J. Brown of Canberra for his invaluable help in establishing the new series.

Department of Chemistry
Princeton University
Princeton, New Jersey

Edward C. Taylor

Research Laboratories
Eastman Kodak Company
Rochester, New York

Arnold Weissberger

PREFACE

Simple heterocyclic systems form the basis of a large part of modern organic chemistry and a critical compilation of their NMR parameters was needed to simplify the mass of data in the literature. In 1962, an eminent chemist commented to me that NMR would have little impact on heterocyclic chemistry because of the few protons normally present in these systems. This book bears out the error in his statement.

Definition of "heterocycle" presented problems. The classic "aromatic" heterocycles had to be included but certain reduced compounds and some of the newer ring systems were often difficult to classify. Natural products, including sugars, were omitted unless they were simple derivatives of an important heterocyclic system. It is hoped that the inclusion of compounds with uncommon heteroatoms does not offend the purist or that the exclusion of more exotic borderline systems will be acceptable to the *avant guarde* chemist. Considerations of space led to the decision to include only compounds with one or two rings and to limit the information given on "mixed" heterocyclic systems. These latter systems will soon warrant a book of this size to themselves.

When this book is published it will be out of date, but an attempt has been made to cover the literature from the first decade of heterocyclic NMR, 1960–1970 inclusive. Over this period of time the fundamental details of the proton spectra of a very large number of heterocyclic systems were firmly established and, although there are sure to be further publications, in most cases these will contain only slight modifications or extensions to the data or principles established earlier. Thus, most of the information in this book will not become out of date for many years to come. The next decade of NMR will emphasize the spectra of nuclei other than hydrogen and undoubtedly this will have a large impact on the heterocyclic field.

I am sure that the reader will find certain references missing from the book. Sometimes these omissions were made on purpose because of inconsistencies in the material contained in these papers, but many references will have been overlooked because of the size of the body of literature covered. It is hoped that the bringing together, on one scale, of data from very diverse and often well-hidden sources will offset any inconvenience caused by omissions.

I am grateful to the many people who have helped me during the writing of this book, particularly to Dr. D. J. Brown who has provided a continual source of inspiration and to Dr. W. L. F. Armarego whose critical comments on the manuscript were most valuable. Without the assistance of my wife, who typed and collated the manuscript, and my daughter, who checked almost every reference, I am sure that this book would never have been completed.

T. J. BATTERHAM

Canberra, Australia
May 1972

CONTENTS

NMR SPECTRA OF SIMPLE HETEROCYCLES

1 INTRODUCTION

A systematic survey of the NMR spectra of simple heterocyclic compounds involves perusal of a large proportion of the chemical and physical investigations of these systems which have been reported over the past ten years. The size and complexity of this undertaking greatly restrict the amount of information which can be given about each individual project and tend to make some areas of this book into literature summaries. Mainly for this reason it is necessary to introduce some fundamental difficulties associated with the recording of NMR data and to give a brief background of a number of the commonly encountered physical and mathematical techniques.

It must be emphasized at this point that the NMR technique has had as big an impact on the heterocyclic field as on any other branch of organic chemistry. Apart from the obvious uses in structural determinations, NMR has been particularly successful in the study of tautomeric systems, and was rapidly utilized to settle many long-standing arguments in this very important area of research. Theoretical chemists have been able to obtain information about the ground state of many systems and to compare the spectral parameters with other molecular properties such as electron distributions, dipole moments, or infrared spectra. Also, interpretation of chemical shift variations with changes of solvent has led to a much better understanding of the intermolecular forces active in solutions. Time-dependent phenomena in the NMR experiment make it possible to study the kinetics of many interconverting binary systems to give data which could not be obtained by other methods.

Before this book was started a decision was made by the author that chemical shifts would be reported in τ units. After a considerable number of tables had been produced it was learned that various international bodies had recommended the use of δ units for the recording of all proton chemical shifts. Conversion of the accumulated τ values into δ values would have been very time-consuming so the τ convention was retained.

The rest of this book will show, in the main, the positive side of the mass of literature which has been published on the NMR of heterocycles. On the negative side, the ease with which spectra can be obtained has tended to produce a publication explosion, and whenever this occurs the average quality of the individual contribution must drop. In the following paragraphs mention is made of a number of practices which I find highly undesirable. The comments are not intended to show that all or even most work in a given field is suspect but rather that certain areas of the NMR of these systems must be looked at in a critical manner. In many cases the objections may be met by the realization that the papers being criticized represent the present state of the art and hence are acceptable. However, just as some investigations can be classed as outstanding, others must be condemned, and it is the author's aim that in the following chapters both of these types of work will be brought to the attention of the reader.

One of the big problems to be overcome in summarizing the available data was to get some idea of the reliability of individual results. It was necessary to determine the extent of variations in spectral parameters obtained by different workers under the *worst* possible conditions where spectra of solutions of unknown concentrations were obtained as part of a routine service and chemical shifts were measured directly from the unexpanded spectrum. Data from the general NMR literature suggest that values obtained in this manner are subject to at least a ± 0.1 ppm error. Work done more accurately with expanded spectra and effective calibration will obviously fall within a much smaller range than this.

Over the past few years it has become almost routine for the programs LAOCOON or NMREN-NMRIT to be used to obtain accurate parameters for complex spin systems. These programs, in most cases, work very well and give essentially identical values of parameters for the same input data. However, at least one case has come to my knowledge where the two programs consistently give completely different results from the same input data. While these programs are of great use, they are much abused. It is most distressing to see workers applying the accuracies obtained in these calculations back to the spin system, and because this happens regularly, I am sure that many chemists carrying out these computations do not really understand the difficulties associated with the use of such programs. "Errors" printed out by the computer (often less than ± 0.01 Hz) are those obtained by iteration onto a given set of frequencies and do not include errors inherent in the determination of the line position. Many complex spectra contain overlapping lines, and it seems to be general practice to assign frequencies to these on a trial-and-error basis to give a reasonable fit to the computation. In very few cases are these decisions checked by calculation of the actual shape of the best-fit spectrum and, in any case, it is doubtful whether the small manipulations of line position would produce much effect on the multiplet envelope. Programs such as DECOMP have been devised to determine the positions of overlapping lines, but the experimental difficulties in obtaining good digitized spectra and the general problems of computer access make these methods of little use to most chemists interested in analyzing a series of spectra. The actual "errors" in these iterative solutions to complex spin systems can best be assessed by comparison of the results obtained by different workers for the same molecule under essentially the same conditions. In the case of pyridine two "accurate" analyses of spectra of the neat liquid have been reported (see Table 2.1) and it is obvious that the values obtained vary by considerably more than the computed "errors." In fact, from these results, it is doubtful whether the sign of J_{26} is really negative or even whether this coupling exists at all. It would be quite reasonable, but probably incorrect, to say that the very small values obtained for this coupling constant were artifacts produced by errors in the measurement and/or assignment of line positions. One important conclusion can be drawn from these considerations: Parameters obtained from iterative computations are not magically correct and must be treated with caution and common sense. Quotation of values to the third decimal place of a hertz is certainly not warranted and, in many cases, even the second decimal place cannot be stated with any certainty.

Another problem, common to all NMR experiments, is the choice of standard signal from which to measure chemical shifts. Rather than discuss this in detail here, the reader is referred to the excellent review by Laszlo (1). For work with protons, the external reference technique, so popular in the early days of NMR, is now seldom used. Tetramethylsilane (TMS) has become almost universally accepted for solvents in which it is soluble, although the old standby, cyclohexane, is still used on occasion. The frequency-independent τ and δ scales, defined with respect to internal TMS, have become the standard method of expressing chemical shift data in a reproducible manner. However, the results of careful analyses of spin systems are often reported in terms of hertz downfield from TMS at the frequency of measurement. This method seems to be more manageable for data of high accuracy but is inconvenient for the reader if a comparison is to be made with data expressed in parts per million. Chemical shifts obtained from the analyses of complex spectra are sometimes reported in terms of the internal chemical shifts, that is, those between the protons within the spin system. This practice greatly reduces the value of the data for comparative purposes and leads to the suggestion that the authors of these papers did not make the effort to calibrate the expanded spectra on which their calculations were made.

The advent of the internally locked spectrometer which locks onto a strong signal in the spectrum being measured brought with it one difficulty which is of particular importance in the heterocyclic field. The solubility of

TMS in dimethyl sulfoxide (DMSO) is not large and, at times, great difficulty is experienced in the use of TMS as a lock signal when using this solvent.

Standards for aqueous solutions are more varied and are much less satisfactory than TMS. Jones *et al.* (2) suggested the use of acetonitrile, dioxan, or *t*-butyl alcohol but these are not chemically inert nor are they isotropic. Further, they are not sufficiently volatile to be removed easily for recovery of the pure sample. The sodium salt of trimethylsilylpropane sulfonic acid (TPS) has been widely used in aqueous solution because the trimethylsilyl group gives a singlet very near the frequency at which TMS would be expected to absorb, and values obtained are assumed to be on the τ or δ scales. Trimethylsilylpropane sulfonic acid is far from satisfactory. It is a solid and cannot be removed from the sample by simple means, and the three methylene groups in the molecule gives peaks in the high-field region of the spectrum. A number of other standards have been used but none of them is really satisfactory in aqueous solutions.

Standards for the measurements of chemical shifts of nuclei other than protons seem to have been chosen almost at random and many of them suffer from considerable disadvantages in terms of their anisotropic nature and their ability to associate with the solute and/or solvent molecules. For ^{19}F shifts, $CFCl_3$, C_6F_6, and trifluoroacetic acid have been extensively used as standards in studies of fluorosubstituted heterocycles (e.g., see Table 2.38). Standards for ^{13}C have normally been CS_2 or benzene, although dimethyl carbonate has been suggested (3) as a more suitable choice; P_4O_6 has been proposed as reference for ^{31}P rather than the more widely used phosphoric acid or trimethyl phosphite (4).

Chemical shifts of nuclei other than protons can be obtained with high accuracy by indirect double-resonance techniques. A standard proton signal, usually TMS, is phase-locked to a reference frequency whose value is exactly that of the nominal proton resonance frequency of the system, for example, 60 or 100 MHz. The various types of "tickling" experiments then make possible the accurate positioning of lines in the spectra of other nuclei and hence the assessment of chemical shift. Although the absolute position of the TMS signal may change by a few hertz from one solution to another, these variations are insignificant when referred to other nuclei where the homonuclear chemical shifts are always larger than those observed for protons.

Correlations between π-electron densities calculated for heteroaromatic molecules and chemical shifts of the ring protons have been attempted many times and with very mixed success. While chemical shifts relative to those of a standard compound can be measured with reasonable accuracy, the determination of electron densities at the ring atoms of these systems is nowhere near as reliable. This has been highlighted in two publications. Black and co-workers (5), in an extremely comprehensive study with many heterocyclic systems, reached the conclusions that correlations between VESCF π-electron densities and proton chemical shifts were only fair, that interpretations of the results were very tentative, and that aspects of the calculation of π-electron densities are not well understood. Fraenkel and his co-workers (6) compared a number of methods for the calculations of electron densities in the pyridine molecule and found that the choice of wave function was critical for reasonable correlations to be obtained between electron densities and proton or ^{13}C chemical shifts. If careful work of this type on parent heterocycles produces correlations which are not good, how much more difficult must it be to get reasonable straight-line relationships between chemical shifts and electron densities for substituted heterocycles using only the simple Hückel approach for the calculations.

Hammett constants (σ) (8) are part of the physical organic chemists' stock in trade and have many legitimate uses. However, they are often abused and it would appear from the wide variety of values for different σ constants now available that one could be found to correlate roughly with almost any measurable molecular property. It would seem that the field has moved a long way from the measurement of equilibrium constants for substituted benzoic acids. Recent work (7) in which Hammett constants were separated into field and resonance components will undoubtedly lead to a "reinterpretation" of many of the correlations obtained between these constants and various NMR parameters.

The problem of "aromaticity" is one which is intimately concerned with the NMR spectra of aromatic systems. The concept of the "aromatic" ring current and its impact on NMR theory are discussed fully in Chapter 2 but a few general comments might be appropriate. Jones, in his excellent review of aromaticity (9), highlights the conflicting nature of different definitions of this concept and gives a full discussion of the ring-current hypothesis. In his summation he criticizes the definition that an aromatic compound is one that can support a ring current, and makes the statement that the anisotropic nature of many cyclic systems is not necessarily due to the presence of a ring current but rather to other factors as yet not well understood. I fully support his feelings on this matter.

Inevitably when one reads through papers on the analysis of complex spin systems one becomes aware of the difficulties inherent in the assignment of multiplets to specific protons. The most usual method for carrying out this operation is to compare values for chemical shifts and particularly coupling constants with those of suitable model compounds. In most cases this approach works well but in strongly coupled systems and in those where the distinctive coupling constants are very small, comparative techniques are often of little use. A typical problem arises in the assignment of peaks to specific protons in benz-fused heterocycles such as quinazoline (see Chapter 2). In these compounds the protons on the nonheterocyclic ring form an "ABMX" system with the two protons "meta" to the ring junction forming a very closely coupled AB system and the two protons adjacent to the ring junction approaching the MX situation. Analysis of the spin system is normally straightforward, but the problem arises of the assignment of M or X to the proton peri to the heteroatom in position 1. The deshielding effect of N-1 is enough to make this decision an easy one in the case of quinazoline, but in substituted compounds and in sulfur heterocycles it can be extremely difficult to determine which end of the ABMX system is which. Perhaps for this reason most workers do not bother to produce analyses for the spectra of the benz-ring protons in this class of compound, although they often expend quite a lot of energy in sorting out the patterns from the heterocyclic ring. The spectrum of quinazoline highlights another problem. Signals from H-2 and H-4 are essentially singlets and occur very close to each other. The original assignment of these depended entirely on the broadness of the peaks, the broader of which was thought to contain the cross-ring coupling J_{48} being assigned to H-4. Specific deuteration experiments showed this choice to be wrong. Unfortunately there are a number of other cases where arguments based on inappropriate model compounds or on the misinterpretation of small coupling constants have led to the wrong assignments, which have later been clarified by the preparation of specifically deuterated compounds. It is my firm view that the relatively simple chemical process of deuteration is far less time-consuming than many of the sophisticated NMR techniques used in attempts to confirm assignments, and the answer obtained is clear-cut and not open to manipulations of logical argument.

The last few years have brought a gradual intensification of interest in spectra of nuclei other than protons. Spectrometers now available commercially are capable of producing spectra of nuclei such as ^{13}C and ^{15}N in natural abundance, and the introduction of Fourier transform techniques greatly reduced the time and effort required for this work. Spectra of "other nuclei" hold tremendous potential for the heterocyclic chemist. The accessibility of the ^{13}C nucleus provides a probe to investigate heterocyclic systems which contain few protons. In nitrogen heterocycles, the ring nitrogen atoms whose natural ^{15}N label can now be routinely observed are usually the active sites of the molecules where tautomerism, association, or reaction has its greatest effect. Heteronuclear NMR spectroscopy is in its infancy, but is rapidly changing from an interesting physical phenomenon to perhaps the most powerful structural tool yet devised for the organic chemist.

REFERENCES

1. Laszlo, "Solvent Effects and Nuclear Magnetic Resonance," in *Progress in Nuclear Magnetic Resonance Spectroscopy*, Vol. 3 (Elmsley, Feeney, and Sutcliffe, Eds.), Pergamon, Oxford, 1967.

2. Jones, Katritzky, Murrell, and Sheppard, *J. Chem. Soc.*, **1962,** 2576.
3. Maciel and Natterstad, *J. Chem. Phys.*, **42,** 2752 (1965).
4. Chapman, Homer, Mowthorpe, and Jones, *Chem. Commun.*, **1965,** 121.
5. Black, Brown, and Heffernan, *Aust. J. Chem.*, **20,** 1305 (1967).
6. Tokuhiro, Wilson, and Fraenkel, *J. Am. Chem. Soc.*, **90,** 3622 (1968); Fraenkel, Adams, and Dean, *J. Phys. Chem.*, **72,** 944 (1968).
7. Swain and Lupton, *J. Am. Chem. Soc.*, **90,** 4328 (1968).
8. Hammett, *Physical Organic Chemistry*, McGraw-Hill, New York, 1970.
9. Jones, *Rev. Pure Appl. Chem.*, **18,** 253 (1968).

2 NITROGEN HETEROCYCLES—SINGLE SIX-MEMBERED RINGS

I. THE PYRIDINE SYSTEM

A. Pyridine and Its Alkyl Derivatives

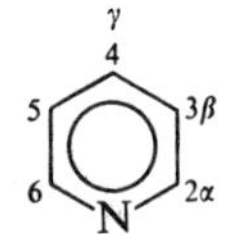

Spectra of the picolines (**1–3**), lutidines (**4–9**), and collidine (**10**), measured at 40 MHz, were first reported in 1955 by Baker (1) who obtained approximate values for chemical shifts and coupling constants. In the following year, spectra of pyridine and certain deutero derivatives were published by Bernstein and Schneider (2) who, together with Pople, later used the simplified spectra of the deutero derivatives to produce a composite analysis of the spectrum of the parent compound (3, 4). Subspectral techniques gave more accurate values for J_{24}, J_{34}, and $J_{23} + J_{25}$ (5), but complete and accurate analysis required iterative computations on spectra sharpened by decoupling from the nitrogen nucleus (6). These accurate analyses have been used to check the validity of subspectral, perturbation, and direct methods for the analysis of the pyridine spectrum (7). Fraenkel and his co-workers (7a) reported full iterative analyses for the spectra of pyridine neat and in a wide range of solvents. It is interesting to note that these results are not quite the same as those obtained previously (6), and the differences reflect the human element in these calculations, an element which few workers consider when carrying out analyses of this type.

The spectrum of pyridine in the nematic phase of anisole-azophenyl-*n*-caproate has been fully analyzed using a modification of the LAOCOON II program called LAOCOONOR (7b). The signs of the direct (D_{ij}) and indirect (J_{ij}) coupling constants were shown to be opposite each other in this molecule and, as the large indirect coupling constants are known to be positive, the direct couplings must take a negative sign. Ratios of interproton distances (r_{ij}/r_{pq}) agree to within experimental error with results obtained from microwave spectroscopy. Data from these analytical studies are summarized in Table 2.1.

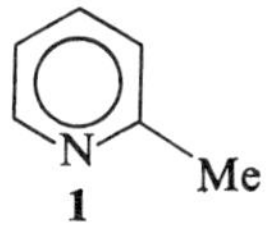

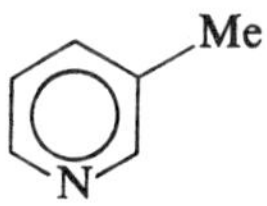

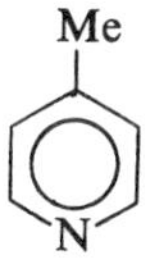

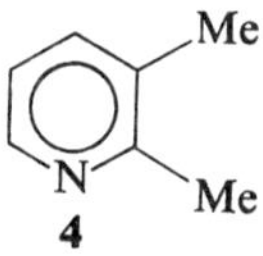

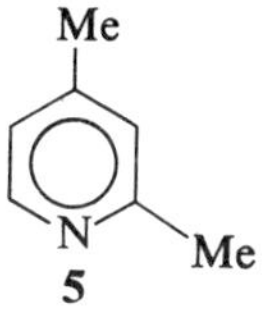

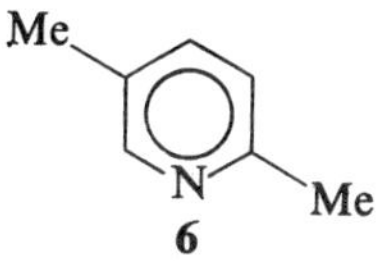

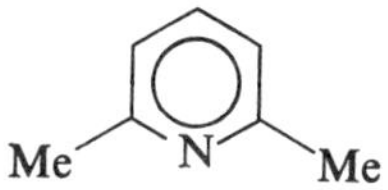

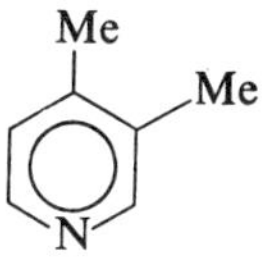

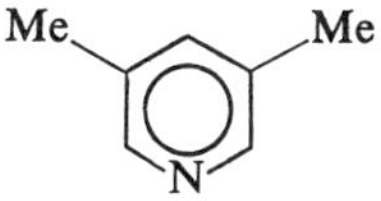

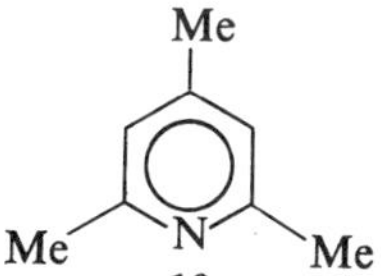

Bernstein and co-workers, in their definitive paper on the theory of AB_2 and ABX systems, used 2,3- and 2,6-lutidines, respectively, as examples of these two types of spin systems and obtained reasonably accurate spectral parameters (8).

Pyridine, with its structural similarity to benzene, has been used to study the contributions made by a ring nitrogen atom to the chemical shifts of protons on the heteroaromatic system. The overall shifts are influenced by a number of specific factors.

1. The aromatic ring current.
2. Distortion of the π-electron distribution by the nitrogen atom.
3. The local magnetic anisotropy of the nitrogen atom.
4. The electrostatic field of the lone-pair dipole.
5. The reaction field of the solvent.
6. Specific solvation.

The general approach to this work has invariably involved benzene as reference and the pyridine ring currents have been considered to be similar but distorted by the electronegativity of the nitrogen atom.

The existence of π-electron ring currents was first postulated about 30 years ago to explain the anisotropic magnetic susceptibility of aromatic

TABLE 2.1 SUMMARY OF PARAMETERS OBTAINED FOR PYRIDINE (IN Hz AT 60 MHz)

A. COMPARISON OF METHODS OF CALCULATION

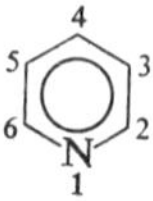

	Method of Analysis					
				Direct		
Parameter (Hz)	Original	Subspectral	Perturbation	abc	ABB′CC′	Iterative
$\delta_{3,4}$		23.247			23.080	23.210
$\delta_{2,4}$		59.086			58.931	58.941
$J_{3,4}$	7.5	7.672			7.612	7.656
$J_{2,4}$	1.9	1.861			1.908	1.850
$J_{2,3}$	5.5		4.920	4.902		4.862
$J_{2,5}$	0.9		0.970	0.988		0.984
$J_{3,5}$	1.6		1.292	1.292	1.316	1.360
$J_{2,6}$	0.4		−0.126	−0.126	−0.185	−0.128
Ref.	3	7	7	7	7	6

B. SOLVENT DEPENDENCE (7a). CONCENTRATION—1.0 *M*

Solvent	$\tau 2$	$\tau 3$	$\tau 4$	J_{23}	J_{24}	J_{25}	J_{26}	J_{35}	J_{34}
Neat	1.31	2.82	2.45	4.76	1.80	0.94	0.05	1.50	7.65
Et_2O	1.45	2.82	2.42	4.82	1.77	0.96	−0.14	1.25	7.64
C_6H_{12}	1.48	2.93	2.54	4.99	1.70	1.16	−0.14	1.15	7.67
CCl_4	1.48	2.84	2.45	4.99	1.69	1.06	−0.07	1.33	7.59
$CDCl_3$	1.40	2.75	2.36	4.93	1.80	1.00	−0.03	1.44	7.66
Me_2CO	1.43	2.69	2.29	4.88	1.76	0.99	−0.02	1.39	7.70
MeOH	1.47	2.60	2.19	5.07	1.72	0.96	0.03	1.47	7.72
2-PrOH	1.48	2.67	2.27	5.03	1.76	0.94	−0.01	1.43	7.70
D_2O	1.50	2.69	2.29	5.02	1.72	0.94	−0.11	1.25	7.67
D_2SO_4/D_2O	1.22	1.91	1.38	5.82	1.69	0.90	0.69	2.16	7.47
Et_2O/R_2Mg	1.49	2.71	2.39	5.18	1.80	0.90	0.13	1.38	7.47

C. PYRIDINE ORIENTED IN NEMATIC PHASE OF ANISOLE-AZOPHENYL-*n*-CAPROATE (7b). MEASURED AT 60 MHz. CHEMICAL SHIFTS AND COUPLING CONSTANTS IN Hz

$\delta_{2,3}$	81.2 ± 0.4	r_{26}/r_{35}	0.962 ± 0.003	J_{23}	4.9
$\delta_{2,4}$	59.1 ± 0.4	r_{23}/r_{35}	0.579 ± 0.003	J_{24}	1.9
D_{23}	−374.6 ± 0.1	r_{25}/r_{35}	1.139 ± 0.003	J_{25}	1.0
D_{24}	−65.0 ± 0.2	r_{34}/r_{35}	0.587 ± 0.004	J_{26}	−0.1
D_{25}	−33.1 ± 0.1	r_{24}/r_{35}	1.009 ± 0.005	J_{34}	7.7
D_{26}	−45.8 ± 0.3	S_{22}	0.0265 ± 0.0003	J_{35}	1.3
D_{34}	−247.2 ± 0.2	S_{33}	0.0484 ± 0.0002		
D_{35}	−40.8 ± 0.3	S_{44}	−0.0749		

hydrocarbons (9), and a considerable amount of effort has been directed toward the quantum mechanical calculation of the magnitude of their effect. Pople replaced the ring current by a point dipole at the center of the aromatic ring (10) and this model was modified by Waugh and Fessenden (11) and Johnson and Bovey (12) who substituted for the magnetic dipole a pair of current loops, above and below the ring plane **11**. The separation between the loops has been assigned values ranging from 0.7 to 1.28 Å with appropriate variation in the magnitude of the predicted proton deshielding (11–13).

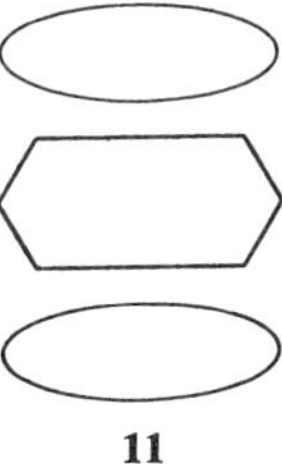

11

Most workers have used the Johnson and Bovey approximation combined with quantum mechanical calculation of the magnitude of the ring current by the methods of Pople (14) or McWeeny (15). Dailey and his co-workers have calculated the chemical shifts for a number of polycyclic hydrocarbons and heterocyclic compounds (16) by this method and found that the predicted chemical shifts for ring protons were invariably too great while the predicted diamagnetic anisotropies were too small. Use of the experimental figure for the diamagnetic anisotropy of benzene gave predicted chemical shifts about 60% greater than those observed, and it has been suggested that the experimental anisotropy be reduced by a factor of 0.6 to accommodate these difficulties. Recent calculations by Pople and Ferguson (17) have shown that almost 30% of the diamagnetic anisotropy in aromatic systems can be ascribed to local paramagnetic terms on the carbon atoms due only to their sp^2 hybridization, in rough agreement with the empirical figure suggested by Dailey, whereas Musher (18) has maintained that the ring-current effect is an artifact of the approximations inherent in the London treatment and has no connection with cyclic delocalization of π-electrons nor with aromatic character.

Theoretical estimates of the ring currents in heteroaromatic compounds (19, 20) suggest that for pyridine, pyrimidine (**12**), and *sym*-triazine (**13**) they are 95–98% that of benzene. However, in these treatments the discrepancies discussed above are all present and the quantitative reliability of the results is open to question.

Most of these difficulties are overcome when the chemical shifts of protons on the pyridine ring are compared with those obtained for the appropriate benzene derivative so that the inadequacies of the mathematical treatments cancel and the effect of the nitrogen atom may be measured. The electronegativity of the heteroatom distorts the π-electron distribution and this is reflected in the π-electron densities at the various ring atoms. With solutions of pyridine in solvents of low dielectric strength, contributions from the reaction field of the solvent (see Ref. 16) and from specific solute–solvent interactions are negligible and factors 3 and 4, which involve local fields, rapidly attenuate as the distance from the nitrogen atom increases. Thus, Schaefer and Schneider (21), working with cyclohexane solutions, found good correlation between the chemical shifts (referred to benzene) of the

TABLE 2.2 π-ELECTRON DENSITIES FOR PYRIDINE AND THE PYRIDINIUM ION

	π-Electron densities (ρ)					
	α-C	β-C	γ-C	N	Method	Ref.
Pyridine		1.014	0.976		Exp.	21
	0.955	1.020	0.974		Exp.	22
	0.965	1.006	0.960	1.908	Exp.	24
	0.952	1.004	0.981	1.107	VESCF	258
	0.951	1.010	0.979		SCF, CI	259
	0.923	1.005	0.923	1.195	HMO	24
	0.86	1.064	0.93			16
Pyridinium ion	0.854	0.915	0.862	1.60	Exp.	21
	0.846	0.915	0.887		Exp.	24
	0.899	0.927	0.829	1.520	VESCF	258
	0.820	1.010	0.880		HMO	24

β- and γ-protons of pyridine and the calculated π-electron densities at these positions (Table 2.2). This was made possible by the approximate linearity, in their hands, of the relationship

$$\delta = k \, \Delta\rho$$

where δ is the chemical shift, $\Delta\rho$ is the local excess charge located on the carbon atom, and k is a constant, 10.7 ± 0.2 ppm per electron. However, they were unable to estimate the magnitudes of effects 3 and 4 and made no comment on the electron density at the α-carbon atoms. This work was criticized by Gawer and Dailey (16) who calculated contributions from the anisotropy of the nitrogen atom but obtained values which were obviously too high. Corrections for the field due to the lone pair were similar to those calculated by other workers (22, 23) (see Table 2.2), but comparison of the corrected chemical shifts with π-electron densities obtained from SCF-MO calculations, using a value for k of 8.08 ppm per electron, was not very satisfactory.

Gil and Murrell (22), working with carbon tetrachloride solutions, made "semiquantitative" calculations of the effects of the *N*-anisotropy and the direct field of the lone-pair dipole (Table 2.3). Subtraction of these contributions from the observed chemical shifts and calculation of π-electron

TABLE 2.3 CALCULATED CONTRIBUTIONS FROM *N*-ANISOTROPY AND LONE-PAIR FIELD EFFECTS (ppm RELATIVE TO BENZENE) TO THE CHEMICAL SHIFTS OF PYRIDINE PROTONS

N-Anisotropy			Lone-pair dipole			
α-H	β-H	γ-H	α-H	β-H	γ-H	Ref.
−1.12[a]	−0.20[a]	−0.02[a]	−0.38	−0.12	−0.08	16
−0.33	0.0	+0.03	−0.52	−0.19	−0.14	22
			−0.43	−0.20	−0.15	23

[a] Discussion of the reasons for these extremely high results can be found in Ref. 25.

densities from the remainders gave results in excellent agreement with theoretical values (Table 2.2). This approach has been extended to the 4-alkylpyridines and the results compared with π-electron densities obtained from simple Hückel molecular orbital calculations (24). Very recently, an extremely comprehensive comparison has been made between π-electron densities calculated by VESCF methods and those obtained from chemical shift data for a large range of heterocycles including pyridine (25). The results are only fair and the authors emphasize that the interpretation is tentative and that aspects of the calculations are not well understood. Work in this field has been very well summarized by Cobb and Memory (25b) in their extensive study of the relationship between π-electron densities and chemical shifts in nitrogen heterocycles.

Fraenkel and his co-workers (7a) carried out a series of calculations and showed that, in one case, the proton and ^{13}C shifts for pyridine could be calculated quite accurately, using atom anisotropies together with a separate π-electron density term. The choice of wave function was critical and the SCF wave function gave by far the best results. Jones and Ladd (25a) reached the conclusion that solvent shifts for pyridine, as solvents were changed over a wide range of permittivities, were dominated by a reaction-field contribution. This effect appears to be inhomogeneous over the dimensions of the solute molecule.

Dewar and Marchand (26), in a study of the contributions to the chemical shift of protons α to the charged group in such species as the *p*-toluidinium ion, used γ-picoline as a model compound where "the effect of the grouping =N– in a conjugated system is due almost entirely to a direct polarization of the π-electrons by the electronegative nitrogen atom." Their choice seems far from ideal.

Evidence for *N*-anisotropy was obtained from the ^{14}N chemical shifts of pyridine and pyridinium ion, measured indirectly by heteronuclear decoupling (27). The unexpected upfield shift from pyridine (-302 ± 10 Hz from NH_3) to pyridinium ion (-179 ± 1 Hz from NH_3) was explained in terms of a large paramagnetic contribution to the *N*-anisotropy from a very low-energy $N \rightarrow \pi^*$ transition in pyridine. The energy of the corresponding $\sigma(N^+\text{–}H) \rightarrow \pi^*$ transition of pyridinium ion will be close to that of the $\sigma(C\text{–}H) \rightarrow \pi^*$ transition of benzene and hence is of little importance in this work (see Ref. 22). Protonation of pyridine leads to displacement to low field of the β- and γ-proton signals by 1.07 and 1.22 ppm, respectively, whereas the α-proton signal is shifted by only 0.25 ppm due to the disappearance of the paramagnetic term [comparing 5% molar solutions in cyclohexane and trifluoroacetic acid (28)].

Attempts have been made to correlate chemical shifts of protons on the pyridinium ion with calculated π-electron densities at the respective carbon atoms (Table 2.2) (21, 22, 24), but these have been less successful than the correlations obtained with the free base. Unfortunately, the cation can only be studied in solvents where factors 5 and 6 become significant and where the additional complication of ion pairing is possible. While correlations between π-electron densities obtained from chemical shifts of the pyridinium protons and those calculated from an "isolated cation" model are classed by the authors as "reasonable," they are not good and the discrepancies observed reflect the inadequacy of the model. Differences are particularly large for the 4-alkylpyridinium ions where simple HMO calculations were used to obtain the theoretical values (24), and in these cases any implied correlation may be coincidental. In a number of solvents, it has been shown that ion pairing between the pyridinium ion and counterion is important but in the highly polar formic acid, solvation of the cation (and/or counterion) predominates (29). Chuck and Randall (24) conclude that in aqueous solutions, the reaction field of the solvent and the polarity of the N^+–H bond cancel and counterions have little effect, presumably because they are solvated. In an earlier paper (30) these authors studied the spectra of 1,4-diethylpyridinium halides (**14**). This model was chosen so that the two ethyl groups, one close to the site of ion pairing and the other as far from it as possible, could be observed simultaneously. Medium effects, though small, were most marked for the

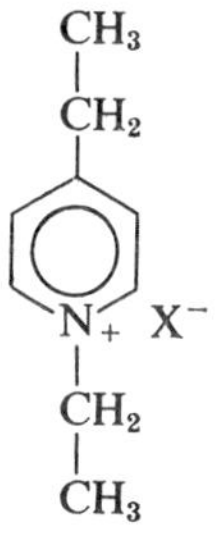

CH_3CH_2–pyridinium–N^+–CH_2CH_3 X^-

14

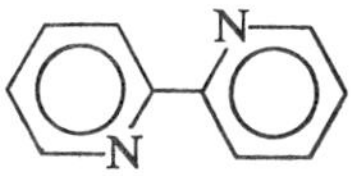

15

α-protons and the 1-methylene group. In deuterochloroform, shifts were anion dependent but not susceptible to concentration changes. This would be expected from the importance of ion pairing in solutions of such low dielectric strength at the concentrations used. In aqueous solution, anion effects were absent but there was a marked concentration effect due to a mobile equilibrium between solvated and unsolvated ions. Spectra of 1-methylpyridinium iodide and bromide have been measured over a range of concentrations in acetone, water, acetonitrile, and DMSO (30a). Chemical shifts, at infinite dilution, were found to be proportional to the square root of the solvent dielectric constant. The anion had an influence (at higher concentrations) in all solvents except water and this was attributed to ion pairing in the less polar solvents.

Proton exchange between alkyl-substituted pyridinium ions and water or methanol has been studied in detail (30b). The effect of methyl substitution was discussed in terms of a proposed mechanism.

Solvent effects on the spectra of pyridine and its alkyl derivatives have been extensively investigated, and it is interesting to note that with few exceptions, authors cannot agree on either the size or nature of the effects causing such shifts. Spectral data are collected in Table 2.4.

One of the characteristic features of pyridine spectra, and indeed those of many nitrogen heteroaromatic compounds, is the broadening of the proton signals, mainly noticeable in those of protons α to the nitrogen atom. That this broadening is due to coupling with the nitrogen atom was shown by heteronuclear decoupling (27), and it has been demonstrated that the actual mechanism causing the broadening is incomplete washing out of the N–H spin–spin coupling by the ^{41}N quadrupolar relaxation (31). In the neutral molecule, the quadrupole moment is caused by distortion of the charge distribution within the nucleus mainly by the lone pair and can be reduced by

TABLE 2.4 THE EFFECT OF SOLVENT ON THE CHEMICAL SHIFTS OF PROTONS IN PYRIDINE AND ITS METHYL DERIVATIVES[a]

Substituent	Solvent						Ref.
None		τ2, 6	τ3, 5	τ4			
	Neat	1.25 $\mp$ 0.02	2.77 $\mp$ 0.02	2.38 $\mp$ 0.01			24, 44
	Pentane	1.43	2.90	2.49			24
	CCl_4	1.44 $\mp$ 0.02	2.79 $\mp$ 0.01	2.38 $\mp$ 0.01			24, 44
	Benzene	1.44	3.22	2.90			44
	Acetone	1.39	2.65	2.21			24
	Methanol	1.41	2.65	2.11			24
	D_2O	1.40	2.52	2.09			24
	Dioxane	1.43	2.76	2.36			68
2-Methyl		τ3	τ4	τ5	τ6	τMe	
	Neat					7.51	44
	Benzene					7.58	44
	CCl_4	3.00	2.54	3.05	1.61	7.46	23, 44
	$CDCl_3$	2.88	2.45	2.93	1.52		23
	Acetone	2.82	2.38	2.87	1.56		23
	Methanol	2.73	2.26	2.79	1.59		23
	D_2O	2.72	2.25	2.79	1.60		23
3-Methyl		τ2	τ4	τ5	τ6	τMe	
	Neat					7.86	44
	Benzene					8.15	44
	CCl_4	1.66	2.62	2.94	1.69	7.65	23, 44
	$CDCl_3$	1.55	2.54	2.84	1.59		23
	Acetone	1.59	2.46	2.73	1.63		23
	Methanol	1.63	2.37	2.71	1.67		23
	D_2O	1.68	2.36	2.70	1.71		23
4-Methyl		τ2, 6	τ3, 5	τMe			
	Neat	1.45 $\mp$ 0.03	2.95 $\mp$ 0.03	7.81 $\pm$ 0.07			24, 42, 44, 238
	Pentane	1.59	3.06	7.75			24
	CCl_4	1.64 $\mp$ 0.01	3.03 $\mp$ 0.01	7.68 $\mp$ 0.03			23, 24, 42, 44, 48
	$CDCl_3$	1.55	2.91	7.66			23
	Acetone	1.55 $\mp$ 0.04	2.84 $\mp$ 0.01	7.68			23, 24
	Methanol	1.60 $\mp$ 0.03	2.76 $\mp$ 0.04	7.63 $\mp$ 0.02			23, 24
	D_2O	1.56 $\mp$ 0.02	2.76 $\mp$ 0.07	7.67 $\mp$ 0.01			23, 24
	Benzene	1.57	3.38	8.16			42

[a] Where two or more values are available the range is given.

TABLE 2.4 (*continued*)

4-Methyl		τ2, 6	τ3.5	τMe	
	Nitrobenzene	1.51	2.92	7.75	42
	t-Butylbenzene	1.64	3.34	8.08	42
	Mesitylene	1.67			42
	N,*N*-Dimethyl-aniline	1.66		8.12	42
	Methyl benzoate	1.49	3.06	7.88	42
	Benzonitrile	1.48	2.97	7.79	42
	Cyclohexane	1.65	3.11		48
	PCl_3	1.56	2.94		48
	$AsCl_3$	1.37	2.51		48
	SO_2	1.50	2.47		48
	CF_3COOH	1.30	2.04		48
	Acetonitrile	1.58	2.83	7.64	23
2,3-Dimethyl		**τ4**	**τ5**	**τ6**	
	Cyclohexane	2.87	3.24	1.81	48
	CCl_4	2.77	3.14	1.81	48
	PCl_3	2.36	3.05	1.69	48
	$AsCl_3$	2.40	2.77	1.55	48
	SO_2	2.36	2.71	1.80	48
2,5-Dimethyl		**τ3**	**τ4**	**τ6**	
	Cyclohexane	2.85	3.16	1.77	48
	CCl_4	2.78	3.11	1.78	48
	PCl_3	2.38	3.05	1.65	48
	$AsCl_3$	2.41	2.81	1.54	48
	SO_2	2.38	2.82	1.83	48
	CF_3COOH	1.50	2.11		48
2,6-Dimethyl		**τ3**	**τ4**	**τMe**	
	Neat	3.19	2.68	7.55	24
	CCl_4	3.17 ∓ 0.01	2.66 ∓ 0.01	7.54	23, 24
	Cyclohexane	3.19	2.72		48
3,4-Dimethyl		**τ2, 6**	**τ5**		
	Cyclohexane	1.80, 1.82	3.20		48
	CCl_4	1.79, 1.81	3.11		48
	PCl_3	1.72, 1.72	3.00		48
	$AsCl_3$	1.50, 1.51	2.55		48

TABLE 2.4 (*continued*)

3,4-Dimethyl		$\tau 2, 6$	$\tau 5$	
	SO_2	1.71, 1.72	2.58	48
	CF_3COOH	1.44, 1.46	2.03	48
3.5-Dimethyl		$\tau 2, 6$	$\tau 4$	
	Cyclohexane	1.72	2.80	48
	CCl_4	1.69	2.85	48
	$AsCl_3$	1.55	2.24	48
	CF_3COOH	0.79	0.99	48
2,4,6-Trimethyl		$\tau 3, 4$		
	Cyclohexane	3.33		48
	CCl_4	3.32		48
	$AsCl_3$	2.81		48
	SO_2	3.09		48
	CF_3COOH	2.44		48

any factor which either counteracts or removes the lone pair. Thus, formation of an *N*-oxide causes significant sharpening of the proton signals.

In solution of low dielectric strength, signals from the α-protons are often so broadened as to show no fine structure but, in aqueous solutions, this broadening disappears and sharp spectra are obtained (23, 24, 32). Such sharpening must be associated with removal of the quadrupole moment of the nitrogen atom and has been interpreted in terms of strong solute–solvent interaction, presumably hydrogen bonding. Support for this hypothesis is obtained with methanol solutions where the sharpness of the signals varies with concentration and is best at 0.5 M fraction of pyridine (24), the concentration at which maximum hydrogen bonding is thought to occur (33). Spotswood and Tanzer (23) in a study of spectra of bipyridyls (**15**) found that the increase in the chemical shift of the α- and β-protons of alkylpyridines on going from carbon tetrachloride to deuterium oxide solutions corresponded almost exactly to the removal of the contribution from the field of the lone pair. This conclusion is supported by calculations of the contributions from the fields of the lone pairs in 2,2′-bipyridyl which again correspond to the differences observed between carbon tetrachloride and aqueous solutions. An extension of the HMO approach to hydrogen bonding has shown that the most likely configuration for water or methanol bonded to pyridine contains the linear group N:–H–O (34). Chemical shifts of ^{13}C calculated from EHT wave functions were in reasonable agreement with experimental values. An NMR investigation of the 2,6-lutidine–water system in the critical region has been reported (34a).

It is interesting to note that while solution in water removes broadening of the ring protons, gradual decrease in the pH of the solution progressively increases the broadening again (32). Decoupling experiments showed that the new broadening was also caused by coupling between the protons and the nitrogen atom, and the effect is best attributed to variation of the nitrogen quadrupole moment by the positive charge.

Most work dealing with solvent or protonation effects on pyridines has concentrated on chemical shift variations. Palmer and Semple (35) found that for a large number of aromatic or heteroaromatic compounds, coupling constants are largely independent of the nature of the solvent and concentration of solution. Merry and Goldstein (32) examined the spectra of pyridine in water and progressively increased the acidity of the solution from pH 9 (neutral molecule) to 3.1 (99% protonation). Over this range, progressive changes occur in five out of six coupling constants. Parameters given in Table 2.5 show how the four couplings involving the α-proton vary with the

TABLE 2.5 PMR PARAMETERS FOR PYRIDINE (32). ALL VALUES IN Hz AT 60 MHz[a]

Solvent	pH	% Protonation	H-2	H-3	H-4	J_{23}	J_{24}	J_{25}	J_{26}	J_{34}	J_{35}
Neat			405.6	316.0	338.1	4.84	1.79	1.00	−0.02	7.62	1.45
H_2O	9.0	0	404.0	335.6	360.3	5.04	1.79	1.00	−0.02	7.73	1.38
H_2O	3.1	99	439.7	395.6	427.5	5.96	1.54	0.81	0.99	7.95	1.38

[a] Chemical shifts are relative to external cyclohexane and are adjusted for bulk susceptibility effects. An uncertainty of ∓2 Hz is quoted for these values.

extent of protonation. By far the largest relative change is that of J_{26} which varies from virtually nothing in the neutral molecule to 1.0 Hz in the cation. Similar but larger variations in coupling constants involving H-2 also occur on *N*-oxide formation (6, 36) and the overall effect appears to involve complete removal of the lone pair. The increase in J_{23} on quaternization of a nitrogen atom in position 1 has been found to be general for many heteroaromatic compounds and has been used to decide the site of protonation in polyaza systems such as cinnoline (**16**) (35). Recently, spectra of a series of 1-substituted pyridines have been examined and the Pauling electronegativity of the first atom of the substituent was correlated with changes in coupling constants throughout the molecules (36). Reasonable "straight-line" relationships were obtained for pyridine, pyridine *N*-oxide, and their protonated

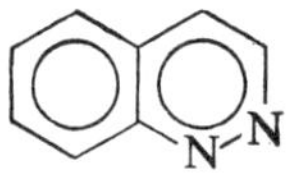

16

species if the pyridine lone pair is assumed to be the shared pair in a hypothetical bond with a substituent corresponding to zero on the electronegativity scale. It is also proposed that, in 1-substituted pyridines (and by inference in other substituted aromatic compounds) charge variation occurs by an inductive mechanism involving σ-bonds where the polarization of the bonds and the transmission of charges alternate along the molecular framework as shown

$$C_2^{\delta\delta\delta+} \rightarrow C_1^{\delta\delta-} \leftarrow A^{\delta+} \rightarrow B_x^{\delta-} \qquad C_2^{\delta\delta\delta-} \leftarrow C_1^{\delta\delta+} \rightarrow A^{\delta-} \leftarrow B_x^{\delta+}$$

Such alternance of charges has been deduced also from theoretical quantum mechanical calculations of the effects of substituents on the dipole moments of saturated and unsaturated molecules (37).

Long-range coupling has been observed between the protons of the methyl group of γ-picoline and protons 3 and 5 and, in the case of β-picoline, between the methyl protons and the 6- as well as the 4-proton (38, 39). The lack of such coupling in α-picoline is thought to involve the low electron density at the α-carbon atom (40), but little is really known about the origins of these couplings.

Variations induced by benzene in the chemical shifts of protons of pyridine and its methyl analogs were reported a number of times in the earlier NMR literature (41–43) but were first studied systematically by Murrell and Gil (44). Shifts obtained by comparing spectra of benzene solutions with those of carbon tetrachloride solutions for protons or methyl groups α to the ring nitrogen atom were negligible but upfield shifts of 0.3 to 0.5 ppm were found for such groups in the β- or γ-positions. The highly specific shifts observed for these compounds can only be caused by the formation of reasonably intransient complexes in which the solvent molecules are bound to the positive ends of the pyridine dipoles as shown (**17**). The structure **18** proposed for paired molecules in neat γ-picoline is similar (42). However, Murrell and Gil were well aware that the chemical shifts quoted were not absolute and did not take into account the solvent shifts of the tetramethylsilane. Similar work was done by Ronayne and Williams (45) who reached essentially the same conclusions and found that successive introduction of *C*-methyl groups caused an algebraic reduction in solvent shifts. The quantitative significance

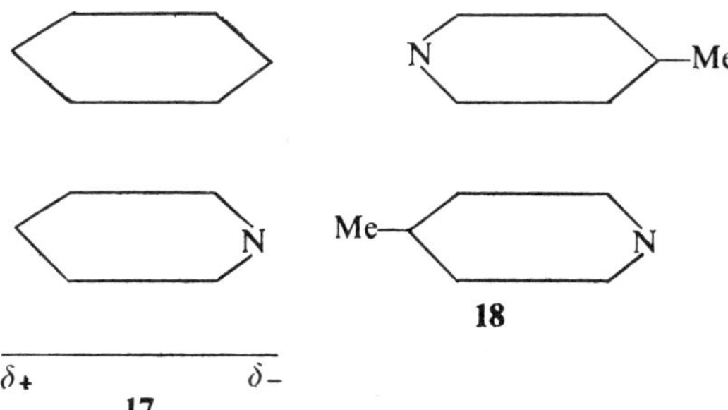

of their work is extremely doubtful because the benzene shifts of the internal standard, tetramethylsilane, have not been taken into account. Calculation of these shifts (46) shows that they may be as high as +0.4 ppm, a value that will greatly increase the observed shifts (see also Ref. 47). The increased shifts would indicate a much stronger (closer) complex and provide no evidence for an angular approach by the benzene molecule.

Perkampus and Krüger (47a) also studied the concentration dependence of the proton resonances of pyridine and its 1-oxide dissolved in C_6D_6. Whereas in the spectrum of the oxide, signals from protons were shifted to higher field as the proportion of benzene increased, for pyridine the α-protons were very little affected. Again, this phenomenon was explained in terms of an "asymmetric" association of pyridine and benzene. Aromatic solvent-induced shifts have been examined for a number of azines and additive CCl_4–C_6H_6 incremental shifts proposed for ring protons in these compounds (47b). These are (in parts per million) −0.07 for α-protons, 0.44 for β-protons, 0.56 for γ-protons, and −0.11 for substituent methyl groups.

The proton chemical shift (relative to internal C_6H_{12}) of chloroform was measured in CCl_4 and pyridine at a variety of concentrations and temperatures (47c). The observed shift was used to calculate thermodynamic parameters for the association between $CHCl_3$ and pyridine. Pyridine-induced solvent shifts (with respect to $CDCl_3$ solutions) have been suggested as an aid in determining structures of hydroxyl-containing compounds (47d). Spectra of a number of asymmetrically substituted pyridines in a range of solvents have been analyzed by the "effective Larmor frequency method" (47e). Anomalous shifts obtained for aqueous solutions were explained in terms of an ordered water structure.

Spectra of methylpyridines in strongly coordinating solvents such as $AsCl_3$, $SbCl_3$, and SO_2 have been measured (48) and solvent shifts with respect to solutions in nonpolar solvents are greatest for the β- and γ-protons. The relative insensitivity of the α-proton is thought to entail removal of the nitrogen lone pair by direct covalent bond formation between the nitrogen atom and the positive end of the solvent dipole. Alkyl substitution decreases the magnitude of these shifts. Similar effects were also observed for pyridine in solutions containing strongly interacting anions (49). Perkampus and Krüger (49a, b) have made a very extensive study of the spectra of complexes between pyridines and Lewis acids and they suggest that three factors determine the shifts of signals from the ring protons which occur when these complexes are formed.

1. Reduction of the overall electron density would occur by donation of the electron pair of the nitrogen atom to the Lewis acid. The effect of this would rapidly diminish as the distance from the nitrogen atom increased.
2. A small reduction of the ring current would occur and this would be felt equally by all ring protons.
3. The paramagnetic shielding of the lone pair would be reduced and this effect would be large for the α-proton but small for the β-proton. The overall result on complex formation was a larger downfield shift of the β-proton than the α-proton.

Complex formation between 4-ethylpyridine and aluminum halides (49c) and 2,4,6-trimethylpyridine and trifluoroiodomethane (49d) has also been studied by NMR techniques. Proton and ^{11}B data have been reported for complexes between pyridines and various boron-containing Lewis acids (49e).

The separate phenomena which contribute to solvent shifts observed for molecules such as pyridine are difficult or impossible to measure accurately and the theoretical basis for their calculation is far from satisfactory. Before these systems can be fully understood, much careful and detailed work will be necessary. The situation is made worse in papers in which sweeping generalizations are made from inadequate data, where earlier work is simply repeated, and where phrases such as "virtually equal," "approximately linear," and "reasonable agreement" are used very loosely. Also, the technique of proposing a possible explanation for experimental data and then using this as established fact to build a large theoretical framework seems to be fashionable.

^{13}C–H coupling constants for pyridine have been measured (50) and from them a self-consistent additivity relationship has been developed where $\zeta'_{i,j}$ represents the contribution from bond i–j, adjacent to the carbon atom in question, and the prime indicates sp^2 hybridization of the i–j bond. The values obtained are

$$\zeta'_{C,C} = 77.5 \text{ Hz}$$

$$\zeta'_{C,N} = 84.5 \text{ Hz}$$

$$\zeta'_{N,C} = 103.0 \text{ Hz}$$

This rule was later extended to many other heterocyclic compounds and seems to be generally applicable (51). Values of $J_{C,H}$ for pyridines studied, together with those reported earlier for 2,4,6-collidine (52), are given in Table 2.6. Attempts have been made to relate ^{13}C chemical shifts of pyridines,

TABLE 2.6 OBSERVED AND CALCULATED ^{13}C–H COUPLING CONSTANTS FOR PYRIDINES (50)

Compound	Carbon atom	J_{CH} observed (Hz)	J_{CH} calculated (Hz)
Pyridine (neat)	2	179	180.5
	3	163	162.0
	4	152	155.0
4-Methyl (neat)	2, 6	174.4	
	3, 5	160.0	
	Me	126.7	
3,5-Dimethyl (neat)	2, 6	176.0	
	4	156.6	
	Me	126.8	
2,6-Dimethyl (neat)	3, 5	161.0	
	4	161.5	
	Me	126.5	
2,4,6-Trimethyl (neat)	3, 5	158.5 ∓ 0.7	
	Me	126.4 ∓ 0.5	
4-Methyl-1-oxide (D_2O)	2, 6	186.5	
	3, 5	163.5	
	Me	128.4	

other heterocyclic compounds, and their cations to calculated π-electron densities (53, 54) and these are discussed in more detail in Section VI.

The ^{14}N chemical shifts for pyridine and pyridinium ion have been measured indirectly by heteronuclear decoupling (27), and it was suggested that the upfield shift on protonation is due mainly to a large paramagnetic term arising from the very low-frequency $N \rightarrow \pi^*$ transition of pyridine (22). More recently the ^{14}N spectra of pyridine neat and in methanol have been measured directly (33). Decrease in the mol fraction of pyridine causes an upfield shift which was attributed to neutralization of the lone-pair dipole by hydrogen bonding with the solvent. The results were used to calculate the contributions of different mesomeric forms to the hydrogen bond using the Coulson-Danielsson approach (55). ^{14}N spectra of pyridine with respect to nitromethane (56) and nitrate ion (57) have also been measured and the ^{15}N–H coupling constant for pyridinium ion in sulfuric acid has been reported (58). High-resolution proton, ^{13}C, and ^{15}N spectra have been reported for ^{15}N-labeled (100%) pyridine and its hydrochloride (58a). All chemical shifts and coupling constants were obtained and the signs of coupling constants, except J_{14} (N–H) and the ^{13}C–^{15}N couplings, were assigned by analogy with spin-tickling experiments carried out on quinoline-^{15}N (58b). Throughout this work, attention was focused on coupling constants which were compared with values for other compounds and discussed in the light of the inapplicability to pyridine of the average energy approximation. The coupling $^1J_{NC}$ is very small for pyridine and this was taken to mean that the assumptions inherent in the relationship between the products of the %*s*-character of each of the orbitals of a C–N bond and $^1J_{NC}$ (58)

$$S_N S_C = 80 J_{NC}$$

do not hold for pyridine. As with unlabeled pyridine, addition of methanol to the neat liquid causes an upfield shift of the nitrogen resonance and protonation, which can be classed as the limiting case of hydrogen-bonding solvation, leads to resonance of the ^{15}N at even higher field. C–N coupling constants over three and four bonds are quite large, in contrast to the absence of similar couplings in saturated systems, and the suggestion was made that the presence of a π-system may be necessary for the transmission of these interactions. Spectra of pyridine-^{15}N in ether and of the corresponding hydrochloride have been obtained at higher fields (100 and 90 MHz for protons) (58c) and all of the data discussed above have been confirmed. Data from nitrogen spectra are summarized in Table 2.7.

Diehl and Leipert (59), in a general study of deuterium spectra, measured that of pentadeuteropyridine and found the relative chemical shifts of the deuterons to be almost identical with those observed for protons:

$$\delta_4 - \delta_2 = 1.16 \text{ ppm}, \qquad \delta_5 - \delta_4 = 0.37 \text{ ppm}$$

B. Bipyridyls

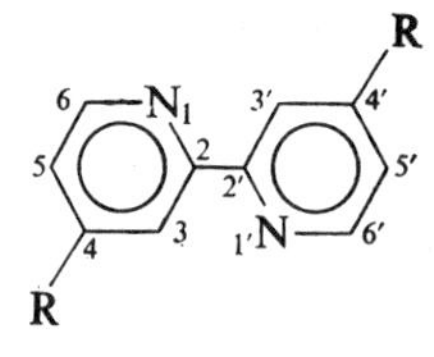

19, R = H
23, R = Me

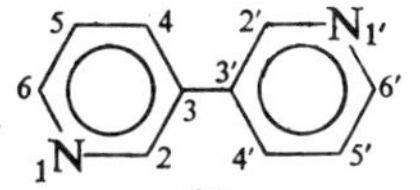

21

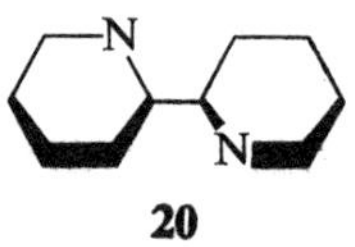

20

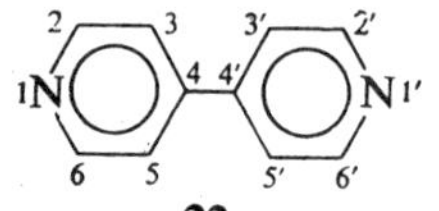

22

TABLE 2.7 NITROGEN SPECTRA OF PYRIDINE AND ITS CATION

Compound	Nucleus	Solvent	Standard	Shift	J_{NH}	Width[a]	Ref.
Pyridine	^{14}N	Neat	NO_3^-	+22 ppm			57
		Neat	NH_3	−302 ± 10 Hz			27
		$MeNO_2$–DMF[b]	$MeNO_2$	+68 ± 2 ppm		260 ± 5 Hz	56
	^{15}N	Neat	$H^{15}NO_3$ (ext.)	56.8 ppm			58a
		Methanol	$H^{15}NO_3$ (ext.)	74.4 ppm			58a

^{15}N–H Coupling constants (58a) (Hz)

J_{12} −10.76; J_{13} −1.53; J_{14} ±0.21

^{15}N–^{13}C Coupling constants (58a). Sign not determined (Hz)

J_{12} 0.45; J_{13} 2.4; J_{14} 3.6

Compound	Nucleus	Solvent	Standard	Shift	J_{NH}	Width[a]	Ref.
Pyridinium ion	^{14}N	TFA[c]	NH_3	−179 ± 2 Hz			27
	^{15}N	H_2SO_4			90.5 ± 1 Hz		58
		CD_3OD	$H^{15}NO_3$ (ext.)	170.1 ppm			58a

^{15}N–H Coupling constants (58a) (Hz)

J_{12} −3.01; J_{13} −3.98; J_{14} ±0.69

^{15}N–^{13}C Coupling constants (58a). Sign not determined (Hz)

J_{12} 12.0; J_{13} 2.1; J_{14} 5.3

[a] Half-height width.
[b] Dimethylformamide.
[c] Trifluoroacetic acid.

The spectrum of 2,2′-bipyridyl (**19**) has been described in general terms (60, 61) and was later analyzed by first-order methods (62). Gil (63) has determined the relative signs of six of the coupling constants from field-sweep double-resonance spectra obtained at 100 MHz and found them all to be positive. Solvent effects were first studied by Murrell *et al.* (64) who surveyed the effects which influence the chemical shifts of the various protons and established that the preferred conformation in inert solvents was approximately trans coplanar. Spectra in more polar solvents were interpreted in terms of higher interplanar angles. A much more extensive study was made by Castellano and his co-workers (65) who measured spectra in 11 single solvents and three binary mixtures. Their analyses of these spectra, carried out with the assistance of the iterative computer program LAOCOON II, are summarized in Table 2.8. In inert solvents the chemical shifts of protons 4, 5, and 6 vary slightly from those observed for pyridine in the same solvent, but for proton 3 a large downfield drift was observed (−1.34 ppm, greater than the shift for proton 5). In agreement with the earlier work, these shifts were interpreted in terms of an essentially trans-coplanar conformation **20** where proton 3 (3′) comes close to the nitrogen atom in the adjacent ring. The shielding of this proton is accounted for by calculations of effects due to the anisotropy of the adjacent ring, the anisotropy and dipolar fields associated with the nearby nitrogen atom, and van der Waals interaction between the proton and the nitrogen lone pair. In methanol and other hydrogen-bonding solvents, signals from protons 4, 5, and 6 move downfield while that from proton 3 moves slightly upfield. This anomalous shift was interpreted in terms of a larger interplanar angle, between 80 and 100°, in the more polar solvents. Spotswood and Tanzer (23) obtained similar spectra in a number of solvents (Table 2.8) but disagreed with the interpretation of the results. They considered that contributions from van der Waals forces to the chemical shift of the 3 (3′)-proton would be small because the required interaction would be fully relieved by a slight twist of the 2–2′ bond. Such a slightly twisted conformation is not inconsistent with the present data and is in agreement with conclusions obtained using other techniques. From electronic spectra, they conclude that the interplanar angles for the bipyridyl in inert or hydrogen-bonding solvents must be similar. The anomalous shifts obtained in the latter solvents were accounted for in terms of attenuation of the field due to the lone-pair dipole by strong solute–solvent interactions, without invoking either van der Waals forces or variation of the interplanar angle. However, they reached the conclusion that 3,3′- and 4,4′-bipyridyls, **21** and **22**, are highly twisted in all solvents or, alternatively, behave as free rotors.

When 4,4′-dimethyl-2,2′-bipyridyl (**23**) is dissolved in acetone with variable proportions of trifluoroacetic acid, the signals from protons 5 and 6 shift continuously to lower field as the concentration of acid increases (64). However, that from H-3 shifts first to low field (τ 1.70 → 1.35 ppm) and then slowly to higher field (τ 1.55 in pure acid). The initial drop was associated with formation of a monocation, and the slow increase to the addition of a second proton. Diprotonation in trifluoroacetic acid was confirmed by comparison of the observed shifts with those obtained for the pyridinium ion and also by the lack of concentration effect (65). In these acid solutions, broadening of all lines of the spectrum takes place and errors in computed parameters, particularly in the coupling constants, are large. The changes observed for most parameters were safely outside experimental error and were interpreted according to the assumption that on protonation the two pyridine rings become more benzene-like (65). Interplanar angles of 25–30° for the monocation and 55–72° for the dication were suggested. Spotswood and his co-workers (66, 67) have made a detailed study of bridged biquaternary 2,2′-bipyridyls (**24**) and concluded that with these compounds the major

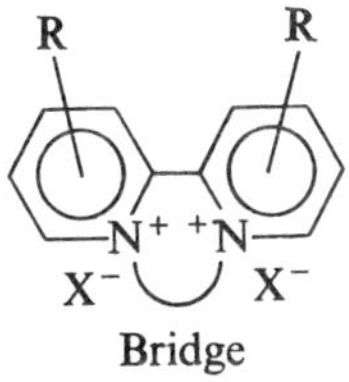

R R

N^+ ^+N

X^- X^-

Bridge

24

contribution which causes chemical shift differences between protons 3 and 5 is the diamagnetic anisotropy of the neighboring pyridinium ring, with significant contributions from the asymmetric distribution of the positive charge on the pyridinium ring. They comment on the inadequacies of their treatment when interplanar angles are small, but obtain reasonable results in other cases. Evidence is provided for the degree of conformational rigidity found in these compounds and the generalization is made that inversion in the bridged bipyridyl series is more difficult than with the corresponding biphenyl. Results are listed in Table 2.9.

C. Substituted Pyridines

Apart from the potentially tautomeric amino- and dihydro-oxopyridines, those substituted with groups other than alkly or aryl have been little studied. By far the most important paper on substituted pyridines is the classic accumulation by Brügel (38) in which spectra of 154 pyridines are presented. These spectral parameters are given in Tables 2.10–2.21 together with other

TABLE 2.8 SPECTRAL DATA FOR BIPYRIDYLS (23, 65)

Junction	Solvent	Molar fraction[a]	Chemical shifts[b]				Coupling constants					
			$\tau 3$	$\tau 4$	$\tau 5$	$\tau 6$	J_{34}	J_{35}	J_{36}	J_{45}	J_{46}	J_{56}
2,2′	CCl_4	0.203	1.50	2.34	2.87	1.40	8.01	1.16	1.00	7.53	1.80	4.75
		0.089	1.54	2.31	2.85	1.42	8.02	1.19	1.00	7.49	1.84	4.84
	$Me_3CC{\equiv}CH$	0.091	1.50	2.30	2.81	1.40	8.06	1.17	0.96	7.49	1.81	4.78
	$CHCl_3$	0.161	1.57	2.27	2.78	1.35	8.04	1.21	0.95	7.51	1.81	4.80
		0.075	1.60	2.21	2.71	1.28	8.02	1.22	0.95	7.54	1.83	4.81
	PrCOOH	0.092	1.66	2.14	2.64	1.23	8.00	1.15	0.91	7.62	1.77	4.97
	MeOH	0.084	1.69	2.13	2.64	1.35	8.03	1.17	0.95	7.50	1.78	4.85
		0.074	1.70	2.13	2.64	1.36	7.98	1.17	0.93	7.57	1.76	4.88
		0.040	1.71	2.10	2.61	1.35	8.03	1.16	0.92	7.57	1.76	4.87
	EtOH	0.100	1.62	2.12	2.64	1.36	8.00	1.17	0.95	7.56	1.78	4.84
	EtCOOH	0.089	1.68	2.10	2.60	1.22	7.99	1.15	0.93	7.59	1.80	4.97
	$CH_2(Cl)CH_2OH$	0.091	1.72	2.05	2.55	1.31	8.02	1.16	0.95	7.62	1.76	4.92
		$\ll$ 0.091	1.66	1.94	2.46	1.27	8.04	1.16	0.92	7.62	1.74	5.02
	MeCOOH	0.100	1.64	1.81	2.32	1.15	8.13	1.16	0.86	7.71	1.73	5.07
	MeCHBrCOOH	0.090	1.45	1.53	2.05	1.03	7.95	1.44	0.66	7.62	1.43	5.31
	CF_3COOH	0.089	1.32	1.12	1.65	0.84	8.12	1.03	0.37	8.09	1.30	5.79
		0.050	1.30	1.10	1.64	0.83	8.12	1.18	0.57	8.02	1.54	5.75
	CCl_4	0.08 *M*	1.55	2.28	2.82	1.42	8.0	1.30		7.6	1.8	4.8
	$CDCl_3$	0.08 *M*	1.51	2.20	2.72	1.32						
	CH_2Cl_2	0.08 *M*	1.58	2.20	2.71	1.36						
	MeI	0.08 *M*	1.58	2.19	2.72	1.37						

	Me_2CO	0.08 *M*	1.51	2.09	2.60	1.33			
	MeOH	0.08 *M*	1.71	2.08	2.58	1.36			
	D_2O	Saturated	2.03	2.09	2.55	1.43			
	$\tau 2$ replaces $\tau 3$ of 2,2′								
3,3′	CCl_4	0.08 *M*	1.23	2.19	2.69	1.45	7.8	1.6	4.9
	$CDCl_3$	0.08 *M*	1.16	2.10	2.59	1.35			
	Me_2CO	0.08 *M*	1.10	1.89	2.51	1.36			
	MeOH	0.08 *M*	1.15	2.01	2.44	1.40			
	D_2O	Saturated	1.45	2.11	2.54	1.50			

			$\tau 2, 6$	$\tau 3, 5$	$J_{23} = J_{56}$
4,4′	CCl_4	0.08 *M*	2.55	1.33	5.0
	$CDCl_3$	0.08 *M*	2.46	1.26	
	Me_2CO	0.08 *M*	2.25	1.28	
	MeOH	0.08 *M*	2.19	1.31	
	D_2O	Saturated	2.15	1.27	

METHYL DERIVATIVES OF 2,2′-BIPYRIDYL[c]

Methyls	Solvent	Molar fraction	$\tau 3$	$\tau 4$	$\tau 5$	$\tau 6$
3,3′	CCl_4	0.08 *M*	(7.84)	2.48	2.90	1.61
	MeOH	0.08 *M*	(7.90)	2.15	2.58	1.57
4,4′	CCl_4	0.08 *M*	1.72	(7.57)	3.00	1.59
	MeOH	0.08 *M*	1.90	(7.55)	2.76	1.54
5,5′	CCl_4	0.08 *M*	1.72	2.53	(7.69)	1.64
	MeOH	0.08 *M*	1.89	2.34	(7.64)	1.59

[a] Concentrations listed as molarities are from Ref. 23.
[b] The accuracy of the chemical shifts has been reduced by conversion to τ values. For more accurate data see references.
[c] Values in parentheses for methyl substituents.

TABLE 2.9 SPECTRAL DATA FOR BRIDGED BIQUATERNARY 2,2′-BIPYRIDYLS. 0.1 *M* SOLUTIONS IN D_2O (66, 67)

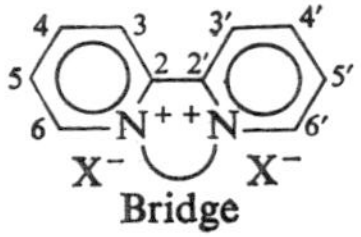

Bridge	Substituents[a]	τ3	τ4	τ5	τ6	J_{34}	J_{45}	J_{56}	J_{35}	J_{46}	J_{36}
$(CH_2)_2$	None	1.02	1.11	1.65	0.81	8.2	7.5	6.0	1.9	1.5	0.9
	3,3′-Di-Me	7.37	1.27	1.80	0.89		8.0	5.7		1.2	
	4,4′-Di-Me	1.24	7.20	1.89	1.08			6.3	1.5		
	5,5′-Di-Me	1.22	1.40	7.36	1.02	8.6				1.4	0.9
$(CH_2)_3$	None	1.48	1.11	1.61	0.75	8.1	7.7	5.9	1.8	1.4	0.7
	3,3′-Di-Me	7.55	1.24	1.75	0.92		8.4	5.7		1	
	4,4′-Di-Me	1.74	7.22	1.87	1.05			6.1	1.6		
	5,5′-Di-Me	1.70	1.34	7.36	0.95	8.2				1.6	
$(CH_2)_4$	None	1.59	1.15	1.53	0.66	8.1	7.7	5.9	1.8	1.5	0.6
	3,3′-Di-Me	7.67	1.27	1.68	0.79		8.3	5.8		1	
	4,4′-Di-Me	1.86	7.24	1.80	0.95			6.3	1.8		
	5,5′-Di-Me	1.81	1.38	7.34	0.84	8.3				1	
CH_2OCH_2	None	1.23	0.91	1.43	0.54	7.8	7.8	5.9	1.7	1.5	0.7
$CH_2C(OD)_2CH_2$	None	1.35	1.02	1.54	0.74	7.8	7.6	5.8	1.9	1.4	
$CH(Me)C(OD)_2CH(Me)$	None	1.41	1.06	1.53	0.77	7.8	7.9	5.8	1.5	1.4	

[a] Chemical shifts are listed for methyl groups in place of the hydrogen atom which they substitute. In all methyl derivatives $J_{MeH} = 0.5–0.7$ Hz

TABLE 2.10 SPECTRAL DATA FOR 2-SUBSTITUTED PYRIDINES. FROM REF. 38 UNLESS INDICATED IN SOLVENT COLUMN

R	Solvent	τ3	τ4	τ5	τ6	J_{34}	J_{35}	J_{36}	J_{45}	J_{46}	J_{56}	Other data[a]
Me	Neat	3.00	2.57	3.08	1.49	8.0	1.1	0.8	7.8	2.0	5.0	Me 7.54
	DMSO	2.73	2.26	2.78	1.33							
Et	Neat	2.83	2.41	2.91	1.32	7.8	1.2	0.8	7.7	2.0	4.8	CH_2 7.18, Me 8.72
	DMSO	2.71	2.33	2.77	1.38							7.14 8.74
CH_2Ph	Neat	2.75	2.62	3.02	1.25	8.7	1.3	1.3	8.3	2.5	5.2	CH_2 6.15, Ph 2.62
	DMSO	2.50	2.33	2.82	1.39							5.83 2.64
CD(Ph)Ac	$CDCl_3$[b]	~2.7	~2.7	~2.7	1.51							
CH=CH	DMSO	2.21	2.05	2.56	1.16	8.2	1.3	0.7	7.9	2.3	4.8	
Ph	CCl_4[c]	2.46	2.54	3.02	1.44	8.01	0.72	0.99	7.86	1.87	4.80	[d]
CH_2NH_2	Neat	2.56	2.32	2.83	1.33	8.0	1.2	0.8	7.7	2.1	5.0	CH_2 6.0, NH_2 7.86
	DMSO	2.42	2.18	2.71	1.36							6.06 7.92
CH_2NHMe	Neat	2.48	2.29	2.90	1.30	8.0	1.3	1.1	7.5	1.9	4.7	CH_2 6.00, Me 7.52
	DMSO	2.37	2.17	2.65	1.33							6.10 7.61
CH=NOH	DMSO	2.22	2.07	2.61	1.25	8.1	1.0	1.4	7.7	1.8	4.9	OH 2.20, =CH 1.54
CN	DMSO	1.74	1.87	2.07	1.02	8.0	1.0	0.4	7.2	1.2	5.0	
	Neat[e]					7.63	1.20	0.99	7.61	1.73	4.64	
CH_2OH	Neat	2.33	2.16	2.73	1.37	8.0	1.0	1.2	7.8	2.0	4.6	OH 2.99, CH_2 4.96
	DMSO	2.25	1.95	2.60	1.35							4.4 5.19
CHO	Neat	1.91	2.07	2.30	1.03	8.3	1.3	1.0	7.7	1.8	4.6	CHO 0.29, $J_{3,CHO}$ 0.3
	DMSO	1.69	1.83	2.12	0.97							0, 24
COMe	DMSO	1.80	1.88	2.23	1.13	7.48						Me 7.13
	Neat	1.86	1.94	2.32	1.14	7.9	1.4	0.9	7.9	1.8	4.9	7.21
	Neat[e]						1.31	0.84	8.02	1.79	4.53	
COPh	DMSO	2.00	1.70	2.30	1.13	8.3	1.0	1.2	7.8	1.6	4.7	
	CCl_4[f]	2.03	2.28	2.70	1.43	7.88	1.24	0.93	7.58	1.77	4.76	[g]
COCO	DMSO	2.18	2.02	2.53	1.13	8.0	1.6	1.3	7.9	1.8	4.9	
$CONH_2$	DMSO	1.57	1.78	2.19	1.11	8.1	1.2	0.8	7.4	2.1	4.5	NH_2 1.72, 1.92
$CONHNH_2$	DMSO	1.73	1.89	2.29	1.22	8.0	1.2	0.9	7.4	1.7	4.5	NH −0.28, NH_2 5.32

TABLE 2.10 (*continued*)

R	Solvent	τ3	τ4	τ5	τ6	J_{34}	J_{35}	J_{36}	J_{45}	J_{46}	J_{56}	Other data[a]
CONHNHCO	DMSO	1.84	1.89	2.23	1.19	8.1	1.2	1.0	7.6	1.8	4.8	NH −0.79
COOH	DMSO	1.65	1.82	2.14	0.99	8.0	1.0	1.0	8.0	2.0	5.0	COOH 2.11
COOBu	Neat	1.66	1.83	2.22	0.87	8.6	1.0	0.8	7.9	2.1	4.4	
	DMSO	1.76	1.86	2.27	1.06							
CH_2S-*n*-Pr	Neat[h]	2.58	2.33	2.88	1.47	8.0	1.8	1.0	6.8	1.9	4.9	
$CSNH_2$	DMSO	1.21	1.88	2.29	1.16	8.0	0.9	0.9	7.4	1.9	4.8	NH_2 −0.56, −0.36
CH_2SO_2Ph	DMSO[i]	~2.6	~2.6	~2.6	1.6							CH_2 5.19, Ph 2.34
NH_2	DMSO	3.30	2.56	3.40	1.89	8.3	1.8	1.0	6.9	1.7	5.0	NH_2 3.79
Piperidyl	DMSO	3.28	2.52	3.45	1.81	9.1	0.8	0.8	7.3	2.0	5.1	NCH_2 6.61, CCH_2 8.56
NHCOMe	DMSO	1.68	2.09	2.82	1.51	9.0	1.0	1.1	8.0	2.1	5.4	NH −0.67, Me 7.78
NHCONH	DMSO	1.99	2.04	2.76	1.48	8.6	1.0	1.2	8.0	1.8	4.8	NH −0.79
NHCOCONH	DMSO	1.82	2.30	2.64	1.44	8.5	1.1	1.0	8.0	1.8	4.7	NH −0.53
NHCOOEt	DMSO	2.03	2.18	2.86	1.62	7.8	1.0	1.0	7.2	1.6	4.8	
NHNHCOOEt	DMSO	1.72	1.81	2.19	1.14	8.0	1.0	1.0	7.2	1.8	4.8	
$NHNO_2$	DMSO	2.28	1.94	2.65	1.82	7.2	0.9	0.5	6.3	1.6	5.2	NH 1.80
NO_2	DMSO	1.53	1.58	1.88	1.15	8.2	1.0	0.9	8.1	1.2	5.2	
OMe	CCl_4[j]	3.38	2.60	3.30	1.94							OMe 6.16
OBu	Neat	3.63	2.52	3.21	1.78	8.5	0.9	0.8	7.3	2.0	5.1	
	DMSO	3.15	2.28	3.02	1.73							
F	Neat	2.83	1.97	2.60	1.39	8.7	0.9	1.0	7.8	2.0	4.9	$J_{3,F}$ 2.8, $J_{4,F}$ 7.4,
	DMSO	2.37	2.17	2.65	1.33							$J_{5,F}$ 2.5, $J_{6,F} < 1$
Cl	Neat	2.51	2.17	2.61	1.42	8.2	1.1	0.8	7.5	2.1	4.9	
	DMSO	2.30	1.96	2.33	1.31							
	Neat[e]					7.75	0.96	0.75	7.22	1.98	4.67	
Br	Neat	2.32	2.23	2.54	1.43	7.8	1.0	0.5	7.3	2.5	4.7	
	DMSO	2.21	2.08	2.43	1.39							
	Neat[e]					7.57	1.13	0.76	7.02	1.94	4.48	

[a] Chemical shifts (τ), coupling constants (Hz).
[b] Ref. 103. [c] Refs. 260.
[d] Phenyl data: τ2 2.02, τ3 2.66, τ4 2.73, J_{23} 7.9, J_{24} 1.25, J_{25} 0.62, J_{26} 1.95, J_{34} 7.41, J_{35} 1.42.
[e] Ref. 239. [f] Ref. 240.
[g] Phenyl data: τ2 1.89, τ3 2.66, τ4 2.54, J_{23} 7.9, J_{24} 1.34, J_{25} 0.59, J_{26} 1.73, J_{34} 7.44, J_{35} 1.30.
[h] Ref. 257. [i] Ref. 241. [j] Ref. 84.

TABLE 2.11 SPECTAL DATA FOR 3-SUBSTITUTED PYRIDINES. FROM REF. 38 UNLESS INDICATED IN SOLVENT COLUMN

R	Solvent	$\tau 2$	$\tau 4$	$\tau 5$	$\tau 6$	J_{24}	J_{25}	J_{26}	J_{45}	J_{46}	J_{56}	Other data[a]
Me	Neat	1.46	2.55	2.87	1.46	2.2	1.0	0.3	7.8	2.2	4.4	Me 7.86
	DMSO	1.43	2.31	2.71	1.43							7.71
CH=CHCOOH	DMSO	0.96	1.73	2.28	1.24	1.6	0.8	0	7.9	1.9	4.7	CH 2.33, 3.33 COOH 3.51
CH_2CN	Neat	1.16	2.11	2.54	1.30	1	0.3	0	8.0	1.6	4.7	CH_2 5.91
	DMSO	1.22	2.08	2.49	1.30							5.84
CH_2COOH	DMSO	1.10	1.85	2.30	1.10				8.0		4.5	COOH 2.52, CH_2 6.08
CH_2NH_2	Neat	1.25	2.24	2.68	1.37	2.0	0.7	0.3	7.8	1.7	4.9	CH_2 6.16, NH_2 7.91
	DMSO	1.25	2.12	2.58	1.41							6.14 7.94
CH_2NHMe	Neat	1.27	2.25	2.71	1.39	2.1	1.0	0.3	7.9	1.7	5.1	CH_2 6.28, Me 7.61
	DMSO	1.25	2.24	2.59	1.40							6.26 7.66
CH=NOH	DMSO	1.02	1.82	2.43	1.26	1.2	0.7	0.3	7.7	1.6	4.7	OH 1.90, CH 1.59
CN	DMSO	0.78	1.53	2.19	0.91	2.1	1.0	0.3	8.1	1.6	5.0	
CH_2OH	Neat	1.17	1.94	2.40	1.36	2.0	1.0	0	7.1	1.4	4.4	OH 2.91, CH_2 5.11
	DMSO	1.23	2.03	2.51	1.38							4.35 5.24
CHO	Neat	0.66	1.59	2.26	0.92	2.1	0.8	0.5	8.1	1.8	5.0	CHO 0.42
	Neat[b]					2.02	0.88	0.00	7.85	1.81	5.00	
COMe	Neat	0.66	1.63	2.42	1.06	1.9	1.0	0.3	8.1	1.6	4.7	Me 7.24

TABLE 2.11 *(continued)*

R	Solvent	$\tau 2$	$\tau 4$	$\tau 5$	$\tau 6$	J_{24}	J_{25}	J_{26}	J_{45}	J_{46}	J_{56}	Other data[a]
	DMSO	0.69	1.57	2.32	1.04							7.27
	Neat[b]					2.12	0.83	0.00	7.99	1.79	4.87	
COPh	DMSO	0.94	1.71	2.25	1.07	2.5	0.7	0	8.2	1.9	4.8	
$CONH_2$	D_2O[c]	1.08	1.76	2.38	1.24	2.3	0.7	0	8.0	1.8	4.9	
$CONHNH_2$	DMSO	0.83	1.68	2.37	1.16	1.6	0.8	0	7.9	1.9	4.7	NH −0.05, NH_2 5.35
$CONEt_2$	D_2O[c]	1.4	2.1	2.38	1.35	1.8			8.0	1.8	5.0	
COOH	NaOD[c]	1.0	1.66	2.42	1.38	2.0			8.0	2.0	5.2	
COOMe	DMSO	0.79	1.65	2.39	1.07	1.5	0.3	0	7.6	1.5	4.9	OMe 6.03
COOBu	Neat	0.65	1.63	2.48	1.08	2.0	0.5	0.3	8.0	1.5	4.6	
	DMSO	0.72	1.59	2.36	1.03							
CH_2SO_2Ph	DMSO[d]	1.65	2.4	2.4	1.40							CH_2 5.18, Ph 2.30
$CSNH_2$	DMSO	0.73	1.58	2.38	1.15	2.4	0.6	0.3	7.9	2.3	4.8	NH_2 2.52, 2.12
NH_2	DMSO	1.47	2.74	2.60	1.77	1.5	1.5	0	7.9	3.2	2.8	NH_2 4.20
NHCOMe	DMSO	1.04	1.75	2.56	1.57	2.4	0.5	0	7.7	1.3	4.5	NH −0.42, COMe 7.82
Piperidyl	Neat	1.47	2.89	2.73	1.74	1.8	1.8	0	9.0	3.1	2.8	NCH_2 6.91, CCH_2 8.51
	DMSO	1.50	2.68	2.51	1.80							6.76 8.40
OH	DMSO	1.44	2.62	2.47	1.65	1.3	1.3	0	9.2	3.0	3.3	OH 0.01
SO_3H	DMSO	0.71	1.11	1.81	0.71	2	0	0	8.5	2	5	
Cl	Neat	1.27	2.27	2.71	1.38	2.5	0.8	0.3	8.2	1.5	4.8	
	DMSO	1.21	2.01	2.43	1.32							
	Neat[b]					2.49	0.71	0.30	8.22	1.52	4.69	
Br	Neat	1.30	2.28	2.92	1.48	1.8	0.8	0	8.3	1.9	4.8	
	DMSO	1.21	1.82	2.28	1.23							
	Neat[b]					2.39	0.81	0.30	7.81	1.44	4.76	

[a] Chemical shifts (τ).
[b] Ref. 117.
[c] Ref. 241. [d] Ref. 86.

TABLE 2.12 SPECTRAL DATA FOR 4-SUBSTITUTED PYRIDINES. FROM REF. 38 UNLESS INDICATED IN SOLVENT COLUMN.

R	Solvent	$\tau 2, 6$	$\tau 3, 5$	$J_{23} = J_{56}$	$J_{25} = J_{36}$	J_{26}	J_{35}	Other data[a]
Me	Neat	1.45	2.96	5.2	0.8	0.4	1.7	Me 7.90
	DMSO	1.40	2.72					7.68
CH_2Ph	Neat	1.48	3.08	5.2	0.9	0.4	1.6	Ph 2.85, CH_2 6.31
	DMSO	1.41	2.77					2.77 6.05
$CH_2CH_2CONH_2$	DMSO	1.35	2.61	5.2	0.8	0.4	1.5	NH_2 2.71
$CH_2CH_2NH_2 \cdot HCl$	DMSO	0.85	1.90	5.2	0.9	0.4	1.6	
CH_2CH_2OH	Neat	1.39	2.61	5.0	0.8	0.4	1.5	
	DMSO	1.37	2.61					
$CH{=}CH_2$	DMSO	1.29	2.49	5.2	0.8	0.4	1.4	
	$CDCl_3$[b]	1.48	2.78					
$CH{=}CH$	DMSO	1.23	2.23	5.1	0.8	0.4	1.5	CH 2.34
CH_2COOMe	$CDCl_3$[c]	1.46	2.81					CH_2 4.93, Me 7.94
CH_2COOEt	DMSO	1.31	2.56	5.4	0.8	0.4	1.5	
CH_2COO—*i*-Pr	$CDCl_3$[c]	1.45	2.80					
CH_2NH_2	Neat	1.34	2.31	5.2	1.0	0.4	1.6	CH_2 5.89, NH_2 7.33
	DMSO	1.40	2.59					6.17 7.61
CH_2NHMe	Neat	1.33	2.66	5.2	1.0	0.4	1.6	CH_2 6.31, Me 7.65
	DMSO	1.38	2.59					6.30 7.69
CH_2NHEt	Neat	1.33	2.65	5.2	1.0	0.4	1.6	
	DMSO	1.39	2.57					
CN	DMSO	0.95	2.00	5.1	0.9	0.4	1.5	
$CH{=}NOH$	DMSO	1.17	2.25	5.2	0.8	0.4	1.5	OH −2.45, CH 1.60
(syn)	D_2O[d]	1.07	2.02					CH 1.37
(anti)	D_2O[d]	0.87	1.68					1.97
$C(Me){=}NNH_2$	DMSO	1.35	2.33	5.3	1.0	0.4	1.4	NH_2 2.97, Me 7.95
CH_2OH	DMSO	1.34	2.48	5.0	0.8	0.4	1.4	OH 4.82, $CH_2$0.41

TABLE 2.12 *(continued)*

R	Solvent	τ2, 6	τ3, 5	$J_{23} = J_{56}$	$J_{25} = J_{36}$	J_{26}	J_{35}	Other data[a]
CH(OH)CH(OH)	DMSO	1.51	2.67	5.0	0.8	0.4	1.4	OH 5.05, CH 3.99
C(OH)(CN)Me	DMSO	1.23	2.33	5.3	1.0	0.4	1.4	Me 8.09
C(OAc)(CN)Me	DMSO	1.17	2.38	5.3	1.0	0.4	1.5	COMe 7.76, Me 8.00
CHO	Neat	0.89	2.04	5.2	1.0	0.4	1.6	CHO −0.38
	DMSO	0.94	2.04					−0.33
COMe	Neat	1.03	2.12	5.1	0.8	0.4	1.7	Me 7.26
	DMSO	1.01	2.04					7.25
COPh	DMSO	1.05	2.20	5.2	1.0	0.4	1.6	Ph 2.20
COOBu	Neat	1.03	2.05	5.2	1.0	0.4	1.6	
	DMSO	1.07	2.08					
$CONHNH_2$	DMSO[e]	1.21	2.20					
$CONHNH_3{}^+I^-$	DMSO[e]	1.03	2.07					
CH_2SPr	Neat[f]	1.47	2.75	4.4	1.6		1.6	
CH_2SPh	CCl_4[g]	1.67	3.00					Ph 2.78, CH_2 6.15
CH_2SO_2Ph	DMSO[h]	1.50	2.80					CH_2 5.18, Ph 2.07
$CSNH_2$	DMSO	1.06	1.94	5.0	0.8	0.4	1.5	NH_2 −0.40
NH_2	DMSO	1.56	3.36					
NMe_2	DMSO	1.75	3.35	5.2	0.9	0.4	1.5	NMe_2 7.07
Piperidyl	DMSO	1.70	3.16	5.0	1.0	0.4	1.4	NCH_2 6.68, CCH_2 8.46
NHCOMe	DMSO	1.46	2.31	5.3	0.8	0.4	1.5	NH −0.48, Me 7.75
OMe	Neat	1.34	3.01	5.4	0.9	0.4	1.5	OMe 6.15
	DMSO	1.39	2.91					6.06
	10% D_2SO_4[i]	1.5	2.5	7.6				6.10
SCH_2Ph	DMSO	1.43	2.58	5.0	1.0	0.4	1.7	Ph 2.48, CH_2 5.58
SPh	Neat	1.49	2.97	5.2	1.0	0.4	1.6	Ph 2.45
	DMSO	1.36	2.78					2.22
Cl	Neat	1.38	2.73	5.4	1.0	0.4	1.6	
	DMSO	1.41	2.57					
Br	DMSO	1.32	2.27	5.4	0.8	0.4	1.8	

[a] Chemical shifts (τ).
[b] Ref. 243. [c] Ref. 103a.
[d] Ref. 111. [e] Ref. 100.
[f] Ref. 257. [g] Ref. 252.
[h] Ref. 241. [i] Ref. 246.

TABLE 2.13 SPECTRAL DATA FOR 2,3-DISUBSTITUTED PYRIDINES. FROM REF. 38 UNLESS INDICATED IN SOLVENT COLUMN

R^2	R^3	Solvent	$\tau 4$	$\tau 5$	$\tau 6$	J_{45}	J_{46}	J_{56}	Other data
Me	$COOCH_2Me$	Neat	1.91	2.85	1.41	7.9	1.8	4.9	CH_2 5.72, Me 7.25, CH_2*Me* 8.86
		DMSO	1.69	2.57	1.25				5.58 7.29 8.63
Me	COMe	Neat	1.79	2.62	1.35	7.9	1.7	4.8	Me 7.26, COMe 7.35
		DMSO	1.68	2.55	1.31				
Me	CN	DMSO	1.82	2.62	1.22	8.2	1.5	4.9	
NH_2	Me	Neat	2.97	3.57	2.06	7.4	1.7	4.8	NH_2 4.20, Me 8.10
		DMSO	2.72	3.46	2.01				4.10 7.91
NH_2	NH_2	DMSO	3.04	3.44	2.42	7.5	1.7	4.9	NH_2 4.78
Cl	Cl	DMSO	1.78	2.39	1.43	7.3	1.7	4.8	
Cl	NH_2	DMSO	2.62	2.85	2.10	7.9	1.9	4.4	NH_2 4.56
Me	OH	DMSO	2.61	2.72	1.94	6.8	0.8?	4.6	OH 1.29, Me 7.53
NO_2	OH	$CDCl_3$[a]	2.24	2.24	1.73		2.8	4.8	
S-n-Pr	Me	Neat[b]	2.90	3.30	1.75	7.5	1.5	5.0	

[a] Ref. 247.
[b] Ref. 257.

TABLE 2.14 SPECTRAL DATA FOR 2,4-DISUBSTITUTED PYRIDINES

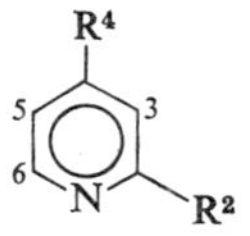

R^2	R^4	Solvent	τ3	τ5	τ6	J_{35}	J_{36}	J_{56}	Other data	Ref.
Me	Me	Neat	3.07	3.10	1.53	1.4	<0.5	4.2	Me 7.57 and 7.89	38
		DMSO	2.89	2.93	1.57				7.54 7.73	38
CMe_3	Me	$CDCl_3$	2.96	3.12	1.59				Me 7.69, Me_3 8.64	103a
SPr	Me	Neat	3.13	3.38	1.77	1.5	0.7	5.2		257
Me	SPr	Neat	3.05	3.16	1.71	1.5	0.7	4.5		257

TABLE 2.15 SPECTRAL DATA FOR 2,5-DISUBSTITUTED PYRIDINES

R^2	R^5	Solvent	$\tau 3$	$\tau 4$	$\tau 6$	J_{34}	J_{36}	J_{46}	Other data	Ref.
Me	Et	DMSO	2.83	2.21	1.32	8.1	0.7	2.4	Me 7.53, CH_2*Me* 8.84	38
Me	$CH{=}CH_2$	DMSO	2.76	2.23	1.24	8.1	0.5	2.0	Me 7.54	38
Me	OH	DMSO	2.85	2.77	1.77	8.7	0.6	2.3	OH 1.21, Me 7.50	38
NH_2	Cl	DMSO	3.37	2.51	1.95	9.0	<0.3	2.5	NH_2 3.80	38
Bu	NO_2	Neat	3.09	1.61	0.96	9.1	<0.3	2.8		38
		DMSO	2.93	1.41	0.82					38
Br	Br	DMSO	2.29	1.92	1.37	8.3	<0.3	2.5		38
Cl	Cl	DMSO	2.32	1.87	1.30	9.0	<0.3	3.0		38
NO_2	Me	$CDCl_3$	1.86	2.14	1.53	8.2	~0.3	2.1	Me 7.48	86
NO_2	OAc	$CDCl_3$	1.71	2.17	1.60	8.8	<1	2.5	Ac 7.62	247
SPr	Me	Neat	3.08	2.85	1.65	7.5	0.9	2.1		257
Me	SPr	Neat	3.03	2.47	1.51	7.4		2.1		257

TABLE 2.16 SPECTRAL DATA FOR 2,6-DISUBSTITUTED PYRIDINES. FROM REF. 38 UNLESS INDICATED IN SOLVENT COLUMN

R^6	R^2	Solvent	$\tau 3$	$\tau 4$	$\tau 5$	J_{34}	J_{35}	J_{45}	Other data
Me	Me	Neat	3.10	2.59	3.10	8.2		8.2	Me 7.50
		DMSO	2.96	2.42	2.96				7.56
CH=NOH	CH=NOH	DMSO	2.01	2.01	2.01				OH −2.00, CH 1.40
CN	CN	DMSO	1.51	1.48	1.51	8		8	
COOBu	COOBu	DMSO	1.62	1.62	1.62				
COMe	COMe	DMSO	1.72	1.72	1.72				Me 6.39
$CSNH_2$	$CSNH_2$	DMSO	1.07	1.71	1.07	7.8		7.8	NH_2 −0.61
NH_2	NH_2	DMSO	4.10	2.77	4.10	7.8		7.8	NH_2 4.46
Me	CH_2OH	Neat	2.48	2.35	2.96	8.1	0.7	8.2	OH 3.37, CH_2 5.23, Me 7.54
		DMSO	2.35	2.19	2.84				4.27, 5.26

Me	CH_2NH_2	Neat	2.80	2.44	2.99	7.3	1.0	7.8	CH_2 6.03, Me 7.50, NH_2 8.02
		DMSO	2.69	2.32	2.87				6.11 8.00
Me	CH_2NHMe	Neat	2.80	2.49	3.03	7.4	0.9	7.9	CH_2 6.15, Me 7.53, NMe 7.61
		DMSO	2.64	2.34	2.88				6.23 7.53
Me	CH=NOH	DMSO	2.20	2.37	2.65	8.1	0.6	8.3	OH −2.00, CH 1.71, Me 7.42
Me	CN	DMSO	1.91	2.05	2.24	7.8	1.2	8.0	Me 7.37
Me	COMe	Neat	2.15	2.19	2.59	7.4	1.3	7.5	Me 7.28 and 7.37
		DMSO	2.05	2.13	2.43				
Me	CHO	Neat	2.04	2.12	2.44	7.6	1.5	7.7	CHO −0.23, Me 7.31
		DMSO	1.93	2.08	2.35				−0.15
Me	COOBu	Neat	2.05	2.20	2.50	8	1	8	
		DMSO	1.95	2.10	2.40				
Me	$COOH{\cdot}H_2O$	DMSO	1.73	1.87	2.18	8	1	8	$COOH{\cdot}H_2O$ 3.81, Me 7.15
Me	$CSNH_2$	DMSO	1.39	1.99	2.39	7.7	1.4	7.9	NH_2 −0.30, Me 7.36
NH_2	NH_2	DMSO[a]	4.33	2.95	4.33				NH_2 4.71
OMe	OMe	CCl_4[b]	3.86	2.70	3.86	8			OMe 6.19
SPr	Me	Neat[c]	3.15	2.83	3.35	7.2	1.0	6.5	

[a] Ref. 88.
[b] Ref. 84.
[c] Ref. 257.

TABLE 2.17 SPECTRAL DATA FOR 3,4-DISUBSTITUTED PYRIDINES. FROM REF 261 UNLESS INDICATED IN SOLVENT COLUMN

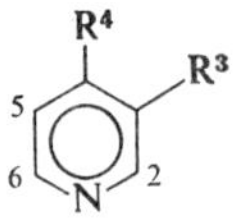

R^3	R^4	Solvent	$\tau 2$	$\tau 5$	$\tau 6$	J_{25}	J_{56}	Other data
CMe_3	Me	$CDCl_3$[a]	1.38	3.05	1.67			Me 7.49, Me_3 8.54
OAc	Me	$CDCl_3$[a]	1.74	2.98	1.74			Ac 7.96, Me 7.86
COOEt	Me	$CDCl_3$	0.95	2.86	1.47		5.05	
Me	NHMe	$CDCl_3$	1.96	3.61	1.85		5.5	Me 7.96, NMe 7.19
COOMe	COOMe	$CDCl_3$	0.92	2.48	1.15	0.5	5.2	Me 7.12
NO_2	OMe	$CDCl_3$	0.99	2.98	1.36		6.0	OMe 5.98
Ac	Et	CCl_4	1.18	2.88	1.54	0.8	6.25	Ac 7.44, Et 7.13, 8.82
CN	Et	CCl_4	1.29	2.73	1.37	0.8	6.4	Et 7.15, 8.68
Me	SPr	Neat[b]	1.77	?	1.70	0.7	5.0	
SPr	Me	Neat[b]	1.45	2.99	1.66	0.7	4.5	
COO^-	Me	NaOD	1.82	2.72	1.95		5.0	Me 7.58
COO^-	COO^-	NaOD	1.02	2.16	1.27	0.7	7.2	
COO^-	NHMe	NaOD	1.42	3.44	1.94		5.4	
COO^-	$NHNH_2$	NaOD	1.40	3.06	1.88	0.4	6.2	
COO^-	Cl	NaOD	1.43	2.45	1.58		5.3	
NH_2	COO^-	NaOD	1.88	1.42	1.08	0.8	5.2	
CO—NH—CO		TFA	0.48	1.37	0.63		6.0	

[a] Ref. 103a.
[b] Ref. 257.

TABLE 2.18 SPECTRAL DATA FOR 3,5-DISUBSTITUTED PYRIDINES

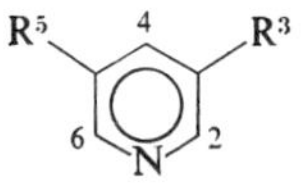

R^3	R^5	Solvent	$\tau 2$	$\tau 6$	$\tau 4$	J_{24}	J_{46}	Ref.
Cl	Cl	DMSO	1.29	1.29	1.77	1.7	1.7	38
Br	Me	Neat	1.63	1.77	2.45	2	2	255
Me	SPr	Neat	1.81	1.65	2.62	1.4	1.8	257

TABLE 2.19 SPECTRAL DATA FOR TRISUBSTITUTED PYRIDINES

A. 2,3,4-TRISUBSTITUTED PYRIDINES

R^2	R^3	R^4	Solvent	$\tau 5$	$\tau 6$	J_{56}	Other data	Ref.
NO_2	OH	Me	$CDCl_3$	2.43	1.92	4.4	Me 7.53	247
Me	OH	NO_2	$CDCl_3$–DMSO	2.33	1.92	5.5	Me 7.47	247
Cl	OH	NO_2	$CDCl_3$	2.05	1.84	5.5		247
OMe	NO_2	OMe	TFA	2.69	1.70	7.5	OMe 5.61, 5.77	262

B. 2,3,5-TRISUBSTITUTED PYRIDINES

R^2	R^3	R^5	Solvent	$\tau 4$	$\tau 6$	J_{46}	Other data	Ref.
NO_2	OMe	OMe	$CDCl_3$	2.96	2.21	3.0	OMe 5.96	244
NO_2	OH	Me	$CDCl_3$	2.53	2.12	1.8	Me 7.59	247

C. 2,3,6-TRISUBSTITUTED PYRIDINES

R^2	R^3	R^6	Solvent	$\tau 4$	$\tau 5$	J_{45}	Other data	Ref.
Me	COMe	Me	Neat	2.03	2.96	8.4	COMe 7.41, Me 7.54	38
			DMSO	1.85	2.76			
Me	COOH	Me	DMSO	1.74	2.73	8.1	COOH −1.30, Me 7.53	38
Me	COMe	Ph	DMSO	1.64	1.85	8.4	Ph 2.26, 1.76, Me 7.15	38
Me	COOH	Ph	DMSO	1.47	1.82	8.8	Ph 2.27, 1.67, COOH −2.03, Me 6.86	38
Me	COMe	OH	DMSO	1.97	3.68	9.7	OH −2.20, COMe 7.39	38
NH_2	NO	NH_2	DMSO	1.79	3.77	9	NH_2 (free) 2.25, NH_2 (bonded) −0.58	88

TABLE 2.19 *(continued)*

C. 2,3,6-TRISUBSTITUTED PYRIDINES–*continued*

R^2	R^3	R^6	Solvent	$\tau 4$	$\tau 5$	J_{45}	Other data	Ref.
NO_2	OH	Me	$CDCl_3$–DMSO	2.56	2.43	8.5	Me 7.51	247
Me	OH	NO_2	$CDCl_3$–DMSO	2.52	1.92	8.7	Me 7.56	247
Cl	OH	NO_2	$CDCl_3$–DMSO	2.42	1.80	8.7		247
NO_2	OH	NO_2	$CDCl_3$–DMSO	2.05	1.49	8.6		247
Cl	NO_2	Cl	D_2SO_4	1.85	2.66	9.0		262

D. 2,4,5-TRISUBSTITUTED PYRIDINES

R^2	R^4	R^5	Solvent	$\tau 3$	$\tau 6$	Other data	Ref.
OMe	$CH_2COCOOEt$	NO_2		2.2	0.9		256

E. 2,4,6-TRISUBSTITUTED PYRIDINES

R^2	R^4	R^6	Solvent	$\tau 3$	$\tau 5$	Other data	Ref.
Br	Me	Me	$CDCl_3$	3.13	3.30	4Me 7.82, 6Me 7.67	251

F. 3,4,5-TRISUBSTITUTED PYRIDINES

R^3	R^4	R^5	Solvent	$\tau 2$	$\tau 6$	Other data	Ref.
CO_2Me	Me	Et	$CDCl_3$	1.22	1.59		242
CO_2Me	CH_2COOMe	Et	$CDCl_3$	1.12	1.54		242
CH_2OH	$(CH_2)_2OH$	Et	$CDCl_3$	1.94	1.94		242
Br	OMe	Br	$CDCl_3$	1.43	1.43	OMe 6.02	86

TABLE 2.20 SPECTRAL DATA FOR TETRASUBSTITUTED PYRIDINES

A. 2,3,4,5-TETRASUBSTITUTED (113, 115)

R^2	R^3	R^4	R^5	Solvent	τ6	2Me	$4CH_2$	$5CH_2$	CMe_3	Other data
Me	O–C(Me)$_2$–OCH$_2$		CH_2OH	CCl_4	2.20	7.67	5.07	5.46	8.48	OH 5.58
Me	O–C(Me)$_2$–OCH$_2$		CH_2OPh	CCl_4		7.63	5.12	4.80	8.48	
Me	O–C(Me)$_2$–OCH$_2$		CHO	CCl_4	1.55	7.54	4.90		8.45	CHO −0.01
Me	O–C(Me)$_2$–OCH$_2$		$CONH_2$	CCl_4	1.62	7.55	4.85		8.43	NH_2 3.6
Me	O–C(Me)$_2$–OCH$_2$		COOMe	CCl_4	1.29	7.54	4.80		8.40	OMe 6.08
Me	OCH_2Ph	CH_2O–C(Me)$_2$–OCH$_2$		CCl_4		7.63	5.28	5.23	8.60	
Me	O-Mesyl	CH_2O–C(Me)$_2$–OCH$_2$		CCl_4	1.92	7.50	5.12	5.23	8.58	SOOMe 6.95
Me	OMe	CH_2OH	CH_2OH	CCl_4	1.94	7.56	5.32	5.23		OH 4.52, OMe 6.24
Me	OAc	CH_2OAc	CH_2OAc	CCl_4	1.52		4.75	4.88		Me 7.63, 7.68, 7.96, 8.04
CN	OAc	CH_2OAc	CH_2OAc	$CHCl_3$	1.27		4.76	4.58		Me 7.53, 7.88, 7.96

B. 2,3,4,6-TETRASUBSTITUTED (84)

R^2	R^3	R^4	R^6	Solvent	τ5	τMe	Other data
Cl	$CH_2CH_2CH_2Cl$	Me	Cl	$CDCl_3$	2.96	7.63	$CH_2CH_2CH_2Cl$ 7.2, 8.2, 6.3
Cl	CH_2CH_2Cl	Me	Cl	CCl_4	3.00	7.61	CH_2CH_2 6.34, 6.81

TABLE 2.20 (*continued*)

B. 2,3,4,6-TETRASUBSTITUTED (84)–*continued*

R^2	R^3	R^4	R^6	Solvent	$\tau 5$	τMe	Other data
Cl	CH_2CH_2CN	Me	Cl	$CDCl_3$	2.88	7.57	CH_2CH_2 6.90, 7.31
OH	CH_2CH_2OH	Me	OH	D_2O	exch.	7.59	CH_2CH_2 6.20, 7.10
Cl	CH_2CH_2COOH	Me	Cl	C_5D_5N	2.98	7.78	CH_2CH_2 6.90, 7.30
				$CDCl_3$	2.95	7.63	6.95, 7.35
Cl	CH_2CH_2COOMe	Me	Cl	CCl_4	2.99	7.62	CH_2CH_2 7.02, 7.46, OMe 6.38
Cl	$CH_2CH_2CH_2Br$	Me	Cl	CCl_4	3.02	7.63	$CH_2CH_2CH_2Br$ 7.2, 7.95, 6.57

C. 2,3,5,6-TETRASUBSTITUTED (244)

R^2	R^3	R^5	R^6	Solvent	$\tau 4$	Other data
NO_2	OMe	OMe	NO_2	$CDCl_3$	2.11	OMe 5.58

D. 2,4,5,6-TETRASUBSTITUTED (251)

R^2	R^4	R^5	R^6	Solvent	$\tau 3$	Other data
Br	Me	Br	Me	$CDCl_3$	2.94	4Me 7.37, 6Me 7.32

TABLE 2.21 SPECTRAL DATA FOR PENTASUBSTITUTED PYRIDINES

4,6-Di-Me, 2,3,5-tri-Br	Solvent—$CDCl_3$	4Me 7.37, 6Me 7.32	Ref. 251

published data on substituted pyridines. Correlations of coupling constants and internal chemical shifts were made.

Substituent-induced chemical shifts, obtained by subtracting the observed shifts from those obtained for pyridine itself, were calculated for a series of γ-substituted pyridines in dioxan (68) and found to be similar to those observed for *p*-disubstituted benzenes in cyclohexane. Moreover, these shifts could be fitted to the additivity relationship for the substituted benzenes (69), for example

$$\delta_2 = d_o(R_1) + \gamma(R_1)d_m(R_4)$$

where δ_2 is the shift of H-2 with respect to benzene, d_o and d_m identify the ortho and meta shifts of the monosubstituted benzenes in cyclohexane at infinite dilution, and $\gamma(R_1)$ is the polarizability of the group R_1. As in the case of benzenes, it was suggested that the π-electron density around the ring of the substituted pyridines was not the dominant factor in determining the chemical shifts of the ring protons but that another factor (or other factors) was of at least equal importance. In a study of substituent effects in α-substituted pyridines the substituent shifts $\Delta\gamma$ of the γ-proton correlated reasonably well with Hammett σ_m values (70). Haigh and Thornton (70a) have recently reexamined the relationships between Hammett constants and chemical shifts for a number of substituted pyridines. They separated the Hammett constants into their field (F) and resonance (R) components and showed that, while substituent shifts did not correlate at all with the F component, a smooth curve was obtained when they were plotted against the R component. Shielding parameters for substituted benzenes in dioxan were applied to spectra of 2,6- and 3,5-disubstituted pyridines (in dioxan), and chemical shifts could be predicted to approximately ± 0.1 ppm (70b).

In compounds with substituents which allow the ready formation of exocyclic double bonds at positions 2, 4, and/or 6 of the pyridine ring, substitution on N-1 causes disruption of the aromatic system. This loss of aromaticity is evident from their NMR spectra (Table 2.22). Thus, the 2-cyclopentadienylidine-1,2-dihydropyridines (**25**) (71) have spectra more akin to those of extended polyene systems than to those of true pyridines. The stable anhydrobases **26** (72) behave similarly. Particularly noticeable are the values of 9.3 and 6.5 Hz, respectively, for J_{34} and J_{56}, which are considerably larger than those observed for simple pyridines and approach those observed for rigidly held *cis-cis*-dienes.

TABLE 2.22 SPECTRAL DATA FOR DIHYDRO-OXOPYRIDINES AND RELATED COMPOUNDS

PYRID-2-ONES

Substituents	Solvent	τ3	τ4	τ5	τ6	J_{34}	J_{36}	J_{45}	J_{46}	J_{56}	Other data	Ref.
None	$CDCl_3$	3.40	2.67	3.80	2.77							249
	DMSO	3.27	2.27	3.57	2.27	10		7	2	7	NH −1.43	38
	D_2SO_4[a]	2.7	1.8	2.7	1.8							246
3-Me	DMSO		2.59	3.80	2.59			6.7			Me 7.95, NH −1.95	38
	D_2SO_4[a]		1.8	2.5	1.8				2.5		Me 7.6	246
5-Me	D_2SO_4[a]	2.8	1.8		2.05						Me 7.6	246
6-Me	$CDCl_3$	3.6	2.60	3.94		9		6			Me 7.65	245
6-OH	DMSO	4.25	2.66	4.33		8.0		8.0			OH + NH −0.08	84
6-OMe	$CDCl_3$	3.80	2.60	4.35		9.5		7.6			OMe 6.18	84
	CCl_4	3.87	2.85	4.15							OMe 6.40	84
1-Me, 6-OH	DMSO	4.33	2.79	4.33		8		8			NMe 6.72	84
1-Me, 6-OMe	CCl_4	4.10	2.90	4.63		9		6.5			OMe 6.18, NMe 6.74	84
	TFA	3.03	1.85	3.20		8.3		8.3			OMe 6.12, NMe 5.78	84
1-Me, 5-Ac	$CDCl_3$	3.41	2.08		1.83	9.6	2.6				COMe 7.55, NME 6.38	248
3-NO_2	DMSO		1.42	3.52	1.97			7.5	1.7	6.6	NH −2.93	38
5-COOH	DMSO	3.47	2.02		1.80	9.7			2.5			38
3-CN, 4,6-di-Me	TFA			3.22							Me 2.59, 2.64	253
3-CN, 4,6-Di-CH_2CH_2Ph	TFA			3.64								253
3-CN, 4-CH_2CH_2Ph, 6-Ph	TFA			3.31							Ph 2.88	253
3-CN, 4-Me, 6-Ph	TFA			3.36							Ph 2.83	253
3-CN, 5-Ph, 6-CH_2CH_2Ph	TFA		2.00									253
3-CN, 5-Ph, 6-Me	TFA		1.94								Ph 2.39–2.86, Me 7.52	253
3-CN, 6-CH_2CH_2Ph	TFA		1.90	3.36								253
3-CN, 6-Me	TFA		1.86	3.22							Me 7.37	253

PYRID-4-ONES

Substituents	Solvent	$\tau 2$	$\tau 3$	J_{23}	Other data	Ref.
None	50% in D_2O	2.02	3.37	7.532	J_{25} 0.260, J_{26} 2.733, J_{35} 1.855	276
	DMSO	1.98	3.48		NH 0.41	38
	D_2SO_4[a]	1.5	2.6	7.6		246
1-Me	D_2O	2.19	3.51	8.4	Me 6.19	276
	D_2SO_4[a]	1.6	2.6	7.6	5.8	246
1-CN	$CDCl_3$	2.41	3.46	8.4		276
1,2,6-Tri-Me	D_2O		3.12		CMe 7.12, NMe 5.90	89
1,3,5-Tri-Me	$CDCl_3$	2.78			7.98 6.38	89
1-Me, 3,5-Di-Br	$CDCl_3$	2.27				89

2-CYCLOPENTADIENYLIDINE-1,2-DIHYDROPYRIDINES

Substituents	Solvent	$\tau 3$	$\tau 4$	$\tau 5$	$\tau 6$	J_{34}	J_{35}	J_{45}	J_{46}	J_{56}	Other data	Ref.
None	Me_2CO	2.37	2.79	3.61	2.48	9.3	1.5	6.5	1.6	6.5	[b]	71a
3-Me	$CDCl_3$										Me 7.42	71
4-Me	$CDCl_3$										7.79	71
5-Me	$CDCl_3$										7.89	71
6-Me	$CDCl_3$										7.60	71
4,6-Di-Me	$CDCl_3$										7.82, 7.59	71

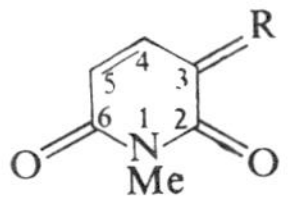

R	Solvent	$\tau 3$	$\tau 4$	$\tau 5$	J_{34}	J_{35}	J_{45}	Other data	Ref.
H_2	$CDCl_3$	6.60	3.27	3.80	3.7	2.0	10.2	NMe 6.81	84
NOH	DMSO		2.29	3.67			10.25	J_{15} 1.9, NH −1.47	88

[a] 10–20% D_2SO_4.
[b] $\tau 1'$ 3.43, $\tau 2'$ 3.79, $\tau 3'$ 3.79, $\tau 4'$ 3.43.

α-Pyridone has long been known to favor the 1,2-dihydro-2-oxopyridine form **27** over the tautomeric hydroxypyridine structure **28**. This is borne out by the similarity between its spectral parameters (Table 2.22) and those of the cyclopentadienyl derivative described above and of 1,2-dihydro-1-methyl-2-oxopyridine (**29**). Low-temperature experiments with 1,2-dihydro-2-oxopyridine-^{15}N showed that, in $CDCl_3$, this form is preferred by about 50:1 over the hydroxy tautomer (73). J_{NH} (90 Hz) was observed at −56°. Similarly, γ-pyridone exists to a large extent in the 1,4-dihydro-4-oxo form **30** and its spectrum compares favorably with that of 1,4-dihydro-1-methyl-4-oxopyridine (**31**) but not with that of 4-methoxypyridine (**32**) (74) (see Table 2.12). The corresponding amino derivatives, though potentially tautomeric,

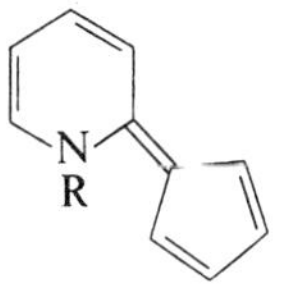

25

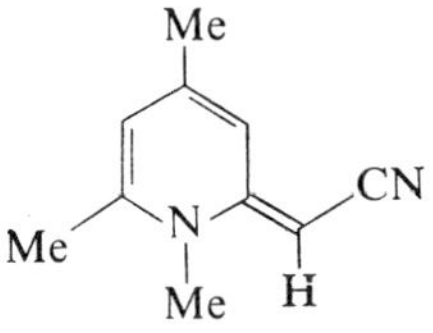

26

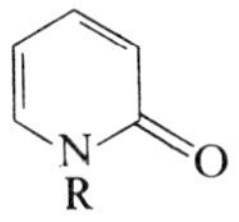

27, R = H
29, R = Me

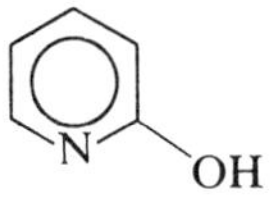

28

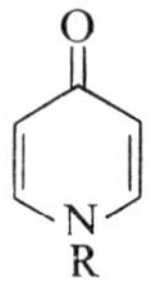

30, R = H
31, R = Me

32

exist mainly in the aromatic amino state (75) and have spectra similar to those of other simple substituted pyridines. Variable-temperature PMR was used to investigate hindered rotation about the exocyclic N–C bond in a number of 2-dimethylaminopyridines (74a). Rates were determined using complete line-shape analysis from the Gutowsky-Holm procedure.

Katritzky and his co-workers have studied the protonation of these pyridones and aminopyridines as well as the corresponding thiopyridones. In concentrated sulfuric acid, 2-pyridones were shown to protonate on the oxygen atom to give the species **33** (76). Proof involves observation of the signal from the 6-proton which, in the spectra of 4-methyl-, 5-methyl-, and 5-chloro-2-pyridone, are, respectively, a triplet, a quartet, and a quartet. These patterns establish the presence of the adjacent N^+–H group whose single proton couples with the 6-proton, $J_{16} = 7$ Hz. Meta couplings between the N^+–H proton and the 3-proton were also observed. Cox and Bothner-By (76a) have examined the spectra of 2-pyridone, its cation, and its anion, and their results confirm the fact that protonation occurs on oxygen. The spectra of 4-pyridones in sulfuric acid were broadened by exchange, and examination of the hydrochlorides in liquid sulfur dioxide was necessary before reasonable resolution was obtained. Again, the 6-proton gave rise to a triplet showing the presence of only one hydrogen on the adjacent nitrogen atom and hence confirming *O*-protonation (77). Spectra of solutions in sulfuric acid were used in a similar manner to show that the two thiopyridones **34** and **35** protonate on the thione grouping to give the species **36** and **37** (78). In the case of the aminopyridines, the position is made more complex by the presence of two nitrogen atoms. However, spectra of the monocations of 4-amino- (**38**), 2-amino-4-methyl- (**39**), and 2-amino-5-chloropyridine (**40**) in liquid sulfur dioxide and in dimethyl

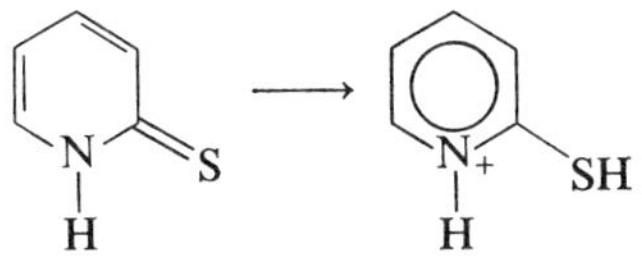

N S H **34** → N+ SH H **36**

S N H **35** → SH N+ H **37**

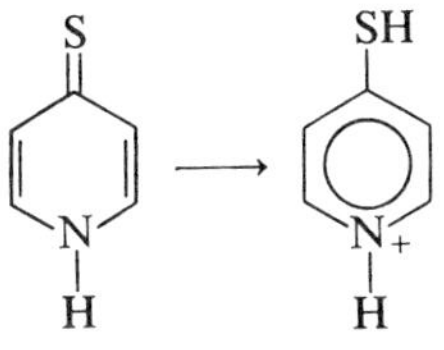

N+ OH H **33**

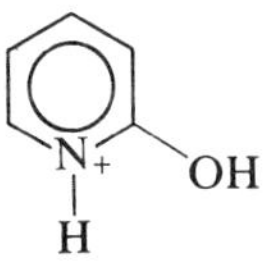

NH_2 N **38**

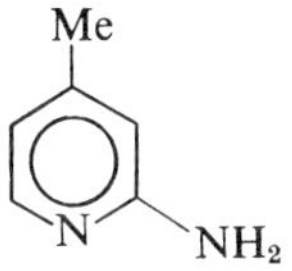

Me N NH_2 **39**

Cl N NH_2 **40**

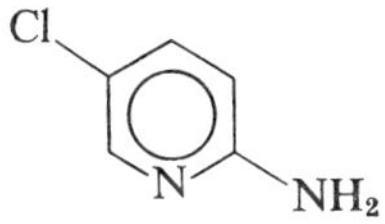

sulfoxide showed preferential protonation of the ring nitrogen atom (Table 2.23) (78), as would be expected from the pK_a values for these compounds (see 79).

Elvidge and Jackman (80) first defined an aromatic compound as one which will sustain an induced ring current. Their calculations, which include all the inherent uncertainties discussed above for pyridine itself, suggest that 2-pyridone has about 35% of the aromatic character of benzene, corresponding to a stabilization of 12–15 kcal/mole. On the other hand, the 1,2-dihydro-2-methylenepyridine (**41**) was found to be nonaromatic. An attempt has been made (81) to relate the observed long-range couplings between methyl and ring protons in a series of substituted picolines and *C*-methylated 2-pyridones and thiones with the aromaticity of the systems.

TABLE 2.23 SPECTRAL DATA FOR CATIONS OF AMINOPYRIDINES (78)

Pyridine	Anion	Solvent	$\tau 2$	$\tau 3$	$\tau 5$	$\tau 6$	J_{23}	J_{35}	J_{56}	J_{12}	τNH	τNH_2	τMe
4-NH_2	Cl^-	SO_2	1.86	2.93	2.93	1.86	7.5			6.0	−2.1	3.8	
	$SbCl_6^-$	DMSO	1.90	3.09	3.09	1.90	7.5				−3.0	2.0	
	Cl^-	DMSO	1.69	3.00	3.00	1.69	7.0				−3.6	1.5	
2-NH_2, 4-Me	Cl^-	SO_2		2.97	3.12	2.20		1.5	6.5		−2.8	3.1	7.55
	$SbCl_6^-$	DMSO		3.04	3.22	2.20			6.5		−2.8	2.2	
	Cl^-	DMSO		3.04	3.19	2.03		1.5	6.5			1.7	
				$\tau 3$	$\tau 4$	$\tau 6$		J_{34}	J_{46}	J_{36}	τNH	τNH_2	
2-NH_2, 5-Cl	Cl^-	SO_2		2.81	2.07	2.17		10.0	2.5		−3.0	2.8	
	$SbCl_6^-$	DMSO		2.96	2.00	1.80		9.0	2.0	1.0		2.1	
	Cl^-	DMSO		2.78	1.89	1.67		9.0	2.0	1.0		0.5	

Comparison of the spectrum of 4-hydroxypyridine-1-oxide (**42**) to those of 1-methoxypyrid-4-one (**43**) and 4-methoxypyridine-1-oxide (**44**) led to the suggestion that it exists in water as a mixture of 30% pyridone and 70% hydroxypyridine, which are rapidly interconverting (74).

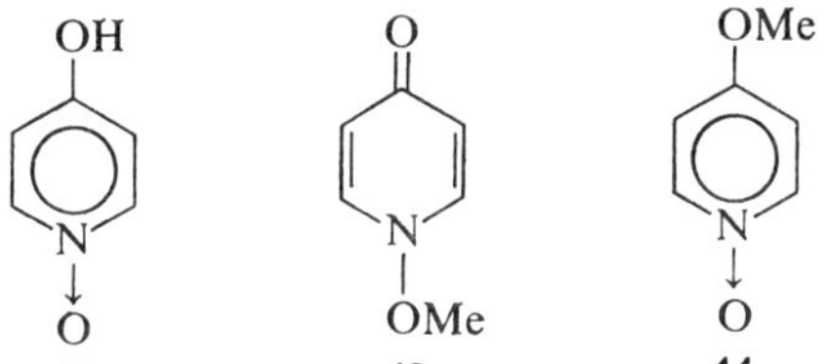

42 **43** **44**

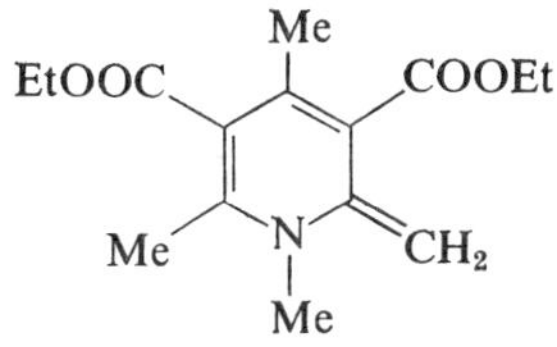

41

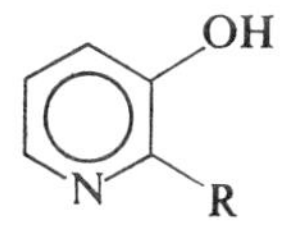

45, R = H
46, R = Alkyl.

The spectra of 3-hydroxypyridines (**45**) have been studied (82) and the π-electron densities obtained from the chemical shifts correlate roughly with those obtained from LCAO calculations. The results were used to determine substitution patterns in the amino- and hydroxymethylation of 2-alkyl-3-hydroxypyridines (**46**) (83).

When two potential oxo groups are present on the pyridine ring the situation becomes more complex. In the case of 2,6-dihydroxypyridine, glutaconimide, the molecule exists as a mixture of the two equivalent 1,2-dihydro-6-hydroxy-2-oxopyridines **47** and **48** in rapid equilibrium, so that the time-averaged spectrum observed seems to be that of a symmetrical molecule. However, the 1-methyl derivative has been shown to exist in deuterochloroform as the stable dioxo form **49** which is presumably in slow equilibrium with the equivalent form **50** through the hydroxypryidines **51** and **52** (84). Such an equilibrium is supported by the rapid exchange of all ring protons except that on C-4 after the solution is shaken with a drop of deuterium oxide, and was also proposed by Katritzky *et al.* (85) from pK_a and ultraviolet data. Spectra of the cations of glutaconimide and some of its derivatives were measured in trifluoroacetic acid (85) and from the data obtained (Table 2.24) it is quite evident that they all exist in the aromatic

TABLE 2.24 GLUTACONIMIDE AND MODEL COMPOUNDS (85). SOLVENT—TRIFLUOROACETIC ACID

	τ3, 5	τ4	J_{34}	τOMe	τNMe
Glutaconimide	3.2	1.8	8		
1,2-Dihydro-2-oxopyridine					
6-OMe	3.1	1.4	8	5.8	
1-Me-6-OMe	3.2	1.8	8	5.8	6.1
1-Me-6-OH	3.1	1.9	8		6.1
2,6-Dimethoxypyridine	3.1	1.5		5.8	
3,3-Dimethylpyridine-2,6(1*H*, 3*H*)-dione	3.6	2.9	10		

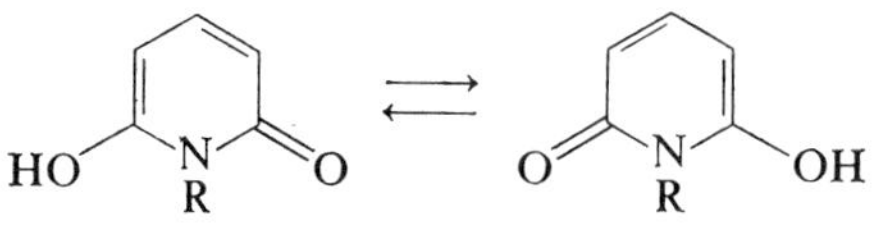

47, R = H
51, R = Me

48
52

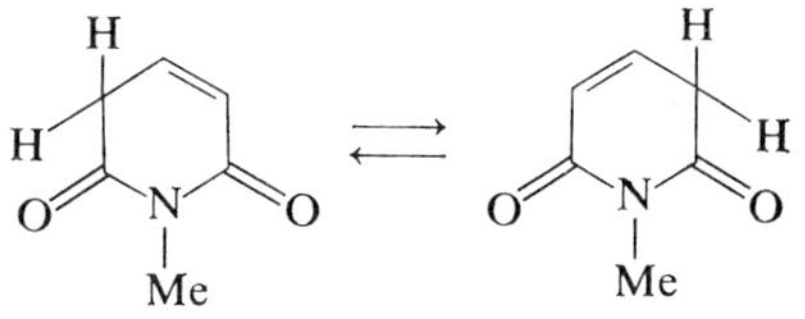

49

50

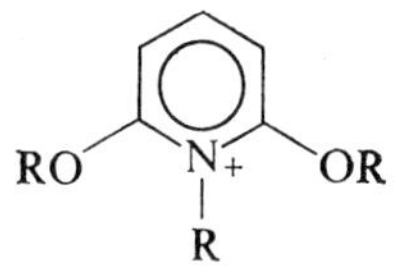

53

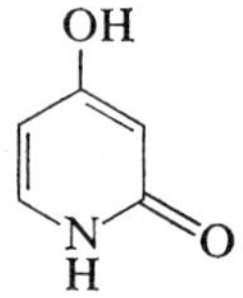

54

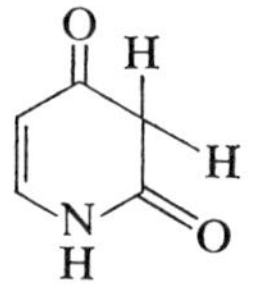

55

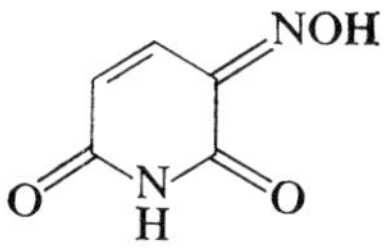

56

form **53**. In deuterated dimethyl sulfoxide the dioxo and monoxo forms are in equilibrium. With 2,4-dihydroxypyridine, the 2-oxo form **54** is known to predominate, although tautomerism between the two oxo compounds, presumably involving the dioxo form **55**, does occur. Thus, the 3-proton exchanges very readily in the presence of deuterium oxide. Interestingly enough, placing an ester group in position 5 slows down this exchange till it is almost unobservable (86). Spectra of a number of fixed 1,2,3,4-tetrahydro-2,4-dioxopyridines have been measured (87). Also, the 3-nitroso derivative appears to exist as the dioxo oxime **56** (88). Data for potential dioxo compounds are collected in Table 2.22.

Nuclear magnetic resonance methods provide a simple way for studying deuteration and have been used to observe the very specific α-deuteration of 1-methyl-4-pyridones in basic media (89). Thus, the 3,5-dimethyl derivative **57** is almost completely deuterated in the 2,6-positions after treatment with D_2O–NaOD at 100° for a short time. However, the 2,6-dimethyl analog **58** gives only the corresponding ditrideuteromethyl compound **59** after similar treatment overnight. Also, the *C*-methyl groups of 2-amino-4-methyl- (**39**), 2-amino-4,6-dimethyl- (**60**), and 1,4-dimethyl-1,2-dihydro-2-iminopyridine (**61**) slowly exchange at room temperature in aqueous solution, pH −0.5, and the last compound at a reasonable speed at pH 15.6 (90). Katritzky and his co-workers have used NMR to study deuteration of the ring protons of aminopyridines (91), pyridine *N*-oxides (92), and pyridones (93). The list of references given on this technique in the pyridine series is far from exhaustive and many further examples may be found in the literature.

The assignments reported by Paudler and Blewitt (94) for spectra of 2-aminopyridines and their cations are not internally consistent and disagree with those suggested by other authors (38, 90). Correct assignments, which follow from a study of substituent effects, are listed in Table 2.25. Small but consistent shifts allowed assignment of peaks to the *C*-methyl protons in the corresponding *N*-methylpyridinium salts (90) (Table 2.25). Rao and Baldeschweiler (95), in a study of the double-resonance spectra of closely coupled three-spin systems, have shown that the coupling constants J_{45}, J_{46}, and J_{56} for 2-amino-3-methylpyridine (**62**) all have the same sign. Also, association of aminomethylpyridines has been studied (96), but the authors were unable to reach firm conclusions about the factors involved in the concentration and solvent effects which they observed. Chemical shifts of

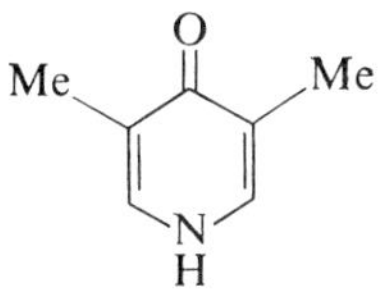

57

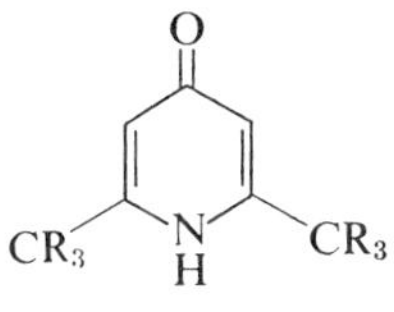

58, R = H
59, R = D

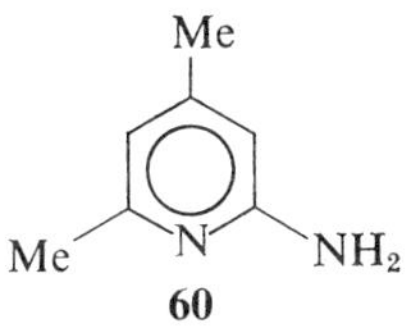

60

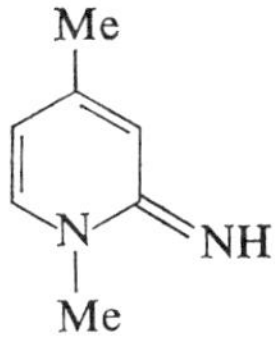

61

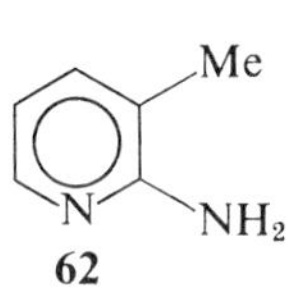

62

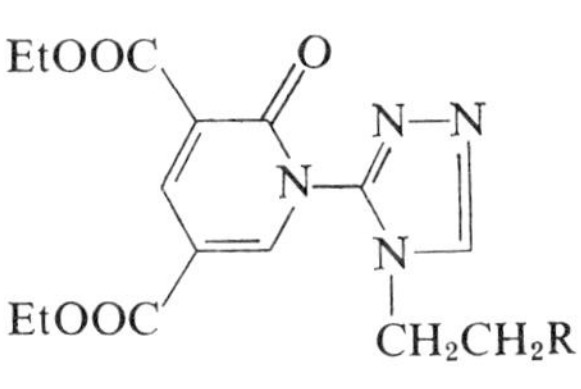

63, R = H
64, R = Ph

TABLE 2.25 CHEMICAL SHIFTS (τ) FOR 2-AMINOPYRIDINES, 2-AMINOPYRIDINIUM IONS, AND 1-METHYL-2-AMINOPYRIDINIUM IONS

Substituent	Solvent	X	τ3	τ4	τ5	τ6	τMe	τNMe	Ref.
None	D_2O	N	3.48	2.72	3.62	1.96			94
	DMSO	N	3.30	2.56	3.40	1.89			38
	D_2O	$N^+Me(Cl^-)$	2.85	2.10	3.02	2.10		6.13	90
3-Methyl	$CDCl_3$	N		2.67	3.35	1.97	7.88		94
4-Methyl	$CDCl_3$	N	3.60		3.50	2.04	7.76		94
	D_2O	$N^+D(Cl^-)$	3.18		3.20	2.28	7.59		94
	D_2O	$N^+D(Cl^-)$	3.06		3.10	2.20	7.53		90
	D_2O	$N^+Me(Cl^-)$	3.03		3.14	2.20	7.53	6.12	90
5-Methyl	$CDCl_3$	N	3.49	2.63		1.99	7.81		94
	D_2O	$N^+D(Cl^-)$	3.10	2.34		2.22	7.78		94
6-Methyl	D_2O	$N^+D(Cl^-)$	3.20	2.10	3.07		7.45		90
	D_2O	$N^+Me(Cl^-)$	3.15	2.25	3.02		7.33	6.17	90
4,6-Dimethyl[a]	$CDCl_3$	N	3.58		3.76		7.63 7.81		94
	D_2O	N	3.65		3.65		7.68 7.87		94
	D_2O	$N^+D(Cl^-)$	3.28		3.28		7.57 7.61		94
	D_2O	$N^+D(Cl^-)$	3.27		3.27		7.52 (6) 7.62 (4)		90
	D_2O	$N^+Me(Cl^-)$	3.19		3.19		7.40 (6) 7.62 (4)	6.27	90

[a] Numbers in parentheses after methyl shifts indicate position.

ring protons of protonated and nonprotonated aminopyridines were shown to correlate reasonably well with calculated π-electron distributions (SCF) (97).

Spectra of the 1-(1,2,4-triazol-3-yl)-2-pyridones **63** and **64** have been measured and differences between the chemical shifts of the upfield doublets from the pyridine protons (presumably from H-4) have been attributed to long-range shielding of the proton by the side-chain phenyl ring of **64** (98) (see Table 2.22).

The spectra of many substituted pyridinium ions have been measured but, with the exception of the cases discussed earlier in the chapter, little detailed work has been done. The general variations of chemical shifts and coupling constants observed on protonation of pyridine hold for most of its simple derivatives and the available data are collected in Table 2.26. Biellmann and Callot (99) have measured the spectra of a large number of 1,4-disubstituted pyridinium ions and, in the case of the 1-ethyl compounds

TABLE 2.26 SPECTRAL DATA FOR 1,4-DISUBSTITUTED PYRIDINIUM SALTS IN D_2O.[a] REFERENCE 99 UNLESS OTHERWISE SPECIFIED

R	R^1	X^-	τ2, 6	τ3, 5	Other data	Ref.
Me	CF_3	I^-	0.91	1.60		
Et	CF_3	I^-	0.86	1.60		
Me	CN	I^-	1.07	1.66		
Et	CN	I^-	0.91	1.58		
Pr	CN	I^-	0.86	1.54		
Allyl	CN	Br^-	0.96	1.60		
i-Pr	CN	I^-	0.82	1.58		
CH_2Ph	CN	Cl^-	0.98	1.74		
CH_2CH_2Ph	CN	Br^-	1.32	1.84		
CH_2COOEt	CN	Br^-	0.94	1.57		
CH_2CH_2COOEt	CN	Br^-	0.79	1.57		
Me	COOH	I^-	1.12	1.66		
Et	COOH	I^-	1.05	1.63		
Me	COOMe	I^-	0.94	1.44		
Et	COOMe	I^-	0.94	1.54		
Me	$CONH_2$	I^-	1.05	1.74		
Et	$CONH_2$	I^-	0.94	1.63		
Me	Cl	I^-	1.35	2.00		
Me	Br	I^-	1.25	1.63		
Me	NHCOMe	I^-	1.66	2.14		
Et	NHCOMe	I^-	1.68	2.25		
Me	Me	I^-	1.50	2.25		
Et	Me	I^-	1.37	2.21		
Et	Et	I^-	1.23	2.05		
Me	H	I^-	1.23	1.96	τ4 1.47	
Et	H	I^-	1.24	2.02	τ4 1.59	
Et	COOEt	I^-	0.88	1.46		
Me	OEt		1.74	2.72	J_{23} 7.5, NMe 5.85, OEt 5.65, 8.49	250
Et	Pyrrolidyl		2.21	3.50	J_{23} 7.5, NEt 6.10, 8.82	250
Me	Synoxime	I^-	0.70	1.33	C*H*N 1.18	111
Me	Antioxime	I^-	0.62	1.04	1.74	111

[a] Data for a number of variously substituted pyridinium ions are included in Tables 2.10 to 2.20 where acids are listed as solvent.

65, they observed couplings between the ^{14}N nucleus and H-3(5), $J = 3.0 \pm 0.2$ Hz, and between the nitrogen nucleus and the side-chain methyl group, $J = 2.4 \pm 0.1$ Hz, which appear to be independent of the nature of the substituent in position 4. Protonation of isonicotinoylhydrazine has been shown to occur on the hydrazino group to give the species **66** (100) (see Table 2.12). A number of broad-line studies of molten pyridinium salts have been reported (100a).

The effects of *N*-oxide formation on the spectrum of the parent pyridine were first reported for 4-aminopyridine (**38**), 4-aminopyridine-1-oxide (**67**), and the corresponding protonated species **68** and **69** (101). Unfortunately,

R, N^+, X^-, CH_2, CH_3 — **65**; $CONHNH_3^+Cl^-$, N — **66**; NH_2, N, ↓, O — **67**; NH_2, N^+, H — **68**; NH_2, N^+, OH — **69**

in this work chemical shifts were measured from the internal water peak which was located with respect to external tetramethylsilane. Thus, the values reported in Table 2.27 cannot be corrected to the τ scale. These results,

TABLE 2.27 *N*-OXIDES OF 4-AMINOPYRIDINES

R = H, Chemical shifts[a]		R = Me, Chemical shifts[a]	
X	α-H	α-H	β-H
N	−3.10	−3.08	−1.49
NO	−2.93	−2.96	−1.73
N^+–H	−3.46	−3.28	−2.16
N^+–OH	−3.65	−3.54	−2.21

[a] Chemical shifts (in parts per million) measured with respect to internal water. Water peak for neutral molecules 5.22 ppm downfield from external TMS, that for cations (20 *N* H_2SO_4) 9.60 ppm downfield from external TMS.

and others which can be obtained from the tables, show that *N*-oxide formation causes shifts of the α-protons to slightly higher fields and of the β-protons to slightly lower fields. As has been mentioned earlier, these shifts are associated with considerable sharpening of the peaks in the spectrum and the appearance of an observable J_{26} (2.3 Hz). Spectra have been reported of 11 pyridine 1-oxides in CCl_4, D_2O, or D_2SO_4 (101a). Shifts of signals from the ring protons were correlated with electron densities. A full analysis of the spectrum of pyridine *N*-oxide has been produced only recently (36). Nuclear magnetic resonance methods have been used by a number of groups to study the rearrangements of *N*-acetylpicolines (102, 103). Data for *N*-oxides and other 1-substituted pyridines are collected in Table 2.28.

Evidence was obtained from NMR measurements for the existence of the ring-open intermediate **70** in the formation of pyridine and formaldehyde from the treatment of *N*-methoxypyridinium perchlorate (**71**) with hydroxide ion (104).

Spectra of pyridinium, *N*-nitrosopyridinium, and *N*-nitropyridinium tetrafluoroborates (**72**) have been measured at $-40°$ in liquid sulfur dioxide (105) and results are summarized in Table 2.29. The general decrease in shielding on going from $H \rightarrow NO \rightarrow NO_2$ is in accord with the relative electron-withdrawing power of the nitroso and nitro groups. Replacement of the BF_4^- ion with the bulkier anions PF_6^-, AsF_6^-, or SbF_6^- had no effect on the spectra.

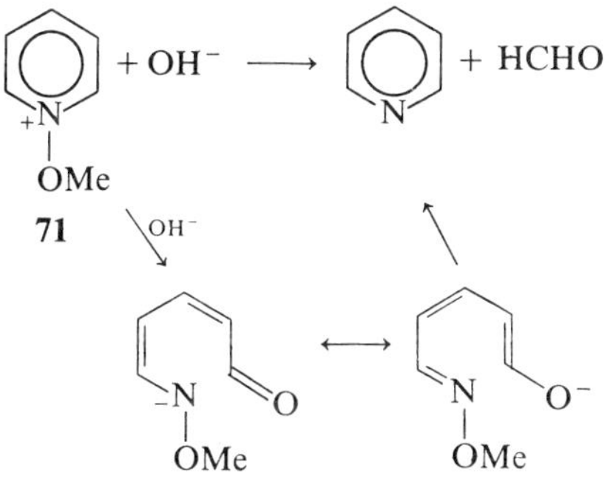

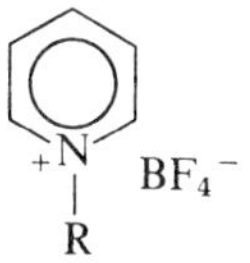

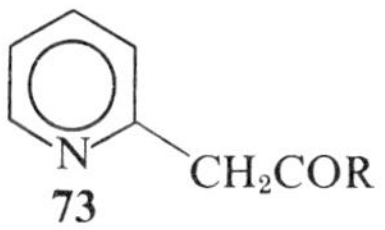

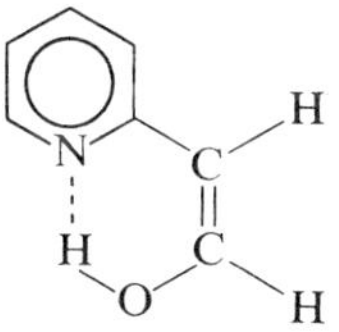

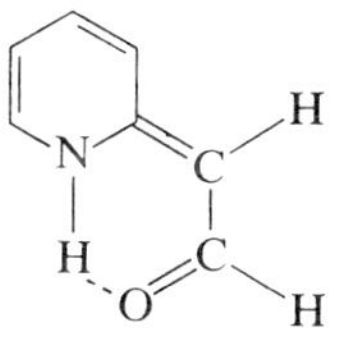

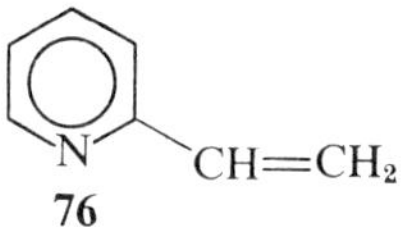

TABLE 2.28 SPECTRAL DATA FOR PYRIDINE *N*-OXIDES

Unsubstituted[a]	Solvent	τ2, 6	τ3, 5	τ4	Ref.
	CCl_4	1.90	2.72	2.92	101a
	D_2O	1.58	2.27	2.27	101a
	D_2SO_4[b]	1.19	1.90	1.51	101a

2-Substituted		τ3	τ4	τ5	τ6	τMe	Ref.
2-Me	CCl_4	2.78	2.78	2.78	1.80	7.62	101a
	D_2O	2.38	2.38	2.38	1.58	7.41	101a
	D_2SO_4[b]	2.04	1.64	2.04	1.26	7.14	101a
2-CH_2Ph	$CDCl_3$		2.6–3.2		1.8		103

3-Substituted		τ2	τ4	τ5	τ6	τMe		Ref.
3-Me	CCl_4	1.94	2.97	2.81	2.20	7.73		101a
	D_2O	1.71	2.35	2.35	1.73	7.58		101a
	D_2SO_4[b]	1.28	1.65	1.97	1.28	7.38		101a
3-CH_2OH	D_2O	1.56	2.23	2.23	1.63			101a
	D_2SO_4[b]	1.11	1.52	1.87	1.18			101a
3-Cl	D_2O	1.65	2.21	2.36	1.73		J_{45} 8.8, J_{56} 6.0 J_{46} 1.6, J_{24} 1.5	279
3-OH	H_2SO_4[c]	1.65	2.15	2.15	1.65			280

4-Substituted		τ2, 6	τ3, 5	τMe	Ref.
4-Me	CCl_4	2.13	3.03	7.70	101a
	D_2O	1.79	2.55	7.56	101a
	D_2SO_4[b]	1.44	2.18	7.34	101a
4-OEt	CCl_4	2.11	3.30		101a
	D_2O	1.78	2.84		101a
	D_2SO_4[b]	1.47	2.64		101a

2,6-Disubstituted		τ3, 4, 5	τMe	Ref.
2,6-Di-Me	H_2SO_4[c]	1.60	7.20	280

3,4-Disubstituted		τ2	τ5	τ6	J_{25}	J_{26}	J_{56}	τMe	Ref.
3-Me, 4-Cl	$CDCl_3$	1.87	2.73	1.94		2.2	6.7	7.68	261
	CCl_4	1.88	2.83	2.05				7.68	101a
	D_2O	1.72	2.39	1.84				7.63	101a
	D_2SO_4[b]	1.35	1.99	1.42				7.45	101a
3-Me, 4-NO_2	CCl_4	2.04	2.04	2.04				7.39	101a
	D_2O	1.53	1.75	1.62				7.39	101a
	D_2SO_4[b]	1.06	1.49	1.14				7.27	101a
3-Me, 4NHMe	D_2O	2.24	3.53	2.15		2.3	7.0		261
3-COO^-, 4-Cl	NaOD	1.58	2.30	1.73		2.5	6.8		261

TABLE 2.28 (*continued*)

3,4-Disubstituted		τ2	τ5	τ6	J_{25}	J_{26}	J_{56}	Ref.
3-COO⁻, 4-NHNH$_2$	NaOD	1.63	2.86	2.05	0.2	2.5	7.4	261
3-COO⁻, 4-NHMe	NaOD	1.64	3.38	2.13		2.6	7.3	261
3-COOH, 4-NO$_2$	TFA	0.65	1.75	0.85		1.8	6.7	261

3,5-Disubstituted	Solvent	τ2, 6	τ4	τMe		Ref.
3,5-Di-Me	CCl_4	2.26	3.27	7.78		101a
	D_2O	2.00	2.59	7.70		101a
	D_2SO_4[b]	1.62	1.96	7.51		101a
	H_2SO_4[c]	1.50	1.85	7.45		280
3,5-Di-OMe	TFA	1.85	2.05		OMe 5.85	244
	H_2SO_4[c]	1.80	2.45		5.90	280
3,5-Di-Cl	MeOH	1.53	2.21			279

2,4,5-Trisubstituted	τ3	τ6	τMe	Ref.
2-(1-Hydroxycyclohexyl), 4-Cl, 5-Me	2.82	2.02	7.73	281

2,4,6-Trisubstituted		τ3, 5	τMe	Ref.
2,4,6-Tri-Me	H_2SO_4[c]	2.15	6.95 (2, 6) 7.15 (4)	280
2,6-Di-Cl, 4-NO$_2$	TFA	1.6		282

3,4,5-Trisubstituted		τ2, 6	Ref.
3,5-Di-Cl, 4-NO$_2$	TFA	1.10	282

2,3,5,6-Tetrasubstituted	τ4	τOMe	Ref.
2,6-Di-NO$_2$, 3,5-Di-OMe	2.11	5.58	244

PYRIDINE *N*-OXIDE—COMPUTED (36)

	Chemical shifts (Hz from TMS at 60 MHz)			Coupling constants (Hz)					
Solvent	H2	H3	H4	J_{23}	J_{24}	J_{25}	J_{26}	J_{34}	J_{35}
10% in Acetone 30°	491.65	444.03	439.08	6.47	1.12	0.63	1.88	7.65	2.13
40°	491.98	443.87	438.93	6.50	1.11	0.69	1.91	7.66	2.16
TFA	530.21	483.96	505.74	6.65	1.10	0.52	2.05	7.83	1.81

[a] Pyridine *N*-Oxide.
[b] D_2SO_4.
[c] 10% H_2SO_4.

TABLE 2.29 DATA FOR *N*-SUBSTITUTED PYRIDINIUM TETRAFLUOROBORATES (105). LIQUID SO_2 AT $-40°$, ppm FROM EXTERNAL TMS

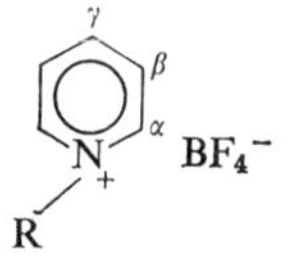

R	α-H	β-H	γ-H
H	8.22	7.54	8.03
NO	8.63	7.77	8.25
NO_2	9.15	7.88	8.52

Tautomerism of 2-picolyl ketones has been studied by Klose and his co-workers (106, 107) who observed the equilibrium between the keto **73** and enol **74** forms but found no evidence for the presence of the 1,2-dihydro-tautomer (**75**).

Proton chemical shifts for cyanopyridines (obtained from Ref. 38) were correlated with electron densities calculated by a simple HMO method and the relationship

$$\tau_i = 14.500q_i - 12.272 \text{ ppm}$$

was obtained with a high correlation coefficient, $r = 0.976$ (108).

The tacticity of polymers of 2-vinylpyridine (**76**) has been studied by NMR methods (109). Polymerization initiated by phenyl magnesium bromide gave an isotactic polymer, while initiation with ammonium persulfate led to a syndiotactic polymer. Initiation with azobisisobutyronitrile gave an intermediate product. The spectrum of poly-4-vinylpyridine could not be resolved.

Olah and Calin (110) have studied the protonation of acetyl- and benzoyl-pyridinium ions **77** at $-60°$ in $FSO_3H–SbF_6–SO_2$ solutions. *O*-Protonation was proven by the presence of hydroxyl peaks in the NMR spectra. With the 3- and 4-substituted compounds, two hydroxyl peaks were observed, indicating the formation of the stereoisomers **78** and **79**, but in the case of the 2-substituted derivatives only one hydroxyl peak was observed, and it was thought that the nearness of the two positive charges would stabilize the form with the hydroxyl bond away from the heterocyclic ring **80**. Results are summarized in Table 2.30.

π-Electron densities have been calculated for a large number of substituted pyridines and for their complexes with various Lewis acids (49a). A linear relationship between the π-electron densities and the proton chemical shifts was obeyed with reasonable accuracy. The results permitted the estimation

TABLE 2.30 ACETYL AND BENZOYLPYRIDINES, PYRIDINIUM IONS, AND PROTONATED PYRIDINIUM IONS (110). CHEMICAL SHIFTS IN ppm DOWNFIELD FROM EXTERNAL TMS

	H-2	H-3	H-4	H-5	H-6	Me	OH^+	$J_{Me,OH}$	NH^+	Ph
2-Acetylpyridine[a]		8.0	7.8	7.5	8.7	2.7				
2-Acetylpyridinium ion[b]		8.9	8.7	8.3	9.2	2.8			14.5	
Protonated 2-acetylpyridinium ion[c]		9.7	9.7	9.4	9.7	4.0	17.4		13.9	
3-Acetylpyridine[a]	9.2		7.5	8.2	8.8	2.7				
3-Acetylpyridinium ion[b]	9.1		8.2	8.8	9.0	2.7			14.8	
Protonated 3-acetylpyridinium ion[c]	10.0		9.0	9.7	10.0	4.0	16.0	0.9	13.2	
							15.0			
4-Acetylpyridine[a]	8.5	7.4		7.4	8.5	2.4				
4-Acetylpyridinium ion[b]	8.8	8.4		8.4	8.8	2.7			14.8	
Protonated 4-acetylpyridinium ion[c]	9.6	9.4		9.4	9.6	4.0	16.5	0.9	13.3	
							16.3			
2-Benzoylpyridine[a]		8.2	8.0	7.9	8.6					7.4
2-Benzoylpyridinium ion[b]		8.6	8.2	8.1	8.9				14.7	7.4
Protonated 2-benzoylpyridinium ion[c]		9.5	9.3	9.2	9.8		14.7		13.5	8.7
3-Benzoylpyridine[a]	9.0		7.1	8.1	8.8					7.6
3-Benzoylpyridinium ion[b]	8.9		8.2	8.8	8.9				15.2	7.6
Protonated 3-benzoylpyridinium ion[c]	9.6		9.0	9.6	9.6		14.2		13.2	8.6
							13.9			
4-Benzoylpyridine[a]	8.7	7.7		7.7	8.7					7.5
4-Benzoylpyridinium ion[b]	8.9	8.2		8.2	8.9				15.1	7.6
Protonated 4-benzoylpyridinium ion[c]	9.6	8.9		8.9	9.6		14.5		13.4	8.7
							14.1			

[a] Solvent is $CDCl_3$ at room temperature.
[b] Solvent is SO_2 at $-60°$.
[c] Solvent is SO_2–SbF_6–FSO_3H at $-60°$.

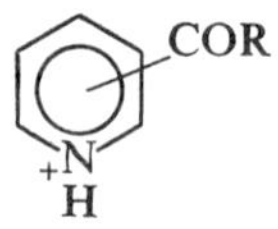

77, R = Me or Ph 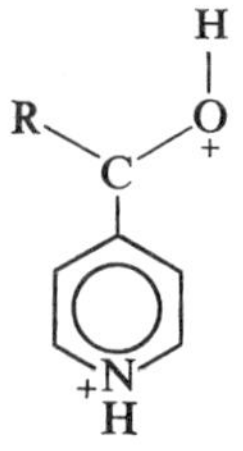 **78** 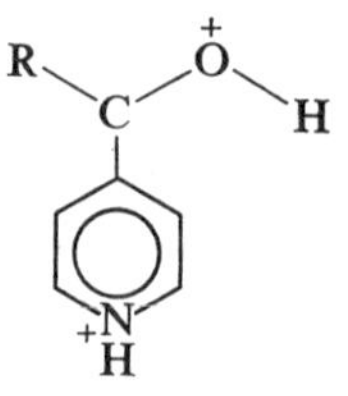 **79** 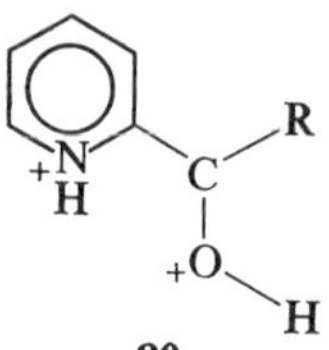 **80**

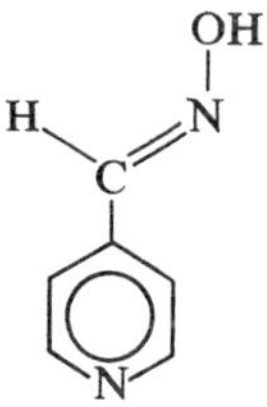

81 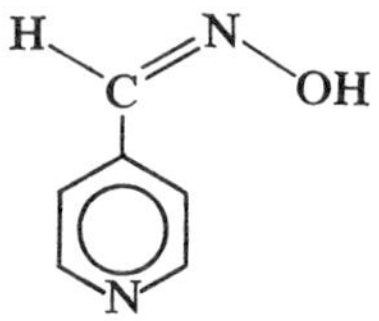 **82** 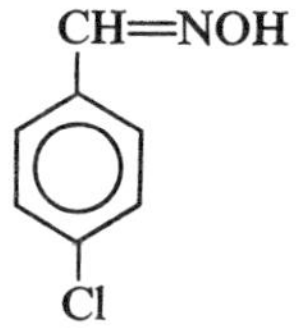 **83**

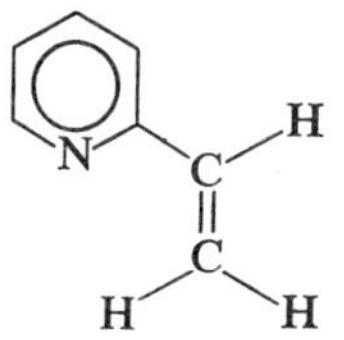

80a 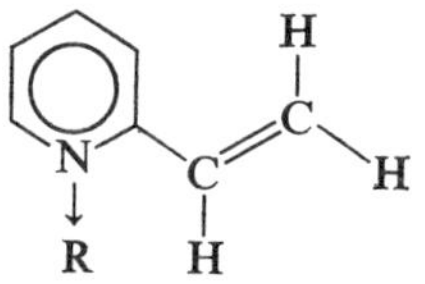 **80b** 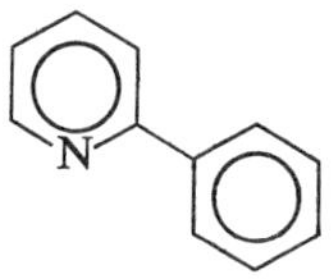 **80c**

of the paramagnetic influence of a nitrogen lone pair on the shift of the α-protons as 0.7 ppm and the variation of chemical shift per unit charge as 9.8 ppm. In a more descriptive paper (49b) the same authors make a number of interesting observations. In general, complexes with $AlBr_3$ are rather unstable and are capable of exchange. Those with BF_3, however, are quite stable 1:1 complexes. As would be expected, α-substitution causes steric hindrance. This is very noticeable for 2-vinylpyridine which normally exists in the predominant conformation **80a** but when complexed prefers the opposite conformation **80b**. Similar effects were observed for 2-phenylpyridine (**80c**).

Diehl and his co-workers (110a) have used 2,6-dibromopyridine as an example of a three-spin system with C_2 symmetry in a study of this system oriented in a nematic phase.

INDOR methods have been used to determine the size and sign of the very small couplings between H-2 and H-6 of a number of substituted pyridines (110b). The values obtained varied between −0.18 and −0.32 Hz.

Spectra of the six isomeric diformylpyridines and of 2,4,6-triformylpyridine have been reported (110c). Long-range couplings of the formyl protons gave no clear diagnosis of preferred conformations. Proton magnetic resonance (PMR) methods were used to determine H^0 and S^0 values for the acid-catalyzed hydration of the formyl group in the three pyridine monoaldehydes (110d). The oxime of pyridine 2-aldehyde reacts readily with organo derivatives of the Group III elements and NMR methods were used extensively to characterize the products (110e). Also, NMR was applied successfully to the determination of p*D* values and tautomeric structures of 3-hydroxypyridine 2- and 4-aldehydes and the related pyridoxamine and pyridoxal (110f).

Gerig and Reinheimer (110g) found that substituent effects on chemical shifts of ring protons of 2-substituted 3- or 5-nitropyridines were substantial but not unusual. Coupling constants were perturbed only slightly when the ring nitrogen was protonated.

Spectra of the isomeric 4-formylpyridine oximes **81** and **82** and their methiodides have been examined (111) and it was shown that, in both species, the hydrogen atom syn to the oxime hydroxyl group gave rise to peaks at lower field than those from the corresponding anti hydrogen atom. This conclusion is in agreement with similar observations on the isomers of *p*-chlorobenzaldoxime (**83**) (112).

An extensive study has been made of the NMR spectra of pyridoxal (**84**), isopyridoxal (**85**), their acetals, and a number of derivatives (113, 114) and it appears that both parent compounds exist in solution as hemiacetals **86** and **87**, respectively, in equilibrium with the free aldehydes. The shielding of every proton in the molecule decreases in a regular fashion as the electron-withdrawing power of the 5-side chain increases. Spectral data for these and other similarly substituted pyridines (115) are collected in Table 2.20.

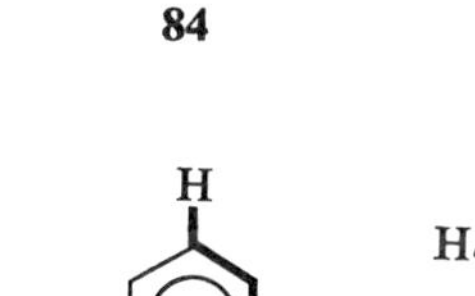

84 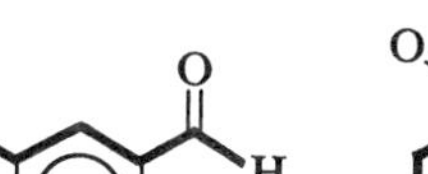 **85** 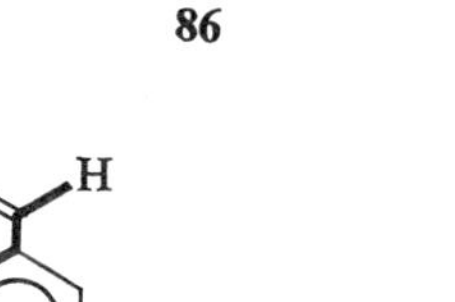 **86** **87**

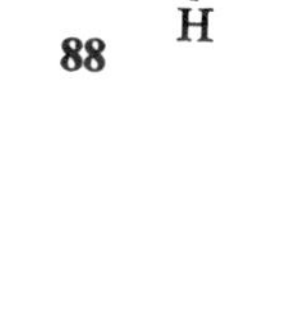

88  **89**  **90**

TABLE 2.31 MEISENHEIMER COMPLEXES OF 3,5-DINITROPYRIDINE (117). SOLVENT—DMSO

R	τ2	τ4	τ6
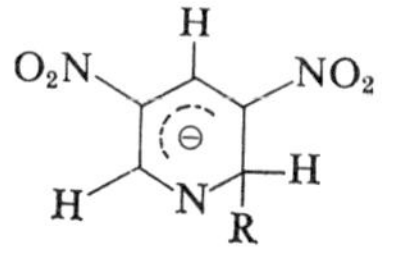 OMe	3.92	1.70	1.38
$N(Et)_2$	4.17	1.70	1.42
CH_2COMe	4.50	1.78	1.62
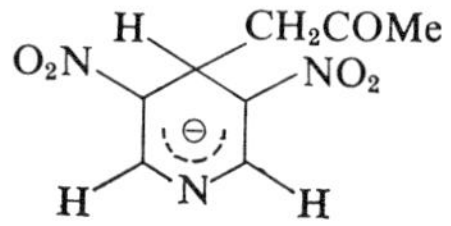 (4-CH_2COMe adduct)	1.85	5.12	1.85

Long-range couplings between the aldehyde hydrogen atoms and ring protons meta to the aldehyde group have been observed for the three pyridine monoaldehydes (116) and were used to estimate the predominant conformation of the aldehyde group, assuming that the "extended *W*" rule holds for these compounds (**88–90**). With the 4-aldehyde, $J_{2,CHO} = 0.35$ Hz, with the 3-aldehyde, $J_{5,CHO} = 0.44$ (CCl_4) and 0.50 (DMSO) Hz, and with the 2-aldehyde, $J_{4,CHO} = 0.5$ (CCl_4) and 0.6 (DMSO) Hz.

Meisenheimer compounds between 3,5-dinitropyridine and a number of nucleophiles have been prepared (117) and data have been collected in Table 2.31.

TABLE 2.31A ^{14}N CHEMICAL SHIFTS OF HALOPYRIDINES (118d). REFERENCE—PYRIDINE

Substituent	Solvent	Chemical Shift (ppm)
2-F	Neat	42.3 ± 0.5
3-F	Neat	−4.0 ± 0.5
2-Cl	Neat	10.0 ± 0.8
3-Cl	Neat	−5.0 ± 0.8
2-Br	Neat	3.0 ± 1.0
3-Br	Neat	−1.0 ± 1.0
2,6-Di-Cl	C_6H_6	18.0 ± 1.5
3,5-Di-Cl	C_6H_6	−11.0 ± 1.0
2,5-Di-Cl	C_6H_6	12.0 ± 3.0
2,3-Di-Cl	C_6H_6	9.5 ± 1.5
2,6-Di-Br	C_6H_6	8.0 ± 2.5
2,5-Di-Br	Dioxan	6.0 ± 2.5

A study has been made of substituent effects on the ^{13}C chemical shifts of a number of 2-substituted pyridines (118). Paramagnetic shifts for C-2 were much smaller than expected in cases where the substituent was a strong electronegative group. Shielding of C-5 (para to the substituent) reflected electron release and withdrawal by the substituent. Substituent effects on the ^{13}C chemical shifts of 3-substituted (118a), 4-substituted (118b), and disubstituted pyridines (118c) were shown to be very similar to those found for substituted benzenes.

^{14}N chemical shifts for a series of mono- and dihalopyridines have been reported (118d). Calculated π-electron densities at the nitrogen atoms correlate satisfactorily with these shifts which are listed in Table 2.31A.

D. Photodimers of 2-Aminopyridines and 2-Pyridones

The dimer produced on ultraviolet irradiation of 2-pyridone was first thought to be a typical cyclobutane-type photoproduct (119). However, the NMR spectrum was not consistent with this formulation but suggested 1,4-cycloaddition had taken place across the 3,6-bonds of the pyridone ring (120, 121). An extensive study of the spectra of a number of substituted 2-pyridone dimers confirmed the 3,6-addition and established the anti-trans structure **91** for most of them (122). This stereochemistry was independently deduced by Taylor and Kan (123) who, after establishing the anti-trans configuration for dimers of 2-aminopyridines, converted the reduced dimer

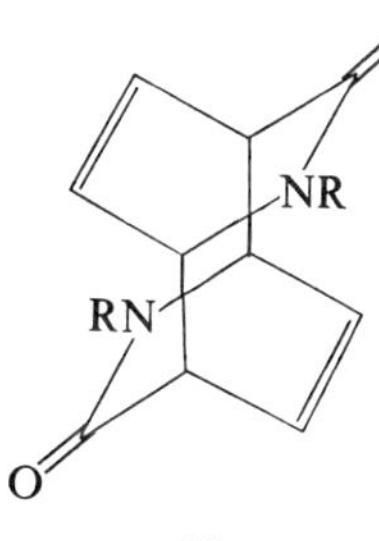

O NR RN O — **91**; NH_2 N N H_2N — **92**; O NH HN O — **93**

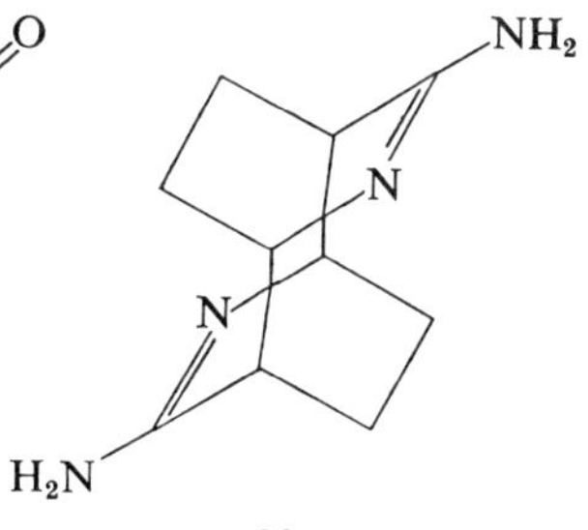

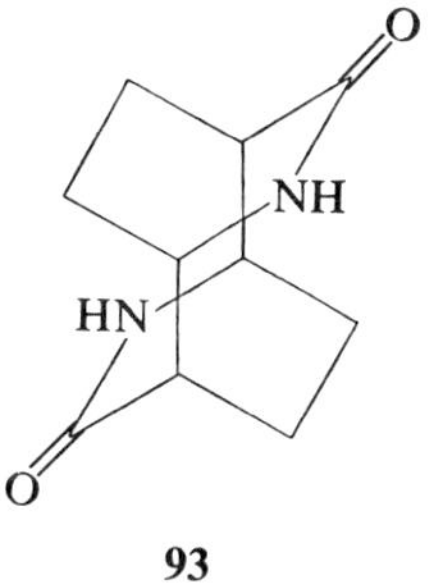

of 2-aminopyridine (**92**) into a reduced dimer of 2-pyridone (**93**), identical in every respect with that obtained by reduction of the 2-pyridone photodimer. Throughout this work arguments were used which, in the light of our present NMR knowledge, are rather inconclusive, but the overall evidence leaves no doubt that such photodimers do exist in the anti-trans configuration. Spectral data are collected in Table 2.32.

TABLE 2.32 SPECTRAL DATA FOR PHOTODIMERS OF 2-AMINOPYRIDINES AND PYRID-2-ONES

DIMER OF 2-AMINOPYRIDINE HYDROCHLORIDE (D_2O)

Substituents	τ3	τ4	τ5	τ6	τMe	τNMe	J_{34}	J_{35}	J_{36}	J_{45}	J_{46}	J_{56}
None	5.89	3.65	3.10	5.32			6.5	1.5	9.5	7.5	1.5	6.5
3-Me		3.94	3.11	5.88	8.30					7.5	1.5	7.5
4-Me	6.21		3.63	5.38	8.08			1.5	10.0			7.5
5-Me	5.97	4.15		5.62	7.96		7.5		10.0		1.5	
6-Me	6.23	3.66	3.38		8.27		7	1.5	8			
5-Cl	5.63	3.66		5.25			7.5		10		2	
1,6-Di-Me (2-imino)	6.15	3.48	3.40		8.17	6.92						
2-$NHCH_2Ph$	5.80	3.73	2.18	5.30			7	1.5	10	8	1.5	6.5

DIMER OF PYRID-2-ONE (TFA)

Substituents	τ3	τ4	τ5	τ6	τMe	τNMe	J_{34}	J_{35}	J_{36}	J_{45}	J_{46}	J_{56}
None	6.08	3.72	3.14	5.50			6.5	1.5	10	8	1.5	6.5
3-Me		4.03	3.21	6.08	8.42					8	1.5	6.5
4-Me	6.41		3.59	5.70	8.15			2	10			6
5-Me	6.20	4.21		5.78	7.95		7		9.5		1.5	
6-Me	6.47	3.71	3.46		8.30		6.5	2		8		
1-Me	5.92	3.68	3.19	5.62		7.30	6	1.5	10	7.7	1.5	
1,6-Di-Me	5.92	4.09	4.02		8.65	7.49						

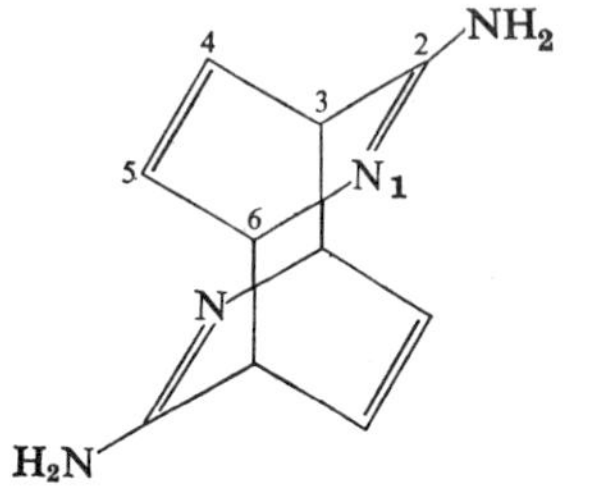

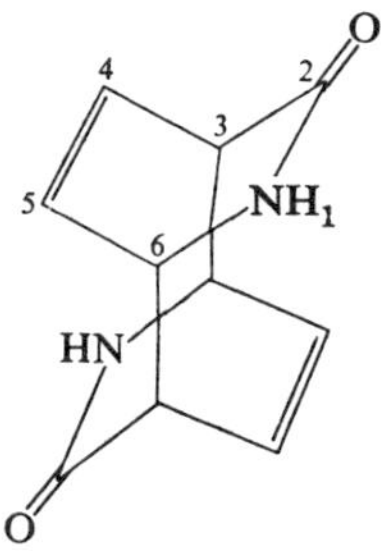

E. 1,2- and 1,4-Dihydropyridines

In recent years there has been a growth of interest in the NMR spectra of 1,4-dihydropyridines, partly because of the occurrence of such a system in dihydronicotinamide-adenine dinucleotide (NADH) and partly because of their formation by ring contraction of azepins. Tautomeric oxo-, thio-, and iminodihydropyridines are not really in the dihydropyridine oxidation state and are discussed in Section I.C.

In 1962, spectra of 1,2- and 1,4-dihydro-1-phenylpyridines **94** and **95** were reported (124). The equivalence of the two 4-protons of the simple NADH analog, 1-benzyl-1,4-dihydronicotinamide (**96**), showed the ring system to be either planar or time-averaged planar (125), in agreement with crystallographic evidence (126). Spectra of NADH and its 4α- and 4β-deutero derivatives, measured in deuterium oxide, again point to a planar or time-averaged planar conformation for the dihydropyridine moiety (see Table 2.33). Although measurements were made in different solvents with different standards, a general and real upfield shift of signals from H-2, H-4, and H-6 is observed on going from the *N*-benzyl compound **96** to NADH **97**, suggesting that the latter is mainly in the folded form **98** (see 127) with the dihydropyridine ring in close juxtaposition to the adenine ring. Since it was established that reversible oxidation–reduction reactions of many pyridine nucleotide dehydrogenases are stereospecific, involving hydrogen transfer from only one specific side of the pyridine ring (see Ref. 128), it has become fashionable to regard the dihydropyridine ring of NADH as preferring the "boat" conformation **98**. Unfortunately, NMR evidence for such a conformer is inconclusive.

More recently, the 1-benzyl-1,4-dihydronicotinamide (**96**) has been found to carry out nonenzymic reductions of haloketones and haloaldehydes to the corresponding alcohols (129). When the 4,4′-dideutero derivative was used, about 40% of the deuterium was lost from the dihydropyridine during the reaction. Biellman and Callot (129a) have extensively studied the spectra of isomeric pairs of 1,2- and 1,4-dihydronicotinamides and a number of similar compounds. The mechanism and kinetics of the primary protonation of 1-methyl-, 1-propyl-, and 1-benzyl-1,4-dihydronicotinamides have been investigated with the aid of NMR methods (129b). Partial deuteration was used to confirm assignments.

The unstable parent compound, 1,4-dihydropyridine (**99**), has been prepared (130), and its spectrum, together with those of the other compounds discussed in this section, is summarized in Table 2.33.

At normal operating temperatures, compound **100** gave a sharp spectrum, but at slightly elevated temperatures, 50° for H-3,5 and 60° for H-2,6,

TABLE 2.33 SPECTRAL DATA FOR 1,4-DIHYDROPYRIDINES

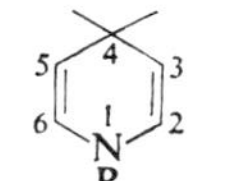

Substituents	Solvent	τ2	τ3	τ4	τ5	τ6	Other Data	Ref.[a]
None	C_6D_6	4.27	5.58	6.85	5.58	4.27		130
1-Si(Me)$_2$	C_6D_6	4.13	5.50	6.95	5.50	4.13		130
1-AlR$_3$	C_5H_5N	3.30	5.70	6.70	5.70	3.30		130
1-Ph	CCl_4	3.73	5.47	7.02	5.47	3.73	J_{23} 9.0, J_{24} 1.6, J_{34} 3.9 Hz	124, 130
1,4,4-Tri-Me	Neat	4.49	5.89		5.89	4.49	4(Me)$_2$ 9.11, N–Me 7.36	130, 270
1-Me, 3-CONH$_2$	Neat ?	3.12			5.48	4.30	NMe 7.22	271
1-CH$_2$Ph, 3-CONH$_2$	$CDCl_3$	2.9		6.9	5.70	4.72	C$\underline{H}_2$Ph 5.77, Ph 2.72	129
		2.86		6.81	5.25	4.28	5.73	124
3,5-Di-COOEt	C_6D_6	3.26		6.40		3.26	Et 9.01, 5.96	272
1-Me, 3,5-di-COOEt	$CDCl_3$	3.15		6.99		3.15	Et 8.80, 5.90, NMe 6.86	272
4-Me, 3,5-di-COOEt	C_6D_6	3.08		5.76		3.08	4Me 8.68, J_{12} 5.6	272
	$CDCl_3$	2.73		6.20		2.73	8.70	272
1-Me, 3,5-di-Ac·H$_2$O	D_2O	2.6		6.60		2.6	NMe 6.20, COMe 7.6	272
1-Me, 3,5-di-CN	$CDCl_3$	3.53		6.85		3.53	NMe 7.02	272
4-Me, 3,5-di-CN	DMSO	3.04		6.74		3.04	4Me 8.73	272
2,4,6-Tri-Me, 3,5-CN	$CHCl_3$			6.75			2,6Me 7.94, 4Me 8.63	272
1,4-Di-Me, 3,5-di-CN	$CHCl_3$	3.28		6.53		3.28	4Me 6.33, NMe 6.7	272
4-CH$_2$CN, 3,5-di-COOEt	?			5.87			CH$_2$CN 7.42	273

[a] See also Refs. 274 and 275 for some other highly substituted derivatives.

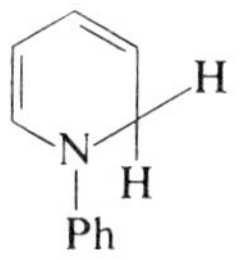

94

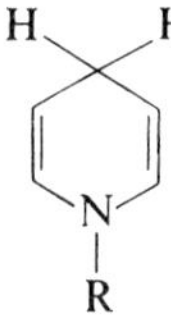

95, R = Ph
96, R = Benzyl
99, R = H

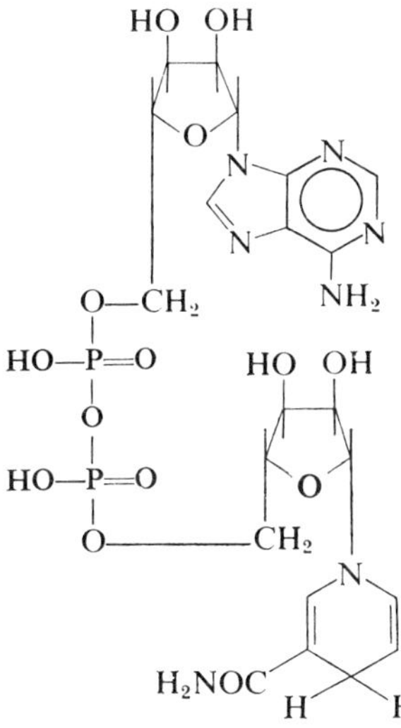

97

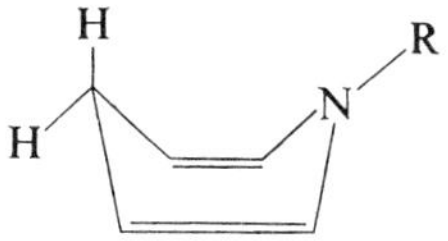

98

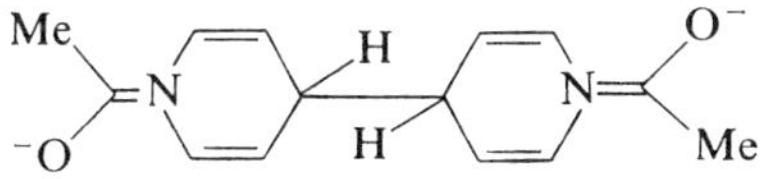

100

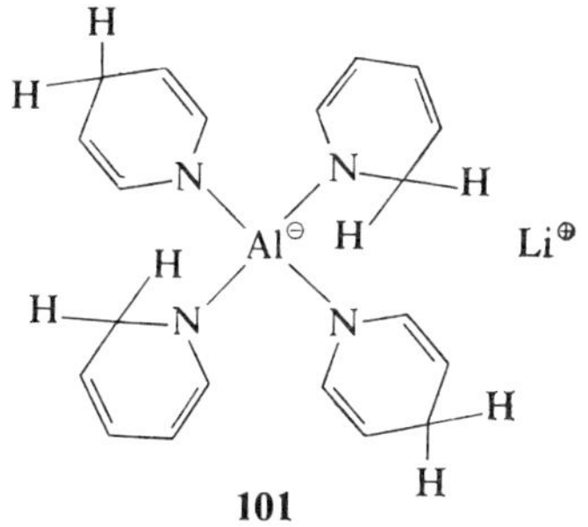

101

coalescence of the spectra of the individual rotamers occurred (131). Use of dimethylformamide as solvent reduced the chemical shift between H-2 and H-6.

Reaction of lithium aluminum hydride with excess of pyridine is thought, from NMR evidence, to produce the ion **101** which acts as a reducing agent by hydrogen transfer from the dihydropyridine systems (132).

F. Piperidines and Tetrahydropyridines

Because of the structural similarity between piperidine and cyclohexane, they exhibit the same types of conformational mobility, and the large amount of data available on the conformational analysis of substituted cyclohexanes can be applied to the heterocyclic system.

In piperidine, as in cyclohexane, the puckered ring allows the ring atoms to have typical sp^3-type bond angles, and the Karplus relationship between vicinal bond angles ϕ and coupling constants J developed for ethanes (133) would be expected to hold. The relationship in its original form,

$$J = k \cos^2 \phi - c$$

where k and c are constants, has been much criticized, and a series of adjusted equations have been used for specific systems. In particular, the electronegativity of substituents on the ethane molecular fragment has an important effect on the size of the observed coupling (see Ref. 134). In spite of the controversy which surrounds the use of the Karplus equation, there is no doubt that it usually allows an unequivocal distinction to be made between possible conformers.

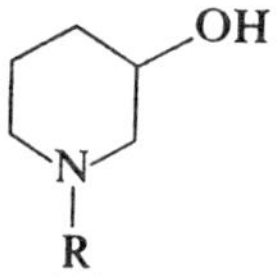

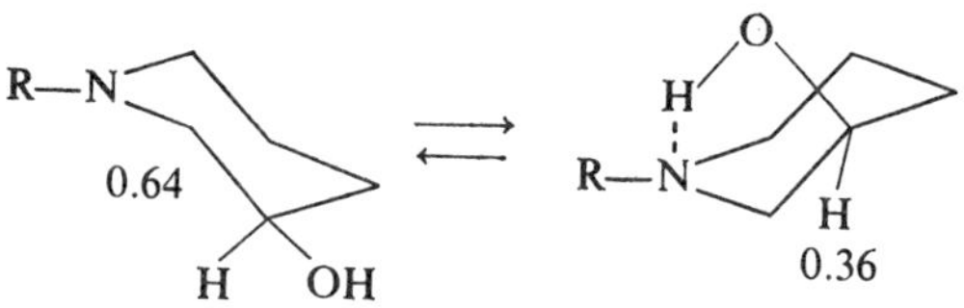

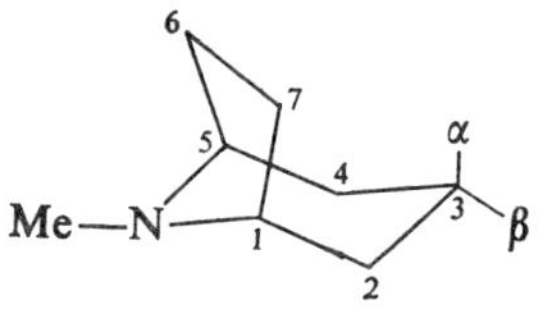

Use of the Karplus equation with a large number of substituted cyclohexanes has produced limiting values for vicinal couplings in the fixed conformers of this ring system (see Ref. 135); 11.2 Hz for trans diaxial protons, 2.5 Hz for trans diequatorial protons, and 4.1 Hz for cis axial-equatorial protons. In many actual spectra these couplings cannot be resolved and the overall width of the signal is used to obtain data about the orientation of the proton concerned. Thus, a single axial proton adjacent to two methylene groups will experience two trans diaxial and two cis axial–equatorial couplings, $2(J_{aa} + J_{ae})$ which, in the limiting case, will total 30.6 Hz. The equatorial alternative will experience a total coupling of only 13.2 Hz. In the case of the piperidin-3-ol (**102**), the separation of the terminal peaks in the multiplet from the carbinol proton, H-3, was found to be 24.4 Hz (136). This result was interpreted in terms of a mixture of conformers containing 64% **103**, axial H-3, and 36% **104** with an equatorial H-3, presumably stabilized by hydrogen bonding between the axial hydroxyl group and the nitrogen atom. These populations correspond to a conformational free-energy difference, $-\Delta G_{OH}{}^{\circ}$, of about 0.37 kcal/mole for the 3-hydroxyl group at 35°. A number of 1- or 4-substituted piperidin-3-ols were also studied. The configurations of *cis*- and *trans*-2,6-dimethylpiperidines have been confirmed by NMR with the aid of spin coupling, and the relative stereochemistry of one isomer of 2,4,6-trimethylpiperidine has been determined (137). Proton spectra at 100 and 220 MHz of different isomers of the 2,3-, 2,4-, and 3,4-dimethylpiperidines have been almost completely assigned (137a). The results were used to determine the configurations of the individual isomers. Many of the spectra are reproduced in the *Spectral Supplement* to Ref. 137a.

The epimeric 3-cyanotropanes (**105**) have also been studied (138). With the β-isomer, the remaining 3-proton would have to be axial if the piperidine ring was to have the expected chair conformation, and this is confirmed by the observed couplings, $J_{aa} = 9$, $J_{ae} = 1.5$ Hz. The 3β-proton of the α-isomer appears as a triplet, $J = 5$ Hz, which is not consistent with an undistorted chair conformation, but rather with a twisted conformation where ϕ_{ae} is approximately 40 or 80°. The *N*-methyl group in both isomers was shown to be equatorial. Data for a few simple tropane derivatives are collected in Table 2.34. Data for some piperidines are collected in Table 2.35.

In 1959, Closs (139) examined the spectra of a number of substituted tropane deutero- and hydrochlorides. An acidified solution of pseudotropine hydrochloride (**106**) in water gave two doublets for the *N*-methyl signals, that at higher field being attributed to the more stable equatorial conformation. Deuterochlorides of tropane derivatives gave corresponding singlets attributable to the two methyl conformers. This method was later extended to salts of 1,2-dimethylpiperidine where conformation **107**, with both methyl groups equatorial, was preferred to **108** with the *N*-methyl group axial (140) (Table 2.36). However, in the piperidine series an *axial* *N*-methyl group normally resonates at *higher* field than its equatorial counterpart, but sometimes this criterion proves unreliable (141). Also, peaks from axial and equatorial *N*-substituents experience quite different solvent shifts, and it has been suggested that spectra should be measured in both chloroform and water before conclusions are drawn about axial–equatorial ratios in these substances (142). House and his co-workers (143), using a 4-*t*-butyl group to fix conformations, studied the axial–equatorial relationships at the nitrogen atom for a series of quaternary piperidinium salts **109** and confirmed that the equatorial substituents gave peaks at lower field than those from the axial isomers. In the course of this investigation, compound **110**, with a

TABLE 2.34 SPECTRAL DATA FOR SOME TROPANE DERIVATIVES (138)

Compound	τ1, 5	τ6, 7	τ2, 4	τNMe	τ3	τOH
Tropinone	6.65	8.0–8.6	7.8	7.59		
2,2,4,4-d_4-Tropinone	6.62	8.0–8.6		7.58		
6β-Hydroxytropinone	6.5–6.7	5.9, 8.0	7.4–7.9	7.35		6.05
Tropine (3α-OH)	7.0	7.95	8.1–8.4	7.84	6.1	5.7
2,2,4,4-d_4-Tropine	7.03	8.0		7.82	6.12	6.44
6,7-d_2-Tropine	7.02	7.95	8.0–8.4	7.82	6.11	6.6
ψ-Tropine (3β-OH)	6.93	8.2–8.5	8.2–8.5	7.78	6.3	6.49
2,2,4,4-d_4-ψ-Tropine	6.91			7.8	6.27	5.21
3α-Cyanotropane	6.83	7.2	7.5–7.8	7.49	6.05	
3β-Cyanotropane	6.90	8.1–8.5	8.1–8.5	7.8		

TABLE 2.35 SELECTED DATA FOR PIPERIDINES

A. *N*-ALKYLPIPERIDINES. SOLVENT—CCl_4 (155)

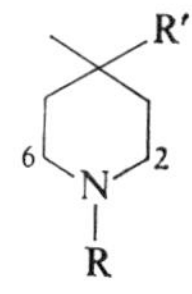

R	R′	τ2, 6
H	H	7.31
Me	H	7.76
Et	H	7.71
CH_2Ph	H	7.71
i-Pr	H	7.61
t-Bu	H	7.55

B. *cis*-2,6-DIMETHYLPIPERIDINE IN $CDCl_3$ (137)

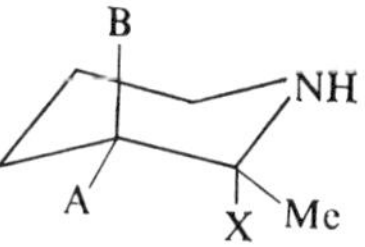

τA	~8.45	J_{AX} 1.9 ± 0.2
τB	~9.00	J_{BX} 10.6 ± 0.2
τX	7.45	J_{XMe} 6.3
τMe	9.05	

TABLE 2.35 (*continued*)

C. 2,6-DIPHENYLPIPERIDIN-4-OLS (148)

Substituent	Solvent	α-Form τ2, 6	α-Form τ4	α-Form τN–Me	β-Form τ2, 6	β-Form τ4	β-Form τN–Me
None	C_6H_6	6.51 (2.6, 11.1)[a]	6.40 (5.1, 11.1)[a]		5.80 (5.7, 9.0)	6.14 (3.0)	
1-Me	$CDCl_3$	6.81 (2.9, 11.2)	6.17 (4.4, 10.8)	8.25	6.39 (7.90)	5.80 (?)	8.20
3,5-Di-Me	$CDCl_3$	6.60 (9.9)	6.97 (9.7)				
	C_6H_6				6.20 (10.3)	6.60 (2.6)	
1,3,5-Tri-Me	$CDCl_3$	7.18 (10.1)	6.77 (9.7)	8.40	6.76 (10.6)	6.25 (2.5)	

D. PIPERIDINES (157)

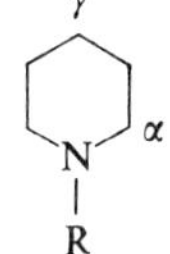

R	Solvent	Parameters (in Hz) $\delta_{ae}(\alpha)$	$J_{ae}(\alpha)$	$\delta_{ae}(\gamma)$	$J_{ae}(\gamma)$
H	MeOH(d_4)	26.1	11.9	24.8	13.1
	C_3H_6	27.5	10.2		
	CH_2Cl_2	29.0	12.3	26.9	13.4
	$C_6H_5Me(d_8)$	32.4	11.2	29.4	12.4
Me	MeOH(d_4)	56.5	11.4	31.1	12.9
	C_3H_6	63.5	11.2		
	CH_2Cl_2	61.4	11.2	34.4	13.1
	$C_6H_5Me(d_8)$	66.1	11.0	39.4	12.6

[a] Coupling constants in parentheses.

TABLE 2.36 CHEMICAL SHIFT DATA FOR *N*-ALKYL GROUPS OF PIPERIDINIUM SALTS

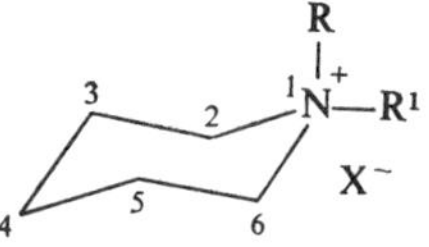

R	R^1	X	Substituents	Solvent	τNMe	τNCH_2	Ref.
Me	Me	I	4-*t*-Bu, 4-OH	D_2O	6.75, 6.88		143
Me	Me	Tosyl	4-*t*-Bu, 4-OH	D_2O	6.88, 7.03		143
Me	Me	I	4-*t*-Bu	D_2O	6.87, 6.78		143
Me	Me	Tosyl	4-*t*-Bu	D_2O	6.91, 7.03		143
CD_3	Me	Tosyl	4-*t*-Bu	D_2O	6.91		143
Me	CD_3	Tosyl	4-*t*-Bu	D_2O	7.03		143
Et	Me	Tosyl	4-*t*-Bu	D_2O	7.08		143
Me	Et	Tosyl	4-*t*-Bu	D_2O	7.16		143
$PhCH_2$	Me	Tosyl	4-*t*-Bu	$CDCl_3$	6.99	5.53	143
Me	$PhCH_2$	Tosyl	4-*t*-Bu	$CDCl_3$	7.09	5.17	143
CH_2CO_2Me	Me	Br	4-*t*-Bu	D_2O	6.64	5.68	143
Me	CH_2CO_2Me	Br	4-*t*-Bu	D_2O	6.76	5.60	143
Me	$CH_2CO_2^-$	Br	4-*t*-Bu	D_2O	6.80	6.08	143
Me	CH_2COOH	Br	4-*t*-Bu	D_2O	6.08	5.67	143
Me	H	Cl	2Me (eq.)	D_2O	7.27		140
H	Me	Cl	2Me (eq.)	D_2O	7.19		140
Me	Et	I	2Me (eq.)	D_2O	7.10		140
Et	Me	I	2Me (eq.)	D_2O	6.97		140
Me	Pr	I	2Me (eq.)	D_2O	7.11		140
Pr	Me	I	2Me (eq.)	D_2O	6.95		140
Me	CH_2Ph	I	2Me (eq.)	D_2O	6.91		140
CH_2Ph	Me	I	2Me (eq.)	D_2O	6.80		140
Me	H	Cl	4-Ph	D_2O	7.13, 7.13		140
Me	Et	I	4-Ph	D_2O	6.98		140, 141
				$CDCl_3$	6.70		142
Et	Me	I	4-Ph	D_2O	6.85		140, 141
				$CDCl_3$	6.62		142

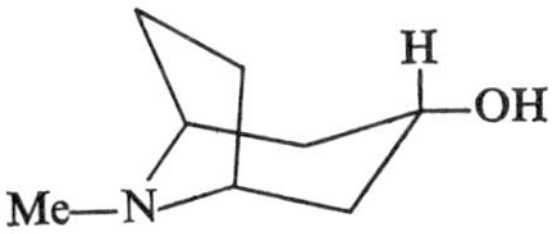

106

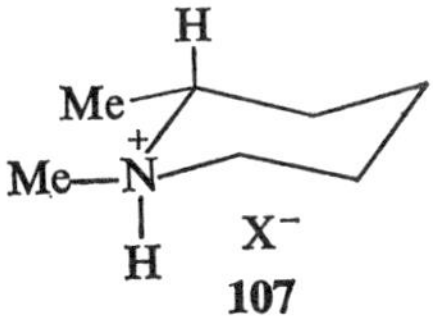

107

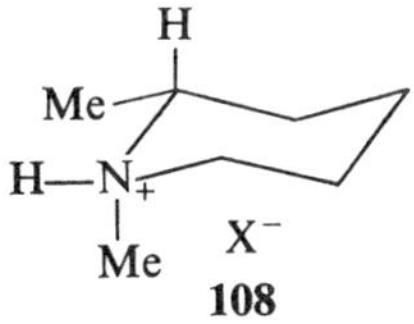

108

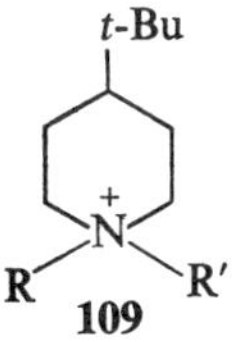

109

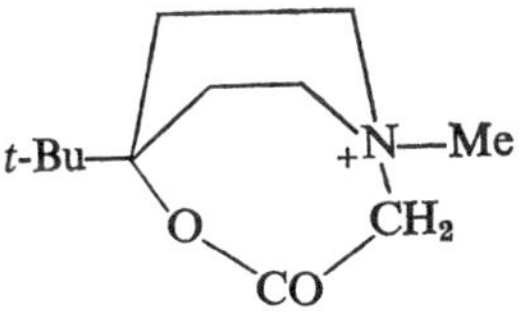

110

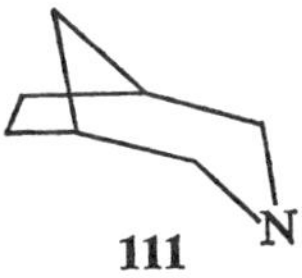

111

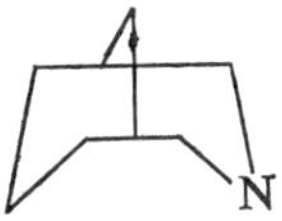

112

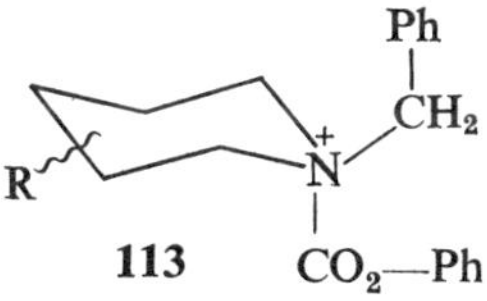

113

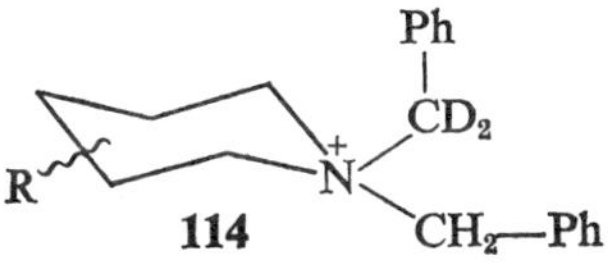

114

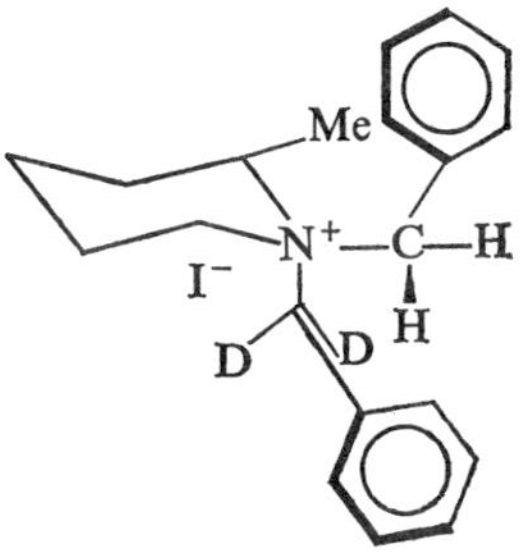

115

fixed boat structure, was obtained and its spectrum measured. These workers also studied *N*-alkyl isomerism for a large number of azabicyclic compounds of types **111** and **112** (144, 145). McKenna *et al.* (146) have repeated work on the direction of methylation of 1-alkyl-4-phenylpiperidines to show that the original kinetic analysis (141) was falsely based because of wrong spectral assignments. After recalculation, it was impossible to say which direction is preferred for alkylation. Reaction of *N*-deuterobenzyl derivatives of variously substituted piperidines with benzyl iodide gave a mixture of the two quaternary salts **113** and **114**, the benzyl methylene protons of the latter absorbing at higher field because of shielding by the other phenyl ring as shown (**115**) (146). The relative configurations of α,α′-disubstituted piperidines have been determined from the structure of the methylene signal from an *N*-benzyl substituent (147). The assignment of configuration of C–Me and N–Me groups in symmetrically disubstituted *N*-methylpyridinium ions has been made from NMR data (147a). One isomer of the *N*,*N*-dimethyl derivative gave two singlets and the other only one singlet in the N–Me region of the spectrum. The former was assumed to be the cis isomer and the latter the trans isomer in which rapid inversion causes time-averaging of the N–Me signals. Conformations of the mono-N–Me derivatives were also studied. The N–Me shifts in 4-alkyl derivatives of N-methyl and *N,N*-dimethylpiperidine salts have been examined (147b). The authors concluded that the shift of an axial N–Me group was affected slightly by a 4-*t*-butyl group and an equatorial N–Me group by an axial methyl group at C-4.

As with cyclohexanes, much of the work on piperidines has involved the use of compounds substituted with one or more groups large enough to fix the conformation. Thus, the 2,6-diphenylpiperidin-4-ols (**116**) were found to exist predominantly in the chair form with the phenyl groups equatorial (**117**) (148). The 2-protons of the β-isomers **118** are deshielded by the axial 4-hydroxyl groups because of their spatial proximity. Similarly, 3-methyl-4-phenyl-piperidin-4-ols (**119**) exist predominantly in the chair form with the phenyl group equatorial. In spectra of the free alcohols and the corresponding esters, a 3-methyl group trans to the phenyl group gives signals at higher field than those from the corresponding cis methyl group. Casy (149) has examined the spectra of the free bases from the analgesics, α- and β-prodine, **120** and **121**, respectively, and used chemical shift data together with the complexity of the phenyl signal to suggest the preferred conformations **122** and **123**, respectively. The spectrum from the reduced ring of the 4-methylenepiperidine (**124**) consists of a singlet (four protons) and an A_2B_2 system (150). Examination of the spectra of model compounds led to the conclusion that the singlet resulted from accidental coincidence of signals from all four protons on the side of the ring adjacent to the phenyl substituent, while the A_2B_2 system arose from the other side of the ring, with protons from each methylene group made time-averaged equivalent by rapid flipping of the piperidine ring.

The effects of various *N*-substituents on the spectra of piperidines and on the conformational equilibrium in these compounds has aroused considerable interest. *N*-Acyl compounds have been investigated by a number of workers (150a–c). With *N*-chloropiperidine it was shown that *N*-inversion has been slowed down until its rate is essentially the same as that of the ring flipping (150d). Preferential shielding of axial and equatorial protons adjacent to the ring nitrogen atom in *N*-nitrosopiperidines has been observed (150b–e). In conformationally mobile systems, cis protons appear 0.3–0.8 ppm to higher field than trans protons but in cyclic compounds possessing a preferred conformation, cis axial protons resonated 2.0–3.0 ppm to higher field than their cis equatorial counterparts. The corresponding difference for trans protons was 0.8–1.4 ppm. A qualitative model of shielding cones was proposed to explain these results (150e). *N*-Nitrosopiperidine was included in a study of the protonation of nitrosoamines (150f).

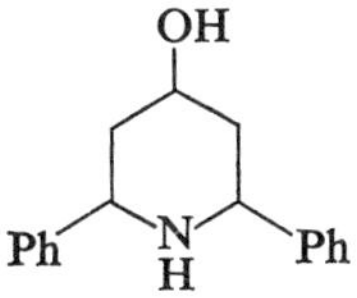

116 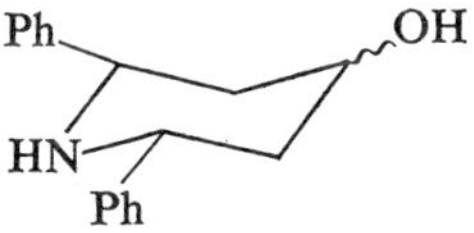 117 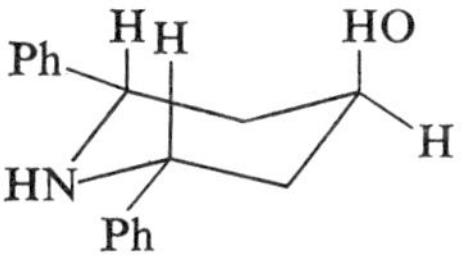 118

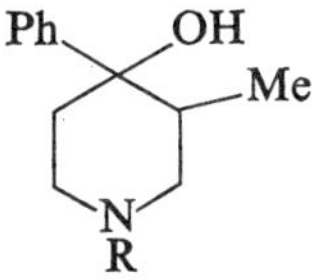

119 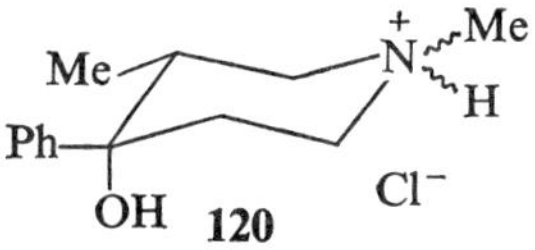 120 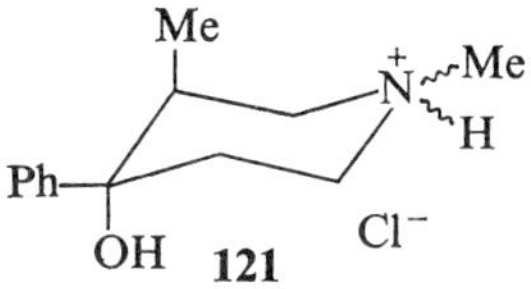 121

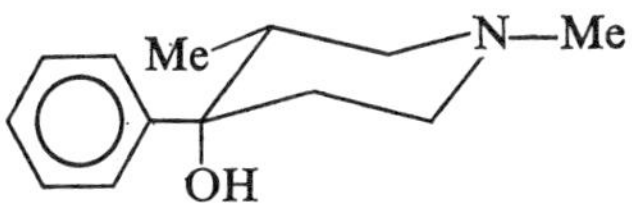

122 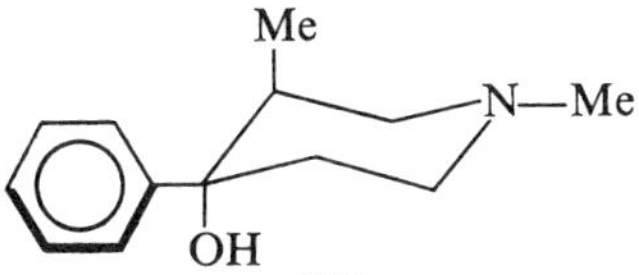 123

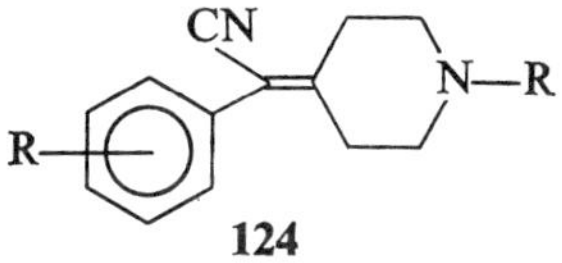

124

Spectra of a number of piperidine nitroxide free radicals have been analyzed (after partial deuteration) to obtain estimates of the paramagnetic shifts induced by the unpaired electron (150g, h). These shifts are proportional to the electron-nucleus hyperfine interaction and were used successfully to predict hyperfine couplings observed in ESR spectra.

Data have been recorded for some acetals and thioacetals of piperid-4-ones and their salts (150i). In the salts the two 4-alkoxy- or 4-alkylthio groups are nonequivalent because of unequal populations of the two chair conformers. Spectra of *N*-alkyl-2,2′,6,6′-tetramethylpiperid-4-ones have been interpreted in terms of rapid inversion between the chair conformers **125** and **126** (151). Also, solvent shifts observed with the 2,6-diphenylpiperidones (**127**) were ascribed to specific solvation rather than to changes in conformation. In each case, boat conformations were considered to be energetically unfavorable and were neglected. When a 1-benzenesulfinyl group is present, signals from the methyl groups of 2,2′,6,6′-tetramethylpiperid-4-one are broadened. Variable-temperature studies and solvent effects were used to show that this was caused by the molecular asymmetry of the sulfinyl group, combined with slow inversion of the nitrogen atom (152). Also, inversion rates for the groups Ph–SO_2–N, Ph–SO–N, and Ph–S–N decreased in that order. From results obtained with a series of *N-t*-butyl-3,5-dimethylpiperid-4-ones, it has been suggested that the axial lone pair offers less steric hindrance than a similarly oriented proton in a cyclohexane (153). Spectra of piperid-2-ones of the general type **128** have been compared with those of the corresponding piperidines to check the effect of the carbonyl group (154). Also, pH variations in spectra of 1,2,6-trimethylpiperidine have been studied (155).

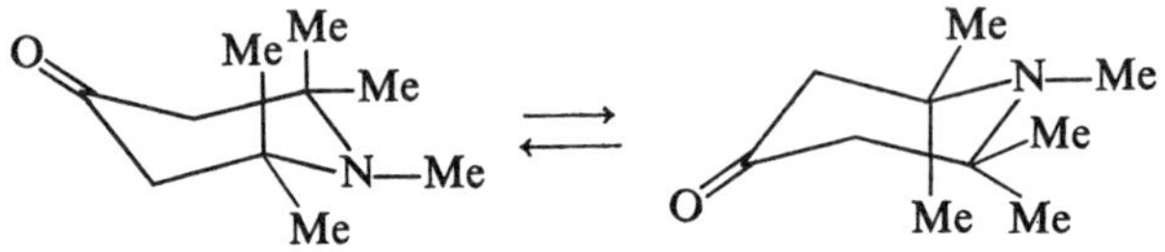

In piperidine itself the 2,6-protons, especially when axial, experience a marked shielding when N–H is converted to *N*-alkyl (156). This effect is attributed to a combination of stereospecific influences, one associated with the orientation of the lone pair and the other with that of the *N*-alkyl group. With carbon tetrachloride as solvent, rapid inversion averaged the axial and equatorial signals, but when trifluoroacetic acid was used, inversion of the nitrogen atom was prevented and, even with small *N*-alkyl substituents such as methyl, the conformation was largely fixed, giving separate signals for axial and equatorial protons.

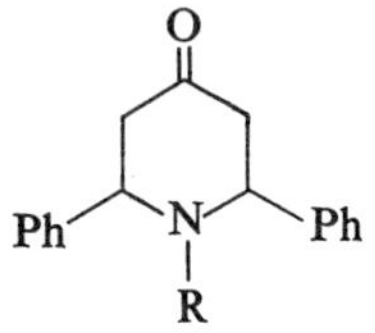

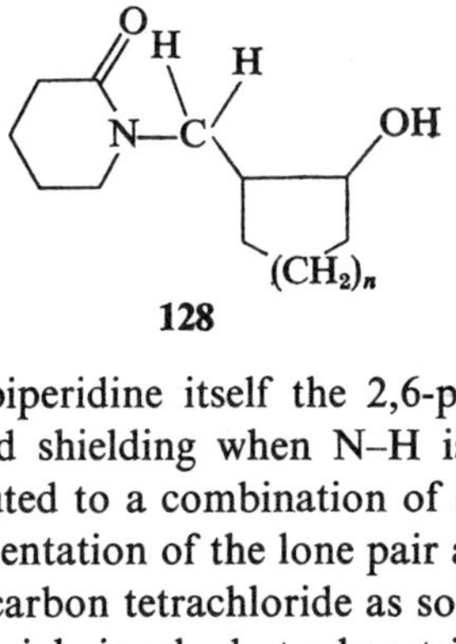

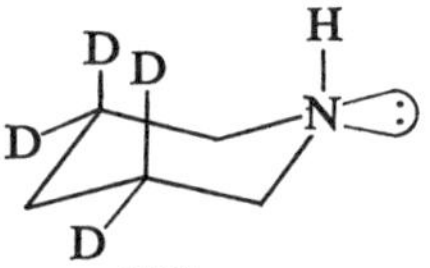

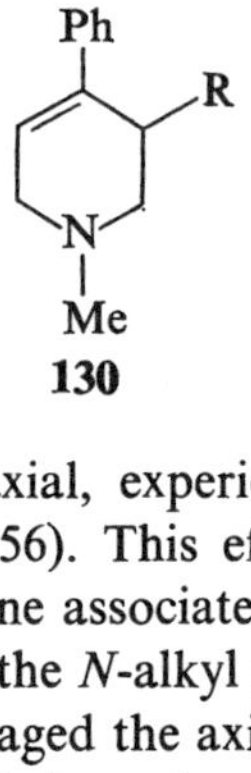

Lambert *et al.* (157), in a study of the isomeric preference of the nitrogen lone pair, measured the spectra of 3,3′,5,5′-d_4-piperidine at temperatures down to −85° and in a number of solvents. They concluded that the lone pair exists predominantly but not completely in the equatorial conformation **129**. However, Robinson (158) has severely criticized this work, asserting that the axial–equatorial shifts $\delta_{a,e}$ for the α-protons cannot, at present, provide valid evidence for the conformation of piperidine. Booth (159) added to the confusion when he suggested that the salt formed between deuterotrifluoroacetic acid and *cis*-dimethylpiperidine consisted of two conformers,

129a and **129b**, fixed on the NMR time scale. From measurements of the complex multiplet from the α-protons the conclusion was reached that the two isomers were present in approximately equal amounts and that the ratio of conformers in the original dimethylpiperidine would be similar to that of the salts. This led to the suggestion that the two conformers of the dimethylpiperidine, one with an axial lone pair and the other with an equatorial lone pair, were energetically very similar. After consideration of the spectra of a number of model compounds, Lambert and Keske (159a) reached the conclusion that the upfield shift of the axial α-protons of *N*-alkylpiperidines was due mainly to the axial lone pair and not to the equatorial *N*-alkyl group as suggested in Ref. 158.

^{19}F spectroscopy has been used to determine the nitrogen inversion rate of 4,4-difluoropiperidine in methanol and chloroform (159b). The conformational mobility of perfluoropiperidine has also been studied (160).

During elucidation of the structure of a number of alkaloids the conformation of the piperidine ring has been extensively studied and pertinent references may be found in Ref. 138. Apart from the brief mention of tropanes above, alkaloids are not included in this discussion.

Published data for piperidines are so extensive and cover so many variables such as solvent and temperature that it is impractical to collect them all here. Data listed in Tables 2.35 and 2.36 are representative, and for specific information the original reference should be consulted.

Tetrahydropyridines are rather rare and very few spectra have been reported. Lyle and Krueger (161) have used the bandwidths of signals from H-3 of 3-substituted 1-methyl-4-phenyl-1,2,3,6-tetrahydropyridine (**130**) to show that the 3-substituent prefers the axial conformation. Casy and his co-workers have studied the stereochemistry of various 5-methyl-1,2,5,6-tetrahydropyridines (162). Spectra obtained of liquid SO_2 solutions of the hydrogen halide salts of glutaronitrile were used to show that the species present had the 2,2-dihalo-6-amino-2,3,4,5-tetrahydropyridinium halide structure **130a** (162a). Other data can be found in papers mainly on the synthesis of this ring system and most of those available are collected in Table 2.37.

G. Fluoropyridines

^{19}F spectra of fluoropyridines have been studied mainly by two groups of workers who unfortunately have used different standards, making the two sets of data not strictly comparable. The most extensive NMR paper is that by Lee and Orrell (163) who systematically examined the range of compounds prepared by the Hazeldine group. Trifluoroacetic acid was used as external reference and susceptibility corrections were found to be small. Chambers,

TABLE 2.37 CHEMICAL SHIFT DATA FOR TETRAHYDROPYRIDINES

1,4,5,6-TETRAHYDROPYRIDINES

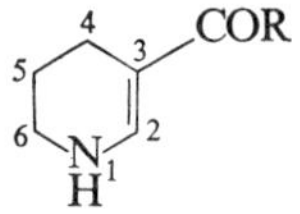

R	Solvent	τ1	τ2	τ4	τ5	τ6	Ref.
$CH_2CH_2CO_2Et$	$CDCl_3$	4.00	2.21	7.58	8.13	6.68	277
CH_2CO_2Et	$CDCl_3$	3.98	2.53	7.71	8.28	6.90	277
OEt	$CDCl_3$	4.96	2.54	7.69	8.20	6.81	277
NH_2	D_2O		2.11	7.34	7.73	6.40	277
Me	$CDCl_3$	2.88	2.52	7.72	8.23	6.77	277
Me	CCl_4		2.51	7.71	8.23	6.77	278

1,2,3,6-TETRAHYDROPYRIDINES[a]

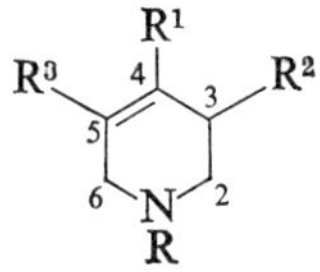

R	R[1]	R[2]	R[3]	τ2	τ3	τ5	τ6	Ref.
Me	Ph	Br	H		4.81	3.92		161
Me(HBr)	Ph	Br	H		4.50	3.93		161
Me	Ph	OH	H		5.62	3.92		161
Me(HCl)	Ph	OH	H		5.15	3.98		161
Me	Ph	OAc	H		4.05	3.69		161
Me(HBr)	Ph	PAc	H		4.00	3.80		161
Me	Ph	Cl	H		5.15	3.98		161
Me(HCl)	Ph	Cl	H		4.74	3.98		161
Ph	NH_2	H	COOEt	6.58	7.72		6.29	290

[a] Many data on 4-aryl-3(or 5)-methyl-1,2,3,6-tetrahydropyridines will be found in Ref. 162.

Musgrave, and their co-workers used trichlorofluoromethane as standard (164, 165). All data are listed in Table 2.38.

In contrast to the proton spectrum of pyridine, internal chemical shifts in the ^{19}F spectrum of pentafluoropyridine are large compared with coupling constants, and spectral analysis is straightforward. The order of shifts, $2 < 4 < 3$, is the same as for the protons of pyridine but the individual shifts are much larger. Shifts are strongly influenced by the nature of substituents (see Table 2.38) and an approximate first-order relationship was found between the meta shifts and Hammett σ_m constants. Geometrical isomerism in propenyl-substituted compounds causes appreciable variation in the shifts of adjacent protons. The ^{19}F shifts of trifluoromethyl groups in positions 2, 3, and 4 are in the reverse order to that observed for fluorine substituents.

The values shown for coupling constants in Table 2.38 are moduli and it was shown that J_{25} (J_{36}) has the opposite sign to J_{23} (J_{56}). Para couplings are usually a little larger than ortho couplings, with meta couplings considerably smaller. The difficulty of determining π-electron densities in this system prevented their correlation with chemical shifts, but it was found that certain couplings were quite sensitive to substitution. Variation in couplings was greatest for meta fluorines coupled across a nonfluorine substituent. The value for J_{26}, 14.8 Hz, seems surprisingly large compared with the negligible value observed for protons. In most spectra strong ^{14}N quadrupole broadening was observed.

It is interesting to compare spectra of 4-hydroxy-2,3,5,6-tetrafluoropyridine and 2,4-dihydroxy-3,5,6-trifluoropyridine with those of the corresponding methoxy compounds **131** and **132** (Table 2.38). The similarities show clearly that the hydroxy compounds exist predominantly as the aromatic hydroxy tautomers **133** and **134** rather than the oxo forms observed for the corresponding unfluorinated pyridines. Introduction of an hydroxyl group into

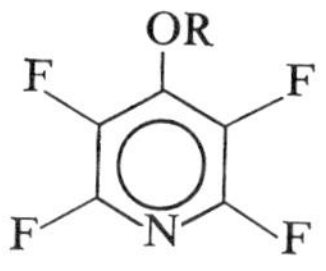

131, R = Me
133, R = H

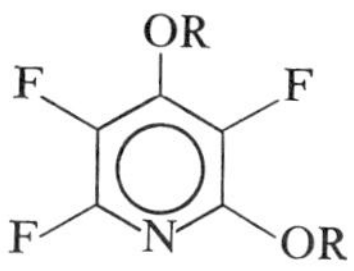

132, R = Me
134, R = H

the ring produces ^{19}F shifts of about +2.0, +5.5, and +10 ppm for nuclei ortho, meta, or para to the hydroxyl group (164).

Large amounts of data are collected in a number of papers describing the synthesis of these fluoro compounds and references for these will be found in Table 2.38. ^{19}F spectra of the three isomeric octafluorobipyridyls have been reported (165).

Emsley and Phillips (165a) have recorded the ^{19}F substituent shifts for a large number of polyfluoropyridines. Comparison of these values with calculated π-electron densities showed little correlation. Empirically calculated values for $^4J_{FF}$ were in good agreement with experiment but $^5J_{FF}$ calculations were less reliable. Substituent shifts for 5-substituted 2-fluoropyridines correlated well with those from para-substituted fluorobenzenes (165d).

TABLE 2.38 ^{19}F SPECTRA OF FLUOROPYRIDINES. CHEMICAL SHIFTS IN ppm, COUPLING CONSTANTS IN Hz. SOLVENTS INDICATED IN PARENTHESES AFTER SUBSTITUENTS[a]

Compound	Standard[b]											Ref.
Perfluoro		F2	F3	F4		J_{23}	J_{24}	J_{25}	J_{26}	J_{34}	J_{35}	
	1	11.77	86.0	57.4		19.8	13.9	25.4	15.6	16.9	1	163
	2	87.63	162.025	134.188								164, 267, 269
	3	−75.1	−1.7	−29.5								268
2-Substituted		F3	F4	F5	F6	J_{34}	J_{35}	J_{36}	J_{45}	J_{46}	J_{56}	
2-H	2	148.8	140.5	157.9	83.9							265
2-CF_3	1	66.8	60.3	74.9	5.7	16.8	9.7	26.6	16.2	18.2	22.7	163
	3	−21.4	−28.0	−13.5	−82.0	17.7	10.1	28.1	17.0	19.8	24.1	279
2-OH	2	164.663	141.305	173.612	91.550							268
2-OMe	3	1.6	−22.4	9.0	−71.5							268
2-Br	3	−7.5	−29.3	−33.2	−82.6	17.2	4.6	22.7	18.2	17.0	25.6	279
3-Substituted		F2	F4	F5	F6	J_{24}	J_{25}	J_{26}	J_{45}	J_{46}	J_{56}	
3-H	2	67.6	99.2	113.1	83.3							265
3-CF_3	1	−10.86	36.6	89.2	1.38	8.1	24.3		16.7	25.2	20.3	163
3-Cl	1	−9.4	35.1	84.9	6.3							263
	2	72.122	114.307	163.903	85.708							164, 267
4-Substituted		F2		F3		J_{23}	J_{25}	J_{26}	J_{35}			
4-H	1	15.61		64.1		21.0	30.4	13.3	1.4			163
	2	92.28		140.97								265
4-Me	2	92.5		144.0								269

$4\text{-}CF_3$	1		11.88			65.6	19.9	31.9	15.5	9.6	163
4-Bu	2		91.9			144.2					269
4-Ph	2		92.83			145.86					264
$4\text{-}CH_2Cl$	2		89.7			143.1					269
$4\text{-}CHCl_2$	2		87.8			140.7					269
$4\text{-}CCl_3$	2		87.1			140.1					269
$4\text{-}CH_2Br$	2		89.75			143.0					269
$4\text{-}CHBr_2$	2		88.1			139.9					269
4-Propenyl	1		15.76			65.1	20.9	29.5	14.4	1	163
	1		7.03			69.7					163
4-CHO	2		90.4			146.8					269
4-COOH (TFA)	1		13.06			63.3	19.2	31.2	13.4	4.1	163
$4\text{-}NH_2$ (TFA)	1		22.8			90.0					163
$4\text{-}NH_2$	2		95.72			165.94					264
	3		−68.2			1.7					268
	2		95.7			165.9					269
$4\text{-}NMe_2$	1		18.42			80.7	22.1	22.1	15.5	9.6	163
$4\text{-}NHNH_2$ (DMF)	1		19.0			86.1					163
$4\text{-}NO_2$	3		−77.5			−16.6					268
4-OH (H_2O)	1		17.0			86.1	20.6	22.2	15.5	7.9	163
	2		94.307			164.663					164
4-OMe	1		16.0			84.2	20.0	22.6	15.5	4.7	163
	2		92.29			161.05					264
	3		−70.0			−1.2					268
2,3-Disubstituted		F4		F5	F6						
2-OH, 5-Cl	2	117.925		173.567	90.200						164
2,4-Disubstituted		F3		F5	F6						
2,4-Di-Me	2	128.65		144.25	90.9						269
2-Bu, 4-Me	2	130.6		144.3	90.7						269

TABLE 2.38 *(continued)*

2,4-Disubstituted		F3	F5	F6	J_{35}	J_{36}	J_{56}	Ref.
2-NH_2, 4-H	2	142.1	155.6	94.4				267
2-NH_2, 4-Me	2	145.8	160.1	94.9				269
2-NH_2, 4-Br	2	135.7	146.3	92.2				266
2-OMe, 4-NH_2	3	2.4	8.26	−65.8				268
2,4-Di-NH_2	2	164.1	175.2	96.4				269
	3	1.9	13.0	−65.8				268
2,4-Di-NMe_2	1	67.5	86.8	15.2	1	25.6	25.6	163
2-NH_2, 4-NO_2	3	−12.7	2.7	−75.3				268
2,4-Dipropenyl	1	48.6	61.5	11.55	1	30.2	27.8	163
	1	53.1	62.2	11.75	1	29.9	27.2	163
	1	52.2	66.7	12.4	1	30.8	28.1	163
2,4-Di-OH (D_2O)	1	94.5	87.4	18.4	2.4	24.4	24.4	163
2,4-Di-OMe	1	91.6	84.6	17.9	1	25.5	22.4	163
	2	168.73	161.73	94.90	0	21.55	25.2	264
2-OMe, 4-Br	2	135.2	145.1	92.2				266
2-OMe, 4-NO_2	3	−15.0	−4.2	−75.5				268
2-OEt, 4-Me	2	144.3	154.1	95.1				269
2,5-Disubstituted		**F3**	**F4**	**F6**				
2-OH, 5-Cl	2	165.883	120.438	75.950				164
3,4-Disubstituted		**F2**	**F5**	**F6**				
3,4-Di-H	2	74.40	147.4	88.69				265
3-Cl, 4-H	1	−4.2	64.7	8.8				263
3-Cl, 4-OH	2	75.950	165.883	92.492				164
3,4-Di-NH_2	3	−68.8	6.3	−58.8				268
3-NH_2, 4-NO_2	3	−82.8	−10.2	−54.6				268

3,5-Disubstituted		F2	F4	J_{24}	F6	Ref.
3,5-Di-H	2	65.05	93.15	19.7		265
3,5-Di-Cl	1	−7.8	16.19	14.3		263
	2	69.858	94.022			164
3-Cl, 5-H	2	68.462	94.63		66.547	267
2,3,5-Trisubstituted		**F4**	**F6**			
2-OH, 3,5-di-Cl	2	99.547	73.317			164
2,4,6-Trisubstituted			**F3**			
2,4,6-Tri-OMe	1		90.5			163
2,6-Di-OMe, 4-Br	2		144.2			266
2,6-Di-OEt, 4-Me	2		152.5			269
3,4,5-Trisubstituted			**F2**			
3,5-Di-Cl, 4-OH	2		74.117			164

Octafluorobipyridyls

Junction	Solvent	Standard[b]	F3	F4	F5	F6	Ref.
2,2′	Acetone	3	−20.9	−25.4	−9.1	−81.2	165
			F2				
3,3′	Acetone	3	−94.4	−52.2	−1.7	−86.6	165
				F3			
4,4′	Acetone	3	−72.3	−23.7			165

[a] Details in most papers are hazy about solvents. Where possible, compounds were measured neat.
[b] Standards: 1—Trifluoroacetic acid; 2—$CFCl_3$; 3—C_6F_6.

Spin-lattice relaxation rates of ^{14}N have been determined by observing spin-echo trains of fluorine in 2-fluoropyridine (165b). The technique is new and is subject to considerable instrumental difficulties. 3,4,5-Trichloro-2,6-difluoropyridine was used as a model compound to study the problem of obtaining coupling constants between spin-$\frac{1}{2}$ and spin-1 nuclei in the fast-exchange limit of quadrupolar relaxation (165c). The value $J_{^{15}NF} = 52.24$ Hz was measured from the ^{15}N satellites and the ^{19}F band shape was then used to give the ^{14}N relaxation times.

II. THE PYRIDAZINE SYSTEM

A. Fully Aromatic Pyridazines

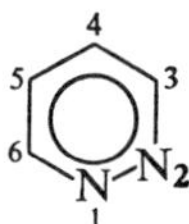

Spectra of compounds with this rather rare ring system have been little studied. That of the parent was first reported in 1960 by Isobe (166) who analyzed the proton spectrum as an A_2X_2 system and obtained the values quoted in Table 2.39. Chemical shift data were later produced by a number of workers (167) and Isobe's analysis was corrected by Tori and Ogata (168) who used ^{13}C satellite spectra to obtain the coupling constants (Table 2.39). These coupling constants have been again corrected by two groups, one using ^{13}C satellites (169) and the other an A_2B_2 analysis of the proton spectrum (170), to give similar results (Table 2.39). The difficulty encountered in obtaining the correct results for this compound serves to emphasize the fact that spectral analysis, particularly of simple systems such as A_2B_2, is not

TABLE 2.39 SPECTRAL DATA FOR PYRIDAZINE

Method	Solvent	Concentration (%)	$\tau 3$	$\tau 4$	J_{34}	J_{35}	J_{36}	J_{45}	$J_{^{13}C,2}$	$J_{^{13}C,3}$	Ref.
A_2X_2					5.1	2.1	1.3	5.0			166
^{13}C	$CDCl_3$	3	0.76	2.46							168
	Neat				4.9	2.0	3.0	8.4	181.5	168.5	168
	$CDCl_3$	10	0.76	2.45	4.9	2.0	3.5	8.4			168
^{13}C	CCl_4		0.83	2.48							169
	Neat		0.31	1.98	5.05	1.85	1.40	8.0	183	168	169
	C_6H_6		1.26	3.58							169
A_2B_2	Neat		0.309	1.967	5.0	2.1	1.4	8.6			170
	$CDCl_3$	20	0.833	2.460	5.2	1.9	1.4	8.6			170
	$CDCl_3$	5–10	0.76	2.50							167

easy, and the task of obtaining a unique solution is often quite a formidable one.

Unusual solvent effects are observed with pyridazine. As the concentration of the compound in deuterochloroform increases, signals from all protons shift to lower field, that is, in the opposite direction to that observed for most aromatic compounds. This was thought to be caused by the high polarity of pyridazine causing unmeshed stacking of the type **135** (170).

The two *C*-methyl derivatives **136** and **137** and the 3-chloropyridazine **138** give deceptively simple ABX-type spectra which required a knowledge of the expected coupling constants before sensible parameters could be derived (168). However, in view of the more recent analyses of the parent compound, the value for J_{36} obtained for the 4-methyl derivative is obviously incorrect and it is difficult to see how satisfactory computation was performed without variation of any of the other parameters. Long-range couplings between the *C*-methyl protons of this compound and the adjacent ring protons, H-3 and H-5, were observed, but such couplings were absent from the spectrum of the 3-methyl isomer, a difference thought to reflect the π-electron densities at C-3 and C-4. An attempt was made to correlate calculated π-electron densities (HMO) with the observed chemical shifts for the parent molecule using the equation and constants suggested by Schaefer and Schneider (21) but the results were not very satisfactory. A more recent correlation was also only fair (25). These rather disappointing results are presumably due in part to difficulties in calculating π-electron densities in such a heteroaromatic system and in part to an inadequate understanding of factors other than electron densities which affect the chemical shifts of protons in these molecules. Temperature-dependent spectra of the pyridazine–iodine complex were interpreted in terms of the presence of a large molecular assembly (170a).

Spectra of *C*-substituted pyridazines have been reported by two main groups of workers (168, 171). Substituent effects and $J_{^{13}C\text{–}H}$ values were discussed in terms of the electronegativity of the substituent and the *s*-character of the carbon atomic orbitals participating in the C–H bonds. Nuclear magnetic resonance data are collected in Table 2.40. Methylthio- and methylsulfonylpyridazines have been studied both as neutral molecules and as cations, but the protonation shifts did not allow the assignment of specific protonation sites (172). Quaternization of 3- or 4-methylpyridazines with methyl iodide in both cases produces a mixture of the N-1 and N-2 methylated isomers whose NMR parameters were extracted from the spectra of the respective mixed methiodides (173).

Possibilities for oxohydroxy tautomerism in maleic hydrazide are considerable and from a study of the possible fixed structures **139** to **142** (Table 2.41), the predominant tautomer for the parent compound in dimethyl

TABLE 2.40 SPECTRAL DATA FOR SUBSTITUTED PYRIDAZINES (168, 171). SOLVENT—$CDCl_3$

	Chemical shifts[a]				Coupling constants[a]				
Substituents	τ3	τ4	τ5	τ6	J_{35}	J_{36}	J_{45}	J_{46}	J_{56}
3-Me	(7.26)	2.62	2.60	0.94	2.2	3.0	8.6	1.8	4.7
4-Me	0.92	(7.60)	2.67	0.96	2.2	3.0	(1.0)		5.0
3-Cl		2.41	2.45	0.83			8.8	1.8	4.7
3-OMe	(5.92)	2.98	2.59	1.12			9.0	1.7	4.5
3-Me, 6-Cl	(7.29)	2.65	2.57				8.8		
3-Cl, 4-Me		(7.54)	2.58	1.01			(1.0)		4.9
4-Me, 6-Cl	1.01	(7.59)	2.60		2.2		(1.0)		
3-Cl, 6-Me		2.57	2.64	(7.30)			8.8		
3-Me, 6-OMe	(7.38)	2.74	3.10	(5.88)			8.8		
3,6-Di-Cl		2.43	2.43						
3-OMe, 6-Cl	(5.99)	3.01	2.62				9.0		
3-Cl, 6-OMe		2.62	3.04	(5.89)			9.1		
3,6-Di-Cl, 4-Me		(7.55)	2.55				(0.9)		
3-Me, 5-Cl, 6-OMe	(7.40)	2.67	(5.82)						

[a] Chemical shifts in parentheses are for the methyl group of the substituent. Coupling constants in parentheses are between the substituent methyl group and the indicated ring proton.

TABLE 2.41 SPECTRAL DATA FOR MALEIC HYDRAZIDE AND MODEL COMPOUNDS (174). SOLVENT—DMSO

Compound	τ4	τ5	J_{45}	τNMe	τOMe
Maleic hydrazide	3.04	3.04			
2,3-Dihydro-3-oxopyridazine					
6-Cl	3.00	2.49	9.9		
2-Me, 6-Cl	2.96	2.46	9.8	6.36	
2-Me, 6-OH	3.13	2.96	9.4	6.51	
2-Me, 6-OMe	3.09	2.84	9.0	6.47	6.22
1,2,3,6-Tetrahydro-3,6-dioxopyridazine	3.09	3.09			
3,6-Dimethoxypyridazine	2.84	2.84			6.01

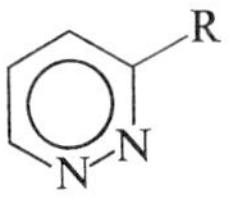

136, R = Me
138, R = Cl

137

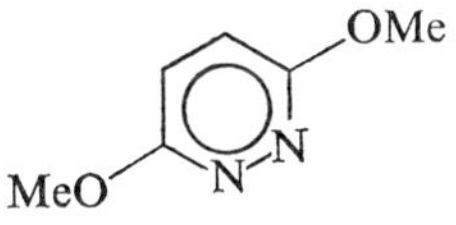

139

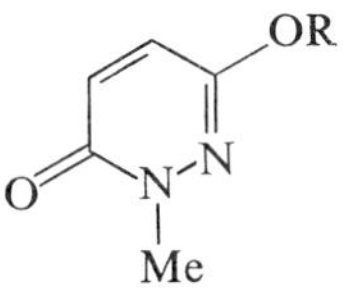

140, R = H
141, R = Me

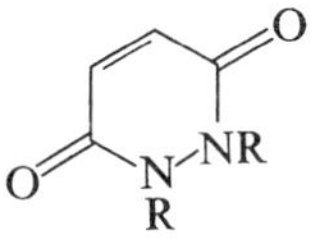

142, R = Me
145, R = H

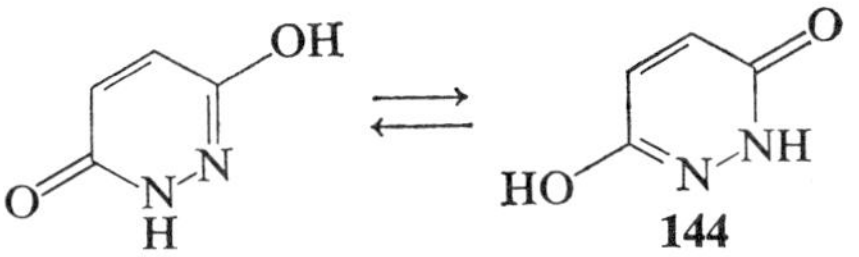

143

144

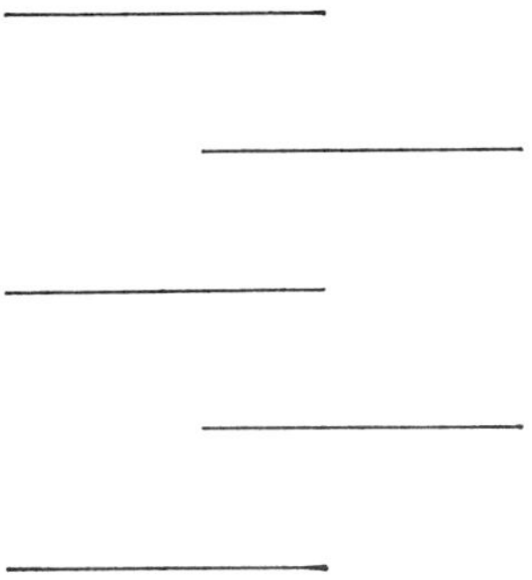

135

sulfoxide was thought to be 2,3-dihydro-5-hydroxy-2-oxopyridazine (**143**), in rapid equilibrium with its equivalent tautomer **144** (174). The earlier misinterpretation of the spectrum (175) where structure **145** was preferred because of the apparent symmetry of the molecule underlines the deceptive nature of "symmetrical" spectra from systems where rapid time-averaging is occurring. Spectra of a number of 6-aryl-2,3-dihydropyridaz-3-ones have also been reported (176). Deuterium exhange in D_2SO_4 solutions of a number of pyridazines has been followed by NMR (176a).

An extremely comprehensive study has been made of the spectra of pyridazine *N*-oxides (177), Table 2.42. In these spectra signals from H-3 and H-6 showed broadening because of incomplete decoupling from the nitrogen nucleus. However, chemical shifts were usually large and first-order

TABLE 2.42 SPECTRAL DATA FOR PYRIDAZINE *N*-OXIDES (167). SOLVENT—$CDCl_3$

A. MONOOXIDES

Substituents	τ3	τ4	τ5	τ6	τMe	J_{34}	J_{35}	J_{36}	J_{45}	J_{46}	J_{56}	τOMe
None	1.46	2.78	2.17	1.74		5.3	2.5	1.0	8.0	1.0	6.5	
3-Cl		2.77	2.25	1.82					8.3	1.0	6.5	
3-Me		3.02	2.42	1.90	7.48				8.2	0.5	6.1	
4-Me	1.65		2.43	1.83	7.64	0.2	2.8	0.5	0.5	0.2	6.2	
5-Me	1.62	3.01		1.92	7.63	5.6	0.2	0.5	0.7	0.7	0.7	
6-Me	1.63	2.88	2.27		7.49	5.6	2.5		8.2			
3,6-Di-Cl		2.78	2.10						8.4			
3-Cl, 4-Me			2.41	1.90	7.61				0.7	0.2	6.2	
3-Cl, 5-Me		2.95		2.01	7.64				0.7	0.7	0.7	
3-Cl, 6-Me		2.83	2.27		7.51				8.3			
4-Cl, 6-Me	1.61		2.32		7.49		3.0					
3,4-Di-Cl, 6-Me			2.23		7.50							
3-OMe		3.31	2.47	2.05					8.6	0.7	5.8	5.98

3-OMe, 6-Me		3.33	2.48		7.55				8.5			5.98
4-OMe, 6-Me	1.92		2.80		7.48		3.7					6.07
3-OMe, 6-Cl		3.28	2.29						8.8			5.96
3,6-Di-OMe		3.27	2.64						8.7			5.92, 6.02
3-OMe, 4-Cl, 6-Me			2.46		7.56							5.92
3,4-Di-OMe, 6-Me			3.05		7.57							5.90, 6.02
4-NO_2	0.70		1.55	1.84			2.3	0.5			7.0	
4-NO_2, 6-Me	0.80		1.55		7.41		3.3					
3-Cl, 4-NO_2, 6-Me			1.66		7.46							
3-OMe, 4-NO_2, 6-Me			1.65		7.49							5.82

B. DIOXIDES

Substituents	$\tau 3$	$\tau 4$	$\tau 5$	$\tau 6$	J_{34}	J_{35}	J_{36}	J_{45}	J_{46}	J_{56}
None	1.85	2.92	2.92	1.85						
3Me	7.44	2.90	2.95	1.86	~0.4		~0.6	8.1	2.4	6.1
4Me	1.93	7.65	3.02	1.89	~0.8	2.4	0.7	0.7		6.8
3,6-Di-Me	7.47	3.04	3.04	7.47	<0.2					<0.2

analyses gave reasonable spectral parameters. As with pyridazine itself, methyl groups in positions 3 or 6 showed no long-range couplings with the ring protons, while those in positions 4 or 5 coupled normally with the ortho protons. LCAO-MO calculations of electron densities predict the order of shielding of protons 3, 4, and 5 in pyridazine-1-oxide, but not that of H-6. The anomalous shift of this proton was explained in terms of the diamagnetic anisotropy of the *N*-oxide bond. Nuclear magnetic resonance parameters have been reported (177a) for a number of 1,2-dioxides of simple pyridazines and these are included in Table 2.42.

Nuclear magnetic resonance data for ^{19}F were recorded for tetrafluoropyridazine and for its mono- and dimethoxy derivatives (177b). In the parent compound F-3,6 resonated 18.1 ppm below hexafluorobenzene and F-4,5 71.7 ppm in the same direction. Methoxyl-substituent shifts from the monomethoxy derivatives were assumed to be additive and the calculated shifts for the dimethoxy compounds were within 2 ppm of the observed values.

B. Reduced Pyridazines

Dihydropyridazines are extremely rare and very few NMR data have been reported. In one paper (178) spectra of a number of 2,5-dihydropyridazines of the general type **146** are reported. Tautomerism between the 4,5-dihydropyridazine (**147**) and the diazepin (**148**) has been observed and structures of the isomers were determined by NMR (179).

Anderson and Lehn (180) have studied conformational changes in various di-, tetra-, and hexahydropyridazine systems (**149**, **150**, and **151**) containing the –N(COOR)–N(COOR)– fragment and have related them to the free enthalpies of activation. Hindered rotation about the N–COOR bonds was observed for the tetra- and hexahydro derivatives and ring "twist" was shown to occur in the dihydropyridazines. This series of compounds is readily accessible through the tetrahydropyridazines (**150**) formed by the action of the dienophile dimethylazidocarboxylate with substituted butadienes. When trans-trans-2,4-hexadiene was used the product was shown to be the cis isomer **152** (181). In the case of the trans diphenyl derivative **153**, variable-temperature NMR showed interconversion between the two possible half-chair conformers **154** and **155** (182). Korsch and Riggs (182a) also studied the slow inversion of compounds of type **150** and NMR data for conformational changes in the corresponding di-*N*-acyl compounds have been reported (182b). Data for reduced pyridazines are collected in Table 2.43.

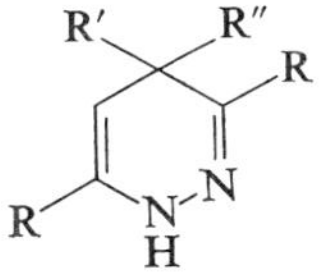

146 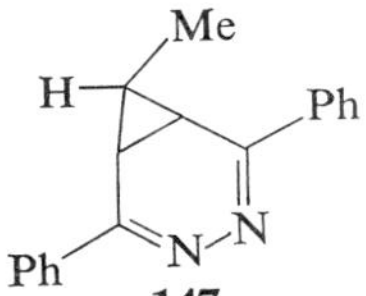 147 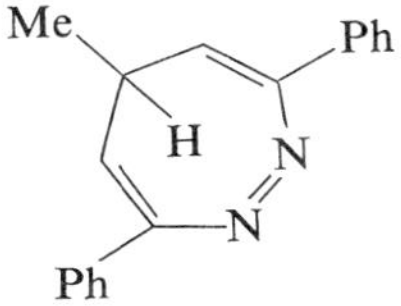 148

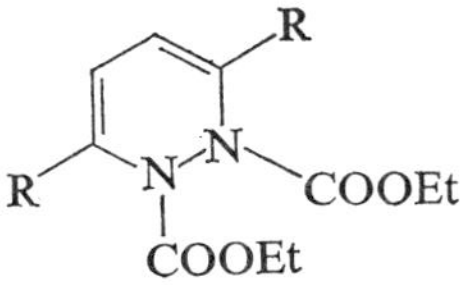

149 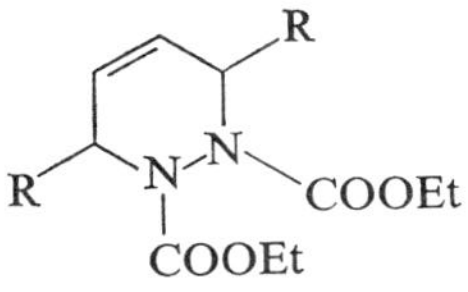 150 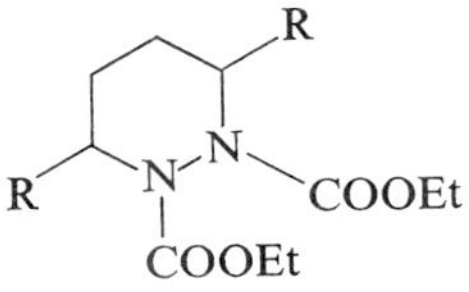 151

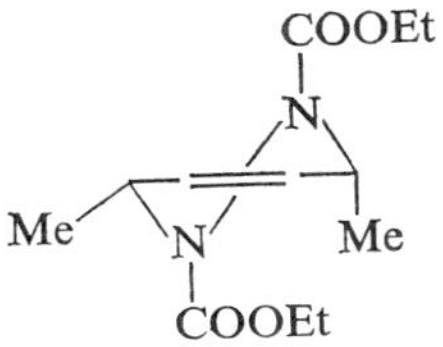

152 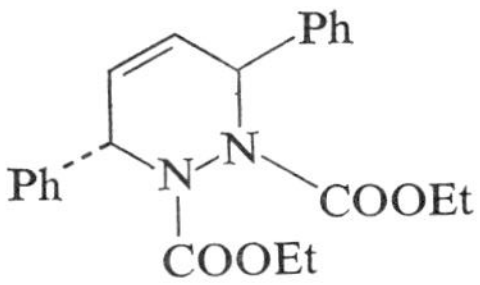 153 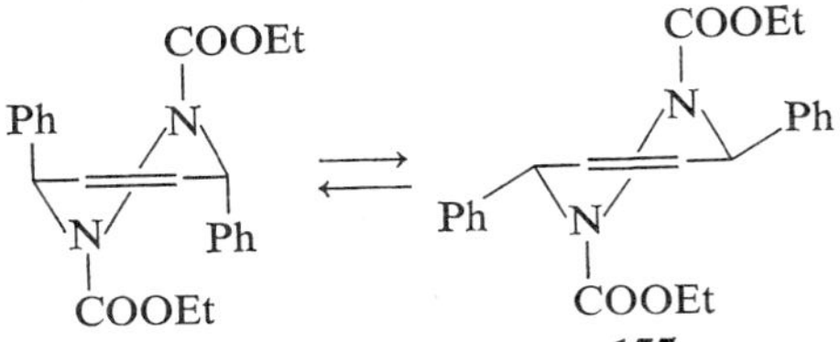 154 ⇌ 155

TABLE 2.43 SPECTRAL DATA FOR REDUCED PYRIDAZINES (180)

1,2-DIHYDRO

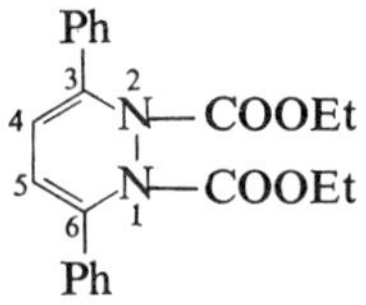

(3,6-diphenyl-1,2-dihydropyridazine-1,2-di(COOEt))

	162° HCB[a]	30° TCE[b]
τ4, 5	3.8	3.8

1,2,3,6-TETRAHYDRO

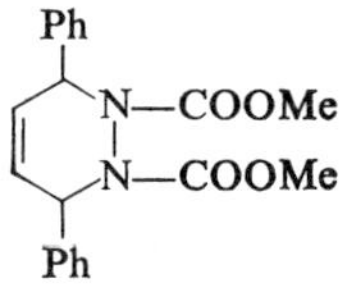

(3,6-diphenyl-1,2,3,6-tetrahydropyridazine-1,2-di(COOMe))

	−60° $CDCl_3$	35° $CDCl_3$	130° TCE[b]	Couplings from $CDCl_3$ 35°
τMe	6.50	6.50	6.45	
	6.46			
τMe	6.13	6.12	6.45	J_{34} 2.2
τ3	4.72	4.68	4.48	J_{45} 10.3
τ4	4.23	4.21	4.07	J_{56} 5.1
τ5	3.70	3.68	4.07	J_{35} 1.5
τ6	4.00	3.98	4.48	J_{46} 1.1
τPh	2.82	2.78	2.80	
	2.54	2.50		

HEXAHYDRO

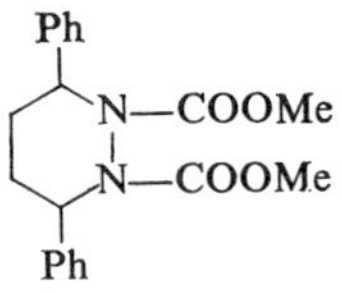

(3,6-diphenylhexahydropyridazine-1,2-di(COOMe))

	95° HCB[a]	32° HCB[a]	−1° HCB[a]
τMe	6.4	6.55, 6.25	6.1–6.7
τ3, 6	5.0–4.7	4.8–5.2	4.7–5.2
τ4, 5	7.9–8.2	7.8–8.3	7.8–8.3

[a] Hexachlorobutadiene.

[b] Tetrachloroethylene. Data are representative only. See text for other references.

III. THE PYRIMIDINE SYSTEM

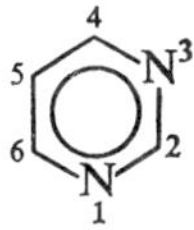

A. Nontautomeric Pyrimidines

The proton magnetic resonance spectra of simple nontautomeric pyrimidines have not been widely studied due probably to the synthetic difficulties involved in obtaining many of the compounds. A large number of the most interesting pyrimidines are highly substituted and contain so few hydrogen atoms that the spectra obtained are of little use in structural assignments. With the free bases, broadening is observed of signals from H-2, H-4, and/or H-6 due to the intermediate relaxation times of the adjacent ^{14}N resonance. Removal of this broadening by heteronuclear decoupling has, to my knowledge, not been reported.

The spectrum of pyrimidine was first reported by Gronowitz and Hoffman (183) who measured chemical shifts at 40 MHz in a number of solvents with respect to either the solvent peak or an external water standard. They established the relative deshielding of the four ring protons, H-2 > H-4 = H-6 > H-5, and by direct measurement obtained values for all coupling constants except J_{46}, which is not observed because of the equivalence of H-4 and H-6. The low value for meta coupling across the nitrogen atoms was noted, compare J_{26} for pyridine −0.13 Hz. (6), but, to the present time no satisfactory explanation has been presented for the size of such couplings. From the relatively constant internal chemical shifts observed for the neat liquid and for dilute solutions in carbon tetrachloride, association effects must have little influence on the magnetic properties of the molecule. Later (184), the spectrum of a deuterochloroform solution was measured with respect to internal tetramethylsilane, confirming the earlier work and giving chemical shift values which are more meaningful by today's standards (Table 2.44). Analysis of the ^{13}C satellites of each multiplet gave J_{46} and the three ^{13}C–H couplings, as well as the known coupling constants (Table 2.44). A rough correlation was obtained between the ^{13}C–H couplings and the chemical shifts of the respective protons. Long-range couplings between the methyl groups and protons in positions ortho or para to them were observed for the three mono-*C*-methylpyrimidines (**156–158**). The concept of a total substituent shift, compared with specific methyl shifts, was used to discuss the transmission of the inductive effect throughout the ring. Spectra of these three methyl pyrimidines have been measured more recently at 60 MHz with complete confirmation of the earlier work (185). The spectrum of pyrimidine has been studied in the nematic phase of 4-methoxybenzylidene-4-amino-α-methylcinnamic acid-*n*-propyl ester at room

TABLE 2.44 SPECTRAL DATA FOR PYRIMIDINE AND ITS *C*-METHYL DERIVATIVES (183, 184)

		Chemical shifts[a]				Coupling constants			
Pyrimidine	Solvent	$\tau 2$	$\tau 4$	$\tau 5$	$\tau 6$	J_{25}	J_{45}	J_{46}	J_{56}
Parent[b]	$CDCl_3$	0.74	1.22	2.64	1.22	1.5	5.0	2.5	5.0
	Me_2CO	0.83	1.20	2.52	1.20	1.45	5.0		5.0
	C_6H_{12}	0.84	1.40	2.92	1.40	1.60	5.0		5.0
	DMSO	0.74	1.13	2.42	1.13				
2-Me	$CDCl_3$		1.37	2.88	1.37	(0.60)	4.9		4.9
	Me_2CO	(7.38)	1.32	2.77	1.32	(0.55)	4.9		4.9
	C_6H_{12}	(7.38)	1.51	3.16	1.51	(0.60)	4.9		4.9
4-Me	$CDCl_3$	0.91		2.79	1.41	1.4	(0.4)		5.1
	C_6H_{12}	1.10		3.02	1.59	1.45	(0.55)	(0.30)	5.16
5-Me	$CDCl_3$	0.96	1.43		1.43				
	Me_2CO	1.01	1.35	(7.68)	1.35				
	C_6H_{12}	1.05	1.55	(7.80)	1.55				

[a] Chemical shifts in parentheses are those of the methyl substituents; coupling constants in parentheses are between methyl protons and the indicated ring proton.
[b] ^{13}C couplings obtained on neat liquid: $J_{^{13}C\text{-}2,H} = 206.0$, $J_{^{13}C\text{-}4,H} = 181.8$, $J_{^{13}C\text{-}5,H} =$ 168 Hz.

temperature (185a). Ratios of interproton distances were found to differ significantly from those obtained from microwave spectroscopy. The molecule, like other aromatics, oriented preferentially with its plane in the direction of the magnetic field. Data reported included the direct couplings, D_{24} -68.2 ± 0.3, D_{45} -412.3 ± 0.3, D_{25} -31.5 ± 0.4, D_{46} -97.8 ± 0.4 Hz, and ratios of interproton distances, r_{46}/r_{45} 1.706 ± 0.004, r_{25}/r_{45} 1.90 ± 0.02, r_{24}/r_{25} 1.62 ± 0.01.

Benzene-induced solvent effects upon the proton chemical shifts of a number of *C*-methylpyrimidines and their *N*-oxides have been described (185b). A relatively fixed collision complex was envisioned with the benzene located near the positive end of the pyrimidine dipole. Larger effects, implying closer association, were noted for the *N*-oxides. In those instances where isomeric *N*-oxides can exist, the observed solvent shifts were useful in determining the oxidation site. Although the qualitative aspects of this work are extremely useful, the assignment of fixed geometrical structures to the collision complexes suffers from the same drawbacks discussed fully for the benzene pyridine case in Section I.A. Data for pyrimidine and its *C*-methyl derivatives are summarized in Table 2.44.

Only one specific study of simple monosubstituted pyrimidines has been made (185) and due to synthetic difficulties, few 4- or 5-substituted derivatives were included. Spectra of 2-substituted pyrimidines (Table 2.45) show a doublet and triplet, typical of an A_2X system, with the H-4,6 doublet considerably broadened by coupling to the adjacent nitrogen atom. Those of 5-substituted derivatives consist of two broad singlets, while 4-substituted pyrimidines give typical ABX patterns with small couplings involving the X nucleus. The para-substituent effects were studied using the corresponding 2- and 5-substituted derivatives, for example, **159** and **160**, to check the reciprocity of transmission of electronic effects between these two nonequivalent para positions. Chemical shifts of the +M 2-substituted pyrimidines follow the mesomeric order $NMe_2 > NH_2 > OMe > Me$ established for other aromatics (186, 187). As in benzenes, the amino group gives a much larger para shift than does the methoxyl group, quite distinct from the almost equal effect of these groups in π-excessive five-membered ring systems such as thiophene. A 5-methoxy substituent causes a para shift of H-2 similar to that in the reverse system, but the 5-cyano group does not. More data are required before substituent effects in these compounds can be understood. Representative data on monosubstituted pyrimidines are collected in Table 2.45.

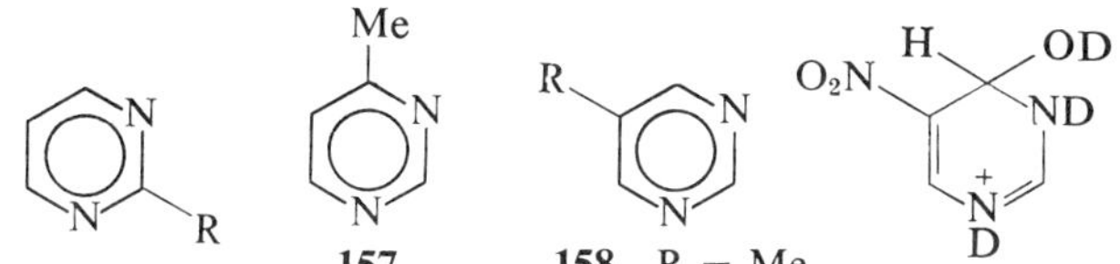

156, R = Me
159

157

158, R = Me
160
161, R = NO_2

162

Decoupling experiments have been used to determine the relative signs of couplings in a number of methyl- and aminosubstituted pyrimidines (188). Coupling constants J_{25} and J_{26} were found to have the same sign as J_{56} and hence were considered to be positive. Coupling constants between substituent *C*-methyl groups and ortho ring protons were negative and those with meta ring protons, though very small, were considered to be positive.

The spectrum of 5-nitropyrimidine (**161**) in deuterium oxide was unexceptional, showing two broad singlets (189) but, after acidification, three singlets were observed—3.51, 1.69, and 1.34 ppm—indicative of the addition of water across one of the C=N bonds (see Ref. 19). From the spectrum of the 2-methyl derivative, the site of hydration was located at C-4, N-3 (or C-6, N-1) giving the adduct structure **162**. These conclusions were confirmed by examination of the ultraviolet spectra and pK_a values for the different species involved.

TABLE 2.45 SPECTRAL DATA FOR MONOSUBSTITUTED PYRIMIDINES UNINVOLVED IN STUDIES OF TAUTOMERIC EQUILIBRIA

A. 2-SUBSTITUTED PYRIMIDINES

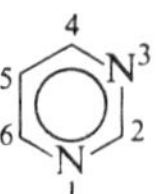

Substituent	Solvent	τ4, 6	τ5	J_{45}	Other data	Ref.
CN	Me_2CO	0.96	2.14	5.10		185
COMe	Me_2CO	1.01	2.31	5.00		185
SCN	Me_2CO	1.16	2.45	5.00		185
	C_6H_{12}	1.47	2.95	4.80		185
Cl	Me_2CO	1.22	2.45	4.90		185
	C_6H_{12}	1.53	2.94	4.75		185
Br	Me_2CO	1.28	2.43	4.80		185
	C_6H_{12}	1.62	2.94	4.80		185
I	Me_2CO	1.42	2.43	4.90		185
	C_6H_{12}	1.72	2.94	4.90		185
OMe	$CDCl_3$	1.28	2.91	6	OMe 5.91	283
OEt	$CDCl_3$	1.38	3.00	5	Et 5.53, 8.57	282
SMe	CCl_4	1.53	3.05	5	SMe 7.48	282
	$CDCl_3$	1.30	2.83	5	SMe 7.43	282
SOMe	$CDCl_3$	0.93	2.28	5	SMe 6.95	282
SO_2Me	$CDCl_3$	0.95	2.32	5	SMe 6.60	282
SO_2F	MeOH	0.80	2.15	5		281
SO_3K	D_2O	0.70	2.24	6.0		281
SPh	$CDCl_3$	1.55	2.50	5	Ph 2.3–2.6	282
SO_2Ph	$CDCl_3$	1.04	2.40	5	Ph 1.8, 2.4	282

B. 4-SUBSTITUTED PYRIMIDINES

Substituent	Solvent	τ2	τ5	τ6	J_{25}	J_{56}	Other data	Ref.
NH_2	DMSO	1.61	3.56	1.96	1.25	6.15		185
NHCOMe	DMSO	1.17	1.98	1.39	1.30	5.80		185
OMe	$CDCl_3$	1.01	3.10	1.35		7	OMe 5.01	283
SMe	$CDCl_3$	0.98	2.75	1.57	1.3	5.6	SMe 7.42	282
	CCl_4	1.10	2.85	1.75	1.0	5.5	SMe 7.47	282
SOMe	$CDCl_3$	0.67	1.83	0.85	1.2	4.2	SMe 7.05	282
SO_2Me	$CDCl_3$	0.45	1.82	0.68	1	5		282

TABLE 2.45 *(continued)*

C. 5-SUBSTITUTED PYRIMIDINES

Substituent	Solvent	τ2	τ4, 6	Other data	Ref.
COOH	DMSO	0.72	1.00		185
CN	Me_2CO	0.56	0.73		185
Br	Me_2CO	0.85	1.04		185
	C_6H_{12}	1.01	1.35		185
OH	Me_2CO	1.31	1.61		185
	DMSO	1.30	1.63		185
OMe	Me_2CO	1.22	1.50	OMe 6.05	185
	C_6H_{12}	1.28	1.75		185
	DMSO	1.13	1.40	OMe 6.07	185
NO_2	CCl_4	0.36	0.33	J_{24} 0.3	189

Spectra of 4-phenylpyrimidine in nonpolar solvents (CCl_4 or $CDCl_3$) and in TFA have been investigated as part of two separate projects (189a, b). Differences between spectra of the neutral molecules and the cations were discussed in terms of delocalization of the nitrogen lone pair.

B. Tautomeric Pyrimidines

Spectroscopic methods have long been used to determine the predominant tautomer in potentially mobile pyrimidines by comparison of the compound with "fixed tautomers," usually specifically methylated derivatives. Nuclear magnetic resonance can also be used in this way, in many cases providing greater sensitivity in these comparisons and in others providing new and highly specific criteria for the predominance of certain forms.

With the simple amino- and "hydroxy"-pyrimidines, PMR spectra have confirmed the predominant forms established by studies of ultraviolet spectra and pK_a values. Thus, with "2-hydroxypyrimidines" the H-5 resonance occurs 0.67 ppm toward higher field than that of 2-methoxypyrimidine (**163**) (185), confirming the 1,2-dihydro-2-oxo structure **164** established by ultraviolet methods. On the other hand, 2-amino-, 2-methylamino-, and 2-dimethylaminopyrimidines have very similar ring-proton shifts (185), reflecting the aromatic nature of these compounds (**165**). Unfortunately, conversion of the fully aromatic structure, as observed in the anion **166** of 2-hydroxypyrimidine, into the dihydroxo form of the free base **164** causes only a small rise in $J_{45}(J_{56})$ and this value is similar to that observed for other nontautomeric pyrimidines (see Table 2.45). The size of this coupling constant (5.3 Hz) is unusual compared with those from more highly substituted 2-oxopyrimidines such as 1-methylcytosine (**167**) where J_{56} = 7.2 Hz (190).

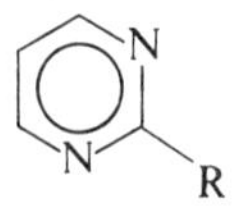

163, R = OMe
165, R = NR_2
166, R = O^-

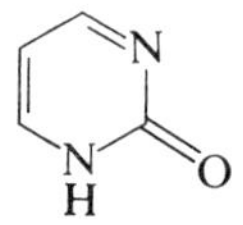

164

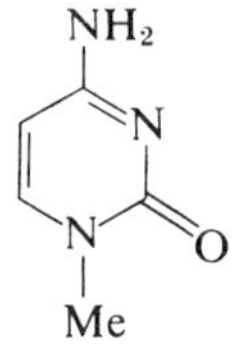

167

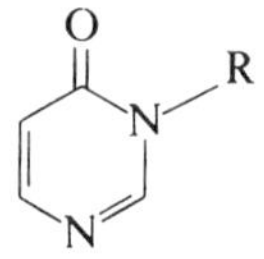

168, R = Me
170, R = H

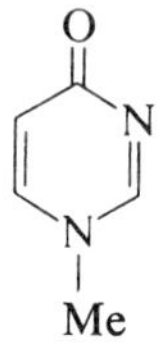

691

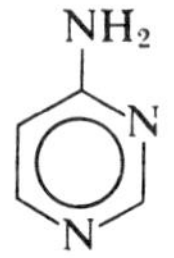

171

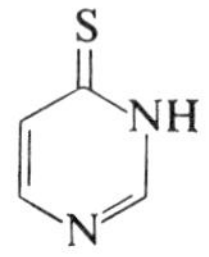

172

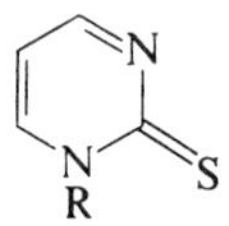

173, R = H
174, R = Me

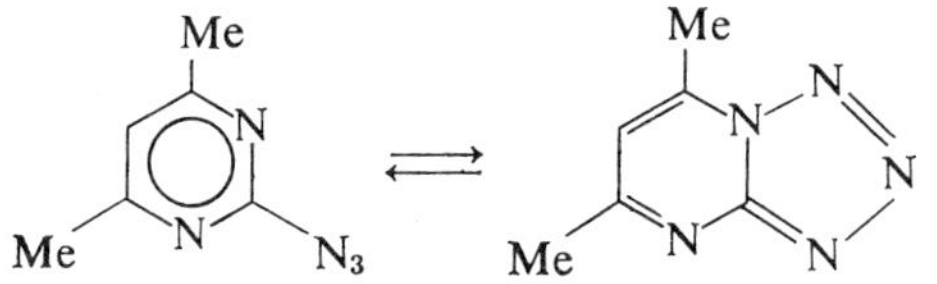

175 ⇌ **176**

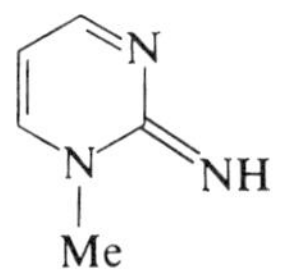

177

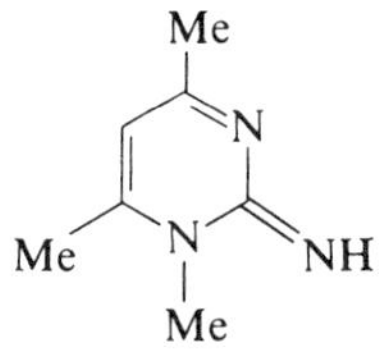

178

In the case of "4-hydroxypyrimidine," the predominant oxo structure in aqueous solution, previously established by Raman and infrared spectroscopy (191), was confirmed (192). Comparison of the spectrum of the dihydro-oxo compound with those of its 1- and 3-methyl derivatives showed the chemical shift of H-5 (τ 3.49) in the parent to be closer to that from the 3-methyl derivative (**168**; τ 3.45) than that from the 1-methyl derivative (**169**; τ 3.67). This evidence was considered sufficient to confirm the 3,4-dihydro-4-oxo structure **170** for the parent pyrimidine. Apparent pK_a values ($pK_1 = 1.8$, $pK_2 = 8.5$) obtained from the plot of chemical shift versus pH and H_0 were in rough agreement with those obtained by the more accurate potentiometric and spectroscopic methods ($pK_1 = 1.69$, $pK_2 = 8.60$) (193). Spectra of 4-aminopyrimidine (**171**) and 3,4-dihydro-4-thiopyrimidine (**172**) have been measured in a number of solvents (**185**) but no discussion of tautomerism is given. Comparison of the chemical shifts of H-5 in 1,2-dihydro-2-thiopyrimidine (**173**) and its 1-methyl derivative **174** shows them to be almost identical (185), confirming the established structure of the thio compound. Data for these monosubstituted tautomeric pyrimidines are summarized in Table 2.46.

TABLE 2.46 PROTON MAGNETIC RESONANCE DATA (185, 191, 194) FOR POTENTIALLY TAUTOMERIC MONO-SUBSTITUTED PYRIMIDINES AND FOR THE MODEL COMPOUNDS USED TO DETERMINE THE PREDOMINANT TAUTOMER

		Chemical shifts[a]				Coupling constants (Hz)		
Pyrimidine	Solvent	$\tau 2$	$\tau 4$	$\tau 5$	$\tau 6$	J_{25}	J_{26}	J_{56}
1,2-Dihydro-2-oxo-	Me_2CO		1.67	3.60	1.67			5.3
	DMSO		1.76	3.66	1.76			
2-Methoxy-	Me_2CO	6.07	1.40	2.93	1.40			4.8
	C_6H_{12}	6.12	1.64	3.30	1.64			4.7
1,2-Dihydro-2-thio-	Me_2CO		1.75	3.16	1.75			
1,2-Dihydro-1-methyl-2-thio-	DMSO		1.50	3.11	1.50			5.35
2-Methylthio-	Me_2CO	(7.50)	1.39	2.86	1.39			4.95
	C_6H_{12}	(7.57)	1.67	3.28	1.67			4.75
2-Amino-	Me_2CO		1.73	3.44	1.73			
	DMSO		1.78	3.46	1.78			
2-Methylamino-	Me_2CO		1.72	3.48	1.72			4.85
	C_6H_{12}		1.85	3.74	1.85			4.70
2-Dimethylamino-	Me_2CO		1.69	3.48	1.69			4.80
	C_6H_{12}		1.86	3.78	1.86			4.65
3,4-Dihydro-4-oxo-[b]	D_2O	1.63		3.49	2.02	1.0	1.2	7.15
	D_2SO_4[c]	0.74		3.15	1.87	0.8	1.7	7.7
	NaOD[d]	1.78		3.70	2.09	1.1	1.0	6.3
	DMSO	1.83		3.71	2.14	1.05	0.7	6.75
3,4-Dihydro-3-methyl-4-oxo-[b]	D_2O	1.58		3.45	2.01	0.8	0.6	6.8
	D_2SO_4	0.54		3.08	1.81	0.6	1.7	7.7
1,4-Dihydro-1-methyl-4-oxo-	D_2O	1.64		3.67	2.21	0.6	2.5	7.5
	D_2SO_4	0.82		3.12	1.91	0.7	2.3	7.9
4-Methoxy-	D_2O	1.32		3.05	1.55	1.2	0.8	6.2
	D_2SO_4	0.77		2.56	1.26	1.0	1.6	7.2
3,4-Dihydro-4-thio-[b]	DMSO	1.74		2.87	2.20			6.05
2-Azido-4,6-dimethyl-	$CDCl_3$		(7.23)	2.65	(7.23)			
	$(CD_3)_2CO$		(7.58)	2.95	(7.23)			
	C_5H_5N		(7.72)	?	(7.72)			
	TFA		(7.23)	2.65	(7.23)			
Tetrazolo-tautomer	$CDCl_3$		(7.22)	2.92	(7.00)			
	$(CD_3)_2CO$		(7.28)	2.65	(7.03)			
	DMSO		(7.31)	2.62	(7.11)			
	C_5H_5N		(7.43)	?	(7.25)			

[a] Values in parentheses are for methyl groups.

[b] Names as a 3,4- instead of a 1,6-dihydropyridimine.

[c] 2.13 N–D_2SO_4. [d] 1.44 N–NaOD.

Spectra of ten pyrimid-4-ones and -4-thiones have been presented and the parameters obtained were interpreted in terms of the amount of aromatic character in the heterocyclic ring (193a). Concentration effects of the N–H shift in pyrimid-4-ones were interpreted in terms of intermolecular hydrogen bonding (193b). Two publications have made extensive use of PMR data for the assignment of structures to compounds of the pyrimid-4-one type (193c, d). Rate constants for the acid-catalyzed deuterium exchange of the 5-proton in a number of pyrimidones have been determined from NMR data (193e).

Nuclear magnetic resonance has been used to follow tautomerism between the 2-azidopyrimidines (**175**) and the corresponding tetrazoles (**176**) (194, 195). In dimethyl sulfoxide the tetrazolotautomer predominates while in trifluoroacetic acid the protonated form of the azidopyrimidine is the major species present. Data are collected in Table 2.46.

A large number of 1,2-dihydro-2-iminopyrimidines (**177**) have been prepared by Brown and his co-workers in studies centered on the Dimroth rearrangement (196, 197). Spectra of these compounds as hydrochlorides, and where possible as free bases, are summarized in Table 2.47. During this

TABLE 2.47 SPECTRAL DATA FOR 1-METHYL-1,2-DIHYDRO-2-IMINOPYRIMIDINES AND THEIR DEUTEROCHLORIDES (90)

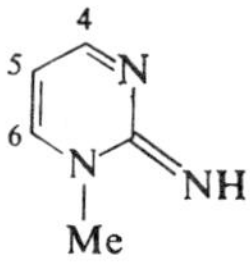

Substituents	Species[a]	τ4	τ5	τ6	τNMe
Nil	Cation	1.12	2.82	1.56	6.03
4-Me	Cation	7.38	2.95	1.73	6.09
	NM	7.72	3.84	2.30	6.53
4,6-Di-Me	Cation	7.34	3.02	7.24	6.15
	NM	7.80	3.91	7.66	6.58

[a] Cation in D_2O/DCl at pH −0.5. Neutral molecule (NM) in D_2O/NaOD at pH 13.5.

work, difficulty was encountered in measuring spectra of 4- and/or 6-methyl derivatives of these iminopyrimidines in deuterium chloride solutions because of rapid deuterium exchange of the methyl protons. The phenomenon was systematically studied (90) and, in the case of 1,2-dihydro-2-imino-1,4,6-trimethylpyrimidine (**178**), the reaction was shown to be catalyzed by D^+ or OD^- ions. Small but consistent shifts allowed the assignment of the downfield *C*-methyl peak in spectra of this compound or its oxo analog to the 6-methyl group (cf. Refs. 198 and 199).

A large amount of effort, much of it overlapping, has gone into the study of cytosine (**179**) and its derivatives. Data are collected in Table 2.48. While interest in a compound of such biological importance is to be expected, much of the work was prompted by a misinterpretation of the spectrum of deoxycytidine (**180**) by Gatlin and Davis (200). Kokko *et al.* (201) first reported the spectrum of cytosine and introduced the use of deuterated dimethyl sulfoxide as a standard solvent for these substances. Quite a lot has been said about the validity of comparisons between aqueous solutions and those in dimethyl sulfoxide but with these tautomeric pyrimidines the similarity of spectra in the two solvents, in cases where they can both be measured, leaves little doubt that the same species are present in both solutions (see also Ref. 202). The zwitterionic structure **181** was suggested for cytosine and structure **182** for its cation. These structures were criticized by Katritzky and Waring (190) who compared the spectra of cytosine with that of its 1-methyl derivative **183** and of the cytosine cation with that of the cation of 1,N^4-dimethylcytosine (**184**). The similarities between the spectra show conclusively that the normally accepted structures for cytosine (**179**) and its cation (**185**) are indeed correct. In the spectrum of the 1,N^4-dimethylcytosine cation, H-5 and H-6 gave rise to *two* AB patterns of different intensities and this was assigned to the presence of the cis–trans isomers **186** and **187**. With solutions of this cation in liquid sulfur dioxide, the N^4-methyl groups of the two isomers had different chemical shifts, and coupling with the adjacent N–H group was observed. However, in deuterium sulfate, the chemical shift difference is extremely small. The cation of 1,N^4,N^4-trimethylcytosine (**188**) similarly shows a chemical shift difference between the two N^4-methyl groups in liquid sulfur dioxide but not in deuterium sulfate. In 4*N*–, but not in concentrated sulfuric acid, 1,N^4-dimethylcytosine shows two peaks for the N^4-methyl group, due to coupling with the adjacent N–H group. Structure **185** for the cytosine cation was also deduced from spectra of solutions in anhydrous trifluoroacetic acid (203).

Undoubtedly the best work on the structure of cytosine derivatives comes from Becker, Miles *et al.* These workers have extensively studied the geometrical isomerism in 4-*N*-methylcytosines (204). Doubtful of the validity of arguments based on the relative chemical shifts expected for different tautomers or on the position and area of peaks from N–H protons, they have relied primarily on the synthesis of ^{15}N analogs, proton spin decoupling, and temperature-variable effects in their extensive study of the nucleoside analog, 1-methylcytosine (**183**) (205, 206). The normal spectrum of this compound in deuterated dimethyl sulfoxide contains two broad peaks (1 proton

TABLE 2.48 SPECTRAL DATA FOR CYTOSINE AND ITS DERIVATIVES (202, 206)[a,b]

Cytosine	Salt	Solvent	τ5	τ6	τ1	τ3	τ7	J_{56}	J_{35}	J_{67}	J_{7Me}	J_{N7}[c]	J_{N5}[c]
Parent		DMSO	4.38	2.64				7.2					
	HCl	DMSO	3.90	2.20				8.0					
	HCl	D_2O	3.85	2.26				8.0					
1-Me		DMSO	4.33	2.38	(6.77)		3.00	7.0				90	
	HCl	DMSO	3.82	1.88	(6.65)		1.15	8.0				92.6	
							0.0					91.4	
	HCl	SO_2	3.72	2.18	(6.45)	−1.44	2.55	7.0	2.5			93.8	0.7
							2.17					96.2	
	HI	SO_2	3.67	2.13	(6.42)	−1.02	2.72	7.5	2.1				
							2.55						
1,3-Di-Me		DMSO	4.30	2.98	(6.80)	(6.82)	2.65	7.0					
	HCl	SO_2	3.58	2.20	(6.42)	(6.35)	2.75	7.5					
							2.20						
	HI	DMSO	3.87	1.95	(6.60)	(6.57)	0.66	8.0					
	HI	SO_2	3.63	2.18	(6.42)	(6.33)	3.00	7.5					
							2.18						

1,7-Di-Me		DMSO	4.25	2.38	(6.73)		(7.20)	7.0			5.0	94.0	1.5
							2.50						
	HCl	DMSO	3.60	1.68	(6.60)		(6.95)	7.7					
							0.46						
		[d]	3.65	1.98	(6.63)		(6.98)	7.7			5.0	94.0	1.5
							−1.02						
	HCl	SO_2	3.62	2.30	(6.46)	−1.09	(6.77)	7.8	2.6	0.7	5.0	93.5	1.5
							1.35						
		[d]	3.48	2.10	(6.48)	−1.54	(6.82)	7.8	2.3				
							1.97						
	D_2SO_4	4N–D_2SO_4	3.78	2.07	(6.56)			8.0					
		[d]	3.54	2.29	(6.54)		(6.90)	8.0					
	HI	DMSO	3.97	2.03	(6.63)	−0.89	(6.98)	7.5			5.0	94.0	
							0.13						
		[d]	3.67	1.75	(6.63)		(7.03)	7.5			5.0	94.0	
							1.61						
	HI	SO_2	3.65	2.23	(6.42)	−1.50	(6.67)	8.0	2.6				
							2.50						
		[d]	3.65	2.03	(6.42)	−0.92	(6.75)	8.0					
							2.77						
1,7,7-Tri-Me	HCl	DMSO	3.52	1.72	(6.50)		(6.50)	8.0					
	HCl	SO_2	3.61	2.10	(6.44)		(6.50)	8.0					
							(6.55)						
	D_2SO_4	4N–D_2SO_4	3.69	2.16	(6.56)		(6.68)	8.0					

[a] In this table the nitrogen atom attached to C-4 is regarded as N-7.
[b] Values in parentheses are for methyl substituents.
[c] From the 7-^{15}N-derivative.
[d] Alternate geometric isomer; see text.

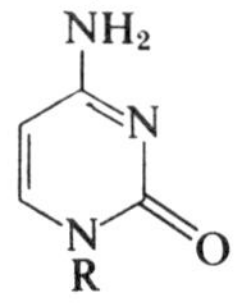

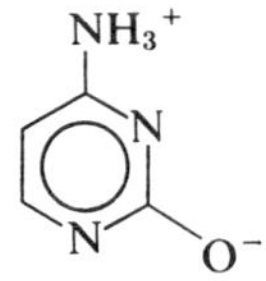

179, R = H
180, R = 2′-Deoxyadenosyl
183, R = Me

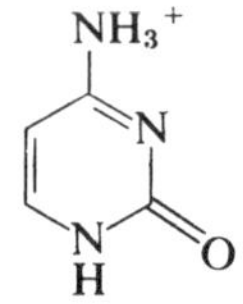

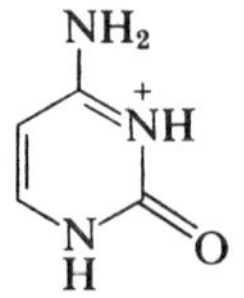

181 **182** **184**

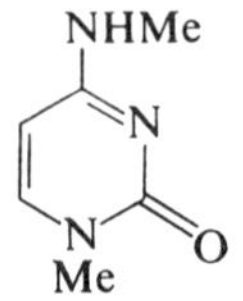

185 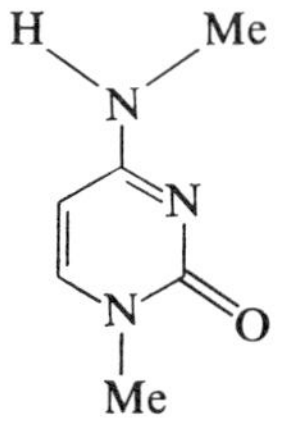 **186** 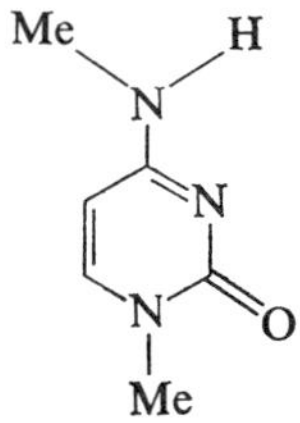 **187** 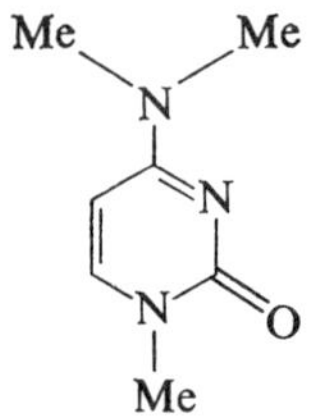 **188**

each) on the low-field side of the doublet from H-6, while that of the exocyclic ^{15}N analog shows each of these signals as doublets, split by the ^{15}N nucleus ($J = 94$ Hz). This information unequivocally proves the amino nature of the 4-substituent. At higher temperatures, these peaks broaden, collapse, and finally reappear as a singlet, consistent with increased rotation around the C(4)–NH_2 bond. Liquid sulfur dioxide was used as solvent for low-temperature studies (−60°) where three N–H signals were observed, two of which were coupled to the ^{15}N nucleus in the labeled analog ($J =$ 94 Hz). Decoupling experiments showed the third downfield signal to be coupled to H-5 ($J_{35} = 2.5$ Hz), eliminating the possibility that protonation had occurred on the oxygen atom. Interconversion of the geometrical isomers of the N^4-methyl compounds was studied and rough thermodynamic data obtained (206).

Complete ^{15}N labeling has since been used (207) to confirm the results above and to provide direct evidence of protonation at N-3. The amino protons from the completely labeled cation, under normal conditions, gave a pair of doublets ($J = 94.2$ Hz) as expected. However, one doublet was further split by 4.4 Hz, assigned to coupling between ^{15}N-3 and one of the amino protons. At low temperature (liquid SO_2, −30°) a clean doublet, absent from spectra measured in dimethyl sulfoxide at room temperature, was present. The value of this splitting, 94.0 Hz, indicates ^{15}N–H coupling and is compatible only with protonation at the N-3-position.

Spectra of 1-methylcytosine are almost identical with those obtained from the pyrimidine ring of cytosine riboside and deoxyriboside and the tautomeric structures of the pyrimidine ring system in both nucleosides are thought to be identical with that of the methylcytosine.

The structures of the stable tautomers of uracil (**189**) and its simple derivatives have always been supported by these PMR studies. Details, often incomplete, of their spectra have been published a number of times, for example, Refs. 201, 203, and 208, and the fine structure is clearly visible in a reported spectrum (185) (40 MHz, DMSO, room temperature). The doublets from H-5 and H-6 are further split by H-1 and, in the former case, by H-3 as well. Thus, H-6 gives a quartet, $J_{56} = 7.8$, $J_{16} = 5.7$ Hz, and H-5 a double triplet, $J_{56} = 7.8$, $J_{15} = J_{35} = 1.4$ Hz. Elevation of the temperature increases the rate of exchange of the amide protons and causes complete decoupling of H-1 and H-3 from H-5 and H-6. These couplings between the N–H protons and the C–H ring protons have been used to show the position of *N*-alkylation in uracils and thymines (208a). Shifts induced by aromatic solvents of protons from 1,3-dimethyluracil and 1,3-dimethylthymine suggest that the solute and aromatic solvent associate in vertical stacks (208b). The 1-methyl group, being twisted out of the ring plane, was affected most by the benzene anisotropy.

The PMR spectrum of the doubly ^{15}N-labeled uracil was extremely complex (207) and the dideutero derivative **190** was prepared to simplify the peak patterns from H-5 and H-6. Thus, the pattern from H-6 reduced to the expected quartet and that from H-5 to an octet from which the H-5–^{15}N splittings of 4.4 and 2.5 Hz were extracted. Assignment of these couplings to specific ^{15}N atoms was not possible. In the conversion of this uracil to similarly labeled cytosine, the correspondingly labeled 2,4-dichloro- (**191**) and dimethoxypyrimidine (**192**) and 1,2-dihydro-4-methoxy-1-methyl-2-oxopyrimidine (**193**) were prepared. Proton magnetic resonance spectra of the three intermediates and the ^{15}N NMR spectrum of the pyrimidine were measured and briefly discussed (207).

Proton magnetic resonance of 15 substituted uracils were used in an attempt to measure the proton mobility within the series and to relate this to biological activity (209). It was shown that the chemical shift of the amide proton

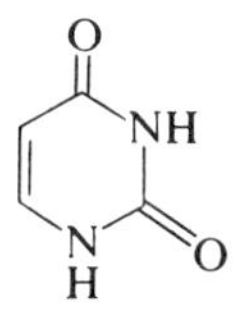

O NH N H O
189

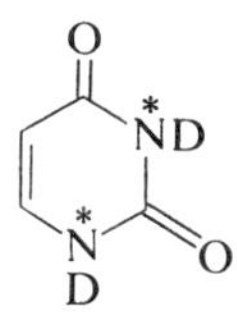

O N*D N* D O
190

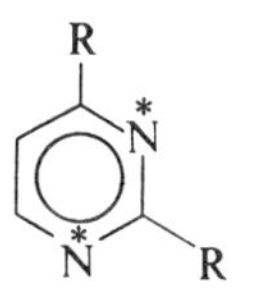

R N* N* R
191, R = Cl
192, R = OMe

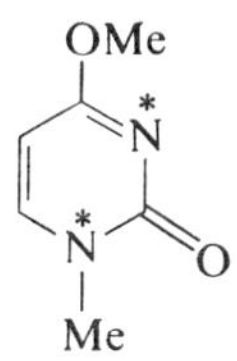

OMe N* N* Me O
193

correlates linearly with both the Hammett σ constant and the group dipole moment of substituents in the 5-position and that the chemical shift of H-6 also correlates linearly with this dipole moment, provided values for a 5-carboxy substituent are ignored. It must be realized, however, that chemical shifts were measured from *external* TMS without compensation for bulk diamagnetic susceptibility and this casts doubts on the numerical validity of this work.

More recently, the structure of the predominant form of 4,6-dihydroxypyrimidine has been the cause of some contention. The first PMR study (191) disproved the suggestion based on ultraviolet spectral evidence (210) that the dioxo form **194** predominated since the spectrum contained only two singlets (each one proton), which could be attributed to ring protons. It is important to note that the ultraviolet spectra were obtained of aqueous solutions while dimethyl sulfoxide was used as solvent for the PMR work. However, addition of 50% deuterium oxide to the dimethyl sulfoxide solutions caused no change in the spectrum of the ring protons, leaving little doubt that both solutions contain the same species in comparable proportions. From a comparison of the spectrum of the dihydroxy compound with those of 4-hydroxy-6-methoxypyrimidine (**195**) and 4,6-dimethoxypyrimidine (**196**) (Table 2.49), either of the dihydroxo tautomers **197** or **198** may predominate. Further comparison with spectra of the 1- and 3-methyl isomers **199** and **200** of **195** led to the conclusion that 3,4-dihydro-6-hydroxy-4-oxopyrimidine **197** was the main species present, presumably in equilibrium with its tautomeric equivalent **201**. On standing with deuterium oxide, particularly with a trace of acid, all protons except H-2 exchange. This suggests that the predominant tautomers are in equilibrium with a small amount of the dioxo form **194**. These conclusions have been criticized by Katritzky *et al.* (211) who, without much practical evidence, suggest the zwitterion **202** as the main species, and by a Russian group (212) who have proposed a number of

TABLE 2.49 CHEMICAL SHIFTS (ppm) FOR 4,6-DIHYDROXYPYRIMIDINE AND ITS METHYL DERIVATIVES (191)

Pyrimidine	Solvent	τNH or OH	τ2	τ5	τOMe	τNMe
4,6-Dihydroxy-	DMSO	−1.8	1.91	4.68		
4,6-Dimethoxy-	D_2O		1.68	3.78	6.04	
4-Hydroxy-6-methoxy-	DMSO	−2.0	1.88	4.43	6.20	
	D_2O		1.81	4.17	6.10	
1,4-Dihydro-6-methoxy-1-methyl-4-oxo-	D_2O		0.85	3.96	6.47	7.14
1,6-Dihydro-4-methoxy-1-methyl-6-oxo-	D_2O		1.69	4.12	6.09	6.47

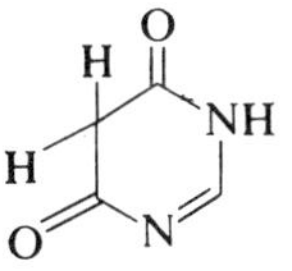

194 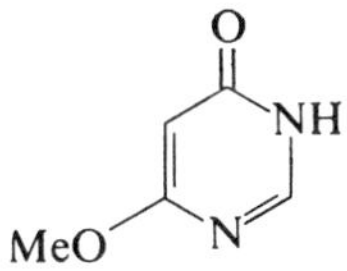 195 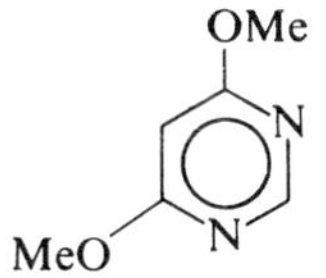 196 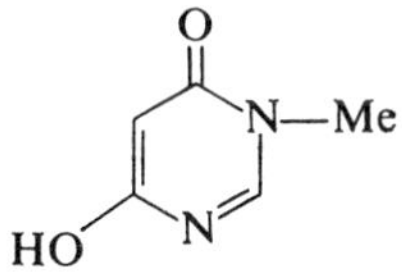 200 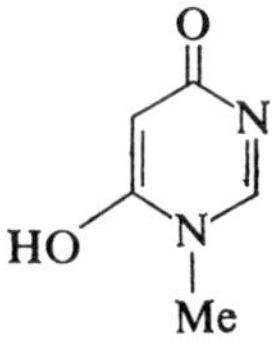 199

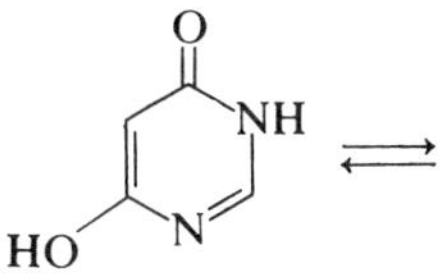

197 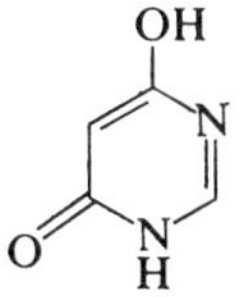 201 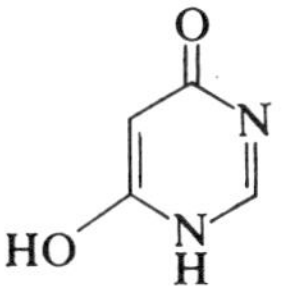 198 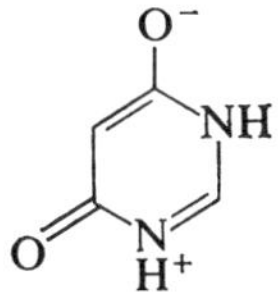 202 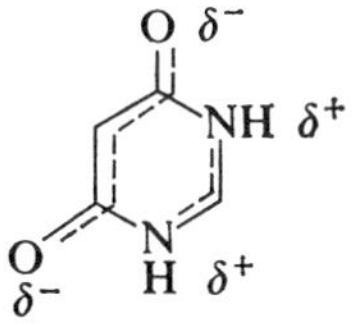 203

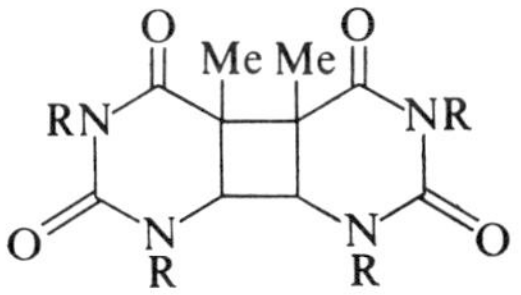

204 *syn.* 203a

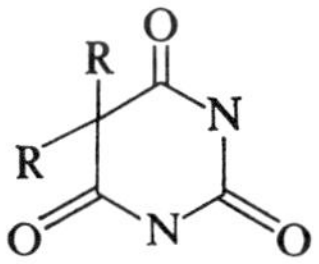

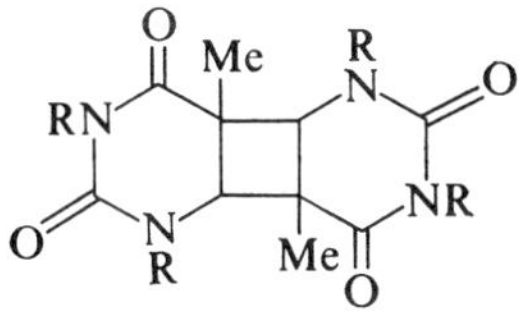

205 *ant.*

structures including the dizwitterion **203** as well as both **197** and **202**. The original PMR interpretation appears to be completely sound.

Snell (212a) has tabulated spectral data for 35 derivatives of "2-amino-4-hydroxypyrimidine" and has indicated the ranges for the chemical shifts of protons from the pyrimid-4-one and 4-hydroxypyrimidine tautomers (in $CDCl_3$). These are, for the pyrimid-4-ones, H-5, τ 4.27–4.48; H-6, τ 2.25–2.40; 6-Me, τ 7.85; and for the hydroxypyrimidine types, 5-H, τ 3.75–4.06; 6-H, τ 1.60–1.94; 6-Me, τ 7.52–7.65.

Studies of line width and chemical shift versus temperature for amide and hydroxyl resonances in a series of barbituric acid derivatives **203a** in anhydrous DMSO led to the conclusion that none of the compounds under investigation exhibited lactim–lactam tautomerism between 20 and 65°C (212b). The amide resonance of uracil was shown to be two closely spaced singlets, due to the nonequivalence of the ring nitrogen atoms. Nuclear magnetic resonance has also been used to identify *N*-methylbarbiturates after their separation by gas chromatography (212c).

An extremely thorough investigation of protonation of amino- and hydroxypyrimidines, including biologically important bases, has been carried out (212d). Four solvents were used, TFA, TFA–SO_2, FSO_3H, and FSO_3H–SbF_5–SO_2, at 27 and −55°C. In TFA, monocations were formed, while in the FSO_3H-based solvents diprotonation occurred. In each case the structure of the protonated species could be derived unambiguously from the chemical shifts of CH, NH, and OH protons and from a consideration of the proton–proton coupling constants. Typical data are given in Table 2.49A.

The work on ^{15}N-labeled cytosine derivatives described above stimulated work on the spectra of a number of 2-substituted pyrimidines, fully labeled in position 1 with ^{15}N (212e). In these compounds, introduction of the label removed the symmetry and the doublet from H-4,6 became a well-resolved

TABLE 2.49A SPECTRAL DATA FOR CATIONS OF AMINO- AND HYDROXYPYRIMIDINES (212c)

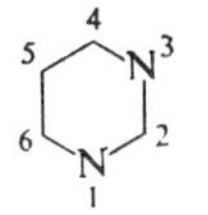

Substituent	Solvent[a]	τ1	τ2	τ3	τ4	τ5	τ6	Coupling constants
None	TFA		0.17		0.49	1.66	0.49	J_{25} 1.1; J_{56} 5.6
	FSO_3H	−4.15	−0.34	−4.15	0.02	0.81	0.02	J_{12} 5.1; J_{16} 6.5; J_{56} 6.2
2-NH_2	TFA		1.95		1.26	2.77	1.26	J_{56} 5.5
	FSO_3H	−1.75	1.82	−1.75	0.74	2.08	0.74	J_{16} 7.6; J_{45} 6.1
2-OH	TFA				1.01	2.76	1.01	J_{45} 6.2
	FSO_3H	−2.75		−2.75	0.45	1.64	0.45	J_{16} 6.5; J_{45} 6.1
4-OH	TFA		0.57			2.94	1.78	J_{26} 1.7; J_{56} 8.0
	FSO_3H	−2.95	0.05	−3.2		1.77	0.70	J_{12} 5.3; J_{16} 5.7; J_{23} 5.3; J_{56} 7.3
2,4-Di-OH	TFA	−0.2				3.77	2.25	J_{16} 5.1; J_{56} 7.9
	FSO_3H	−1.63		−2.34		2.46	1.16	J_{16} 5.3; J_{56} 7.4
2-OH, 4-NH_2	TFA	−0.45				3.59	2.13	J_{56} 7.5
	FSO_2H	−0.79		−0.79	1.81	2.77	1.74	J_{16} 6.4; J_{56} 7.7
2-NH_2, 4-OH	TFA		1.90			3.68	2.21	J_{56} 7.5
	FSO_3H	−0.75	2.13			2.81	1.29	J_{16} 6.3; J_{56} 7.4

[a] TFA solutions at 27°, FSO_3H solutions at −55°.

16-peak multiplet. The triplet from H-5 was doubled by coupling to the ^{15}N nucleus. Complete analyses of proton and ^{15}N spectra were carried out and ^{13}C–^{15}N coupling constants were measured. As would be expected, protonation, which placed a positive charge on the nitrogen atoms, caused very large variations in the N–H coupling constants and hence in the spectral patterns. Data obtained were discussed fully in terms of the structures of the species being investigated and of current ideas on the mechanisms causing chemical shifts and couplings in this type of compound. Representative data are collected in Table 2.49B.

Thymine photodimers have been extensively studied in recent years and ^{13}C–H satellite spectra were used to differentiate between syn and anti configurations, **204** and **205**, respectively (213).

C. Reduced Pyrimidines

Spectra of a number of dihydropyrimidines have been reported. Silversmith (214) was unable to distinguish between the 1,4- and 3,4-dihydro structures **206** and **207**. Mehta *et al.* (215) examined the spectrum of the 2,5-dihydropyrimidine (**208**) and showed by decoupling that J_{25} has the unexpectedly large value of 5.5 Hz. A number of 1,4- (or 3,4-) dihydro-2-oxopyrimidines of the type **209** have been examined (216) and data are collected in Table 2.50.

Nofre and his co-workers (217–219) have carried out an extensive investigation of the preferred conformations of dihydrouracil (**210**) and dihydrothymine (**211**) derivatives. Spectra were analyzed to obtain coupling constants which were then used with the aid of the Karplus relationship to determine the predominant configuration. Coupling between H-1 and H-6 could be used as a diagnostic tool. In the case of dihydrouracil and its 1-methyl and 1,3-dimethyl derivatives rapid flipping between the two half-chairs **212** and **213** was occurring at room temperature. In 5-hydroxy-, 1-methyl-5-hydroxy-, and 6-methyl-dihydrouracil and 1- and 5-methyl and 1,3-dimethyldihydrothymine and *C*-substituent preferred the equatorial position. The italicized substituents in the following compounds are predominantly axial: *5-bromo-*, 1-methyl-*5-bromo-*, 1,3-dimethyl-*5-bromo-*, 1-methyl-*6-hydroxy-*, 1-methyl-5,*6*-di*hydroxy-*, 1,3-dimethyl-5,*6*-di*hydroxy*dihydouracil, and *5-bromo-*, *6-hydroxy-*, and *trans-5-bromo-6-hydroxy*dihydrothymine. Thus, in the absence of substituents in positions 5 or 6 rapid inversion is possible, but substitution on either of these carbon atoms serves to "freeze" the conformation. The essential details of this work on dihydrouracils have been confirmed (219a). 6-Phenyldihydrouracil (**213a**) (219) was thought to behave in an anomalous manner but the results were shown to be those expected from the ring-opened compound, β-phenyl-β-ureopropionic acid (**213b**) (219b). Spectra of both the cyclic and the open-chain structures have been discussed at some length. Some of the data for these compounds are collected in Table 2.51.

TABLE 2.49B SPECTRAL DATA FOR ^{15}N-LABELED PYRIMIDINES (212e)

I II III IV

Type	Substituents	Solvent	τ^A	τ^B	τ^M	δX^a	J_{AX}	J_{MX}	J_{BX}	J_{AB}	J_{AM}	J_{BM}
I	R = SMe	CCl_4	1.59	1.59	3.15	265.53	−11.04	−1.02	0.79	2.58	4.82	4.82
		Me_2CO	1.49	1.49	2.95	264.53	−11.23	−1.30	0.43	2.51	4.80	4.80
	R = Cl	$CDCl_3$	1.39	1.39	2.75	272.47	−11.43	−1.40	0.85	2.65	4.81	4.81
	R = S^-	NaOD	1.69	1.69	3.84	262.79	−10.00	−1.40	0.40	2.71	4.89	4.89
	R = O^-	NaOD				226.97						
II	R = H, X = S	DMSO	1.73	1.73	3.17	226.82	−7.13	−2.30	0.68	2.52	5.31	5.31
	R = Me, X = S	DMSO	1.50	1.50	3.10	290.31	−12.05	−1.10	0.52	2.28	4.26	6.52
	R = H, X = 0	DMSO				202.94						
IV	R = H, X = S	DCl	1.27	1.27	2.66	179.52	−2.89	−3.98	0.83	2.18	6.00	6.00
	R = Me, X = S	DMSO	1.19	0.91	3.06	167.23	−3.81	−3.60	0.51	2.40	6.00	6.11
III		DMSO	1.27	1.27	2.62	264.59	−11.32	−1.40	0.30	2.55	4.75	4.76

[a] δX in parts per million downfield from NH_4^+.

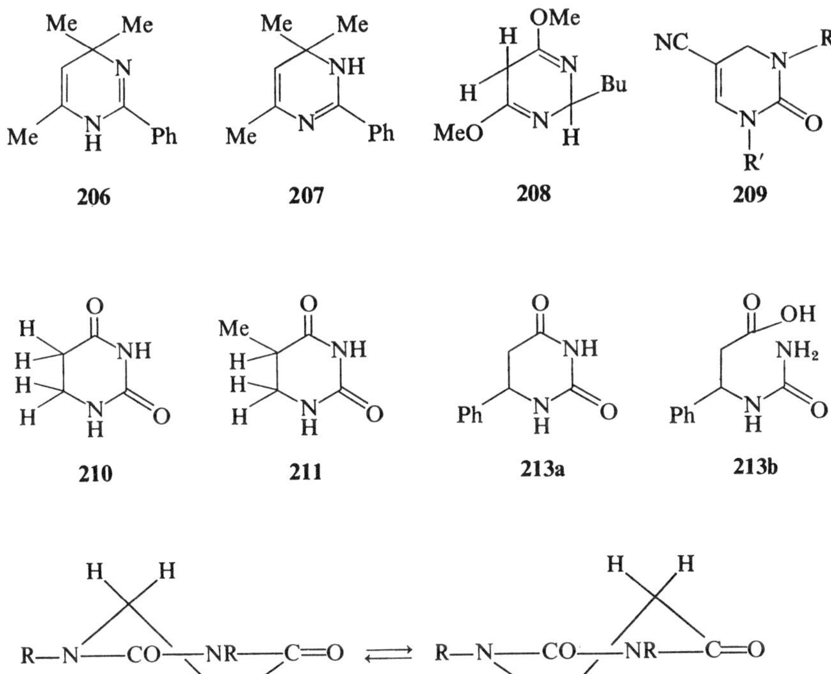

TABLE 2.50 SPECTRAL DATA FOR 1,4- OR 3,4-DIHYDRO-2-OXOPYRIMIDINES (216). SOLVENT—$CDCl_3$[a]

R	R′	τ4	τ6	τ1	τ3
H	Me	5.95	3.15		7.08
Me	H	5.87	3.20	6.88	
Me	Me	5.93	3.17	6.87	7.07
Ac	Ac	5.56	2.05	7.32	7.40

[a] J_{46} in all cases is approximately 1.0 Hz.

TABLE 2.51 SPECTRAL DATA FOR DIHYDROURACIL, DIHYDROTHYMINE, AND DERIVATIVES (217–219). SOLVENT—DMSO

	Chemical shifts[a]				Coupling constants (J)						
	τ1	τ3	τ5	τ6	5*e*, *a*	6*e*, *a*	5*a*, 6*a*	5*e*, 6*a*	5*e*, 6*e*	5*a*, 6*e*	1, 6
Dihydrouracil	2.58	0.10	7.56*ae*	6.75*ae*			7.0	7.0	7.0	7.0	2.5*ea*
1-Me	(7.16)		7.47*ae*	6.65*ae*			7.0	7.0	7.0	7.0	
1,3-Di-Me	(7.10, 7.03)		7.33*ae*	6.65*ae*			6.9	6.9	6.9	6.9	
5-OH	2.50	−0.05	5.86*a*	6.95*a*		12.6	10.3			6.1	3.8*e*
											1.4*a*
1-Me, 5-OH	(7.01)		5.87*a*	6.76*a*		12.2	10.1			5.54	
				6.56*e*							
1-Me, 6-OH	(6.98)		7.00*a*	5.04*e*	16.0				1.7	4.3	
			7.46*e*								
1,3-Di-Me, 6-OH	(7.04, 7.40)		7.10*a*	5.10*e*	15.8				1.9	4.2	
			7.43*e*								
1-Me, 5,6-Di-OH											
(cis)	(7.13)		5.61*a*	5.25*e*						3.6	
(trans)	(7.07)		6.24*e*	5.35*e*					3.0		
1,3-Di-Me, 5,6-di-OH											
(cis)	(7.13, 7.08)		5.77*a*	5.33*e*						3.8	
(trans)	(7.10, 7.07)		6.18*e*	5.45*e*					3.0		
5-Br	2.17	−0.4	5.28*e*	6.22*a*		14.4		3.4	3.4		4.2*a*

1-Me, 5-Br	(7.07)		5.19*e*	5.98*a*		14.2		3.3	3.3		
				6.52*e*							
1,3-Di-Me, 5-Br, 6-OH											
(trans)	(6.95, 6.92)		5.45*e*	5.01*e*					2.2		
5,5-Di-Br, 6-OH	1.33	−0.7		5.00*e*							4.5*e*
6-Me	2.5	0.1	7.80*a*	(8.9*e*)	15.8		9.54	4.1			
			7.66*e*	6.42*a*							
Dihydrothymine	2.77	0.45	(8.97*e*)	7.12*a*		11.4	9.83			5.77	3.7*e*
			7.55*a*	6.82*e*							1.5*a*
1-Me	(7.11)	0.2	(8.90*e*)	6.81*a*		11.8	9.57			7.23	
			7.38*a*	6.68*e*							
1,3-Di-Me	(7.05, 7.03)		(8.83*e*)	6.76*a*		11.6	9.72			6.68	
			7.29*a*	6.65*e*							
5-OH	2.48	0.1	(8.75*ae*)	6.13*ae*							2.8*ae*
6-OH	1.90	0.15	(8.95*e*)	5.32*e*						3.6	4.2*e*
			7.30*a*								
5,6-Di-OH											
(cis)	1.78	−0.2	(8.73*a*)	5.67*e*							4.9*e*
5-Br	2.0	−0.4	(7.84*e*)	6.54*a*		14					4.2*e*
											1*a*
5-Br, 6-OH											
(trans)	1.50	−0.5	(7.84*e*)	5.22*e*							4.8*e*
5-Cl, 6-OH	1.50	−0.5	(8.37*e*)	5.32*e*							4.9*e*

[a] Chemical shifts in parentheses are for substituent methyl groups. The orientation of a group is indicated by *a* (axial) or *e* (equatorial) following the value.
[b] The orientation of H-6 is indicated by *a* or *e* after the value.

Spectra of a number of tetrahydropyrimidines have been reported (220–223).

Ring inversion of 1,3-dialkylhexahydropyrimidines has been studied and thermodynamic activation constants were derived (224, 225). At low temperatures, signals from axial and equatorial protons α to nitrogen atoms are separated by approximately 1 ppm.

D. Fluoropyrimidines

^{19}F spectra of a number of fluoropyrimidines have been measured and used to check their structures (226, 227). It is interesting that the values obtained by the two groups for tetrafluoro- and 2,4,6-trifluoropyrimidines are rather different. Spectral parameters are summarized in Table 2.52.

TABLE 2.52 DATA FOR FLUOROPYRIMIDINES (266, 227)[a]

Substituent	F-2	F-4	F-5	F-6	J_{25}	J_{56}	J_{26}	Other data
None	−30.6	−3.8	99.0	−3.8	26.0	17.9		
	−31.3	−6.3	95.2	−6.3				
2-H		2.0	93.4	2.0				
4-H	−28.0		83.0	−1.5				
5-H	−35.2	−24.0		−24.0				$J_{2,5H}$ 1.1, $J_{4,5H}$ 1.8
	−35.6	−22.9		−22.9				
4,6-H_2	−24.8		69.9					
4-NH_2	−27.6		103.2	11.2	25.7	17.5		
	−28.0		103.0	10.7				
4-NHMe	−28.4		105.6	14.6	26.5	17.5		$J_{5,NH}$ 1.7, $J_{NH,H}$ 4.8
4-NMe_2	−30.0		98.4	9.8	26.1	17.5		$J_{5,H}$ 2.2
4-NHPh	−29.8		101.2	10.0	26.6	16.9	2.8	$J_{5,NH}$ 2.8
4,6-$(NH_2)_2$	−24.8		107.2		27.1			
4,6-$(NHMe)_2$	−26.6		110.2		27.4			$J_{5,NH}$ 1.7
4,6-$(NMe_2)_2$	−28.0		92.2		27.0			$J_{5,H}$ 3.0
4,6-$(NHPh)_2$	−28.6		100.8		28.2			
4-OH	−25.4		103.0	7.8	24.8	16.4	4.6	
4-OMe	−29.6		101.8	5.6	25.9	16.9		$J_{5,H} \leqslant 0.3$
4,6-$(OMe)_2$	−29.8		102.6		26.7			$J_{5,H} \leqslant 0.3$
2,4,6-$(OMe)_3$			111.2					$J_{5,H} \leqslant 0.3$
5-Cl	−33.8	−21.0		−21.0				

[a] ^{19}F shifts in parts per million from external trifluoroacetic acid. Coupling constants in hertz.

IV. THE PYRAZINE SYSTEM

The proton spectrum of pyrazine consists of a single peak, and its chemical shift has been reported for a range of solvents (16, 44, 228, 229). Tori and

Ogata (168) analyzed the ^{13}C–H satellite spectra of this singlet to obtain values for the proton–proton coupling constants. Unfortunately, the satellite spectra were deceptively simple and only average couplings were obtained for the two large couplings, J_{23} and J_{25}. Complete analyses of the spectra of a number of monosubstituted pyrazines have been reported (229a). Substitution effects were found to follow those exhibited by monosubstituted benzenes and pyridines. While J_{23} and J_{25} were of the same sign (positive) the four-bond couplings across the nitrogen atom (J_{26}) were shown by tickling experiments to be negative. From the chemical shift data it appears that 2-hydroxypyrazine prefers the oxo form **213c** while the 2-amino derivative exists as such. This work was extended to cover di- and trisubstituted pyrazines (229b) but recent information (229c) based on results with specifically deuterated compounds shows that the original assignments for these compounds were incorrect. It will be interesting to see the complete results from these labeling studies when they are published. Data for pyrazine and some of its derivatives are collected in Table 2.53.

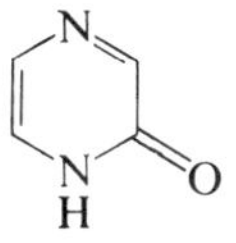

120°40′ ± 20′

213c  **214** 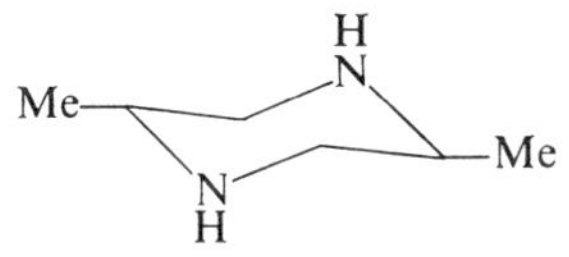 **215**

Benzene-induced solvent shifts in the spectra of pyrazine and its 1-oxide led to the assignment to the collision complexes of structures in which the benzene molecule aligns itself close to the positive ends of the molecular dipoles (185b). Double and triple irradiation experiments aided in the measurement of couplings between the ^{14}N nuclei of pyrazine methiodides and the ring protons (229d). Coupling constants for these systems are included in Table 2.53. The spectrum of pyrazine in a nematic phase has also been measured (230) and the direct couplings used to determine the C–C–H bond angle (**214**), 120°40′ ± 20′.

Ring inversion of *N,N′*-dialkylpiperazines and their salts has been extensively studied in an effort to obtain accurate thermodynamic activation parameters for the change (231, 232). With the dimethyl derivative, ring inversion could be slowed down enough at low temperatures (e.g., −56°) for axial and equatorial ring protons to be differentiated, but the *N*-methyl groups still gave a time-averaged singlet. However, in acid solution protonation stops *N*-inversion and the two orientations of the *N*-methyl groups can be distinguished (232). Sudmeier (233) has analyzed spectra of the free base, monocation, and dication of trans-2,5-dimethylpiperazine to show that they all exist predominantly (98%) in the diequatorial conformation **215**.

TABLE 2.53 SPECTRAL DATA FOR PYRAZINES

A. Neutral Molecules (229a)

Substituent	Solvent	$\tau 2$	$\tau 3$	$\tau 5$	$\tau 6$	J_{23}	J_{25}	J_{26}	J_{35}	J_{36}	J_{56}
None	Neat	1.37 (all positions)				1.8[a]	1.8[a]	0.5[a]	0.5[a]	1.8[a]	1.8[a]
2-Me	$CDCl_3$	7.43	1.43	1.62	1.55	0.21	0.45	0.70	−0.20	1.48	2.55
2-$CONH_2$	DMSO	1.72, 2.10	0.75	1.10	1.24				−0.01	1.51	2.49
2-CO_2Me	DMSO	6.00	0.78	1.07	1.14				−0.29	1.49	2.43
2-NH_2	DMSO	3.68	2.06	2.29	2.10				−0.29	1.54	2.77
2-OMe	DMSO	6.04	1.69	1.79	1.78				−0.35	1.39	2.86
2-OH (oxo)	DMSO		2.08	2.70	2.60				−0.05	1.34	3.91
2-F	DMSO		1.27	1.30	1.58	−8.17	4.72	−1.43	−0.46	1.33	2.67
2-Cl	DMSO		1.29	1.40	1.54				−0.40	1.43	2.61

B. Methiodides (229d)

	Coupling constants						
	^{14}N–H				H–H		
Substituent	J_{12}	J_{13}	J_{15}	J_{16}	J_{25}	J_{26}	J_{56}
None	1.05	2.8					
3-Me	1.0		2.9	1.15	1.1	1.1	3.6
3-NH_2	1.0		3.0	1.25	1.0	1.5	4.0

[a] Average values.

Axial and equatorial protons exhibit almost equal protonation shifts. It is interesting that vicinal couplings increase and geminal couplings decrease on protonation. This is the opposite of the usual trend and suggests that in these compounds β-effects are greater than α-effects. Analyses of spectra of the three species are summarized in Table 2.54.

TABLE 2.54 SPECTRAL DATA FOR TRANS-2,5-DIMETHYLPIPERAZINE AND ITS SALTS (233). VALUES OBTAINED FROM LAOCOON II. SOLVENT—D_2O

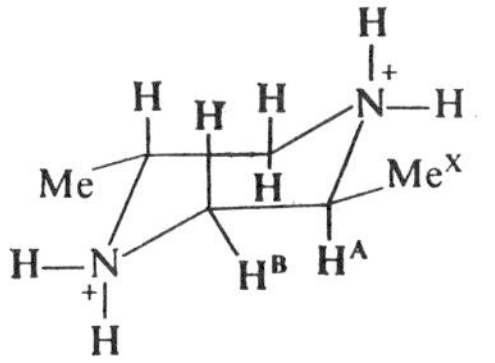

	pH 12.8	pH 7.7	pH 1.3
	Free base	Monocation	Dication
δA $\mp$ 0.003 ppm	2.644	3.103	3.693
δB $\mp$ 0.003 ppm	2.859	3.248	3.728
δC $\mp$ 0.003 ppm	2.293	2.700	3.239
δX $\mp$ 0.003 ppm	0.960	1.180	1.422
J_{AB} $\mp$ 0.2 Hz	2.9	3.0	3.4
J_{AC} $\mp$ 0.2 Hz	10.8	11.5	12.2
J_{AX} $\mp$ 0.2 Hz	6.4	6.2	6.1
J_{BC} $\mp$ 0.2 Hz	−12.5	−13.5	−14.2
J_{BX} $\mp$ 0.2 Hz	−0.1	−0.1	−0.2
J_{CX} $\mp$ 0.2 Hz	−0.2	−0.3	−0.3

A study of the spectrum of the cis-2,4 isomer of 1,2,4,6-tetramethylpiperazine over the pH range 0–10.75 yielded information about the rate of *N*-inversion of the individual nitrogen atoms (233a). Two groups of workers (233b, c) have studied the spectra of 1,4-dinitrosopiperazine. In solution, both syn and anti conformers are present with a preponderance of the latter. The size of 1.5 for the *R* value (J_{trans}/J_{cis}), which depends only on the conformation of the molecule and is independent of the nature of the atoms attached to the $-CH_2-CH_2-$ fragment, is diagnostic for flattened conformations such as boats or very flattened chairs. Proton and ^{11}B data are available for some mono- and bisborane adducts of 1,4-dimethylpiperazine (233d). Data for a number of polyfluoropyrazines (233e) are collected in Table 2.54A. A number of partly reduced systems have been reported (234).

TABLE 2.54A ^{19}F DATA FOR POLYFLUOROPYRAZINES (233e)

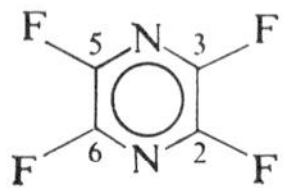

Reference Compound—C_6F_6.Solvent—Ether

Substituent	^{19}F Shifts (ppm)			F–F Coupling constants		
				Ortho	Meta	Para
None		−68.4				
2-NH_2	−67.4	−62.3	−49.3	16.2	10.8	49.8
2-Br	−85.4	−71.1	−70.7	17.5	6.9	39.9
2-Cl	−82.0	−71.8	−70.0	17.2	5.6	44.6
2-OH	−68.6	−63.7	−55.7	16.5	12.0	51.0
2-OMe	−70.1	−63.5	−56.0	15.1	12.3	51.2
2-Me[a]	−79.2	−68.6	−69.8	19.7	7.9	43.9

[a] Proton data: τMe 7.5; J_{MeF} 2.2, 2.2.

V. THE TRIAZINE SYSTEMS

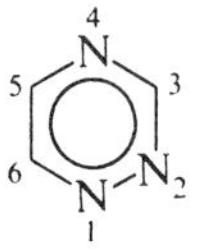

1,2,4-
as

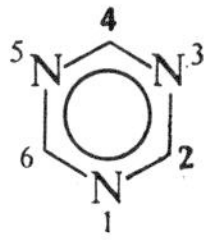

1,3,5-
sym

Spectra of a number of 1,3,5-triazines have been measured (228) and it was found that substituent effects of electron-releasing groups were greatly enhanced at the meta positions. Structural assignments for 2,4-dialkyl-1,3,5-triazines have been made from spectral data (234a). ^{19}F shifts for certain fluorinated *s*-triazines have been reported (234b). Nuclear magnetic resonance has also been used to determine the structures of 1,2,4-triazine derivatives (235). Compounds of the type **215a** were shown, from their spectra, to contain the exocyclic alkylidene groups instead of the usual alkyl groups (235a). The 3-amino analog exists in the ethyl form **216** as the neutral molecule in DMSO but in TFA solution, in which annular *N*-protonation takes place, the ethylidene iminium form **216a** is present in about 30% concentration. Data are collected in Table 2.55.

TABLE 2.55 NMR DATA FOR TRIAZINES (228, 235)

A. 1,3,5-TRIAZINE

Substituents	Solvent	$\tau 2$	$\tau 4$	$\tau 6$
None	$CDCl_3$	0.75	0.75	0.75
	CCl_4	0.82	0.82	0.82
2,4-Di-Me	CCl_4	7.46	7.46	1.20
2,4-Di-OMe	CCl_4	5.99	5.99	1.49
2-N(Et)$_2$, 4-OMe	CCl_4	6.59, 8.81	6.04	1.81
2,4-Di-N(Et)$_2$	CCl_4	6.46, 8.85	6.46, 8.85	2.02
2-N(Et)$_2$	CCl_4	6.37, 8.80	1.60	1.60
2,4-Di-Ph	CCl_4	1.45, 2.59	1.45, 2.59	0.93
2-Ph	CCl_4	1.52, 2.54	0.87	0.87

B. 1,2,4-TRIAZINE

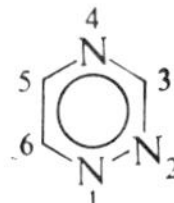

Substituents	Solvent	$\tau 3$	$\tau 5$	$\tau 6$	J_{56}	J_{36}
None	CCl_4	0.37	1.47	0.76	2.7	2.2
	$CDCl_3$	0.12	1.16	0.52	2.7	2.2
3-CO_2Et	$CDCl_3$		1.07	0.48	2.5	
3-NH_2	TFA		1.38	0.93	1.80	
3-NH_2, 5-Ph	Dioxan	3.65	2.69, 1.88	0.87		
3-NH_2, 6-Ph	Dioxan	3.55	1.40	2.59, 2.04		

^{19}F shifts have been obtained for six trifluoromethyl-substituted 1,2-dihydro-1,3,5-triazines (**216b**) (235b). Use has been made of vicinal couplings in assigning relative stereochemistry in 5,6-disubstituted 1,4,5,6-tetrahydro-*as*-triazines (**216c**) (235c).

Ring inversion of a series of tri-*N*-alkylhexahydro-1,3,5-triazines (**217**) and tetramethylhexahydrotetrazine has been extensively studied (236) and activation parameters were obtained.

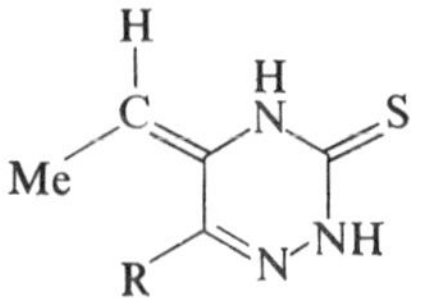

215 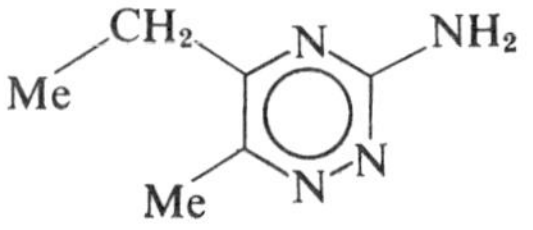 216 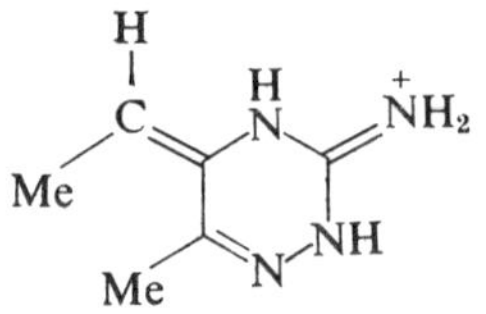 216a 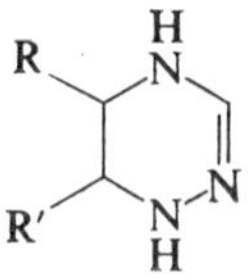 216c

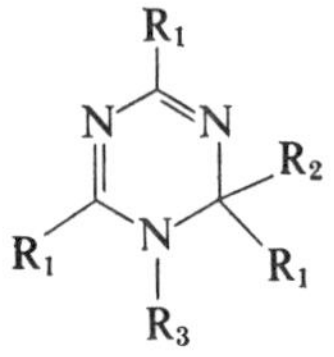

216b 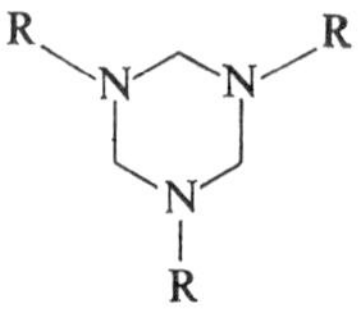 217

VI. GENERAL

While most topics have been adequately discussed in the appropriate sections of this chapter, there have been a number of comparative studies involving all or most of the ring systems discussed.

Thus, π-electron densities of a large number of heterocyclic systems have been calculated and compared with those obtained from chemical shift data (16, 25). In both cases correlation for single six-membered *N*-heterocyclic systems was found to be poor.

Malinowski's additivity rule (Section I.A) for predicting ^{13}C chemical shifts in heterocyclic systems has been shown to be valid for a whole series of compounds including pyridine, pyridazine, pyrimidine, pyrazine, 1,3,5-triazine, and certain derivatives (51). Also, ^{13}C chemical shifts and ^{13}C–H coupling constants have been separately measured for these compounds and for *s*-tetrazine (53). Recently, Pugmire and Grant (54) have studied ^{13}C chemical shift data for these azines and their protonated species and found that simple additivity parameters accurately accounted for monoprotonation shifts. A detailed examination of ^{13}C chemical shift calculations indicated that the shifts were critically dependent on both charge-transfer features and variation in the bond-order parameters. Extended HMO wave functions were used to estimate the ^{13}C shifts and reasonable agreement was obtained between predicted and experimental results. ^{13}C data for these ring systems are collected in Table 2.56.

A review on the use of NMR to study conformational equilibria in rapidly interconverting six-membered ring systems (237) apparently stimulated a large interest in the thermodynamic activation parameters for inversion of

TABLE 2.56 ^{13}C SPECTRA OF AZINES

	Standard	Chemical shifts (ppm)			$J_{C,H}$ (Hz)			Ref.[a]
		C-2, 6	C-3, 5	C-4	C-2, 6	C-3, 5	C-4	
Pyridine	1	−21.72	4.59	−7.42				54
	2	43.1	69.2	57.3	180	162	160	53
					179	163	152	51
					178.4	162.3	162	280
		C-3, 6	C-4, 5		C-3, 6	C-4, 5		
Pyridazine	1	−24.31	0.85					54
	2	41.1	66.0		186	174		53
					181.5	168.5		51
					182.3	170.5		280
		C-2	C-4, 6	C-5	C-2	C-4, 6	C-5	
Pyrimidine	1	−30.99	−28.96	6.42				54
	2	34.5	36.1	71.1	211	182	171	53
					206	181.8	168	184
					202.7	182.5	168.8	280
		C-2, 3, 5, 6			C-2, 3, 5, 6			
Pyrazine	1	−17.14						54
	2	46.6			184			53
					183			51
					182.0			280
		C-2, 4, 6						
s-Triazine	2	26.0			206			53
					207.7			51
		C-3, 6						
s-Tetrazine	2	31.6			214			53

[a] These references also contain data on a number of methyl derivatives and cationic species.

fully reduced heterocycles. Unfortunately the results are marred by the confusion of terms and symbols characteristic of this field and also by the fact that values obtained under different experimental conditions are not strictly comparable. For these reasons, values are not tabulated and the reader should consult the original reports, referred to in the appropriate sections of this chapter.

REFERENCES

1. Baker, *J. Chem. Phys.*, **23,** 1981 (1955).
2. Bernstein and Schneider, *J. Chem. Phys.*, **24,** 469 (1956).
3. Schneider, Bernstein, and Pople, *Can. J. Chem.*, **35,** 1487 (1957).
4. Schneider, Bernstein, and Pople, *Ann. N.Y. Acad. Sci.*, **70,** 806 (1958).
5. Diehl, Jones, and Bernstein, *Can. J. Chem.*, **43,** 81 (1965).
6. Castellano, Sun, and Kostelnik, *J. Chem. Phys.*, **46,** 327 (1967).
7. Diehl and Chuck, *Mol. Phys.*, **14,** 183 (1968).

7a. Tokuhiro, Wilson, and Fraenkel, *J. Am. Chem. Soc.*, **90,** 3622 (1968); Fraenkel, Adams, and Dean, *J. Phys. Chem.*, **72,** 944 (1968).

7b. Diehl, Khetrapal, and Kellerhals, *Mol. Phys.*, **15,** 333 (1968).

8. Bernstein, Pople, and Schneider, *Can. J. Chem.*, **36,** 65 (1957).
9. London, *J. Phys. Radium*, **8,** 397 (1937).
10. Pople, *J. Chem. Phys.*, **24,** 1111 (1956).
11. Waugh and Fessenden, *J. Am. Chem. Soc.*, **79,** 846 (1957).
12. Johnson and Bovey, *J. Chem. Phys.*, **29,** 1012 (1958).
13. Slater, *Phys. Rev.*, **36,** 57 (1930).
14. Pople, *Mol. Phys.*, **1,** 175 (1958).
15. McWeeny, *Mol. Phys.*, **1,** 311 (1958).
16. Dailey, Jonathan, and Gordon, *J. Chem. Phys.*, **36,** 2443 (1962); Gawer and Dailey, *ibid.*, **42,** 2658 (1965).
17. Pople and Ferguson, *J. Chem. Phys.*, **42,** 1560 (1965).
18. Musher, *J. Chem. Phys.*, **43,** 4081 (1965).
19. Gordon, Thesis, Columbia University, New York, 1962, in Ref. 16.
20. Hall, Hardisson, and Jackman, *Discuss. Faraday Soc.*, **34,** 15 (1962); Veillard, *J. Chim. Phys.*, **59,** 1056 (1962).
21. Schaefer and Schneider, *Can. J. Chem.*, **41,** 966 (1963).
22. Gil and Murrell, *Trans. Faraday Soc.*, **60,** 248 (1964).
23. Spotswood and Tanzer, *Tetrahedron Lett.*, **1967,** 911.

23a. Spotswood and Tanzer, *Aust. J. Chem.*, **20,** 1227 (1967).

24. Chuck and Randall, *J. Chem. Soc., B*, **1967,** 261.
25. Black, Brown, and Heffernan, *Aust. J. Chem.*, **20,** 1305 (1967).

25a. Jones and Ladd, *Mol. Phys.*, **19,** 233 (1970); see also Ladd and Jones, *Spectrochim. Acta*, **23A,** 2791 (1967).

25b. Cobb and Memory, *J. Chem. Phys.*, **50,** 4262 (1969).

26. Dewar and Marchand, *J. Am. Chem. Soc.*, **88,** 354 (1966).
27. Baldeschweiler and Randall, *Proc. Chem. Soc.*, **1961,** 303.
28. Smith and Schneider, *Can. J. Chem.*, **39,** 1158 (1961).
29. Kotowycz, Schaefer, and Boch, *Can. J. Chem.*, **42,** 2541 (1964).
30. Chuck and Randall, *Spectrochim. Acta*, **22,** 221 (1966).

30a. Reynolds and Priller, *Can. J. Chem.*, **46,** 2787 (1968).

30b. Cocivera, *J. Phys. Chem.*, **72,** 2515, 2520 (1968).

31. Kintzinger and Lehn, *Mol. Phys.*, **14,** 133 (1968).
32. Merry and Goldstein, *J. Am. Chem. Soc.*, **88,** 5560 (1966).
33. Saito, Nukuda, Kato, Yonezawa, and Fukui, *Tetrahedron Lett.*, **1965,** 111.
34. Adam, Grimison, Hoffmann, and de Ortiz, *J. Am. Chem. Soc.*, **90,** 1509 (1968).

34a. Arata and Fukumi, *Mol. Phys.*, **19,** 135 (1970).

35. Palmer and Semple, *Chem. Ind.*, **1965,** 1766.
36. Castellano and Kostelnik, *J. Am. Chem. Soc.*, **90,** 141 (1968).

37. Pople and Gordon, *J. Am. Chem. Soc.*, **89,** 4253 (1967).
38. Brügel, *Z. Electrochem.*, **66,** 159 (1962).
39. Batterham, unpublished results.
40. Tori and Ogata, *Chem. Pharm. Bull.*, **12,** 272 (1964), and references cited therein.
41. Schaefer and Schneider, *J. Chem. Phys.*, **32,** 1224 (1960).
42. Hatton and Richards, *Mol. Phys.*, **5,** 153 (1962).
43. Nakagawa and Fujiwara, *Bull. Chem. Soc. Japan*, **34,** 143 (1961).
44. Murrell and Gil, *Trans. Faraday Soc.*, **61,** 402 (1965).
45. Ronayne and Williams, *J. Chem. Soc.*, *B*, **1967,** 805.
46. Becker, *J. Phys. Chem.*, **63,** 1379 (1959).
47. Barbier, Delman, and Bene, *J. Chim. Phys.*, **58,** 764 (1961).
47a. Perkampus and Kruger, *Z. Phys. Chem.*, **55,** 202 (1967).
47b. Laszlo and Soong, *J. Chem. Phys.*, **47,** 4472 (1967).
47c. Findlay, Keniry, Kidman, and Pickles, *Trans. Faraday Soc.*, **63,** 846 (1967).
47d. Demarco, Farkas, Doddrell, Mylari, and Wenkert, *J. Am. Chem. Soc.*, **90,** 5480 (1968).
47e. Chatterjee and Bose, *Mol. Phys.*, **12,** 341 (1967).
48. Katritzky, Swinbourne, and Ternai, *J. Chem. Soc.*, *B*, **1966,** 235.
49. Fratiello and Christie, *Trans. Faraday Soc.*, **61,** 306 (1965).
49a. Perkampus and Kruger, *Ber. Bunsenges. Phys. Chem.*, **71,** 447 (1967).
49b. Perkampus and Kruger, *Ber. Bunsenges. Phys. Chem.*, **71,** 439 (1967).
49c. Wilson and Worrall, *J. Chem. Soc.*, *A*, **1968,** 2389; Huckerby, Wilson, and Worrall, *ibid.*, **1969,** 1189.
49d. Larsen and Allred, *J. Phys. Chem.*, **69,** 2400 (1965).
49e. Mooney and Qaseem, *J. Inorg. Nucl. Chem.*, **30,** 1638 (1968); Mooney and Qaseem, *Spectrochim. Acta*, **24A,** 969 (1968); McLauchlan and Mooney, *ibid.*, **23A,** 1227 (1967); Nainan and Ryschkewitsch, *Inorg. Chem.*, **7,** 1316 (1968).
50. Malinowski, Pollara, and Larmann, *J. Am. Chem. Soc.*, **84,** 2649 (1962).
51. Tori and Nakagawa, *J. Phys. Chem.*, **63,** 3163 (1964).
52. Hutton, Reynolds, and Schaefer, *Can. J. Chem.*, **40,** 1758 (1962).
53. Lauterbur, *J. Chem. Phys.*, **43,** 360 (1965).
54. Pugmire and Grant, *J. Am. Chem. Soc.*, **90,** 697 (1968).
55. Coulson and Danielsson, *Ark. Fys.*, **8,** 239, 245 (1954).
56. Witanowski and Januszewski, *J. Chem. Soc.*, *B*, **1967,** 1062.
57. Holder and Klein, *J. Chem. Phys.*, **23,** 1956 (1955).
58. Binsch, Lambert, Roberts, and Roberts, *J. Am. Chem. Soc.*, **86,** 5564 (1964).
58a. Lichter and Roberts, *J. Am. Chem. Soc.*, **93,** 5218 (1971).
58b. Tori, Ohtsura, Aono, Kawazoe, and Ohnishi, *J. Am. Chem. Soc.*, **89,** 2765 (1967).
58c. Batterham and Bigum, unpublished results.
59. Diehl and Leipert, *Helv. Chim. Acta*, **47,** 545 (1964).
60. Freymann and Freymann, *Arch. Sci. (Geneva)*, **13,** 506 (1960).
61. Freymann, Freymann, and Libermann, *Compt. Rend.*, **250,** 2185 (1960).
62. Kramer and West, *J. Phys. Chem.*, **69,** 673 (1965).
63. Gil, *Mol. Phys.*, **9,** 97 (1965).
64. Murrell, Gil, and Van Duijneveldt, *Rec. Trav. Chim.*, **84,** 1399 (1965).
65. Castellano, Günther, and Ebersole, *J. Phys. Chem.*, **69,** 4166 (1965).
66. Calder, Spotswood, and Tanzer, *Aust. J. Chem.*, **20,** 1195 (1967).
67. Spotswood and Tanzer, *Aust. J. Chem.*, **20,** 1213 (1967).
68. Wu and Dailey, *J. Chem. Phys.*, **41,** 3307 (1964).
69. Martin and Dailey, *J. Chem. Phys.*, **37,** 2575 (1962); Martin and Dailey, *ibid.*, **38,** 2582 (1963).

70. Perkampus and Kruger, *Chem. Ber.*, **100,** 1165 (1967).
70a. Haigh and Thornton, *Tetrahedron Lett.*, **1970,** 2043; see also Smith and Roark, *J. Phys. Chem.*, **73,** 1049 (1969).
70b. Wu, *J. Phys. Chem.*, **71,** 3089 (1967).
71. Boyd and Singer, *J. Chem. Soc., B*, **1966,** 1017.
71a. Crabtree and Bertelli, *J. Am. Chem. Soc.*, **89,** 5384 (1967).
72. Boyd and Ezekiel, *J. Chem. Soc., C*, **1967,** 1866.
73. Coburn and Dudek, *J. Phys. Chem.*, **72,** 1177 (1968).
74. Jones, Katritzky, and Lagowski, *Chem. Ind.*, **1960,** 870.
74a. MacNicol, *Chem. Commun.*, **1969,** 933.
75. Craig and Pearson, *J. Heterocycl. Chem.*, **5,** 631 (1968); Katritzky and Tiddy, *Org. Magnetic Resonance*, **1,** 57 (1969).
76. Katritzky and Reavill, *J. Chem. Soc.*, **1963,** 753.
76a. Cox and Bothner-By, *J. Phys. Chem.*, **73,** 2465 (1969).
77. Katritzky and Jones, *Proc. Chem. Soc.*, **1960,** 313.
78. Katritzky and Reavill, *J. Chem. Soc.*, **1965,** 3825.
79. Perrin, *Dissociation Constants of Organic Bases in Aqueous Solution*, Butterworths, London, 1965.
80. Elvidge and Jackman, *J. Chem. Soc.*, **1961,** 859.
81. Bell, Egan, and Bauer, *J. Heterocycl. Chem.*, **2,** 420 (1965).
82. Lezina, Bystrov, Smirnov, and Dyumaev, *Teor. i Eksp. Khim., Akad. Nauk Ukr. SSR*, **1,** 281 (1965); through *Chem. Abstr.*, **63,** 13049*f* (1965).
83. Smirnov, Lezina, Bystrov, and Dyumaev, *Izv. Akad. Nauk SSSR, Ser. Khim.*, **1965,** 1836.
84. Spinner and Yeoh, unpublished results.
85. Katritzky, Popp, and Rowe, *J. Chem. Soc., B*, **1966,** 562.
86. Batterham, unpublished results.
87. Ostercamp, *J. Org. Chem.*, **30,** 1169 (1965).
88. Cox, Elvidge, and Jones, *J. Chem. Soc.*, **1964,** 1423.
89. Beak and Bonham, *J. Am. Chem. Soc.*, **87,** 3365 (1965).
90. Batterham, Brown, and Paddon-Row, *J. Chem. Soc., B*, **1967,** 171.
91. Bean, Johnson, Katritzky, Ridgewell, and White, *J. Chem. Soc., B*, **1967,** 1219.
92. Bean, Brignell, Johnson, Katritzky, Ridgewell, Tarhan, and White, *J. Chem. Soc., B*, **1967,** 1222.
93. Bellingham, Johnson, and Katritzky, *J. Chem. Soc., B*, **1967,** 1226.
94. Paudler and Blewitt, *J. Org. Chem.*, **31,** 1295 (1966).
95. Rao and Baldeschweiler, *J. Chem. Phys.*, **37,** 2473 (1962).
96. Dhingra, Govil, Khetropal, and Ramiah, *Proc. Indian Acad. Sci.*, **62A,** 90 (1965).
97. Martin, Dorie, and Peradejordi, *J. Chim. Phys.*, **64,** 1193 (1967).
98. Spickett and Wright, *J. Chem. Soc., C*, **1967,** 503.
99. Biellman and Callot, *Bull. Soc. Chim. Fr.*, **1967,** 397.
100. Craig, Garnett, and Temple, *J. Chem. Soc.*, **1964,** 4057.
100a. Reinsborough, *Aust. J. Chem.*, **23,** 1473 (1970); Matthews and Gilson, *Can. J. Chem.*, **48,** 2625 (1970), and references cited therein.
101. Katritzky and Lagowski, *J. Chem. Soc.*, **1961,** 43.
101a. Abramovitch and Davis, *J. Chem. Soc., B*, **1966,** 1137.
102. Ford and Swan, *Aust. J. Chem.*, **18,** 867 (1965), and references cited therein.
103. Traynelis and Pacini, *J. Am. Chem. Soc.*, **86,** 4917 (1964).
103a. Traynelis and Gallagher, *J. Am. Chem. Soc.*, **87,** 5710 (1965).
104. Eisenthal and Katritzky, *Tetrahedron*, **21,** 2205 (1965).
105. Olah, Olah, and Overchuk, *J. Org. Chem.*, **30,** 3373 (1965).

106. Klose and Arnold, *Mol. Phys.*, **11,** 1 (1966).
107. Klose and Uhlemann, *Tetrahedron*, **22,** 1373 (1966).
108. Kuthan, *Collect. Czech. Chem. Commun.*, **31,** 3593 (1966).
109. Lubiculu and David, *Polymer*, **7,** 63 (1966).
110. Olah and Calin, *J. Am. Chem. Soc.*, **89,** 4736 (1967). [See also Olah and Calin, *ibid.*, **90,** 943 (1968).]
110a. Diehl, Khetrapal, and Lienhard, *Mol. Phys.*, **14,** 465 (1968).
110b. Kowalewski, de Kowalewski, and Ferra, *J. Mol. Spectrosc.*, **20,** 203 (1966); de Kowalewski and Ferra, *Mol. Phys.*, **13,** 547 (1968).
110c. Queguiner and Pastour, *Bull. Soc. Chim. Fr.*, **1968,** 4117.
110d. Cabani, Conti, and Ceccarelli, *Gazz. Chim. Ital.*, **98,** 923 (1968).
110e. Pattison and Wade, *J. Chem. Soc., A*, **1968,** 2618.
110f. Gansow and Holm, *Tetrahedron*, **24,** 4477 (1968).
110g. Gerig and Reinheimer, *Org. Magnetic Resonance*, **1,** 239 (1969).
111. Poziomek, Kramer, Mosher, and Michel, *J. Am. Chem. Soc.*, **83,** 3916 (1961).
112. Lustig, *J. Phys. Chem.*, **65,** 491 (1961).
113. Korytnyk, Kris, and Singh, *J. Org. Chem.*, **29,** 574 (1964).
114. Korytnyk and Singh, *J. Am. Chem. Soc.*, **85,** 2813 (1963).
115. Bedford, Katritzky, and Wuest, *J. Chem. Soc.*, **1963,** 4600.
116. Karabatsos and Vane, *J. Am. Chem. Soc.*, **85,** 3886 (1963); Kowalewski and de Kowalewski, *J. Chem. Phys.*, **36,** 266 (1962).
117. Fyfe, *Tetrahedron Lett.*, **1968,** 659.
118. Retcofsky and Friedel, *J. Phys. Chem.*, **72,** 2619 (1968).
118a. Retcofsky and Friedel, *J. Phys. Chem.*, **72,** 290 (1968).
118b. Retcofsky and Friedel, *J. Phys. Chem.*, **71,** 3592 (1967).
118c. Retcofsky and McDonald, *Tetrahedron Lett.*, **1968,** 2575.
118d. Hensen and Messer, *Chem. Ber.*, **102,** 957 (1969); Herbison-Evans and Richards, *Mol. Phys.*, **8,** 19 (1964).
119. Taylor and Paudler, *Tetrahedron Lett.*, **No. 25,** 1 (1960).
120. Ayer, Hayatsu, de Mayo, Reid, and Stothers, *Tetrahedron Lett.*, **1961,** 648.
120a. Slomp, MacKellar, and Paquette, *J. Am. Chem. Soc.*, **83,** 4472 (1961).
121. Taylor, Kan, and Paudler, *J. Am. Chem. Soc.*, **83,** 4484 (1961).
122. Paquette and Slomp, *J. Am. Chem. Soc.*, **85,** 765 (1963).
123. Taylor and Kan, *J. Am. Chem. Soc.*, **85,** 776 (1963).
124. Saunders and Gold, *J. Org. Chem.*, **27,** 1439 (1962).
125. Meyer, Mahler, and Baker, *Biochim. Biophys. Acta*, **64,** 353 (1962).
126. Karle, *Acta Crystallogr.*, **14,** 497 (1961).
127. Weber, *Nature*, **180,** 1409 (1957); Veitch, in *Light and Life*, Johns Hopkins Press, Baltimore, Maryland, 1961, p. 108.
128. Levy and Vennesland, *J. Biol. Chem.*, **228,** 85 (1957); Vennesland, *Fed. Proc.*, **17,** 1150 (1958).
129. Dittmer and Fouty, *J. Am. Chem. Soc.*, **86,** 91 (1964).
129a. Biellman and Callot, *Bull. Soc. Chim. Fr.*, **1968,** 1154; Biellman and Callot, *ibid.*, **1968,** 1159.
129b. Choi and Alivisatos, *Biochem.*, **7,** 190 (1968).
130. Cook and Lyons, *J. Am. Chem. Soc.*, **87,** 3283 (1965).
131. Neilsen, Moore, Muha and Berry, *J. Org. Chem.*, **29,** 2175 (1964).
132. Lansbury and Peterson, *J. Am. Chem. Soc.*, **85,** 2236 (1963).
133. Karplus, *J. Chem. Phys.*, **30,** 11 (1959).
134. Anteunis, *Bull. Soc. Chim. Belges*, **75,** 413 (1966).
135. Booth, *Tetrahedron*, **20,** 2211 (1964).

136. Lyle, McMahon, Krueger, and Spicer, *J. Org. Chem.*, **31**, 4164 (1966).
137. Booth, Little, and Feeney, *Tetrahedron*, **24**, 279 (1968).
137a. Feltkamp, Naegele, and Wendisch, *Org. Magnetic Resonance*, **1**, 11 (1969).
138. Bishop, Fodor, Katritzky, Soti, Sutton, Swinbourne, *J. Chem. Soc., C*, **1966**, 74.
139. Closs, *J. Am. Chem. Soc.*, **81**, 5456 (1959).
140. Becconsall, Jones, and McKenna, *J. Chem. Soc.*, **1965**, 1726.
141. Imbach, Katritzky, and Kolinski, *J. Chem. Soc., B*, **1966**, 556.
142. Bottini and O'Rell, *Tetrahedron Lett.*, **1967**, 423, 429.
143. House, Tefertiller, and Pitt, *J. Org. Chem.*, **31**, 1073 (1966).
144. House and Pitt, *J. Org. Chem.*, **31**, 1062 (1966).
145. House and Tefertiller, *J. Org. Chem.*, **31**, 1068 (1966).
146. Brown, Hutley, McKenna, and McKenna, *Chem. Commun.*, **1966**, 719.
146a. Brown, McKenna, and McKenna, *J. Chem. Soc., B*, **1967**, 1195.
147. Hill and Chan, *Tetrahedron*, **21**, 2015 (1965).
147a. Kawazoe, Tsuda, and Ohnishi, *Chem. Pharm. Bull.*, **15**, 51 (1967).
147b. Stien, Ottinger, Reisse, and Chiurdoglu, *Tetrahedron Lett.*, **1968**, 1521.
148. Chen and Le Fevre, *J. Chem. Soc.*, **1965**, 3467.
149. Casy, *Tetrahedron*, **22**, 2711 (1966).
150. Lee, Beckett, and Sugden, *Tetrahedron*, **22**, 2721 (1966).
150a. Johnson, *J. Org. Chem.*, **33**, 3627 (1968).
150b. Chow, Colon, and Tam, *Can. J. Chem.*, **46**, 2821 (1968).
150c. Paulsen, Toth, and Ripperger, *Chem. Ber.*, **101**, 3365 (1968).
150d. Lambert and Oliver, *Tetrahedron Lett.*, **1968**, 6187.
150e. Chow and Colon, *Can. J. Chem.*, **46**, 2827 (1968).
150f. Kuhn and McIntyre, *Can. J. Chem.*, **44**, 105 (1966).
150g. Kreilick, *J. Chem. Phys.*, **46**, 4260 (1967).
150h. Briere, Lemaire, Rassat, Rey, and Rousseau, *Bull. Soc. Chim. Fr.*, **1967**, 4479.
150i. Branch and Casy, *J. Chem. Soc., B*, **1968**, 1087.
151. Aroney, Chen, Le Fevre, and Singh, *J. Chem. Soc., B*, **1966**, 98.
152. Murayama and Yoshioka, *Tetrahedron Lett.*, **1968**, 1363.
153. Brignell, Katritzky, and Russell, *Chem. Commun.*, **1966**, 723.
154. Baumann, Franklin, and Möhrle, *Tetrahedron*, **24**, 589 (1968); see also Kessler and Möhrle, *Z. Naturforsch.*, **24b**, 301 (1969).
155. Booth and Little, *Tetrahedron*, **23**, 291 (1967).
156. Delpuech and Deschamps, *Chem. Commun.*, **1967**, 1188; Delpuech and Gay, *Tetrahedron Lett.*, **1966**, 2603.
157. Lambert, Keske, Cahart, and Jovanovich, *J. Am. Chem. Soc.*, **89**, 3761 (1967).
157a. Lambert and Keske, *J. Am. Chem. Soc.*, **88**, 620 (1966).
158. Robinson, *Tetrahedron Lett.*, **1968**, 1153.
159. Booth, *Chem. Commun.*, **1968**, 802.
159a. Lambert and Keske, *Tetrahedron Lett.*, **1969**, 2023.
159b. Yousif and Roberts, *J. Am. Chem. Soc.*, **90**, 6428 (1968).
160. Reeves and Wells, *Discuss. Faraday Soc.*, **34**, 177 ((1962); Lee and Orrell, *Trans. Faraday Soc.*, **63**, 16 (1967).
161. Lyle and Krueger, *J. Org. Chem.*, **32**, 3613 (1967).
162. Casy, Beckett, and Iorio, *Tetrahedron*, **22**, 2751 (1966); Beckett, Casy, and Iorio, *ibid.*, **22**, 2745 (1966).
162a. Duquette and Johnson, *Tetrahedron*, **23**, 4517 (1967).
163. Lee and Orrell, *J. Chem. Soc.*, **1965**, 582.
164. Chambers, Hutchison, and Musgrave, *J. Chem. Soc.*, **1964**, 5634.
165. Chambers, Lomas, and Musgrave, *J. Chem. Soc., C*, **1968**, 625.

165a. Emsley and Phillips, *J. Chem. Soc., B*, **1969,** 434; Emsley, *J. Chem. Soc., A*, **1968,** 2735.
165b. Alger and Gutowsky, *J. Chem. Phys.*, **48,** 4625 (1968).
165c. Cunliffe and Harris, *Mol. Phys.*, **15,** 413 (1968).
165d. Giam and Lyle, *J. Chem. Soc., B*, **1970,** 1516.
166. Isobe, *Bull. Chem. Res. Inst. Nonaqueous Solutions, Tohoku University*, **9,** 115 (1960).
167. Kawazoe and Natsume, *Yakugaku Zasshi*, **83,** 523 (1963); Coad, Coad, and Wilkins, *J. Phys. Chem.*, **67,** 2815 (1963); Elvidge, Newbold, Senciall, and Symes, *J. Chem. Soc.*, **1964,** 4157.
168. Tori and Ogata, *Chem. Pharm. Bull.*, **12,** 272 (1964).
169. Gil, *Mol. Phys.*, **9,** 443 (1965).
170. Elvidge and Ralph, *J. Chem. Soc., B*, **1966,** 249.
170a. Dratler and Laszlo, *Chem. Commun.*, **1970,** 180.
171. Declerck, Degroote, de Lannoy, Nasielski-Hinkens, and Nasielski, *Bull. Soc. Chim. Belges*, **74,** 119 (1965).
172. Barlin and Brown, *J. Chem. Soc., B*, **1967,** 648.
173. Bale, Simmonds, and Trager, *J. Chem. Soc., B*, **1966,** 867.
174. Katritzky and Waring, *J. Chem. Soc.*, **1964,** 1523.
175. Gompper and Altreuther, *Z. Anal. Chem.*, **170,** 205 (1959).
176. Leclerc and Wermuth, *Bull. Soc. Chim. Fr.*, **1967,** 1307.
176a. Katritzky and Pojarlieff, *J. Chem. Soc., B*, **1968,** 873.
177. Kawazoe and Ohnishi, *Chem. Pharm. Bull.*, **11,** 243 (1963).
177a. Suzuki, Nakadate, and Sueyoshi, *Tetrahedron Lett.*, **1968,** 1855.
177b. Chambers, MacBride, and Musgrave, *J. Chem. Soc., C*, **1968,** 2989.
178. Sauer, Mielert, Lang, and Peter, *Chem. Ber.*, **98,** 1435 (1965).
179. Maier, *Chem. Ber.*, **98,** 2446 (1965).
180. Anderson and Lehn, *Tetrahedron*, **24,** 137 (1968).
181. Daniels and Roseman, *Tetrahedron Lett.*, **1966,** 1335.
182. Breliere and Lehn, *Chem. Commun.*, **1965,** 426.
182a. Korsch and Riggs, *Tetrahedron Lett.*, **1966,** 5897.
182b. Price, Sutherland, and Williamson, *Tetrahedron*, **22,** 3477 (1966).
183. Gronowitz and Hoffman, *Ark. Kemi*, **16,** 459 (1960).
184. Reddy, Hobgood, and Goldstein, *J. Am. Chem. Soc.*, **84,** 336 (1962).
185. Gronowitz, Norrman, Gestblom, Mathiasson, and Hoffman, *Ark. Kemi*, **22,** 65 (1964).
185a. Khetrepal, Pantakar and Diehl, *Org. Magnetic Resonance*, **2,** 405 (1970).
185b. Paudler and Humphrey, *Org. Magnetic Resonance*, **3,** 217 (1970).
186. Spiesecke and Schneider, *J. Chem. Phys.*, **35,** 731 (1961).
187. Gronowitz and Hoffman, *Ark. Kemi*, **16,** 539 (1960).
188. Rodmar, Rodmar, Khan, and Gronowitz, *Ark. Kemi*, **27,** 87 (1967).
189. Biffin, Brown, and Lee, *J. Chem. Soc., C*, **1967,** 573.
189a. Lynch and Poon, *Can. J. Chem.*, **45,** 1431 (1967).
189b. Ditchfield and Gil, *J. Chem. Soc., A*, **1969,** 533.
190. Katritzky and Waring, *J. Chem. Soc.*, **1963,** 3046.
191. Albert and Spinner, *J. Chem. Soc.*, **1960,** 1221.
192. Inoue, Furutachi and Nakanishi, *J. Org. Chem.*, **31,** 175 (1966).
193. Brown and Short, *J. Chem. Soc.*, **1953,** 331.
193a. Bauer, Wright, Mikrut, and Bell, *J. Heterocycl. Chem.*, **2,** 447 (1965).
193b. Nishiwaki, *Tetrahedron*, **23,** 2657 (1967).
193c. Elvidge and Zaidi, *J. Chem. Soc., C*, **1968,** 2188.
193d. Promel, Cardon, Daniel, Jacques, and Vandersmissen, *Tetrahedron Lett.*, **1968,** 3067.

193e. Katritzky, Kingsland, and Tee, *J. Chem. Soc.*, *B*, **1968,** 1484.
194. Temple and Montgomery, *J. Am. Chem. Soc.*, **86,** 2946 (1964).
195. Temple and Montgomery, *J. Org. Chem.*, **30,** 826 (1965); Temple, Coburn, Thorpe, and Montgomery, *ibid.*, **30,** 2395 (1965).
196. Brown, England, and Harper, *J. Chem. Soc.*, *C*, **1966,** 1165.
197. Brown and Paddon-Row, *J. Chem. Soc.*, *C*, **1967,** 903; Brown and Paddon-Row, *ibid.*, **1966,** 164; Brown and Paddon-Row, *ibid.*, **1967,** 1928; Brown and England, *ibid.*, **1967,** 1922.
198. Paudler and Kuder, *J. Org. Chem.*, **31,** 809 (1966).
199. Paudler and Blewitt, *J. Org. Chem.*, **30,** 4081 (1965).
200. Gatlin and Davis, *J. Am. Chem. Soc.*, **84,** 4464 (1962).
201. Kokko, Goldstein, and Mandell, *J. Am. Chem. Soc.*, **83,** 2909 (1961).
202. Batterham, Ghambeer, Blakley, and Broenson, *Biochem.*, **6,** 1203 (1967).
203. Jardetzky, Pappas, and Wade, *J. Am. Chem. Soc.*, **85,** 1657 (1963).
204. Shoup, Miles, and Becker, *J. Am. Chem. Soc.*, **89,** 6200 (1967).
205. Miles, Bradley, and Becker, *Science*, **142,** 1569 (1963).
206. Becker, Miles, and Bradley, *J. Am. Chem. Soc.*, **87,** 5575 (1965).
207. Roberts, Lambert, and Roberts, *J. Am. Chem. Soc.*, **87,** 5439 (1965).
208. Jardetzky and Jardetzky, *J. Am. Chem. Soc.*, **82,** 222 (1960).
208a. Nollet, Koomen, Grose, and Pandit, *Tetrahedron Lett.*, **1969,** 4607.
208b. Rosenthal, *Tetrahedron Lett.*, **1969,** 3333.
209. Kokko, Mandell, and Goldstein, *J. Am. Chem. Soc.*, **84,** 1042 (1962).
210. Brown and Teitei, *Aust. J. Chem.*, **17,** 567 (1964).
211. Katritzky, Popp, and Waring, *J. Chem. Soc.*, *B*, **1966,** 565.
212. Kheifets, Khromov-Borisov, and Kol'tsov, *Dokl. Akad. Nauk SSSR*, **166,** 635 (1966); through *Chem. Abstr.*, **64,** 11207 (1966); Kheifets, Khromov-Borisov, and Kol'tsov, *Zh. Org. Khim.*, **2,** 1516 (1966); Kheifets and Khromov-Borisov, *ibid.*, **2,** 1511 (1966); Kheifets and Khromov-Borisov, *J. Gen. Chem.*, **34,** 3134 (1964); Kheifets, Khromov-Borisov, Kol'tsov, and Volkenstein, *Tetrahedron*, **23,** 1197 (1967).
212a. Snell, *J. Chem. Soc.*, *C*, **1968,** 2358.
212b. Glasel, *Org. Magnetic Resonance*, **1,** 481 (1969).
212c. Neville, *Anal. Chem.*, **42,** 347 (1970).
212d. Wagner and von Philipsborn, *Helv. Chim. Acta*, **53,** 299 (1970).
212e. Batterham and Bigum, *Org. Magnetic Resonance*, in press (1972).
213. Hollis and Wang, *J. Org. Chem.*, **32,** 1620 (1967), and references cited therein.
214. Silversmith, *J. Org. Chem.*, **27,** 4090 (1962).
215. Mehta, Miller, and Mooney, *J. Chem. Soc.*, **1965,** 6695.
216. Takamizawa, Hirai, Sato, and Tori, *J. Org. Chem.*, **29,** 1740 (1964).
217. Rouiller, Delmau, Duplan, and Nofre, *Tetrahedron Lett.*, **1966,** 4189.
218. Chabre, Gagnaire, and Nofre, *Bull. Soc. Chim. Fr.*, **1966,** 108.
219. Rouillier, Delmau, and Nofre, *Bull. Soc. Chim. Fr.*, **1966,** 3515.
219a. Katritzky, Nesbit, Kurtev, Lyopova, and Pojarlieff, *Tetrahedron*, **25,** 3807 (1969).
219b. Pojarlieff, *Bull. Soc. Chim. Fr.*, **1968,** 5033.
220. Evans and Shannon, *J. Chem. Soc.*, **1965,** 1406.
221. Evans, *Aust. J. Chem.*, **20,** 1643 (1967).
222. David and Sinay, *Bull. Soc. Chim. Fr.*, **1965,** 2301.
223. Suzuki, Inoue, and Goto, *Chem. Pharm. Bull.*, **16,** 933 (1968).
224. Riddell, *J. Chem. Soc.*, *B*, **1967,** 560; Riddell and Lehn, *Chem. Commun.*, **1966,** 375.
225. Farmer and Hamer, *Tetrahedron*, **24,** 829 (1968).

226. Schroeder, Kober, Ulrich, Ratz, Agahigian, and Grundmann, *J. Org. Chem.*, **27**, 2580 (1962).
227. Banks, Field, and Hazeldine, *J. Chem. Soc.*, *C*, **1967**, 1822.
228. Elvidge, Newbold, Senciall, and Symes, *J. Chem. Soc.*, **1964**, 4157.
229. Kamei, *J. Phys. Chem.*, **69**, 2791 (1965).
229a. Cox and Bothner-By, *J. Phys. Chem.*, **72**, 1646 (1968).
229b. Cox and Bothner-By, *J. Phys. Chem.*, **72**, 1642 (1968).
229c. Bramwell, Riezebos, and Wells, *Tetrahedron Lett.*, **1971**, 2489.
229d. Goto, Isobe, Ohtsura, and Tori, *Tetrahedron Lett.*, **1968**, 1511.
230. Diehl and Khetrapal, *Mol. Phys.*, **14**, 327 (1968).
231. Reeves and Stromme, *J. Chem. Phys.*, **34**, 1711 (1961); Harris and Spragg, *Chem. Commun.*, **1966**, 314; Spragg, *J. Chem. Soc.*, *B*, **1968**, 1128.
232. Sudmeier and Occupati, *J. Am. Chem. Soc.*, **90**, 154 (1968).
233. Sudmeier, *J. Phys. Chem.*, **72**, 2344 (1968).
233a. Delpuech and Martinet, *Chem. Commun.*, **1968**, 478.
233b. Harris, *J. Mol. Spectrosc.*, **15**, 100 (1965).
233c. Lambert, Gosnell, Bailey, and Henkin, *J. Org. Chem.*, **34**, 4147 (1969).
233d. Getti and Wartik, *Inorg. Chem.*, **5**, 2075 (1966).
233e. Allison, Chambers, MacBride, and Musgrave, *J. Chem. Soc.*, *C*, **1970**, 1023.
234. Lamchen and Mittag, *J. Chem. Soc.*, *C*, **1966**, 2300.
234a. van der Plas, Haase, Zuurdeeg, and Vollering, *Rec. Trav. Chim.*, **85**, 1101 (1966).
234b. Young and Dressler, *J. Org. Chem.*, **32**, 2237 (1967).
235. Declerck, Degroote, de Lannoy, Nasielski-Hinkens, Nasielski, *Bull. Soc. Chim. Belges*, **74**, 119 (1965); Paudler and Barton, *J. Org. Chem.*, **31**, 1720 (1966).
235a. Adams and Shepherd, *Tetrahedron Lett.*, **1968**, 2747.
235b. Moore, *J. Org. Chem.*, **31**, 3910 (1966).
235c. Trepanier, Wagner, Harris, and Rudzik, *J. Med. Chem.*, **9**, 881 (1966).
236. Riddell and Lehn, *Chem. Commun.*, **1966**, 375; Lehn, Riddell, Price, and Sutherland, *J. Chem. Soc.*, *B*, **1967**, 387; Gutowsky and Temussi, *J. Am. Chem. Soc.*, **89**, 4358 (1967); Farmer and Hamer, *Chem. Commun.*, **1966**, 866; Anderson and Roberts, *J. Am. Chem. Soc.*, **90**, 4186 (1968), and references cited therein.
237. Anderson, *Q. Rev.*, **19**, 426 (1965).
238. Biddiscombe, Herington, Lawrenson, and Martin, *J. Chem. Soc.*, **1963**, 444.
239. Kowalewski and de Kowalewski, *J. Chem. Phys.*, **37**, 2603 (1962).
240. Castellano and Bothner-By, *J. Chem. Phys.*, **41**, 3863 (1965).
241. Golding, Katritzky, and Kucharska, *J. Chem. Soc.*, **1965**, 3090.
242. Wenkert, Dave, and Haglid, *J. Am. Chem. Soc.*, **87**, 5461 (1965).
243. Bhacca, Johnson, and Shoolery, *NMR Spectra Catalog*, Varian Associates, Palo Alto, California, 1962.
244. Johnson, Katritzky, and Viney, *J. Chem. Soc.*, *B*, **1967**, 1211.
245. Streith and Sigwalt, *Tetrahedron Lett.*, **1966**, 1347.
246. Bellingham, Johnson, and Katritzky, *J. Chem. Soc.*, *B*, **1967**, 1226.
247. De Selms, *J. Org. Chem.*, **33**, 478 (1968).
248. Neuhoff and Harris, *Naturwiss.*, **51**, 291 (1964).
249. Dinan and Tieckelmann, *J. Org. Chem.*, **29**, 892 (1964).
250. Baker, Katritzky, and Moynehan, *J. Chem. Soc.*, **1964**, 6138.
251. Potts and Burton, *J. Org. Chem.*, **31**, 251 (1966).
252. Bauer and Gardella, *J. Org. Chem.*, **28**, 1323 (1963).
253. Boatman, Harris, and Hauser, *J. Org. Chem.*, **30**, 3593 (1965).
254. Butt, Elvidge, and Foster, *J. Chem. Soc.*, **1963**, 3069.

255. Van der Does and den Hertog, *Rec. Trav. Chim.*, **84,** 951 (1965).
256. Frydman, Despuy, and Rapoport, *J. Am. Chem. Soc.*, **87**, 3530 (1965).
257. Bauer and Hirsch, *J. Org. Chem.*, **31,** 1210 (1966).
258. Brown and Heffernan, *Aust. J. Chem.*, **12,** 543 (1959).
259. McWeeny and Peacock, *Proc. Phys. Soc. (London), Sect. A*, **70,** 41 (1957).
260. Gunther and Castellano, *Ber. Bunsenges. Phys. Chem.*, **70,** 913 (1966).
261. Batterham and Rao, unpublished results.
262. Johnson, Katritzky, Ridgewell, and Viney, *J. Chem. Soc., B*, **1967,** 1204.
263. Banks, Haszeldine, Latham, and Young, *J. Chem. Soc.*, **1965,** 594.
264. Chambers, Hutchison, and Musgrave, *J. Chem. Soc.*, **1964,** 3736.
265. Chambers, Drakesmith, and Musgrave, *J. Chem. Soc.*, **1965,** 5045.
266. Chambers, Hutchison, and Musgrave, *J. Chem. Soc.*, **1965,** 5040.
267. Chambers, Hutchison, and Musgrave, *J. Chem. Soc.*, **1964,** 3574.
268. Chambers, Hutchison, and Musgrave, *J. Chem. Soc., C*, **1966,** 220.
269. Chambers, Iddon, Musgrave, and Chadwick, *Tetrahedron*, **24,** 877 (1968).
270. Kosower and Sorensen, *J. Org. Chem.*, **27,** 3764 (1962).
271. Hutton and Westheimer, *Tetrahedron*, **3,** 73 (1958).
272. Brignell, Eisner, and Farrell, *J. Chem. Soc., B*, **1966,** 1083.
273. Brignell, Bullock, Eisner, Gregory, Johnson, and Williams, *J. Chem. Soc.*, **1963,** 4819.
274. Weber, Slates, and Wendler, *J. Org. Chem.*, **32,** 1668 (1967).
275. Childs and Johnson, *J. Chem. Soc., C*, **1966,** 1950.
276. Batterham and Batts, unpublished results.
277. Quan and Quin, *J. Org. Chem.*, **31,** 2487 (1966).
278. Freifelder, *J. Org. Chem.*, **29,** 2895 (1964).
279. Anderson, Feast, and Musgrave, *Chem. Commun.*, **1968,** 1433.
280. Gil and Pinto, *Mol. Phys.*, **19,** 573 (1970).
281. Brown and Hoskins, *J. Chem. Soc., Perkin Trans I*, **1972,** 522.
282. Ford, Ph.D. Thesis, Australian National University, Canberra, Australia, 1968.
283. Brown and Lee, *Aust. J. Chem.*, **21,** 243 (1968).

3 NITROGEN HETEROCYCLES—SINGLE THREE-, FOUR-, FIVE-, AND SEVEN-MEMBERED RINGS

I. THREE-MEMBERED RINGS

A. The Azirine System

2 3 N 1

This rather unstable system has been studied very little, but a certain amount of data has been collected in a number of recent papers. A comparison of the chemical shifts of the 3-proton in the azirine (**1**) (τ 0.2) and the corresponding aziridine (**2**) (τ 2.6) suggests that a substantial azirine ring current exists (1). Similarly, H-2 in the series **3** absorbs close to τ 0 and is much deshielded (by 2.0–2.6 ppm) with respect to the methine proton of the open-chain aldimine (**4**). In compounds with a $-CH_2-$ attached to C-3, long-range coupling between the 2-proton and the methylene substituent was observed (2). The value of J_{23} for 3-phenylazirine was shown to be about 2 Hz (3). Data for azirines are collected in Table 3.1.

The reduced azirines or aziridines (**5**) were among the earliest compounds studied by NMR spectroscopy. Some confusion exists in the use of the terms cis and trans with respect to these compounds. These terms will be used here to describe only the stereochemistry of the C–C fragment while the terms syn and anti, respectively, will replace them when referring to the C–N bonds.

Mortimer (4), using ^{13}C satellite spectra as an aid to analyzing symmetrical spin systems, produced an analysis of aziridine or ethylenimine and obtained values for all parameters except J_{gem}. Quite a lot of work has gone into the analysis of the spectra of 2-phenylaziridines or styrenimines (**6**). Brois and

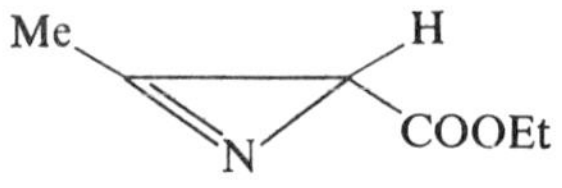

1 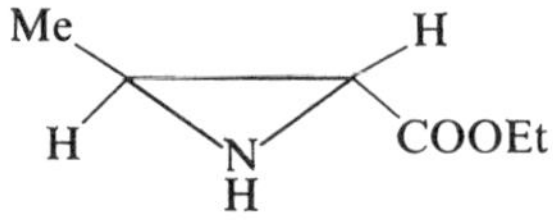 **2** 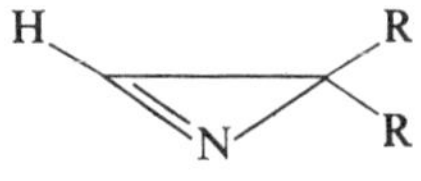 **3**

$MeCH{=}NCH_2CH_2Me$

4

TABLE 3.1 SPECTRAL DATA FOR AZIRINES

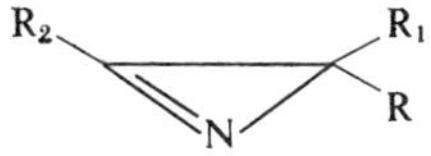

R	R_1	R_2	Solvent	τR_1	τR_2	Ref.
n-Pr	H	H	CCl_4		−0.1	2
$PhCH_2CH_2$	H	H	CCl_4		0.3	2
Et	Et	H	CCl_4		−0.2	2
Ph	Et	H	CCl_4		0.3	2
Ph	Ph	H	CCl_4		0.1	2
Ph	H	H	CCl_4	7.2	0.4	3
COOEt	H	Me	$CDCl_3$	7.8		1
COOEt	Me	Me	$CDCl_3$	8.65	7.7	1

his co-workers have completely analyzed the spectrum of the parent compound (5, 6) and finally have produced models of known stereochemistry to prove that J_{cis} is always larger than J_{trans} (7). They showed that all couplings between the three aziridine protons had the same sign and predicted J_{gem} for the parent aziridine as 1 Hz. These results have been used to assign stereochemistry in the isomeric 3-deuteroaziridines (7) (8). More recently, a full analysis of spectra of a series of *N*-substituted 2-phenylaziridines has been reported (9). Coupling constants between the geminal protons correlated linearly with the electronegativity of the *N*-substituent. In the case of the 1-methyl derivative, long-range couplings between the *N*-methyl group and the ring protons were stronger when these were syn to each other than when they were anti (see Table 3.2). ^{14}N–H coupling, observed as broadening, was used to assign the position of the lone pair.

The analysis of the ^{13}C satellite spectra of a number of *N*-alkylaziridines (10), which at room temperature have fixed conformations (11), shows a number of remarkable effects which are attributed to localization of the lone pair. The value of J_{cis} for two protons syn to the alkyl group is about 7 Hz, whereas for protons anti to the alkyl group it is only about 5.3 Hz. Similarly,

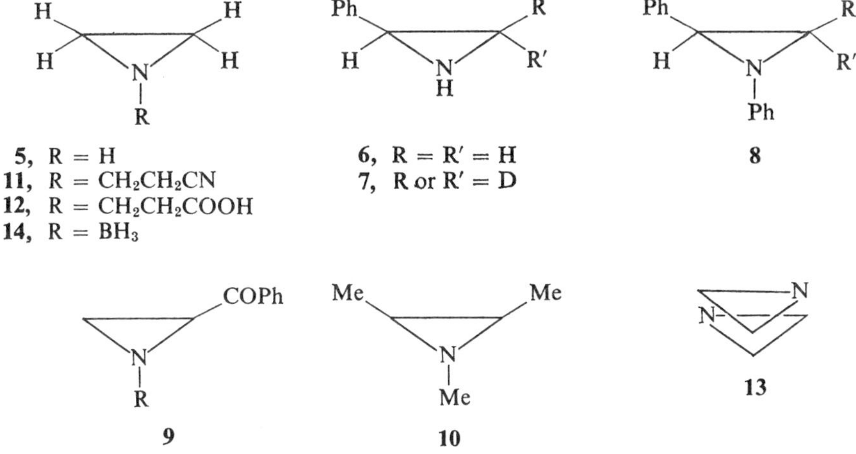

$J_{^{13}C-H_{syn}}$ is 161 Hz while $J_{^{13}C-H_{anti}}$ is 171 Hz, a difference which reflects decisively the nonequivalence of these two types of geminal protons. This nonequivalence is also evident from data obtained for the ^{15}N isomer of 2-(α-naphthyl)aziridine (11a) (see Table 3.2).

Spectra of a number of 3-substituted 1,2-diphenylaziridines (**8**) were measured as models for a 3-chloro-3-methyl derivative of unknown stereochemistry (12). Again couplings of J_{cis} 5–6 Hz and J_{trans} 2–2.7 Hz were found for a large range of compounds. Brois (12a) showed that ring protons cis to an *N*-alkyl bond are shifted to higher field than the corresponding trans protons. Earlier work was discussed in relation to this finding and van der Waal's interactions were invoked to explain why the shifts occurred. Chemical shift data for a series of *C*-alkylaziridines have recently been published (12b).

A great deal of the work on aziridines has been centered around the relatively slow inversion of the nitrogen atom in 1-substituted derivatives of this ring system. Unfortunately, almost all of the information on this aspect of aziridines appears in short communications without adequate experimental descriptions and beset by the difficulties now known to be inherent in coalescence temperature determinations of thermodynamic data. For this reason I refer the reader to the original references (11, 13–25c) and only discuss points particularly related to the NMR spectra. The effects of various *N*-substituents can be summarized; substituents able to delocalize the electron pair enhance the rate (19, 20), severe steric effects also enhance the rate (13), and *N*-halogeno substituents reduce the rate so dramatically that in certain cases the two isomers have been isolated (22–24). Recently, Jautelat and Roberts (26) have used a complete line-shape method for the analysis

TABLE 3.2 SPECTRAL DATA FOR AZIRIDINES

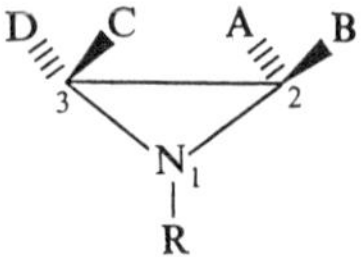

A. STEREOCHEMISTRY UNKNOWN OR UNIMPORTANT

Substituents	Solvent	τA	τB	τC	τD	τNR	J_{AB} (gem)	J_{AC} (trans)	J_{BC} (cis)	Ref.
None	Neat	8.63								28
	CCl_4	8.52								28
	$CDCl_3$	8.39								28
1-Me							1.0	3.8	7.0 (syn) 5.3 (anti)	10[a]
1-$(CH_2)_3$CN	Neat	8.35	8.84	8.84	8.35		1.0	3.5	6.0	28
	CCl_4	8.28	8.83	8.83	8.28					28
	$CDCl_3$	8.16	8.74	8.74	8.16					28
	CD_3OD	8.20	8.62	8.62	8.20					28
1-$(CH_2)_3$COOH	CCl_4	8.40	8.96	8.96	8.40		1.3	3.4	6.0	28
	$CDCl_3$	8.24	8.84	8.84	8.24					28
	CD_3OD	8.27	8.69	8.69	8.27					28
1-*t*-Bu, 3,4-Di-Me, 2-oxo		8.55	8.55			8.67				248

B. COMPOUNDS WITH FIXED STEREOCHEMISTRY

	Substituents and their chemical shifts (τ)					Coupling constants			
Solvent	A	B	C	D	R	J_{AB} (gem)	J_{AC} (trans)	J_{BC} (cis)	Ref.
Neat	H 8.58	H 8.58	H 8.58	H 8.58	Ph				5
$CDCl_3$	H 7.37	H 7.73	H 7.23	CO_2Me 6.22	Ph	1.70	3.06	6.34	221
$CDCl_3$	H 7.40	H 7.79	H 7.30	CO_2Me 6.22	*p*-MeOPh	1.83	3.27	6.45	221
$CDCl_3$	H 7.39	H 7.78	H 7.28	CO_2Me 6.18	*p*-MePh	1.96	3.26	6.39	221
$CDCl_3$	H 7.36	H 7.72	H 7.24	CO_2Me 6.20	*p*-ClPh	1.65	3.21	6.39	221
$CDCl_3$	H 7.21	H 7.54	H 7.02	CO_2Me 6.17	*p*-NO_2Ph	1.55	3.12	6.19	221
$CDCl_3$	H 7.17	H 7.84	Me 8.67	CO_2Me 6.37	Ph	1.25			221
$CDCl_3$	H 6.56	CO_2Me 6.31	H 6.56	CO_2Me 6.31	*p*-OMePh				221
$CDCl_3$	CO_2Me 6.22	H 6.95	H 6.95	CO_2Me 6.22	*p*-OMePh				221
$CDCl_3$	H 7.46	H 7.68	H 7.36	CN	Ph	1.07	3.10	6.08	221
$CDCl_3$	H 6.75	PhCO	PhCO	H 6.75	Me				25
$CDCl_3$	H 6.62	PhCO	PhCO	H 6.62	$PhCH_2$				25
$CDCl_3$	H 6.62	PhCO	PhCO	H 6.62	C_6H_{11}				25
C_6H_6	7.25			7.25					27

TABLE 3.2 (*continued*)

Solvent	Substituents and their chemical shifts (τ)					Coupling constants			Ref.
	A	B	C	D	R	J_{AB} (gem)	J_{AC} (trans)	J_{BC} (cis)	
$CDCl_3$	H 6.08	PhCO	H 6.08	PhCO	Me				25
$CDCl_3$	H 5.90	PhCO	H 5.90	PhCO	$PhCH_2$				25
C_6H_6	5.90		5.90						27
$CDCl_3$	H 5.93	PhCO	H 5.93	PhCO	C_6H_{11}				25
C_6H_6	5.82		5.82						27
CCl_4[b]	H 8.42	H 7.96	H 7.15	Ph	H	0.76	3.32	5.96	9
C_6H_6[b]	8.57	8.24	7.38			0.87	3.29	6.12	6
$CDCl_3$	8.52	8.14	7.34			0.6	3.1	6.0	5
CCl_4[b]	H 8.33	H 7.87	H 6.97	$C_{10}H_7$[c]		0.7	3.3	5.9	9
CCl_4	H 8.33	H 8.55	H 7.90	Ph	Me	1.2	3.1	6.3	9
CCl_4	H 8.20	H 8.46	H 7.71	$C_{10}H_7$[c]	Me	1.1	3.1	6.3	9
CCl_4	H 7.77	H 7.66	H 7.03	Ph	Ph	1.4	3.3	6.3	9
CCl_4	H 7.88	H 7.42	H 6.66	Ph	Ac	0.8	3.5	6.3	9
$CDCl_3$	H 6.35	COPh	H 6.35	Ph	CH_2Ph				27
C_6H_6	6.55		6.30						27
$CDCl_3$	H 6.62	COPh	Ph	H 6.77	CH_2Ph				27
C_6H_6	7.10			7.10					27
$CDCl_3$	H 6.42	COPh	H 6.42	Ph	C_6H_{11}				27
C_6H_6	6.57		6.37						27
$CDCl_3$	H 6.78	COPh	Ph	H 6.78	C_6H_{11}				27
C_6H_6	7.18			7.18					27
CCl_4	Cl	Me 8.54	H 6.90	Ph	Ph				12
CCl_4	Cl	H 5.69	H 6.80	Ph	Ph			5.0	12
CCl_4	H 5.84	OR	H 6.66	Ph	Ph		2.5		12
CCl_4	Me	CN	H 6.35	Ph	Ph				12
CCl_4	Me	SPh	H 6.47	Ph	Ph				12
CCl_4	H 6.18	SPh	H 6.8	Ph	Ph		2.5		12
CCl_4	H 7.66	Me 8.92	H 7.25	Ph	Ph		2.7		12
CCl_4	Me 8.97	H 7.66	H 6.90	Ph	Ph			6.0	12
	H	H	H	Me	Cl (syn)	2.3	5.7	6.7	24
	H	H	H	Me	Cl (anti)	2.4	5.4	6.7	24

C. 2-(α-NAPHTHYL)AZIRIDINE-^{15}N (11a)

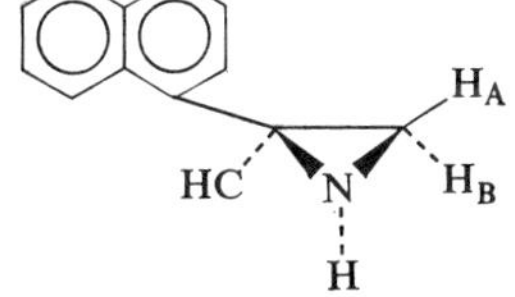

Solvent	τA	τB	τC	J_{AB}	J_{AC}	J_{BC}	J_{NA}	J_{NB}	J_{NC}
$CDCl_3$	8.22	7.74	6.54	0.7	3.6	5.9	−3.6	−1.4	−1.7
C_6D_6	8.65	8.32	7.01	1.1	3.4	6.0	−4.6	−0.8	−1.1
CD_3CN	8.46	7.76	6.53	1.2	3.5	6.1	−4.8	−0.4	< -0.6
DMSO	8.52	7.78	6.53	1.4	3.3	6.1	−4.9	< -0.3	< -0.7

[a] ^{13}C Coupling constants: $J_{^{13}C-H}$ (syn) 161, (anti) 171 Hz.

[b] Plus a trace of acid.

[c] $C_{10}H_7 = \beta$-naphthyl.

of the temperature-dependent proton spectra of 1,2,2-trimethylaziridine (**10**) as neat liquid, and as solutions in acetone, benzene, and chloroform. This method for the evaluation of kinetic parameters is far superior to the coalescence temperature methods and the reader is referred to this paper for an excellent discussion of the problem. Previous estimates for the nitrogen inversion process were shown to be too low and a solvent effect on the Arrhenius parameters, but not on the rate, was noted.

In spite of the amount of work done on this group of compounds, little attention has been paid to the effects of concentration and solvent on their spectra. Brois (11) was the first to comment on the very strong concentration and solvent effects in spectra of the *N*-alkylaziridines. $CDCl_3$–benzene shifts have been reported for a series of 2-acylaziridines (**9**) (27) and it was suggested that they may be of use in structural assignments. The only real theoretical paper (28) involves the use of spectra of aziridine, its 1-propionitrile (**11**), and its 1-propionate (**12**) to study the anisotropy of the nitrogen atom and the effect of hydrogen bonding to the nitrogen atom. As neat liquid, or in concentrated solutions, the aziridines were assumed to be stacked (**13**), and magnetic interaction between adjacent molecules, shown up in dilution studies, was termed the intermolecular anisotropy effect. This effect has been investigated for many three-membered rings (29, 30) and is thought to be caused by ring currents, in a manner similar to the more popular ideas of benzene anisotropy. Assignment of syn and anti proton signals from the *N*-substituted compounds **11** and **12** was made with the aid of these intermolecular anisotropy effects; high-field peaks were assigned to syn and low field to anti protons. It was also found that downfield shifts of the aziridine protons occur on dilution with $CDCl_3$ or CD_3OD. These were interpreted as due to the partly ionic character of hydrogen bonds between the solvent molecules and the nitrogen lone pair.

^{11}B and ^{1}H NMR spectra were used to determine the structure of aziridine bornane (**14**) and certain of its dimers and trimers (31). Data for aziridines are collected in Table 3.2.

B. The Diazirine System

Simple diazirines are rare and little is known of their NMR spectra. Proton and ^{19}F spectra were used to assign structures to certain 3-alkyl-3-fluorodiazirines (**15a**) (31a). Also, the ring system is found in a number of complex structures and, in the case of the spiro compounds **15** (32), the powerful anisotropy of the diazirine ring has been used to differentiate between *axial* and equatorial protons on the six-membered ring.

TABLE 3.3 SPECTRAL DATA FOR DIAZIRIDINES (33)

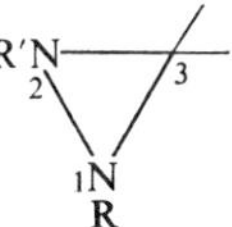

Substituents	Solvent	τR	τR′	τ3	Other data
1-*t*-Bu	CCl_4			7.41, 7.69	J_{gem} 6.0
	$(Ph)_2O$	9.05	8.55	7.64, 7.82	
1,2-Di-Me	Neat			7.81	$J_{^{13}CH}$ 173.1
	CCl_4	7.71	7.71	7.81	
1,2-Di-Et	CCl_4	7.61, 8.01	7.61, 8.01	7.83	
1,2-Di-*i*-Pr	CCl_4			7.83	
1-*i*-Pr; 3,3-di-Me	$CHCl_3$		8.0	8.61, 8.63	
1-Me, 2-CONHPh; 3,3-di-Me	$CDCl_3$	7.40		8.54, 8.65	
	$(Ph)_2O$	7.64		8.77, 8.81	
1-Me, 2-CH(OH)CCl_3; 3,3-di-Me	$CDCl_3$	7.41		8.60	
	$(Ph)_2O$	7.63		8.63, 8.70	

Mannschreck *et al.* (33) have reported the spectra of a number of diaziridines (**16**) and found that, with respect to an *N*-substituent, the diaziridine ring was asymmetric. Only if inversion of the ring nitrogen atoms was very slow could the rapidly rotating R–CH_2 substituents exhibit the nonequivalence observed. Similarly the methyl groups of isopropyl substituents showed nonequivalence and in the case of mono-*N*-substituted derivatives **17** the ring methylene protons were also nonequivalent. Thermodynamic parameters were also calculated. Data for diaziridines are collected in Table 3.3.

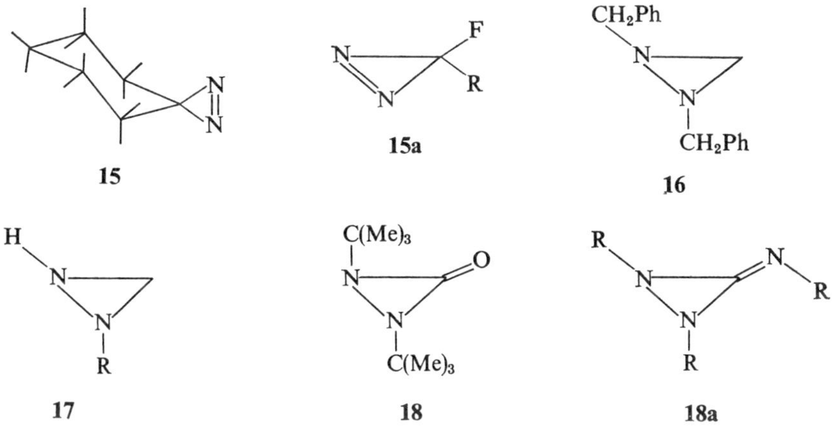

TABLE 3.4 SPECTRAL DATA FOR AZETINES, AZETIDINES, AND SIMILAR COMPOUNDS

1-AZETINES (35). SOLVENT—CCl_4

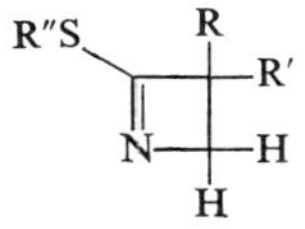

R	R′	R″	τ4	J_{gem}	J_{vic} Cis	J_{vic} Trans
Ph	Et	Et	6.15, 6.12	9.9		
C_6H_{12}	H	Et	6.54, 6.29	10.0	2.0	4.0

AZETIDINES (36). SOLVENT—$CDCl_3$

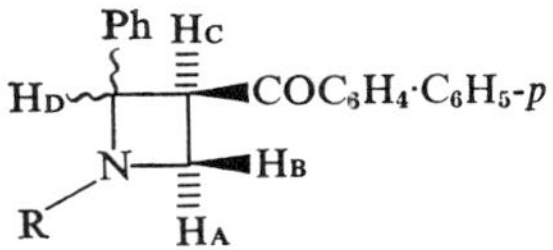

R	H_C–H_D configuration	τA	τB	τC	τD	J_{AB}	J_{AC}	J_{BC}	J_{DC}
$(Et)_3C$	Cis	6.40	6.04	5.61	4.97	6.8	7.8	3.5	9.5
	Trans	6.15	6.42		4.95	6.0			5.2
t-Bu	Cis	6.65	6.05	5.73	5.05	6.9	8.0	3.4	9.5
	Trans	6.45	6.45		5.30				6.5
i-Pr	Cis	6.95	5.85	5.69	5.40	7	7	3.0	9.0
	Trans	6.81	6.23		5.60	6.8			7.0
C_6H_{11}	Cis	6.94	5.81	5.61	5.36	6.8	7.4	3.0	9.0
	Trans	6.80	6.23		5.59	6.8			7.2

Et	Cis	6.99	5.84	5.61	5.42	6.5	7.5	2.8	8.5
	Trans	6.79	6.17		5.64	6.5			7.3
Me	Cis	6.79	5.85	5.61	5.47	6.6	7.4	2.8	7.5
	Trans	6.79	6.16		5.71	6.8			7.3

AZETIDIN-2-ONES. SOLVENT—$CDCl_3$

Substituents	$\tau3$	$\tau3'$	$\tau4$	$\tau4'$	$\tau1$	$J_{33'}$	$J_{44'}$	J_{34} Cis	J_{34} Trans	J_{13}	$J_{13'}$	Ref.
None	6.92	6.92	6.58	6.58	3.0							37
3-Ph	(2.73)	5.61	6.73	6.34	3.17		−5.6	5.9	2.6		1.1	37
4-Ph	7.19	6.37	(2.74)	5.30	3.20	−15.0		5.4	2.7	0.9	2.4	37
1-Me, 4-Ph	7.31	6.67	(2.74)	5.55	(7.3)	−14.2		5.0	2.4	0.9	0.46	37
1,3-Di-Me, 4-Ph	(9.24)	6.52	(2.77)	5.38	(7.21)			5.5			0.40	37
1,3′-Di-Me, 4-Ph	7.04	(8.65)	(2.77)	6.00	(7.29)				2.2	0.7		37
1-CH_2Ph	7.13	7.13	6.95	6.95	(5.72, 2.73)							37
1-CH_2Ph, 3-Me	(8.73)	7.00	7.32	6.80	(5.60, 2.73)		−5.5	5.1	2.8			37
1-CH_2Ph, 4-Me	7.58	7.06	(8.92)	6.51	(5.96, 5.54) (2.75)	−14.3		4.9	2.2	0.6	0.35	37
1-Ph	6.92	6.92	6.43	6.43	(2.73)							37
1,4′-Di-Ph	6.4	7.15	5.04					5.6	2.3			39
1,4′-Di-Ph, 3-Me		6.90	5.43						2.4			39
1,4′-Di-Ph, 3′-Me	6.34		4.82					5.8				39
1,3,4′-Tri-Ph		5.79	5.09						2.5			39

The spectrum of di-*t*-butyldiaziridinone (**18**) is a sharp singlet between −40 and +150° (34). This suggests that, as in other three-membered azacyclic systems, interchange between the two optical antipodes takes place by "slow" inversion around the nitrogen atoms. Recently data have become available for a number of diaziridinimines (**18a**) (34a).

II. FOUR-MEMBERED RINGS

A. The Azete System

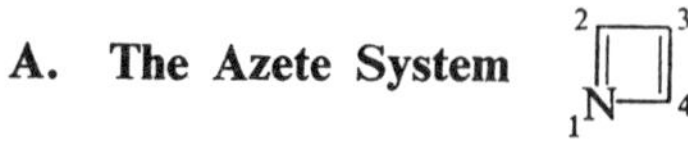

To my knowledge, no spectra have been reported for simple azetes, but some information is available for certain of the reduced systems. The only azetines to be dealt with in any detail are the 1-isomers **19** (35) in which the geminal coupling between the protons on C-4 was compared with similar couplings for the corresponding azetidin-2-ones (**20**) and azetidine-2-thiones (**21**) and for a number of azetidines (**22**). The observed ranges for the last three classes 5.2–5.7, 6.9–7.2, and 7.4–8.0 Hz, respectively, correlate reasonably well with the electron-withdrawing power of the atom attached to C-2. The larger values 9.9–10.0 Hz, for the azetines reflect the presence of the adjacent double bond.

The spectra of azetidines were first reported by Bottini and Roberts (13) who noted the temperature dependence of *N*-substituted derivatives. Much more recently, the inversion of the nitrogen atom in a series of compounds of general structure **23** has been investigated (22). The stereochemistry of 1-alkyl-2-phenyl-3-aroylazetidines (**24**) has been studied in some detail (36). The cis vicinal couplings in these rings are generally larger than the trans vicinal couplings (see Table 3.4), the actual values being influenced by the electronegativity of the substituents and quite strongly by differences in dihedral angle caused by variations in the ring puckering. The amount of pucker appears to decrease with increase in the size of the *N*-alkyl substituent. Chemical shifts of the azetidine ring protons at C-2 and C-4 in the cis compounds are sensitive to the steric requirement of the *N*-alkyl group. Any shielding or deshielding by the *N*-alkyl substituent is thought to involve mainly intramolecular van der Waals dispersion effects. In the corresponding trans compounds, the anisotropy of the aryl group trans to the *N*-alkyl group completely overrides any effect of the latter.

Barrow and Spotswood (37) in a study of β-lactams have reported the spectra of a group of azetidin-2-ones (**25**). In these compounds, J_{cis} was greater than J_{trans}. The vicinal coupling constants had the same sign, opposite to that of the geminal couplings. Comparatively large cross-ring couplings were observed between the *N*-substituents and the protons on C-3. In the

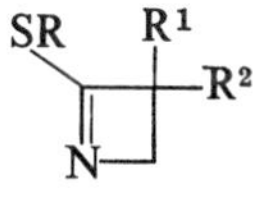

SR R1 -R2 N— 19

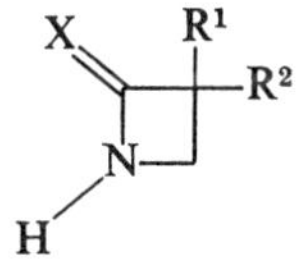

X R1 -R2 N— H 20, X = O 21, X = S

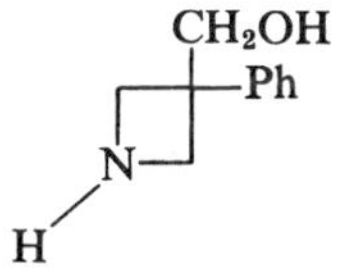

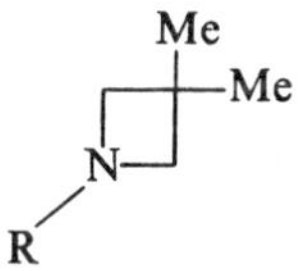

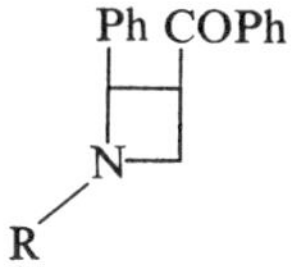

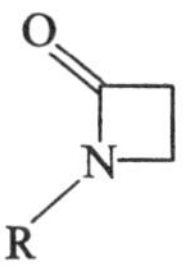

CH₂OH -Ph N— H 22

Me -Me N— R 23

Ph COPh N— R 24

O N— R 25

O R -H N— -Ph Ph H 26, R = H 27

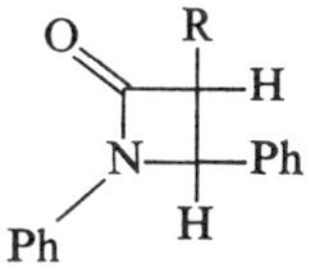

N-benzyl compounds, the benzylic protons showed marked nonequivalence and couplings between these protons and the 3-protons were stereospecific, suggesting a highly preferred conformation of the benzyl groups. The ABX system in 1,4-diphenylazetidin-2-one (**26**) has been thoroughly investigated (38) and spin-tickling experiments showed the geminal and vicinal couplings to have opposite signs. Nuclear magnetic resonance has also been used to distinguish between cis and trans isomers in compounds of the type **27** (39). Data for reduced azetes are collected in Table 3.4.

B. The Diazete System

The spectrum of 1,2-diphenyl-1,2-diazetidin-3-one (**28**) has been reported (40). In acetone solution at $-50°$, the 4-methylene group shows a typical AB-type structure (τ 4.75, 5.43; $J_{AB} = 13$ Hz) and this nonequivalence has been attributed to relatively slow inversion around the "nonamide" nitrogen atom, N-1. Similar results were obtained for a series of highly substituted 1,2-diazetidin-3-ones (40a).

A complete analysis of the ABX pattern from the ring protons of compound **28a** in $CDCl_3$ gave the following results: τA 5.1, τB 5.8, τX 4.2 ppm; J_{AB} 9.8, J_{AX} 5.7, J_{BX} 3.7 Hz (40b).

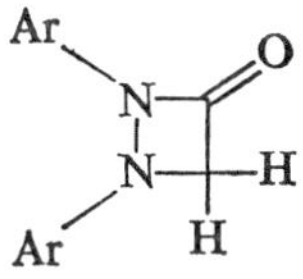

Ar N— O N— H Ar H

28

OAr H_X N—CO₂Me H_A N—CO₂Me H_B

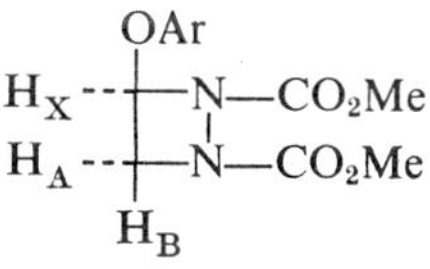

28a

III. FIVE-MEMBERED RINGS

A. The Pyrrole System

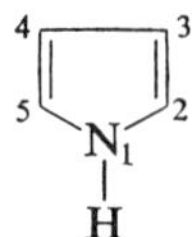

1. *Aromatic Pyrroles*

The spectrum of pyrrole was first reported by Reeves in 1957 (41) and this was followed by a series of measurements (42–45) of the spectra of pyrrole and its alkyl derivatives. Analysis of the proton spectrum of pyrrole is complicated by quadrupolar broadening effects associated with the nitrogen nucleus and by the fact that all couplings have similar values. Replacement of H-1 by deuterium and simultaneous decoupling of the remaining protons from ^{14}N and ^{2}D produced a sharp AA′BB′-type spectrum (46) which, although it was amenable to analysis, still allowed a number of possible solutions (see also Ref. 47). However, analysis of the fine structure in the ^{13}C satellite peaks of the proton resonances provided unambiguous values for all coupling constants. These were checked by measuring the spectra of *C*-deuterated pyrroles. The use of similar methods in a later publication led to much the same results (48). Proton data obtained for ^{15}N-labeled pyrrole (48a) were similar to those discussed above for the ^{14}N isotopomer but the differences in coupling constants were large enough to prompt a very careful analysis of the spectrum of the labeled compound (48b). The parameters from this work were very similar to those reported by Katekar and Moritz for the unlabeled molecule. Data obtained are summarized in Table 3.5.

TABLE 3.5 SPECTRAL DATA FOR PYRROLE AND VARIOUS ISOTOPIC DERIVATIVES (46, 48a). 60 MHz—NEAT LIQUID

Isotopic derivative	Chemical shifts (Hz from TMS)[a]								
	α-H	β-H	NH	J_{12}	J_{13}	J_{23}	J_{24}	J_{25}	J_{34}
Pyrrole	381.6 (3.64)	369.3 (3.85)	438.8 (2.69)	2.67	2.50	2.73	1.46		
^{13}C-2				2.53		2.75	1.39	1.90	
^{13}C-3					2.40	2.70	1.43		3.35
^{2}H-1	382.7 (3.62)	368.8 (3.86)				2.75	1.42		
^{2}H-1, ^{13}C-2						2.71	1.42	1.86	
^{2}H-1, ^{13}C-3						2.64	1.45		3.34
Random ^{2}H				2.60	2.45	2.70	1.45	1.85	3.31
^{15}N				2.58	2.46	2.70	1.44	1.87	3.35

[a] τ in parentheses.

The aromatic nature of pyrrole has been studied using the "ring-current" criterion for aromaticity but agreement was not reached between the two groups of workers (49, 50). However, the predominant aromatic character of pyrrole can be inferred from the similarity between the observed ^{13}C–H couplings and those of benzene (51). An effort has been made to correlate proton chemical shifts of pyrrole and a number of annulated benzene derivatives with π-electron densities calculated by a VESCF method (52). The estimated excess charges do not correlate well with chemical shifts and it was noted that benzene may not be a satisfactory reference compound for these systems. Also it was suggested that in pyrrole and other azoles, the role of the σ-electrons in determining electron densities needed closer investigation.

Solvent effects and self-association are extremely important for pyrrole and its derivatives. A number of spectra of pyrrole in associating solvents have been reported (41, 44, 45), and it is interesting to note that benzene shifts for the α-protons are greater than those observed for the β-protons, the reverse of the situation with pyridine. Happe (53) suggests that the self-association of pyrrole molecules is of the "π-donor type" **29**, whereas pyrrole–pyridine association is of the "n-donor type" **30**. Autocomplexing of pyrrole in CCl_4, Freon 112, and cyclohexane solutions, and complexing between pyrrole and acetone in cyclohexane have been studied, after sharp spectra were obtained by decoupling the ^{14}N nucleus (54). Thermodynamic data were calculated for the complex formation. The chemical shifts of the N–H proton were compared with the wave numbers for the NH vibrations for pyrrole hydrogen bonded to various electron donors (55). Linear correlations were found between τ(NH), $\tilde{\nu}$(NH), and the pK values of the electron donors. A comparison of the chemical shifts of the ring protons of pyrrole and other five-membered heterocycles in nine different solvents has been made (56). Hydrogen bonding between pyrrole and dimethyl sulfoxide has been studied by NMR (57). The concentration and temperature dependence of the chemical shift of the N–H resonance of pyrrole shows that the pyrrole–dimethyl sulfoxide interaction is probably of the "n-donor type." The equilibrium between the complex and the pyrrole monomer plus solvent was studied and kinetic data obtained for the system. While the chemical shift of the NH proton in the monomer was assumed to be temperature independent, the chemical shift of this group in the complex shows a regular trend toward lower field with increasing temperature. Interaction of pyrrole with benzene (58), and with a number of substituted benzenes (58a) has also been studied. The ready association of pyrrole with other molecules led to its use as a solvent for ketones where it produces specific shifts similar to but much larger than those induced by benzene (59). The spectrum of pyrrole trapped in a nickel cyanide–ammonia clathrate has been investigated in the temperature range 20–300° (60). Observed line widths were interpreted in relation to molecular motions.

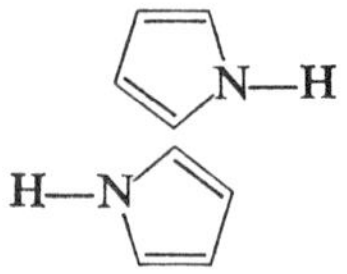

29

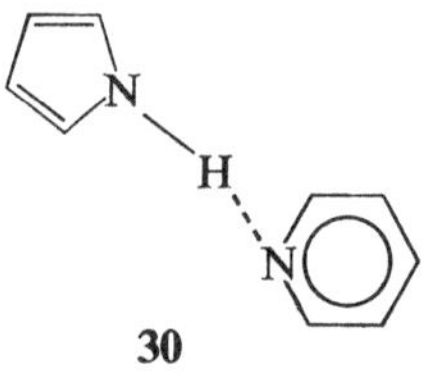

30

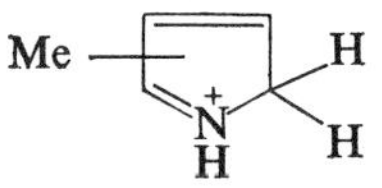

31

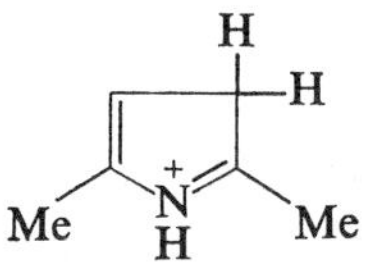

32

Protonation of alkylpyrroles has been studied using hydrochloride salts (61) or sulfates (62). Methylpyrroles are usually protonated in the α-position to give ions of type **31** but competitive β-protonation was also observed for 2,5-dimethylpyrrole, giving the ion **32**. Nuclear magnetic resonance has also been used in a study of the protonation of *N*-phenylpyrroles (63). Competitive protonation was observed and both α- and β-protonated isomers are evident in the published spectra. Data for a number of protonated species are collected in Table 3.6.

Many substituted pyrroles are not readily available and so the spectra of relatively few of these compounds have been reported, in many cases with only chemical shift data given. Where coupling constants are reported, the values must be viewed with caution because the first-order methods often used to obtain them are not always applicable to these systems. Chemical shifts reported in earlier work cannot be related to any modern scale and, where acetone is used as internal standard, are open to serious interaction errors. Thus, in Table 3.7, the only chemical shift data included come from more modern papers where present-day standard methods of spectral calibration were used.

Chemical shifts of various alkyl derivatives have been measured many times (e.g., see Ref. 64). These compounds normally exhibit two complex multiplets in dimethyl sulfoxide in the aromatic region of the spectrum, one at τ 3.3–3.8 attributable to H-2,5 and the other at τ 4.0–4.5 from H-3,4 (65). The largest group of substituted pyrroles for which coupling constants

TABLE 3.6 SPECTRAL DATA FOR PROTONATED PYRROLES

A — B

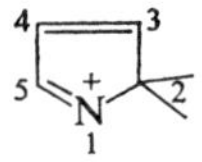

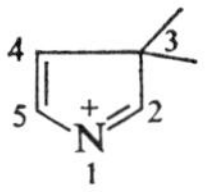

Pyrrolium ion	Chemical shifts[a] R2	R3	R4	R5	Ref.
	Hz from internal $N(Me)_4^+$ in 12 *M* H_2SO_4				62
2,5-Di-Me A	−111 (+94)	−287	−223	(+26)	
2,5-Di-Me B	(+26)	−50	−190	(+54)	
	Hz from internal dioxan in conc. HCl				61
2,3,5-Tri-Me A	−78 (+134)	(+65)	−180	(+86)	
3,4,5-Tri-Me A	−61	(+70)	(+108)	(+93)	
2,3,4,5-Tetra-Me A	−57 (+138)	(+71)	(+107)	(+97)	

[a] Values in parentheses are for methyl substituents.

are given was associated with a study of pyrrole thiocyanation by Gronowitz and his co-workers (66, 67). Data for substituted pyrroles are collected in Table 3.7.

Long-range couplings between substitutent methyl protons in both α- and β-positions with both ortho- and meta-type ring protons have been observed (66). For pyrrole-2-aldehyde, a small coupling between the aldehyde proton and H-5 led to the suggestion of the extended *W* conformation **33** for the aldehyde in dioxan solutions (68). As part of a study on the conformations of 1-*t*-butyl-2-formylpyrrole (**34**) (69), this long-range coupling was measured and compared with that for *N*-alkyl derivatives (Me, Et, *i*-Pr). Since profound changes were not observed the authors concluded that the conformations of these compounds were essentially the same, although infrared and ultraviolet data showed that there was a variation in the conjugation between the formyl group and the ring. The barrier to the rotation of the aldehyde group in 2-formyl-1-methylpyrrole has also been measured by NMR methods (69a). Partial "virtual coupling" has been observed between H-4 and the C*H*O proton in 2-formyl-1-methylpyrrole (70). I would like to see use of the somewhat magical term "virtual coupling" discontinued since such effects, caused by second-order coupling effects, are commonplace in the theoretical treatment of many spin systems. In a study

TABLE 3.7 SPECTRAL DATA FOR *C*-SUBSTITUTED PYRROLES

A. PARENT AND 2-SUBSTITUTED PYRROLES

Substituents	Solvent	τ1	τ2	τ3	τ4	τ5	J_{13}	J_{14}	J_{15}	J_{34}	J_{35}	J_{45}	Ref.
None	CCl_4	2.3	3.38	3.95	3.95	3.38							64
	Et_2O		3.27	3.86	3.86	3.27							249
	DMSO	−0.8	3.17			3.17							71
2-Me	CCl_4	2.8	7.84	4.28	4.11	3.64							64
	Dioxan							2.8[a]	2.4[a]	3.40	1.50	2.45	66
2-CN	Me_2CO						2.25	2.40	2.70	3.74	1.45	2.63	43
2-CHO	Me_2CO						2.47	2.26	2.45	3.73	1.36	2.68	43
							2.53	2.49	2.95	3.84	1.47	2.37	261
	CCl_4		0.48	3.02	3.68	2.77				3.92	1.40	2.44	80
	$CDCl_3$		0.40	2.96	3.58	2.78							80
	$MeNO_2$		0.48	2.94	3.62	2.73							80
	$C_2H_4Cl_2$		0.48	2.97	3.65	2.78							80
	DMSO	−2.2	0.51	3.01		2.83							71
2-COMe	CCl_4		7.58	3.17	3.85	2.94				3.75	1.50	2.45	80
	$CDCl_3$		7.52	3.00	3.65	2.84							80
				3.11	3.75	2.91	2.50	2.75	3.0	3.75	1.35	2.40	250
	$MeNO_2$		7.61	3.01	3.70	2.87							80
	Dioxan						2.50	2.45	3.00				66
$2\text{-}CO_2H$	DMSO	−1.8		3.16		3.16							71
	Me_2CO						2.56	2.33	2.87	3.64	1.52	2.63	261
$2\text{-}CO_2Me$	CCl_4		6.17	3.16	3.82	3.09				3.84	1.46	2.56	80
	$CDCl_3$		6.13	3.02	3.70	3.02							80
	$MeNO_2$		6.16	3.08	3.72	2.90							80

	DMSO	−2.0		2.90		3.15							71
2-$CONH_2$	DMSO	−1.4		3.22		3.22							71
2-$CONHNH_2$	DMSO	−1.4		3.14		3.14							71
2-NO_2	$CDCl_3$			2.89	3.71	2.95	2.5	2.75	3.1	4.2	1.7	2.8	250
2-SMe	Dioxan		7.75	3.77	3.90	3.28	2.6	2.85	2.85	3.35	1.50	2.85	67
2-SCN	Dioxan			3.47	3.85	3.10	2.4	2.45	2.85	3.60	1.50	2.90	67
	DMSO			3.43	3.81	2.94	2.4	2.45	2.80	3.50	1.55	2.85	67
	C_6H_{12}			3.47	3.85	3.10							67

B. 3-SUBSTITUTED PYRROLES

Substituents	Solvent	$\tau 2$	$\tau 3$	$\tau 4$	$\tau 5$	J_{12}	J_{14}	J_{15}	J_{24}	J_{25}	J_{45}	Ref.
3-Me	CCl_4	3.72	7.95	4.15	3.58							64
3-COMe	$CDCl_3$	2.59		3.32	3.23				1.60	1.90	2.98	251
		2.57		3.36[b]	3.22[b]	3.0	2.6	3.3	1.7	1.9	3.0	250
3-CO_2Me	$CDCl_3$	2.48		3.22	3.22				1.60	2.10	2.89	251
	Dioxan								1.40	1.95	2.80	66
3-NO_2	$CDCl_3$[c]	2.34		3.25	3.25	3.5	2.65	2.65	1.9	1.9		250

C. POLYSUBSTITUTED PYRROLES

Substituents	Solvent	$\tau 1$	$\tau 2$	$\tau 3$	$\tau 4$	$\tau 5$	Coupling constants[d]	Ref.
2,3-Di-Me	Neat						2.6 (14), 2.6 (15), 2.6 (45)	43
	CCl_4	2.9	7.98	8.04	4.18	3.72		64
2-Me, 3-CO_2Et	Dioxan						3.1 (45)	66
2-CHO, 3-Me	DMSO	−1.8	0.31			2.85		71
2,4-Di-Me	CCl_4	3.0	7.93	4.43	8.00	3.92		64
2-Br, 4-COMe	$CDCl_3$			3.41		2.66	3.12 (12), 2.56 (14), 1.80 (24)	251
2-*i*-Pr, 4-COMe	$CDCl_3$			3.66		2.60	1.68 (24)	251

TABLE 3.7 *(continued)*

Substituents	Solvent	$\tau 1$	$\tau 2$	$\tau 3$	$\tau 4$	$\tau 5$	Coupling constants[d]	Ref.
2-Br, 3-CO_2Me	$CDCl_3$			3.45		2.67	2.90 (12), 2.58 (14), 1.72 (24)	251
2-*i*-Pr, 3-CO_2Me	$CDCl_3$			3.64		2.67	1.64 (24)	251
2,5-Di-Me	CCl_4	2.9	7.87	4.43	4.43	7.87		64
2-Me, 5-CHO	$CDCl_3$		7.65	3.12	3.97	0.62	4 (34)	67
2-Me, 5-SCN	Dioxan			4.13	3.56		0.8 (23), 0.35 (24), 3.55 (34)	67
2-SMe, 5-CHO	Dioxan		7.59	3.78	3.12	0.69		67
2,5-Di-SMe	Dioxan		7.77	3.85	3.85	7.77	2.60 (13), 2.60 (14)	67
3,4-Di-Me	CCl_4		3.73	8.05	8.05	3.73		64
	Me_2CO						2.3 (15)	43
3-Me, 4-CO_2Et	Dioxan						3.05 (15), 1.0 (23), 2.2 (2.5)	66
3-Me, 4-COOH	DMSO		3.42			2.66		71
2,3,4-Tri-Me	CCl_4		7.92	8.16	8.08	3.85		64
2,4-Di-Me, 3-CHO	DMSO			0.11		3.50		71
2,4-Di-CO_2Et, 3-Me	DMSO					2.51		71
2,3,5-Tri-Me	CCl_4		7.98	8.13	4.58	7.92		64
2,3,4,5-Tetra-Me	CCl_4	3.2	8.02	8.20	8.20	8.02		64

[a] For neat liquid.
[b] Values reverse of those in the reference.
[c] Plus a little acetone.
[d] The respective nuclei indicated in parentheses.

of the bipyrrole precursor of prodigiosin (71), the NMR spectra of a series of substituted pyrroles were examined but complete analyses are not given. Emphasis was placed on signals from the pyrrole N–H protons and from *C*-formyl substituents. Pyrroloethylenes (**35**), in $CDCl_3$, give a singlet from the ethylenic protons at τ 3.1 (72). Acetonepyrrole (**36**), formed by acid-catalyzed condensation of acetone and pyrrole, gives rise to a singlet from the dimethyl group (τ 8.47), a doublet (τ 4.0, J 2.8 Hz) from the β-protons, and a broad signal (τ 2.9) from the N–H protons ($CDCl_3$ solvent) (73). The spectrum of 2-vinylpyrrole (**37**) has been reported by Brugel *et al.* in their general study of the vinyl group (74).

Recently, a detailed study has been made of the spectra of 1-substituted derivatives of pyrrole to check the effect of 1-aryl substituents on the aromaticity of the pyrrole ring (75). No evidence was obtained to support any appreciable change of ring current with substitution. Linear correlations with Hammett σ-constants were obtained for the chemical shifts of α- and β-protons and *C*-methyl groups of 1-*p*-substituted arylpyrroles. These were

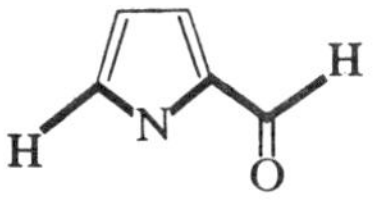

H N H O

33

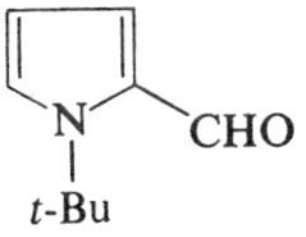

N CHO *t*-Bu

34

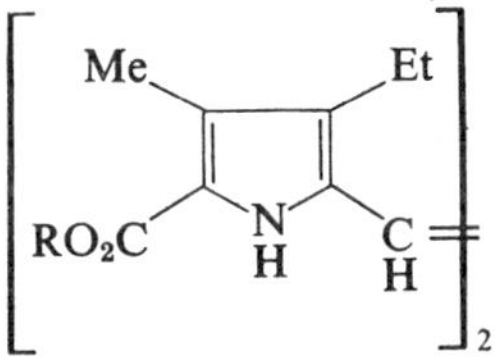

Me Et RO_2C N H C H]$_2$

35

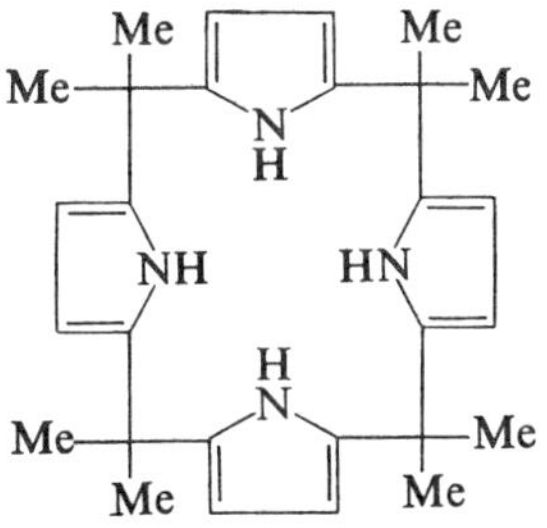

Me Me Me Me N H NH HN H N Me Me Me Me

36

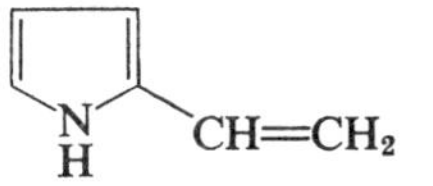

N H $CH{=}CH_2$

37

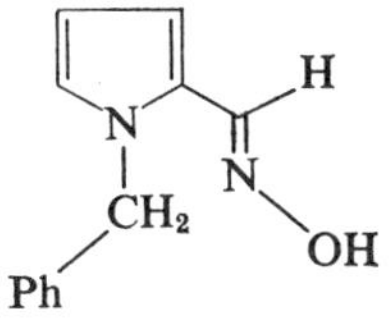

N H N CH_2 OH Ph

38

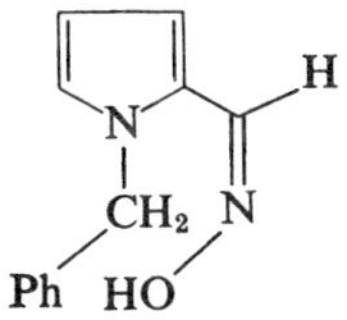

N H CH_2 N Ph HO

39

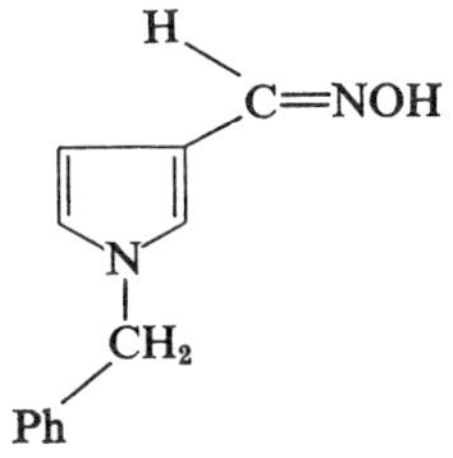

H C=NOH N CH_2 Ph

40

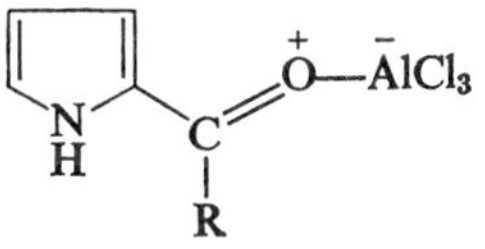

N H C $\overset{+}{O}$—$\overset{-}{A}lCl_3$ R

41

interpreted in terms of a conjugative interaction of the para substituent with the benzene ring but essentially an inductive interaction between the pyrrole nitrogen and the aryl ring. In a study of ring-substituted 1-benzylpyrroles it was shown that electron withdrawal in the pyrrole ring was transmitted more effectively to the benzyl methylene group from the 2-position than from the 3-position (76). The oxime of 1-benzyl-2-formylpyrrole showed syn and anti isomers (**38** and **39**) and the benzyl methylene protons were more strongly shielded by the anti form. However, with the oxime of 1-benzyl-3-formylpyrrole (**40**) only one isomer was obtained. Spectra of 1-methylpyrrole and a number of its 2-aryl derivatives have been reported (77, 78) and it was noted that, as in other heterocycles, the addition of halogen caused only minor variations in chemical shifts. Incomplete data have also been published for some *N*-anilinopyrroles (79).

A study has been made of the effect of complexing with Lewis acids on the spectrum of various 1-carbonylpyrroles (80). Formation of species such as **41** caused a general shift of the ring protons to lower field, but with much greater shifts of the signals from the 3- and 5-protons. This agrees with the preferential deactivation of these sites on complex formation, shown by substitution experiments and predicted by HMO calculations. Data for 1-substituted pyrroles are collected in Table 3.8.

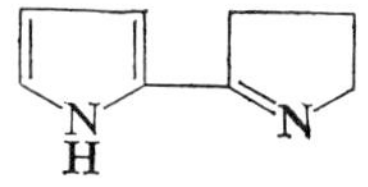

42 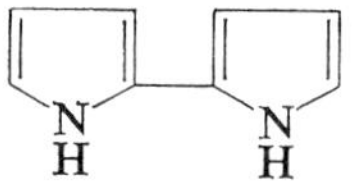 **43** 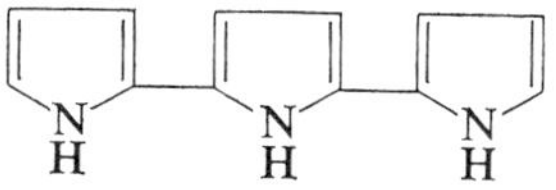 **44**

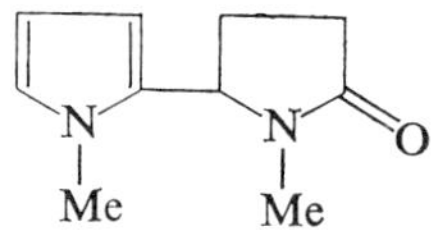

42a 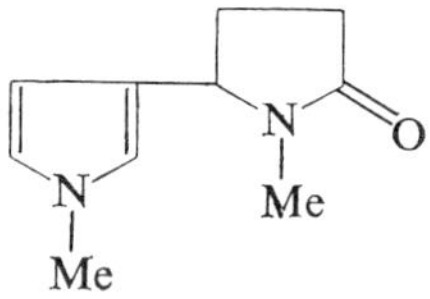 **43a** 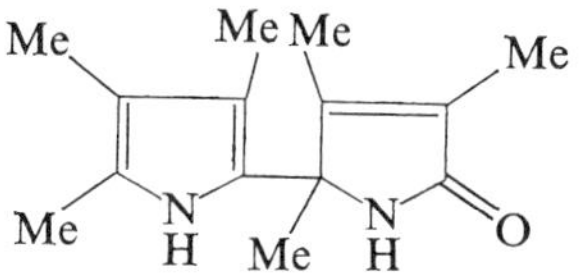 **45**

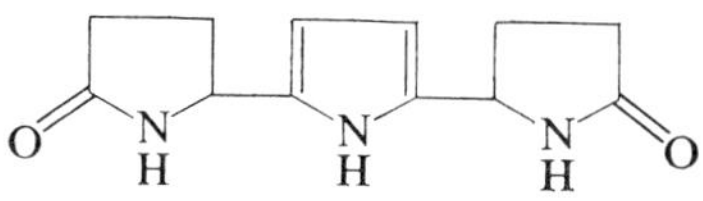

44a 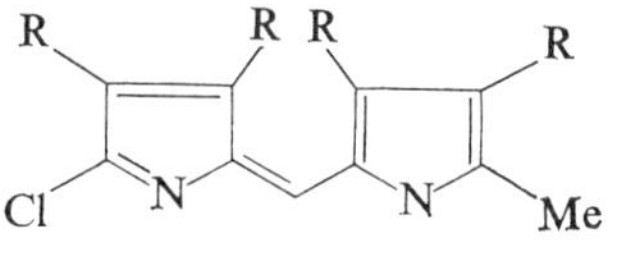 **46**

TABLE 3.8 SPECTRAL DATA FOR *N*-SUBSTITUTED PYRROLES

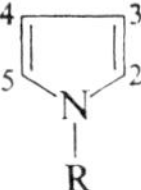

R	*C*-Substituents	τ2	τ3	τ4	τ5	J_{24}	J_{25}	J_{34}	J_{45}	τR[b]	Ref.
Me	None	3.63	4.08	4.08	3.63					6.40	64
	2-Me	7.84	4.33	4.23	3.70					6.52	64
	2,5-Di-Me	7.90	4.48	4.48	7.90					6.73	64
	2,3,5-Tri-Me	7.95	8.11	4.59	7.90					6.72	64
	2,3,4,5-Tetra-Me	7.98	8.18	8.18	7.98					6.73	64
Et	None	3.54	4.07	4.07	3.54						75
	2,5-Di-Me	7.84	4.49	4.49	7.84						75
CH_2Ph	None	3.50	3.99	3.99	3.50						75
	None	3.51	3.92	3.92	3.51						76
	2,5-Di-Me	7.91	4.37	4.37	7.91						75
	2-CHO		3.13	3.82	3.13	1.8		3.86	2.55	4.50	76
	2-CHNOH (DMSO)		3.50	3.77	2.90	1.8		3.70	2.65	4.47 (syn) 4.57 (anti)	76
	3-CHNOH (DMSO)	3.09		3.52	3.52	1.6			2.8	4.81	76
	2-CO_2Me		3.27	4.08	3.42	1.75		3.85	2.56	4.64	76
	3-CO_2Me	2.8		3.52	3.52	1.75	2.25		2.86	5.01	76
	3-NHCOMe ($CDCl_3$)	2.62		3.97	3.48	1.70			2.75	5.03	76
	2-NO_2		2.96	3.87	3.18	2.20		4.14	2.80	4.48	76
	3-NO_2	2.52		3.47	3.37	1.86	2.48		3.19	4.92	76
	3-Br	3.47		3.91	3.47	1.77	2.39		2.78	5.05	76
	2,5-Di-Br		3.75	3.75						4.76	76
	2,3,5-Tri-Br			3.61						4.72	76
	Tetra-Br									4.75	76
COMe	None	2.82	3.83	3.83	2.82						75
	2,5-Di-Me	7.63	4.32	4.32	7.63						75
	3,4-Di-Me	3.11	8.06	8.06	3.11						75
COPh	None	2.81	3.77	3.77	2.81						75
	2,5-Di-Me	7.96	4.28	4.28	7.96						75
Ph	None	3.05	3.81	3.81	3.05						75
	2,5-Di-Me	8.02	4.34	4.34	8.02						75
	3,4-Di-Me	3.33	7.99	7.99	3.33						75
p-MePh	None	3.11	3.84	3.84	3.11						75
	2,5-Di-Me	8.04	4.36	4.36	8.04						75
	3,4-Di-Me	3.37	8.01	8.01	3.37						75
4-Pyridyl	None	2.93	3.74	3.74	2.93						75
	2,5-Di-Me	8.00	4.33	4.33	8.00						75

[a] Solvent—CCl_4, unless otherwise indicated.
[b] For benzyl substituents only the $-CH_2-$shift is given.

TABLE 3.9 SPECTRAL DATA FOR DIMERIC AND TRIMERIC PYRROLES

A. 2,2′-BIPYRROLES[a]

Substituents	Solvent	τ3(3′)	τ4(4′)	τ5(5′)	τNH	Ref.
None	SO_2	3.78	3.78	3.34		81
3-Me	$CDCl_3$	7.8, 3.8–4.0	3.8–4.0	3.4		83
4-Me	$CDCl_3$	3.90	4.0 (7.9)	3.4, 3.6		83
5-Me	$CDCl_3$	4.1–4.5	4.1–4.5	3.7 (7.8)		83
4,4′-Di-Me	$CDCl_3$	3.94	(7.84)	3.5	1.03	85
5,5′-Di-Me	$CDCl_3$	4.05	4.15	(7.7)		83
3,3′,4,4′-Tetra-Me	$CDCl_3$	(8.19)	(8.06)	3.5	2.96	85
3,3′-Di-Et,4,4′-di-Me	$CDCl_3$	(7.61, 8.99)	(7.89)	3.6	2.93	85
5-COOEt	$CDCl_3$		3.2–3.9			83
1-Me, 5-COOEt	$CDCl_3$	3.4–4.0	3.4–4.0	3.1	(6.2)	83
3-Me, 4-COOEt	$CDCl_3$	3.8 (7.6)	3.8	3.3–2.9		83
3,3′,4,4′-Tetra-Me, 5,5′-COOEt	$CDCl_3$	(7.94)	(7.66)	(5.68, 8.63)	0.95	85

B. 2,2′-(1′-PYRROLINYL)PYRROLES. SOLVENT—$CDCl_3$

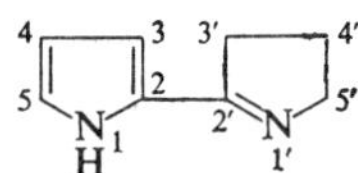

Substituents	τ3	τ3′	τ4	τ4′	τ5	τ5′	τ1	Ref.
None	3.58	6.50	3.93	8.83	3.08	7.78	−0.32	81
5-Me	3.5	7.1	4.1	8.1	(7.8)	6.0		83
5′-Me	3.8	7.2	4.1	7.8–8.5	3.4	5.9 (8.7)		83
3-Me	(7.9)	7.0	3.9	7.6	3.1	6.0		83
4-Me	3.6	7.1	(7.8)	7.6	3.2	5.9		83
5,5′-Di-Me	3.8	7.2	4.2	8.0–8.5	(7.8)	5.9 (8.8)		83
5′-COOEt	3.4	6.9	3.8	7.7	3.1	5.2		83
3-Me, 4-COOEt	(7.7)	6.9		7.8	2.5	6.0		83
1-Me, 5′-COOEt	3.3	6.9	3.8	7.7	3.1	5.0	(5.9)	83

C. 2,2′-PYRROLIDINYLPYRROLE. SOLVENT—$CDCl_3$ (81)

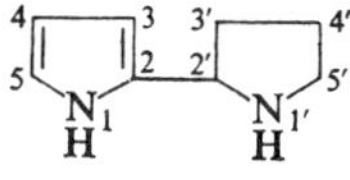

Substituents	τ1	τ3	τ4	τ5	τ3′	τ4′	τ5′
None	−0.50	3.78	3.78	3.20	5.75	7.90	6.88

TABLE 3.9 (*continued*)

D. 2,2′-(PYRROLONYL)PYRROLE. SOLVENT—$CDCl_3$ (88)

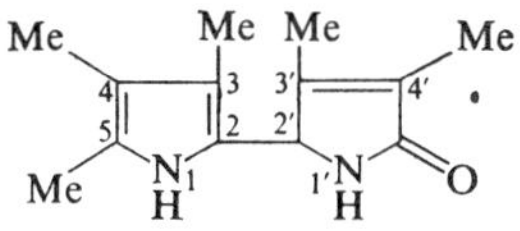

τ2′	τ3	τ3′	τ4	τ4′	τ5
(8.13)	(8.25)	(8.30)	(8.21)	(8.05)	(7.91)

E. 2,2′-DIPYRROLONE. SOLVENT—$CDCl_3$ (88)

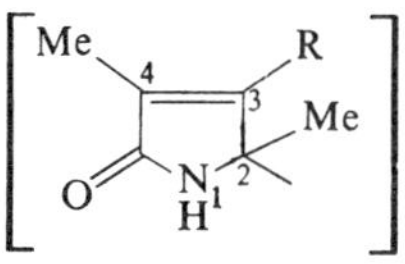

R	τ1	τ2	τ3	τ4
Me	0.96	(8.19)	(8.35)	(8.14)
Et	1.04	(8.22)	(7.64, 8.80)	(8.07)

F. 2,2′-PYRROLIDONYLPYRROLE. SOLVENT—DMSO (87)

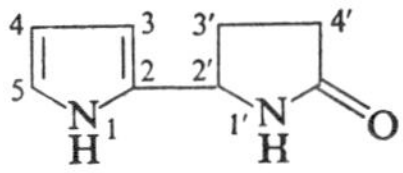

Substituents	τ1	τ3	τ4	τ5	τ1′	τ3′ + τ4′	τ5′
None	−0.8	4.06	4.06	3.33	2.15	7.5–8.4	5.35
5-(2′-Pyrrolidonyl)	−0.6	4.17	4.17		2.17	7.4–8.5	5.37

[a] Further data for highly substituted derivatives are in Ref. 85.

The spectra of 2,2′-(1′-pyrrolinyl)pyrroles (**42**), 2,2′-bipyrroles (**43**), and 2,2′:5′,2″-terpyrroles (**44**) have been reported by Rapoport and his co-workers (81–83), and those of a number of highly substituted 2,2′-bipyrroles by Grigg and Johnson (84, 85). Smith and Jensen (86) give the spectra of the isomeric 5,2′- and 5,3′-(1′-methyl-pyrrolyl)-1-methyl-pyrrolid-2-ones (**42a** and **43a**). One of the oxidation products of pyrrole has been shown by NMR methods to be a hydrate of 2,5-bis(pyrrolidin-2-on-5-yl)pyrrole (**44a**), and data for this and similar compounds were reported (87). The autoxidation products of alkylpyrroles were shown to be mainly pyrrolylpyrrolones (**45**) (88). Also, some work has been done on a number of highly substituted dipyrrylmethenes (**46**) (89). Data on these di- and trimeric compounds are collected in Table 3.9.

Measurement of ^{13}C parameters for pyrrole has kept pace with instrumental developments in the field. The original one-bond ^{13}C–H coupling constants were obtained from ^{13}C satellites of the proton signals (90). Later, these values were confirmed by direct measurement from low-resolution, dispersion-mode ^{13}C spectra (51) and, only recently, high-resolution ^{13}C spectra have furnished long-range ^{13}C–H coupling data (91). In this latter case, broadening of the signal from the α-protons occurs because of quadrupole relaxation effects of the adjacent ^{14}N nucleus and a nonzero ^{13}C–^{14}N coupling. Estimates of some of the long-range ^{13}C–H couplings were obtained from ^{13}C satellites of proton spectra decoupled from ^{14}N and, if deuterated at N-1, also decoupled from ^{2}H (46). A very extensive study of the ^{13}C chemical shifts for the five-membered nitrogen heterocycles, their cations and anions, has been reported (92). The coupling constants $^{1}J_{^{14}N,H}$ for pyrrole and its 2,3,4,5-tetradeutero derivative, have been reported ($J = 69.5$ and 68.6 Hz, respectively) (93). Proton spectra of ^{15}N-pyrrole have been very carefully analyzed and values for coupling constants between the ring protons and the ^{15}N nucleus are available (48a, b). ^{13}C and ^{15}N data for pyrrole are given in Table 3.10. Also, a short note has appeared containing information on the spectrum of 2,6-di-*t*-butylpyrrole-^{15}N (93a).

TABLE 3.10 ^{13}C AND ^{15}N DATA FOR PYRROLE

A. ^{13}C (91)

	Chemical shift[a]	J_{C-H-2}	J_{C-H-3}	J_{C-H-4}	J_{C-H-5}
C-2	74.1	182		7.6 av	
C-3	85.1	7.8	170	4.6	7.8

B. ^{15}N (48b)

$^{1}J_{NH}$ −96.53; $^{2}J_{NH}$ −4.52; $^{3}J_{NH}$ −5.39 Hz

[a] Parts per million upfield from CS_2.

2. *Reduced Pyrroles*

Reduced pyrroles are not commonly encountered, and hence few have been extensively studied. By far the most important contributions to this field come from Robertson and his co-workers in their investigation of the spectra of proline (**47**) derivatives. The only spectra reported for Δ^1-pyrrolines (**48**) are those in various reduced di- and terpyrroles synthesized by Rapoport and his co-workers (81–83) (see Table 3.9). Δ^2-Pyrrolines (**49**) are enamines and hence extremely unstable unless stabilized by the presence of a keto or thione grouping in the molecule. Spectra of a number of *N*-alkyl-4-thio derivatives have been measured (94). Bordner and Rapoport (95) investigated the spectra of Δ^3- and Δ^4-pyrrolin-2-ones, **50** and **51**, respectively. The position of the double bond was elegantly established by reduction with diimide(d_2) and examination of the spectra of the resulting pyrrolid-2-ones (**52** and **53**). This diimide reduction is not susceptible to the deuterium scrambling which occurs during catalytic deuteration. In the spectrum of compound **54**, a long-range coupling of 2.3 Hz was observed

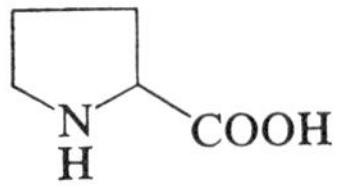

47 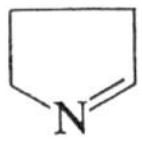 **48** 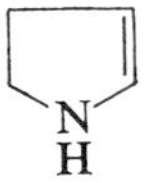 **49** 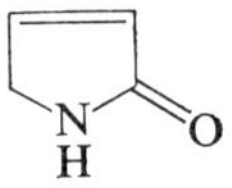 **50** 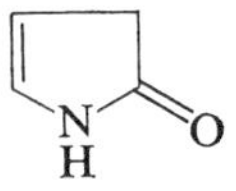 **51**

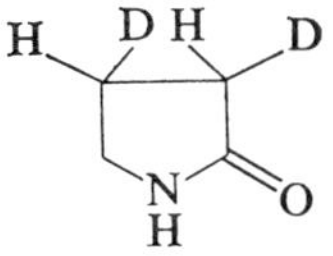

52 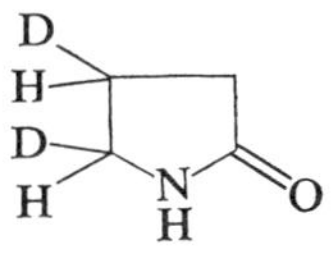 **53** 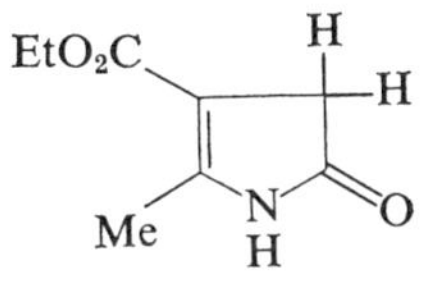 **54** 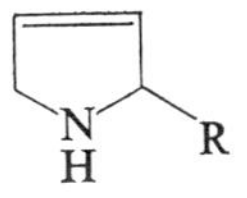 55, R = H
56, R = $CONH_2$

between the 5-methyl group and the 2-proton (96). Spectra of a number of other Δ^3-pyrrolin-2-ones have been reported (90, 97).

Δ^3-Pyrrolines (**55**) include 3,4-dehydroproline, derivatives of which have been studied in great detail (98–100). The amide (**56**) in $CDCl_3$ gives spectra where coupling constants of equal value but opposite sign cancel, leading to deceptive simplicity. In D_2O, different deceptive simplicity is caused by the accidental coincidence of chemical shifts of the 5-protons. Spectra of this compound proved very difficult to analyze and a combined approach, using analytical tables for the ABXYM approximation and computation (LAOCOON III) where the system was treated as ABCDE, was necessary before accurate parameters could be obtained. It is extremely doubtful whether either of these approaches would have been successful on its own, and the results obtained bear witness to the power of the method. Also, progress in this work was dependent on the development of better spectrometers, and satisfactory spectra were unobtainable until internal-locking, high-resolution spectrometers became available. Usually, spectra measured at 100 MHz are more easily analyzed than the corresponding spectra at 60 MHz. However, with 3,4-dehydroprolinamide in $CDCl_3$ at 100 MHz, accidental coincidences caused by chemical shift variations between 60 and 100 MHz remove most of the fine structure evident in the 60-MHz spectrum, making *ab initio* analysis of the 100-MHz spectrum quite impossible. The results obtained in this analysis, together with data from other pyrrolines, are summarized in Table 3.11. The most noticeable feature of these spectra

TABLE 3.11 SPECTRAL DATA FOR PYRROLINES AND PYRROLONES

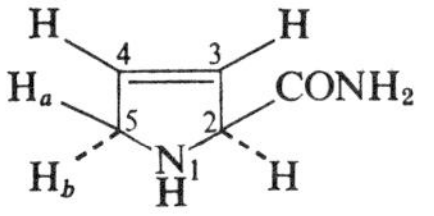

A 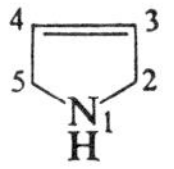   B, C 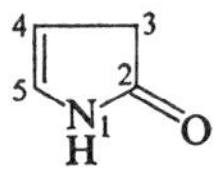   D

A. DEHYDROPROLINAMIDE (99). SOLVENT—$CDCl_3$

τ2 5.45, τ3 4.10, τ4 4.14, τ5a 6.21, τ5b 6.09 ppm; J_{23} 2.10, J_{24} −2.54, J_{25a} 5.81, J_{25b} 3.76, J_{34} 5.86, J_{35a} −2.47, J_{35b} −2.21, J_{45a} 1.99, J_{45b} 1.86, J_{5a5b} −14.73 Hz

B. Δ^3-PYRROLINES (94). SOLVENT—DICHLOROTETRAFLUOROACETONEDEUTEROHYDRATE

Substituents	τ1	τ2	τ3	τ5
1,2-Di-Me, 3-NH_2, 4-CN	7.12	8.44, 5.7	3.58	5.93
1,2,2-Tri-Me, 3-NH_2, 4-CN	7.18	8.44	3.53	5.95
1,2-Di-Me, 3-SH, 4-$SCNH_2$	7.01	8.35, 5.6		5.6

C. Δ^3-PYRROLIN-2-ONES (95)

Substituents	Solvent	τ3	τ4	τ5
None	D_2O	3.8	2.4	5.8
4-Me	$CDCl_3$	4.6	7.8	6.0

D. Δ^4-PYRROLIN-2-ONES (95)

Substituents	τ3	τ4	τ5
None	6.8	3.9	2.6
4-CO_2Et	6.6		2.5

are the extremely large cross-ring couplings between protons on C-2 and C-5 (five-bond through C–C, four-bond through N). Also, the respective values for J_{25} (cis) and J_{25} (trans) provide a diagnostic tool for the relative stereochemistry of the Δ^3-pyrrolines at C-2 and C-5.

Fully reduced pyrroles, the pyrrolidines (**57**), are also rather rare and the only NMR studies have involved proline (**58**) and its derivatives (Table 3.12). Abraham and his co-workers (101–104) produced complete analyses of *cis*- and *trans*-4-hydroxyproline, **59** and **60**, respectively, and a number of other derivatives. The analyses were complicated because at the time there was no way of differentiating between the α- and β-protons of the methylene groups. Analyses for all couplings were made and modifications of the Karplus equation were used to assign the cis and trans relationships of vicinal protons. The problems of using the Karplus equation in detail on such systems are formidable. Uncertainties arise due to ring strain, hybridization, presence of the heteroatom, and the presence and orientation of electronegative and charged substituents. The conclusion was reached that, in neutral solution, the 4-hydroxyprolines exist in the envelope conformation with angles of buckle of 70° for the cis and 53° for the trans isomer. Such extreme buckling seems unlikely, particularly since the X-ray structure of cystalline *trans*-4-hydroxyproline showed only a 17° buckle (105). The difficulty in assigning configuration by Karplus-type relationships is apparent in the epimers of 3-methylproline (**61**) (106, 107), 3-hydroxyproline (**62**) (49), and 4-substituted prolines (109, 110). Robertson and his co-workers (109, 110) have analyzed the spectra of a number of 4-oxoprolines (**63**) and cis- and trans-4-substituted prolines (**64**). From the large number of compounds studied, correlations were obtained which permitted assignment of configuration from spectral data. The difficulties encountered in deducing precise conformations from spectral parameters are emphasized. Spectra of the *p*-nitrophenyl ester of *N*-benzyloxycarbonyl-L-proline (**65**) have been used to demonstrate restricted rotation about the urethane linkage (111).

Differences between the chemical shifts of R and R′ in salts of *N*-substituted 2-iminopyrrolidines led to the conclusion that with $CDCl_3$ as solvent the positive charge must be localized as in **66**, but when D_2O was used delocalization occurred to give ions of the type **67** (112). At 100 MHz, the spectrum of 1-methyl-2-cyclopentadienylidenepyrrolidine (**68**) is approximately first order, and on this basis, a full analysis has been reported (113). Variable-temperature studies permitted an estimation to be made of the size of the barrier to rotation about the exocyclic double bond. Pyrrole-2,5-diacetic acid derivatives **69** have been shown to exist mainly in the tautomeric form **70** (114) although the pyrrole forms could be isolated. The *N*-methyl half-ester **71** was shown to exist as a mixture of the cis and trans isomers **72**

TABLE 3.12 SPECTRAL DATA FOR PROLINES (110)

A. Cis-4-SUBSTITUTED PROLINES. SOLVENT—D_2O

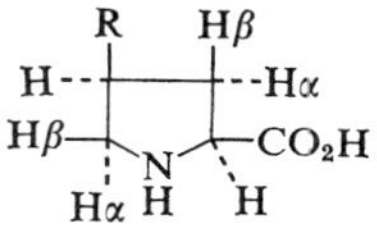 R	$\tau 2$	$\tau 3\alpha$	$\tau 3\beta$	$\tau 4$	$\tau 5\alpha$	$\tau 5\beta$	$J_{23\alpha}$	$J_{23\beta}$	J_{33}	$J_{3\alpha 4}$	$J_{3\beta 4}$	$J_{45\alpha}$	$J_{45\beta}$	J_{55}
Cl[a]	5.43	7.12	7.32	5.14	6.22		3.3	10.5	−15.0	5.6	1.4	3.0		
Br	5.64	7 02	7.28	5.18	6.13		4.1	9.9	−14.6	5.9	2.1	3.8		
NH_2	5.92	7.36	8.09	6.16	6.54	6.80	7.5	8.9	−13.4	6.3	6.1	5.8	6.0	−11.5
OH[b]	5.79	7.52	7.78	5.43	6.58	6.64	3.5	10.5	−14.2	4.7	2.1	0.9	4.6	−12.5

B. Trans-4-SUBSTITUTED PROLINES. SOLVENT—D_2O

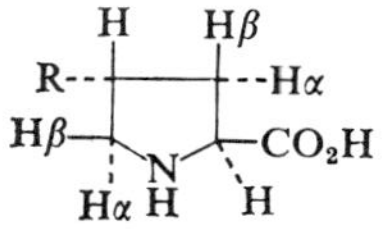

R	$\tau 2$	$\tau 3\alpha$	$\tau 3\beta$	$\tau 4$	$\tau 5\alpha$	$\tau 5\beta$	$J_{23\alpha}$	$J_{23\beta}$	J_{33}	$J_{3\alpha 4}$	$J_{3\beta 4}$	$J_{45\alpha}$	$J_{45\beta}$	J_{55}
Cl	5.45	7.22	7.45	5.07	6.15	6.25	7.5	10.7	−15.2	1.9	5.1	4.4	2.2	−13.2
Br	5.49	7.27	7.36	5.15	6.07	6.22	7.0	10.6	−14.5	1.7	5.3	4.6	2.3	−13.4
OH[b]	5.66	7.60	7.88	5.35	6.52	6.66	7.7	10.4	−14.1	1.4	4.3	4.1	1.2	−12.7

[a] Spectrum of HBr salt at 5°.
[b] From Ref. 103.

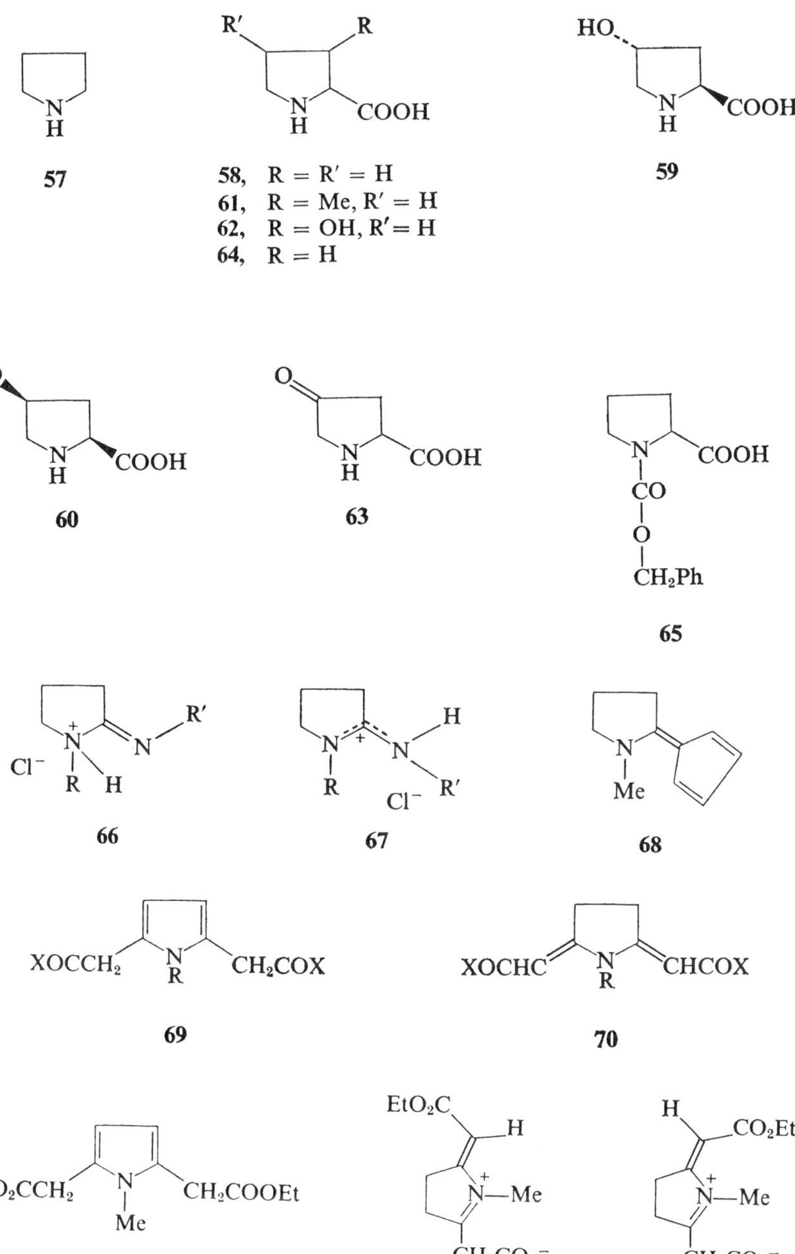

57
R′
R
COOH
58, R = R′ = H
61, R = Me, R′ = H
62, R = OH, R′ = H
64, R = H
HO
COOH
59
HO
COOH
60
O
COOH
63
COOH
CO
O
CH2Ph
65
R′
Cl−
R
H
66
H
R
Cl−
R′
67
Me
68
XOCCH2
R
CH2COX
69
XOCHC
R
CHCOX
70
HO2CCH2
Me
CH2COOEt
71
EtO2C
H
N—Me
CH2CO2−
72
H
CO2Et
N—Me
CH2CO2−
73

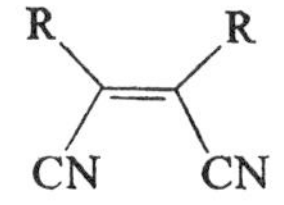

74, (R = H)
79

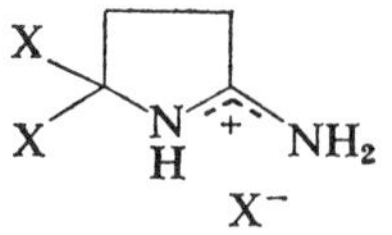

75

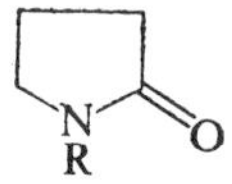

76

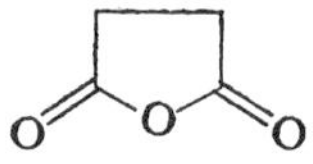

77

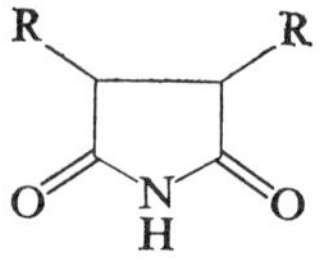

78, R = H
80

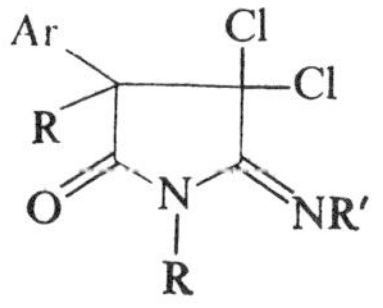

81

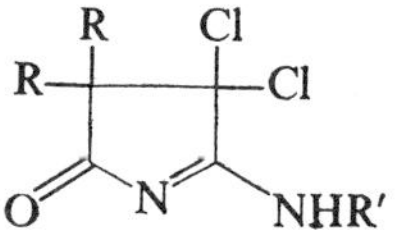

82

and **73**, respectively. Also, the hydrogen halide salts of substituted succinonitriles (**74**) have been shown to be 2-amino-5,5-dihalo-1-pyrrolinium halides **75** (115).

An extensive study of the spectra of *N*-substituted pyrrolid-2-ones (**76**) has been reported (116). The spectrum of the parent compound is broadened and complicated by molecular motion but *N*-acetylation is enough to produce a reasonable AA′BB′ pattern. Many of the compounds studied contained methylene protons in the *N*-substituents but nonequivalence of these protons was not general. In contrast to many other lactams and "asymmetric" ring systems, this nonequivalence was shown to be due entirely to the presence of adjacent asymmetric centers in the side chains. Chemical shifts of the ring protons were shown to vary greatly with the nature of the *N*-substituent, but attempts to correlate the shifts with the electronegativity of the substituent failed.

A number of papers have appeared describing the spectra of maleic anhydride **77** and the corresponding maleimides or pyrrolidine-2,5-diones (**78**). Morel and Foucaud (117, 118), working with succinic dinitriles (**79**), used the spectrum of the corresponding maleimides (**80**) to determine the relative stereochemistry of the dinitriles. In the case of the 4-substituted 2-iminopyrrolidin-5-ones (**81**), the tautomeric form has been shown to depend on the nature of the 4-substituent (119). If this is an aryl group, the iminoxo tautomer **81** is thought to predominate, but with alkyl substitution the amino form **82** is most important.

It is interesting to compare two papers which have appeared on the solvent effects in spectra of maleimides and similar compounds. The second of these (120), written after the style of Williams (58), reports the large upfield shift of the pyrrole protons in benzene solutions, with respect to values for CCl_4 solutions, and attributes the phenomenon to the formation of a 1:1 complex between the solute and solvent. This much is reasonable, but to go on to calculate the actual mutual positions of the two molecules from the raw solvent-shift data seems to me to be quite untenable. Solvent effects on the TMS signal are dismissed by a statement that the choice of standard "minimized" contributions from everything except the stereospecific anisotropy of the complexing molecules. The effects of time-averaging on the spectrum are largely neglected and the calculated "shifts" for the various models do not correlate at all well with the actual values. Many such papers have appeared in the literature over the past few years and, while the usefulness of these solvent shifts as a diagnostic tool is indeed great, we could well have done without the associated theoretical calculations and structures. At this stage it was refreshing indeed to turn to the earlier of these publications (121) which leaves no doubts about the competence of the author as a physical chemist. Solvent effects were measured for maleimide in benzene, *n*-hexane, cyclohexane, acetonitrile, acetone, DMSO, and dioxane. The results were interpreted in terms of a specific interaction with the π-cloud of benzene, preferential solvation of the lone-pair electrons of the polar solvents with electron-deficient centers in the solutes, and some anisotropic shieldings by polar solvents of solvent molecules. Of importance is the realization of the real complexity of the problem and the difficulty of knowing just what is really being measured. Of great interest is the statement concerning solvent shifts between CCl_4 and cyclohexane solutions, "The van der Waals shifts of the solutes are generally assumed to be the same as that of the internal standard. The assumption hardly holds however, when one compares a pair of polar and nonpolar solutes in the same highly polarizable solvent." If this applies to these solvents, how much more must it apply to the CCl_4–benzene pair. Reference to the validity of solvent shifts is given in Chapter 1. Data for reduced pyrroles are collected in Table 3.11.

B. The Pyrazole System

1. *Aromatic Pyrazoles*

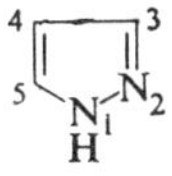

The ease with which pyrazoles can be synthesized is reflected in the large number of these compounds whose spectra have been reported. One review-type paper from the pen of the most prolific workers in the field (122) lists spectra of 180 pyrazoles, mainly 1-substituted, and many other publications contain considerable amounts of data.

Apart from normal structural determinations, NMR has been used in the pyrazole field to study two main problems. The first and most fundamental of these involves the attachment of the single hydrogen to the two nitrogen atoms to give structures in which H-3 and H-5 are equivalent. Of the two proposed mechanisms, normal tautomerism **83** and "mesohydric tautomerism" **84**, NMR evidence has usually been interpreted to support the former, although, as will be seen below, the arguments put forward are often far from conclusive. The second major problem involves assignment of specific peaks to H-3 and H-5 in *N*-substituted pyrazoles. Attempts to make this assignment are studded with arbitrary rules which are often contradictory and not at all general.

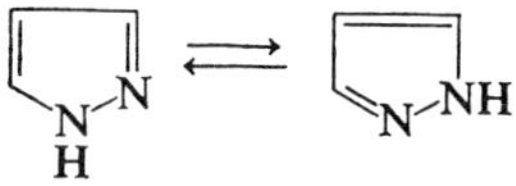

83  **84**

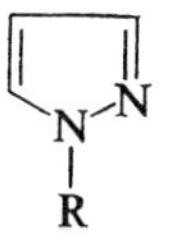

85, R = H
86, R = Me
89, R = Ac

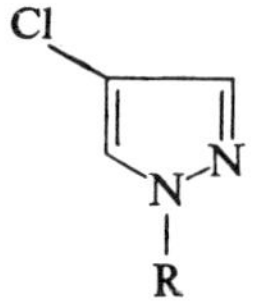

87, R = H
90, R = Ac

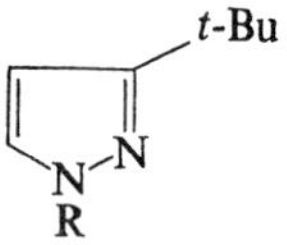

88, R = H
91, R = Ac

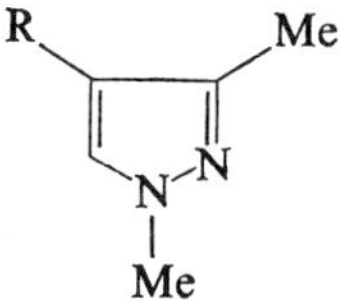

95, R = H
96, R = Ph

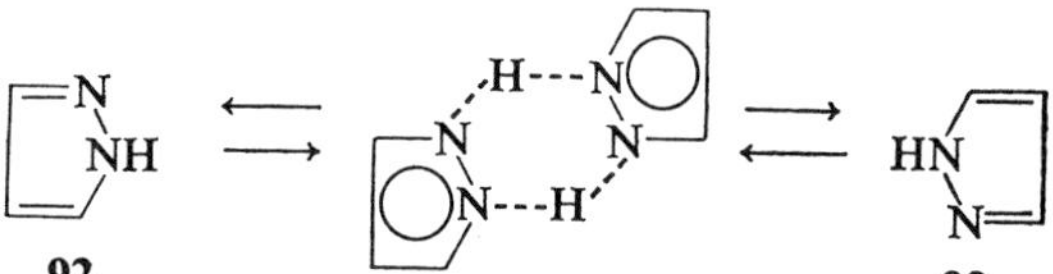

92 **94** **93**

The spectra of pyrazole (**85**) and its 1-methyl derivative **86** in D_2O were first reported in 1959 (123) to demonstrate that the heterocyclic ring present in a natural amino acid was a pyrazole and not an imidazole. However, no real structural work was attempted until 1964 when Williams (124) examined the spectra of pyrazole, 4-chloropyrazole (**87**), 3-*t*-butylpyrazole (**88**),* and their *N*-acetyl derivatives **89–91** in CCl_4. He concluded that pyrazole consisted of the tautomers **92** and **93**, interconverting through a cyclic dimer **94** so rapidly that averaging of all signals occurred, making H-3 and H-5 magnetically equivalent. After acetylation, H-3 and H-5 were nonequivalent and H-5, adjacent to the *N*-acetyl group, was assumed to resonate at lower field with $J_{45} > J_{34} > J_{35}$.

* Pyrazoles in which tautomerism at N-1 and N-2 can occur have positions 3 and 5 interchangeable. Often the notation 3(5)- is used to describe substituents in these positions. Throughout this chapter only the lower number is used in naming the tautomeric mixtures.

Moore and Habraken (125, 126) prepared authentic samples of 1,3-dimethylpyrazole (**95**) and 1,3-dimethyl-4-phenylpyrazole (**96**) and used these to show that in mixtures of the 1-alkyl-3- and 5-methyl derivatives, the 5-methyl group gave signals at higher field than the 3-methyl group and H-3 resonated at lower field than H-5. Comparison of peak spacings for these compounds and for the N–CH_2COOEt derivatives led to the conclusion that in the N–H compounds, the tautomer with its methyl group in position 5 is preferred. They pointed out, however, that their results did not rule out the presence of dimers or rapid interconversion between the two tautomeric forms. Spectra of pyrazole and certain 1-alkyl derivatives were examined by Cola and Perotti (127) but they were unable to assign specific signals to H-3 and H-5. The available data for pyrazole are collected in Table 3.13.

TABLE 3.13 SPECTRAL DATA FOR PYRAZOLE

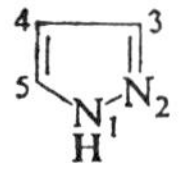

Solvent	τ1	τ3	τ4	J	Ref.
CCl_4	−3.60	2.26	3.60	2.1	124
	−3.65	2.45	3.75		133
		2.65	3.93	2.05	127
	−3.58	2.48	3.74		135
		2.46	3.76	2.0	122
$CDCl_3$	−2.52	2.38	3.67		246
	−2.64	2.39	3.69	1.9	122
	−3.13	2.47	3.75		137
		2.35	3.64		139
DMSO		2.36	3.71	2.0	122
		2.40	3.75		247
$(Me)_2CO$		2.39	3.72	1.9	122
Mesitylene		2.69	3.92	1.9	122
Collidine		2.30	3.92	1.9	122

Solvent effects on pyrazoles have been little studied. Nuclear magnetic resonance results with CCl_4 solutions support the existence of monomers, together with cyclic dimers and trimers (128), conclusions similar to those obtained from infrared data (129–131). In chloroform solutions solute–solvent interactions appear to be important. Variable-temperature studies led to the conclusion that in solutions of normal NMR strength in CCl_4, pyrazole exists mainly as cyclic dimers and trimers but, in very dilute solutions the main species present were thought to be the monomer and an "open-chain dimer" (132). Finar and Mooney (133) have examined the spectra of a number of N–H, N–Me, and *N*-arylpyrazoles in CCl_4, $CDCl_3$, and trifluoroacetic acid and agree with the presence of dimers and trimers in solutions of pyrazole and its 3,5-dimethyl derivative. The difference in chemical shift between the N–H protons of these two compounds (τ −3.65 and −1.49, respectively) is attributed to the higher basicity of the latter. In *N*-aryl derivatives, as in most of the pyrazoles, H-4 gives peaks upfield from those of the other ring protons, reflecting the higher proton density at C-4. Also, in $CDCl_3$, the signal from H-5 for a number of *N*-phenyl derivatives appears at 2.22 ∓ 0.06. Given the definition

$$\Delta\tau = \tau(CDCl_3) - \tau(CF_3COOH),$$

$\Delta\tau_3$ is normally but not exclusively greater than $\Delta\tau_5$. In a recent paper (134) an extremely interesting solvent phenomenon was observed. Pyrazole, as well as other azoles, in acetone appears to exist in equilibrium with the acetone adduct **97** whose formation is thermodynamically favored by lowering the solution temperature. Thus, at −90°, the spectrum shows peaks due to the 1-substituted pyrazole as well as those due to the original N–H tautomers. However, the claim that the results prove the normal tautomeric structure for pyrazoles and discredit the concept of "mesohydric tautomerism" seems to be rather rash.

The ever-present problem of the assignment of peaks to H-3 and H-5 in 1-substituted pyrazoles, while occurring in some of the references above and also in those below dealing with 1-aryl compounds, has led to a number of publications specifically dealing with these assignments. In the spectrum of 1,4-dimethylpyrazole (**98**) in CCl_4, the signal at τ 2.66 was attributed to H-3 and that at τ 2.80 to H-5 ($J_{34} = 1.8$, $J_{45} = 2.4$ Hz) (135). This difference in coupling constants appears to have been used widely to assist in assignment (e.g., Refs. 122, 134). More recently, Albright and Goldman (136), studying isomerism in the *N*-methyltetrahydroindazole moieties (**99**, **100**) of various natural products, found that the difference between the chemical shifts of a given proton in an N–H pyrazole and in the corresponding *N*-methyl derivative, $\Delta = \delta_H - \delta_{NMe}$, was characteristic of the position of methylation. These rules appear to hold quite well for complex pyrazoles. Also, with compounds **101** and **102**, that with the 5-methyl group shows a small coupling between the methyl protons and H-4, characteristic of the group CH–CMe. Elguero *et al.* (122) point out the obvious weaknesses of using trifluoroacetic acid as solvent in these determinations (see Ref. 133) and suggest the use of hexamethylphosphorotriamide (HMPT) or benzene as solvents to spread the proton signals of 1-substituted pyrazoles. Unfortunately, they did not use deuterated solvents so that much important data were obscured by the solvent peaks. They were able to show that with

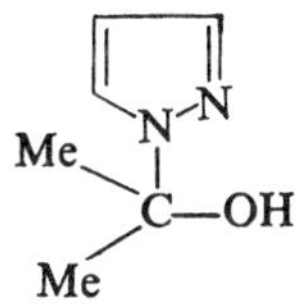

97 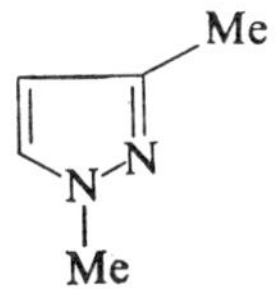 98 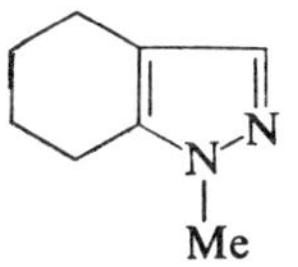 99 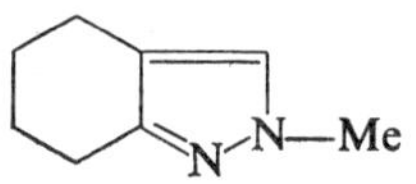 100

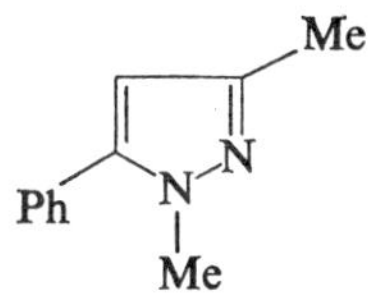

101 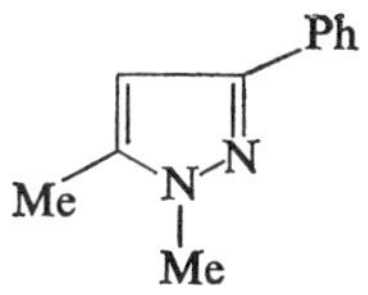 102 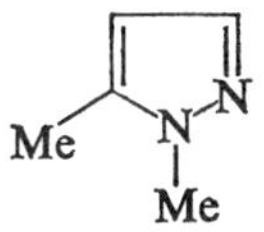 103 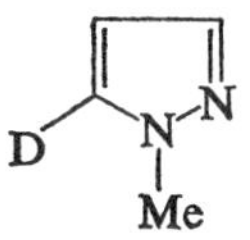 104

benzene as solvent, the various methyl groups could be readily distinguished (see Table 3.14) and their results agree qualitatively with those published earlier (127, 135). However, the method of Albright and Goldman (136) does not appear to work for all of the substituted pyrazoles listed. In the case of 1,3- and 1,5-dimethylpyrazoles (**98** and **103**), the signal from the 5-methyl group usually appears downfield from that of the 3-methyl group and H-3 absorbs upfield from H-5 (126). Insertion of strongly electron-withdrawing and anisotropic groups can cause reversal of this order (137). Thus, a 4-nitro substituent does not exercise a uniform deshielding effect on protons or methyl groups in positions 3 and 5. This variation is thought to involve electron withdrawal from the pyrazole ring by the substituent with the associated partial positive charge located mainly at N-1, resulting in preferential deshielding at position 5.

Elguero and Jacquier (138) have summarized the empirical methods which permit the assignment of specific signals to H-3 and H-5 in spectra of the groups of compounds which they have studied. The first of these, discussed above, involves the use of model compounds. The next approach is derived from their work on *N*-arylpyrazoles (122) and is really an internally consistent verification of their assignments for various nitrophenyl derivatives. Variation of the *N*-substituent from a 4-nitrophenyl to a 2,4-dinitrophenyl to a 2,4,6-trinitrophenyl group produces different but predictable shifts of the pyrazole protons, allowing a 3-H series to be distinguished from a 5-H series. The third method relies on the dramatic shifts observed for the signal from H-5 of simple *N*-methyl derivatives when the solvent is varied from benzene, through $CDCl_3$, DMSO, to HMPT. In contrast, the chemical shift of a 3-proton is almost independent of this solvent. The final method depends on the fact that J_{45} is usually greater than J_{34}, regardless of the nature of the

TABLE 3.14A SPECTRAL DATA FOR 1-*H*-PYRAZOLES

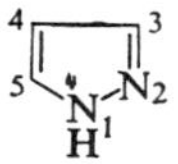

R3	R5	R4	τ1	τ3	τ5	τ4	J	Solvent[a]	Ref.[b]
Me	H	H		7.68	2.52	3.94	1.7		
				7.68	2.53	3.95			139
				7.91		4.09		C_6H_6	
Et	H	H		8.74, 7.27	2.48	3.92	1.7		
t-Bu	H	H		8.63	2.56	3.94	1.9		
Ph	H	H				3.40	2.2		
				2.23	2.48	3.47			139
					1.80	2.92	2.2	TFA	
CO_2Et	H	H		5.49, 8.60	2.20	3.14			
Cl	H	H			2.47	3.76	1.9		
H	H	Me		2.64	2.64	7.91			
			−3.04	2.74	2.74	8.05		CCl_4	137
H	H	Ph		1.60	1.60			TFA	
H	H	OEt	−0.96	2.75	2.75	6.08, 8.64			137
H	H	Cl		2.43	2.43				
			−2.75	2.47	2.47				137
H	H	Br		2.42	2.42				
			−2.23	2.40	2.40				137
H	H	I		2.35	2.35				
Me	H	Me		7.80	2.74	8.02			
Me	H	Cl		7.70	2.54				
Me	H	Br		7.69	2.50				

Me	H	I		7.70	2.47			
Me	H	NO_2		7.33	1.77			
Et	H	Cl		7.31, 8.73	2.54			
Et	H	Br		7.30, 8.73	2.51			
Me	H	Ph		7.54	2.29			
				7.34	1.82		TFA	
Ph	H	Me			2.29	7.80		
					1.99	7.58	TFA	
Ph	H	Br			1.75		TFA	
CO_2Et	H	Me		5.80, 8.62	2.39	7.72		
CO_2Et	H	Br		5.52, 8.56	2.23			
Br	H	Me			2.62	7.95		
Br	H	Br			2.36			
Me	Me	H		7.79	7.79	4.24		
				7.73	7.73	4.18		139
				7.85	7.85	4.31	C_6H_6	
t-Bu	*t*-Bu	H		8.70	8.70	4.13		
Me	Et	H		7.72	7.34, 8.77	4.14		
Me	*n*-Pr	H		7.75	8.37, 9.07	4.20		
Me	CO_2Et	H		7.64	5.64, 8.68	3.44		
			−3.20	7.63	5.6, 8.7	3.44		139
Me	OMe	H	−1.2	7.80	6.15	4.53		154
Me	Ph	H		7.80		3.68		
			−1.2	7.75	2.27	3.68		139
				7.38		3.16	TFA	
Ph	Ph	H	−0.3	2.30	2.30	3.21		139
						3.17		
Me	Cl	H		7.59		4.02		
			−3.1	7.62		4.03		139

TABLE 3.14A (*continued*)

R3	R5	R4	τ1	τ3	τ5	τ4	J	Solvent[a]	Ref.[b]
Me	Br	H		7.59		3.92			
Ph	Cl	H				3.66			
						2.98		TFA	
Ph	CO_2Et	H	−3.5	2.27	5.85, 8.85	3.07			139
					5.32, 8.46	2.44		TFA	
Ph	NH_2	H		2.4, 2.6	4.9	4.06			139
Me	Me	Me		7.80	7.80	8.13			
			−2.36	7.82	7.82	8.12		CCl_4	137
Me	Ph	Me		7.91		7.91			
				7.45		7.71		TFA	
Me	CO_2Et	Me		7.82	5.98, 8.96	7.93			
Me	Me	Cl		7.75	7.75				
Me	Me	Br		7.71	7.71				
Me	Me	I		7.74	7.74				
Me	Me	NH_2		7.82	7.82				
Me	Me	NO		7.33	7.33				
Me	Me	NO_2		7.37	7.37				
Me	Cl	Br		7.62					
Me	Br	Br		7.58					
Ph	Me	Br			7.37				
Me	CO_2Et	Br		7.68	5.61, 8.64				

[a] $CDCl_3$ unless otherwise indicated.
[b] Reference 122 unless otherwise indicated.

TABLE 3.14B SPECTRAL DATA FOR 1-METHYLPYRAZOLES

R3	R5	R4	τ1	τ3	τ5	τ4	J_{34}	J_{45}	Solvent[a]	Ref.[b]
H	H	H	6.63	2.63	3.17	3.98			Mesitylene	
				2.48	3.26	3.96			C_6H_6	140
			6.19	2.70	2.78	3.90	2.0	2.3	CCl_4	
			6.27	2.53	2.64	3.58			Collidine	
				2.34	2.53	3.73			C_5H_5N	140
			6.12	2.51	2.65	3.78			$CDCl_3$	
			6.19	2.63	2.60	3.82			$MeNO_3$	
			6.20	2.64	2.58	3.82			MeCN	
			6.17	2.60	2.52	3.80			MeOD	
				2.64	2.48	3.83			Me_2CO	
			6.14	2.56	2.43	3.79			MMF	
			6.14	2.62	2.40	3.82			DMF	
				2.59	2.34	3.79			DMSO	
			6.05	2.39	2.24	3.61			$(NH_2)_2$	
				2.69	2.15	3.89			HMPT	
			5.74	1.89	1.95	3.17			TFA	
Me	H	H	6.20	7.77	2.78	4.05		2.0		
			6.72	7.76	3.28	4.11			C_6H_6	
			6.30	7.72		4.01			C_5H_5N	
			6.28	7.88	2.53	4.05			DMSO	
			6.21	7.87	2.33	4.10			HMPT	
			5.80	7.44	2.11	3.42			TFA	
H	H	Me	6.20	2.72	2.88	7.96				
			6.30	2.88	2.94	8.00				137
			6.70	2.82	3.44	8.10			C_6H_6	

TABLE 3.14B (*continued*)

R3	R5	R4	$\tau 1$	$\tau 3$	$\tau 5$	$\tau 4$	J_{34}	J_{45}	Solvent[a]	Ref.[b]
			5.80	2.11	2.16	7.74			TFA	
			6.63	2.86		8.10			Mesitylene	
			6.29			8.01			C_5H_5N	
			6.24	2.80	2.62	8.02			DMSO	
			6.22	2.93	2.44	8.04			HMPT	
H	Me	H	6.27	2.64	7.78	4.02	2.0			
			6.82		8.32	4.15			C_6H_6	
			5.88	2.02	7.45	3.37			TFA	
			6.69	2.75	8.16	4.20			Mesitylene	
			6.36		7.94	4.01			C_5H_5N	
			6.33	2.78	7.80	4.05			DMSO	
			6.28	2.83	7.76	4.08			HMPT	
H	H	Cl	6.28	2.73	2.77					137
H	H	Br	6.12	2.54	2.60					
			6.15	2.57	2.62					137
H	H	CHO	6.09	2.20	2.13	0.30				137
H	H	NO_2	5.97	1.92	1.73					137
H	H	Ph	6.08	2.25	2.42					143
CO_2Me	H	H	6.01	6.13	2.54	3.32		2.5	CCl_4	140
H	CO_2Me	H	5.81	2.62	6.13	3.24	2.5		CCl_4	140
CO_2H	H	H	6.07		2.16	3.27		2.5	DMSO	140
H	CO_2H	H	5.88	2.48		3.15	2.0		DMSO	140
Me	H	Me	6.24	7.83	2.98	8.04				
			6.74	7.82	3.48	8.16			C_6H_6	
			6.67	7.90	3.48	8.16			Mesitylene	
			6.33	7.80		8.09			C_5H_5N	
			6.37	7.98	2.80	8.12			DMSO	
			6.32	7.98	2.60	8.12			HMPT	
			5.92	7.59	2.32	7.86			TFA	
Me	H	Ph	6.40	7.63	2.67	2.68				143
Me	H	Cl	6.18	7.75	2.74					
			6.96	7.82					C_6H_6	
Me	H	Br	6.22	7.80	2.64					

			6.96	7.80			C_6H_6	
			5.84	7.51	2.08		TFA	
Me	H	NO_2	6.17	7.39	1.32		DMSO	253
CO_2H	H	NO_2	6.05		1.17		DMSO	253
$CONH_2$	H	NO_2	6.09		1.17		DMSO	253
Me	Me	H	6.34	7.83	7.83	4.22		
			6.83	7.76	8.32	4.27	C_6H_6	
			6.43	7.72	7.95	4.21	C_5H_5N	
			6.00	7.54	7.54	3.64	TFA	
Me	Ph	H	6.20	7.70	2.57	3.91		139
Me	OEt	H	6.54	7.91	6.08, 8.72	4.78	Neat	153
OEt	Me	H	6.58	5.91, 8.78	7.98	4.65	Neat	153
Me	CO_2H	H	5.87	7.71		3.22		139
Me	Br	H	6.20	7.77		3.94		
			6.72	7.89		4.16	C_6H_6	
Me	Cl	H	6.25	7.80		4.04		137
Ph	Me	H	6.20	2.22	7.70	3.70		139
CO_2H	Me	H	6.11	−0.9	7.69	3.37		139
Ph	CO_2Me	H	5.83	2.18	6.14	2.91		139
Ph	CO_2H	H	5.84	2.18	−0.7	2.75		139
Ph	NH_2	H	6.37	2.30	6.40	4.18		139
CO_2Me	Ph	H	5.80	6.10	2.55	3.15		139
CO_2H	Ph	H	6.00	−1.6	2.50	3.07		139
					2.12		DMSO	254
Br	Me	H	6.28		7.77	3.98		
			7.03		8.53	4.23	C_6H_6	
H	Me	Me	6.26	2.81	7.84	8.04		
			6.77		8.40	8.20	C_6H_6	
			6.70	2.90	8.27	8.16	Mesitylene	
			6.38		8.08	8.08	C_5H_5N	
			6.37	2.94	7.90	8.12	DMSO	
			6.32	3.01	7.86	8.12	HMPT	
			5.93	2.22	7.59	7.84	TFA	
H	Me	Ph	6.30	2.50	7.81	2.74		143

TABLE 3.14B (*continued*)

R3	R5	R4	$\tau 1$	$\tau 3$	$\tau 5$	$\tau 4$	J_{34}	J_{45}	Solvent[a]	Ref.[b]
H	Me	Br	6.21	2.61	7.76					
			6.99		8.44				C_6H_6	
H	Me	NO_2	6.17	1.87	7.40				DMSO	253
H	CO_2H	Me	5.96	2.65		7.80			DMSO	254
Br	H	Me	6.17		2.89	8.01				254
			6.22		2.48	8.09			DMSO	254
			6.77		3.63	8.17			Mesitylene	254
Me	Me	Me	6.32	7.86	7.88	8.11				
			6.76	7.80	8.36	8.24			C_6H_6	
			6.42	7.80	8.18	8.08			C_5H_5N	
Me	Me	Ph	6.26	7.75	7.81	2.70				143
Me	Me	Br	6.29	7.78	7.76					
			6.94	7.76	8.38				C_6H_6	
CO_2Et	Me	Br	6.11	5.60, 8.56	7.68					
			7.04	5.80, 8.93	8.50				C_6H_6	
Me	CO_2Et	Br	5.88	7.76	5.60, 8.59					
			6.24	7.80	5.98, 8.98				C_6H_6	
CO_2H	Me	NO_2	6.17		7.50				DMSO	253
Me	CO_2H	NO_2	6.12	7.66					DMSO	253
Me	Cl	Me	6.28	7.85		8.19				137
Me	Cl	Ph	6.30	7.75		2.71				143
Br	CO_2H	Me	5.95			7.86			DMSO	254
Me	Br	Br	6.16	7.76						
			6.88	7.86					C_6H_6	
			5.82	7.45					TFA	
Br	Me	Br	6.20		7.72					
			7.16		8.57				C_6H_6	
			5.86		7.44				TFA	

[a] $CDCl_3$ unless otherwise indicated.
[b] Reference 122 unless otherwise indicated.

TABLE 3.14C SPECTRAL DATA FOR 1-CARBOXAMIDOPYRAZOLES (122)

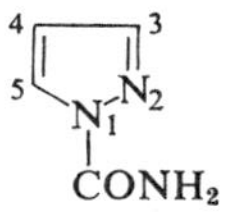

R3	R4	R5	$\tau 3$	$\tau 4$	$\tau 5$	J_{34}	J_{45}	J_{35}	Solvent
H	H	H	2.34	3.60	1.74	1.5	2.7	0.7	$CDCl_3$
			2.25	3.50	1.70	1.6	2.7	0.6	DMSO
Me	H	H	7.71	3.81	1.89		2.6		$CDCl_3$
			8.02						C_6H_6
Me	Br	H	7.73		1.89				$CDCl_3$
			8.07						C_6H_6
H	Br	H	2.17		1.57				DMSO
H	H	Me	2.58	3.91	7.41	1.5			$CDCl_3$
					7.60				C_6H_6
Me	Me	H	7.81	8.00	2.12		1.0		$CDCl_3$
			7.88	8.07	2.11		1.0		DMSO
Me	H	Me	7.80	4.09	7.46				$CDCl_3$
			8.01		7.57				C_6H_6
			7.87	4.01	7.57				DMSO
Me	Br	Me	7.78		7.43				$CDCl_3$
			8.01		7.59				C_6H_6
H	Me	Me	2.67	8.00	7.50				$CDCl_3$
Me	Me	Me	7.86	8.11	7.54				$CDCl_3$
			8.07	8.53	7.62				C_6H_6
			7.91	8.17	7.64				DMSO

TABLE 3.14D SPECTRAL DATA FOR 1-ARYLPYRAZOLES

R–Aryl	R3	R4	R5	τ1	τ3	τ4	τ5	J_{34}	J_{45}	Solvent	Ref.
	H	H	H		2.28	3.54	2.13	1.9	2.5	$CDCl_3$	122
					2.33	3.57	1.10	1.6	2.6	HMPT	122
					2.60	3.90	3.04	1.7	2.5	Mesitylene	122
	Me	H	H	2.3, 2.8	7.65	3.77	2.12	1.1	2.9	$CDCl_3$	139
	H	H	Me	2.57	2.44	3.81	7.66			$CDCl_3$	139
	H	Cl	H		2.34		2.08			$CDCl_3$	122
	H	Br	H		2.34		2.09			$CDCl_3$	122
	OH	H	H		−0.24	4.11	1.75			DMSO	257
	OAc	H	H		7.69	3.64	2.16			$CDCl_3$	257
	OMe	H	H		6.05	4.14	2.33			$CDCl_3$	257
	Me	H	Me	2.61	7.73	3.99	7.70			$CDCl_3$	139
					7.69	4.18	8.08			C_6H_6	122

	Me	H	CO_2H	2.58	7.64	3.13				$CDCl_3$	139
	CO_2H	H	Me	2.54		3.20	7.67			$CDCl_3$	139
	Me	Ph	H		7.54		2.11			$CDCl_3$	143
	H	Ph	Me	2.52	2.20	2.59	7.59			$CDCl_3$	143
	Me	H	OEt	2.1–3.0	7.74	4.50	5.86, 8.59			$CHCl_3$	153
	Me	Me	OEt	2.1–3.0	7.90	8.13	6.12, 8.82			CCl_4	153
	OMe	H	Me	2.63	6.13	4.34	7.90			Neat	153
4′-Br	H	H	H	2.45	2.30	3.54	2.13	1.6	2.4	$CDCl_3$	122
				1.86, 2.37	2.31	3.54	1.02	1.6	2.4	HMPT	122
	H	Br	H	2.47	2.33		2.11			$CDCl_3$	122
	H	Br	Br	2.39, 2.59	2.30					$CDCl_3$	122
				2.08, 2.37	1.89					HMPT	122
2′, 3′, 5′, 6-Tetra-F	Me	H	Me	2.71	7.69	3.88	7.81		0.6	$CDCl_3$	122
				3.65	7.84	4.26	8.23		0.7	C_6H_6	122
Penta-F	H	H	H	[1]	2.14	3.44	2.28	1.8	2.5	$CDCl_3$	144
	H	Br	H	[2]	2.21		2.31			$CDCl_3$	144
	H	NO_2	H	[3]	1.65		1.54			$CDCl_3$	144
	Me	H	Me	[4]	7.83	3.95	7.71			$CDCl_3$	144
	Ph	H	Ph	[5]	2.1–2.75	3.13	2.68			$CDCl_3$	144

FLUORINE RESONANCES (ppm TO HIGH FIELD OF $CFCl_3$)

	Ortho	Meta	Para	$J_{o,m}$	$J_{m,p}$
[1]	147.7	162.5	154.8	16.4	21.2
[2]	147.2	160.5	152.5	15.8	21.5
[3]	146.4	159.2	149.6	15.4	21.5
[4]	145.3	161.2	152.3	15.6	21.2
[5]	144.2	160.8	151.7	15.8	21.5

TABLE 3.14E SPECTRAL DATA FOR 1-(*p*-NITROPHENYL)-PYRAZOLES (122)

R3	R4	R5	τPNP	τ3	τ4	τ5	J_{34}	J_{45}	Solvent
H	H	H	2.27, 2.85	2.54		2.82			Mesitylene
			2.10, 1.64	2.17	3.43	1.94	1.8	2.7	$CDCl_3$
			1.88, 1.60	2.18	3.38	1.46	1.5	2.4	Me_2CO
			1.88, 1.60	2.10	3.32	1.26	1.5	2.4	DMSO
			1.65, 1.42	2.14	3.38	0.73			HMPT
			2.02, 1.40	2.16	2.86	1.46	2.5	2.5	TFA
Me	H	H	2.20, 1.71	7.60	3.66	2.09		2.5	$CDCl_3$
				7.82					C_6H_6
H	Me	H	2.18, 1.71	2.39	7.82	2.20			$CDCl_3$
					8.20				C_6H_6
H	H	Me	2.30, 1.64	2.36	3.70	7.52			$CDCl_3$
						8.20			C_6H_6
H	Br	H	2.15, 1.65	2.25		1.94			$CDCl_3$

Me	Me	H	2.25, 1.73	7.72	7.92	2.28	$CDCl_3$
				7.91	8.28		C_6H_6
			2.11, 1.43	7.34	7.68	1.76	TFA
Me	H	Me	2.33, 1.69	7.70	3.91	7.56	$CDCl_3$
				7.78		8.18	C_6H_6
			2.19, 1.63	7.70	3.81	7.58	DMSO
			2.18, 1.41	7.40	3.36	7.52	TFA
H	Me	Me	2.32, 1.66	2.46	7.92	7.64	$CDCl_3$
					8.26	8.32	C_6H_6
			2.16, 1.38	1.84	7.71	7.56	TFA
Ph	H	Ph	2.47, 1.80		3.15		$CDCl_3$
H	Ph	Me	2.24, 1.61	2.14	2.57	7.47	$CDCl_3$
						8.08	C_6H_6
			1.98, 1.44	1.86	2.42	7.43	DMSO
H	Br	Me	2.31, 1.62	2.32		7.57	$CDCl_3$
						8.29	C_6H_6
Me	Br	Me	2.32, 1.64	7.68		7.56	$CDCl_3$
				7.78		8.25	C_6H_6
Me	Br	Br	2.10, 1.42	7.35			TFA
Me	Me	Me	2.34, 1.69	7.74	8.00	7.64	$CDCl_3$
				7.85	8.27	8.32	C_6H_6
			2.17, 1.40	7.41	7.75	7.57	TFA

TABLE 3.14F SPECTRAL DATA FOR 1-(2,4-DINITROPHENYL)-PYRAZOLES (122)

R3	R4	R5	τDNP	τ3	τ4	τ5	J_{34}	J_{45}	Solvent
H	H	H	2.54, 2.16	2.64	3.98	2.98			Mesitylene
			2.16, 1.43, 1.28	2.20	3.42	2.14	1.8	2.7	$CDCl_3$
			1.91, 1.38, 1.24	2.23	3.38	1.70	1.8	2.7	Me_2CO
			1.86, 1.38, 1.16	2.12	3.30	1.46			DMSO
				2.14	3.33	1.05			HMPT
			1.88, 1.18, 0.78	1.48	2.87	1.64	2.6	2.6	TFA
Me	H	H	2.19, 1.52, 1.36	7.67	3.64	2.32		2.7	$CDCl_3$
				7.90					C_6H_6
			1.94, 1.28, 1.20	7.78	3.52	1.54		2.7	DMSO
				7.74					C_5H_5N
			1.88, 1.18, 0.78	7.28	3.03	1.72		2.9	TFA
H	Me	H	2.22, 1.52, 1.36	2.38	7.82	2.44			$CDCl_3$
			3.35, 2.40, 2.02	2.73	8.31	3.26			C_6D_6
			1.96, 1.44, 1.20	2.30	7.88	1.73			DMSO
H	H	Me	2.28, 1.42, 1.18	2.38	3.72	7.66	1.8		$CDCl_3$
						8.26			C_6H_6
			1.93, 1.31, 1.07	2.35	3.62	7.64			DMSO
			1.49, 1.26, 0.96	2.41	3.58				HMPT
						7.66			C_5H_5N
			1.82, 1.09, 0.69	1.58	3.06	7.52	2.7		TFA
Ph	H	H	2.16, 1.50, 1.32		3.12	2.07		2.6	$CDCl_3$
H	Ph	H	1.80, 1.14, 0.76	1.26		1.45			TFA
H	H	Ph	2.57, 1.64, 1.29	2.18					$CDCl_3$

Cl	H	H	2.11, 1.41, 1.24		3.48	2.26	2.5	$CDCl_3$
H	Cl	H	2.20, 1.46, 1.28	2.28		2.18		$CDCl_3$
H	Br	H	2.20, 1.44, 1.28	2.24		2.16		$CDCl_3$
H	NO_2	H	1.72, 1.29, 1.11	1.48		0.37		DMSO
CO_2Et	H	H	2.05, 1.42, 1.20	8.58, 5.57	2.92	2.22	2.4	$CDCl_3$
Me	Me	H	2.22, 1.56, 1.40	7.76	7.93	2.53		$CDCl_3$
				8.06	8.35			C_6H_6
			2.05, 1.50, 1.28	7.89	7.98	1.88		DMSO
Me	H	Me	2.32, 1.49, 1.25	7.79	3.93	7.74		$CDCl_3$
				7.91		8.23		C_6H_6
			1.98, 1.42, 1.16	7.87	3.82	7.80		DMSO
H	Me	Me	2.31, 1.46, 1.22	2.49	7.92	7.77		$CDCl_3$
					8.34	8.30		C_6H_6
			1.99, 1.38, 1.15	2.48	7.97	7.74		DMSO
Ph	Me	H	2.18, 1.54, 1.37		7.57			$CDCl_3$
Ph	H	Me	2.29, 1.50, 1.22		3.40	7.66		$CDCl_3$
H	Ph	Me	2.22, 1.40, 1.15	2.14	2.58	7.58		$CDCl_3$
			3.52, 2.54, 1.97	2.28		8.16		C_6D_6
			1.87, 1.33, 1.10	1.99		7.54		DMSO
			1.77, 1.10, 0.73	1.50		7.49		TFA
Me	Ph	H	2.16, 1.52, 1.37	7.58	2.60	2.21		$CDCl_3$
			3.34, 2.49, 2.04	7.84		2.85		C_6D_6
			1.85, 1.38, 1.17	7.60		1.26		DMSO
			1.81, 1.17, 0.80	7.19		1.67		TFA
Me	H	Ph	2.58, 1.68, 1.34	7.65	3.62			$CDCl_3$
H	Me	Ph	1.72, 1.36	2.33	7.86			$CDCl_3$
Ph	H	Ph	1.64, 1.27		3.10			$CDCl_3$
CO_2Et	Me	H	2.06, 1.44, 1.24	8.58, 5.58	7.64	2.43		$CDCl_3$
				8.98, 5.89	7.80			C_6H_6
CO_2Et	H	Me	2.19, 1.37, 1.07	8.61, 5.60	3.18	7.72		$CDCl_3$
				8.99, 5.89		8.36		C_6H_6
CO_2Me	Ph	H	2.01, 1.41, 1.18	6.10	2.54	2.20		$CDCl_3$

TABLE 3.14F *(Continued)*

R3	R4	R5	τDNP	τ3	τ4	τ5	J_{34}	J_{45}	Solvent
CO_2Et	H	Ph	2.32, 1.52, 1.22	8.56, 5.54	2.90				$CDCl_3$
CO_2Et	Br	H	2.08, 1.38, 1.16	8.58, 5.56		2.10			$CDCl_3$
Me	Br	H	2.22, 1.50, 1.31	7.72		2.26			$CDCl_3$
				8.00					C_6H_6
			1.88, 1.14, 0.77	7.35		1.74			TFA
Me	NO_2	H	2.21, 1.36, 1.16	7.38		1.44			$CDCl_3$
				7.77					C_6H_6
			1.94, 1.28, 1.08			0.38			DMSO
Me	H	Cl	2.16, 1.46, 1.18	7.69	3.68				$CDCl_3$
				8.00					C_6H_6
				7.77					C_5H_5N
Me	H	Br	2.16, 1.44, 1.18	7.66	3.61				$CDCl_3$
Br	Me	H	2.16, 1.49, 1.33		7.90	2.50			$CDCl_3$
					8.34				C_6H_6
			2.01, 1.26, 0.98		7.76	2.26			TFA
			1.93, 1.36, 1.14			1.61			DMSO
					8.06				C_5H_5N
Cl	H	Me	2.25, 1.42, 1.15		3.76	7.72			$CDCl_3$
						8.44			C_6H_6
						7.71			C_5H_5N

Br	H	Me	2.25, 1.41, 1.14		3.68	7.70	$CDCl_3$
H	Br	Me	2.28, 1.38, 1.09	2.36		7.71	$CDCl_3$
Ph	Br	H	1.48, 1.29			2.09	$CDCl_3$
H	Br	Ph	2.51, 1.60, 1.29				$CDCl_3$
Br	Br	H	2.16, 1.44, 1.25			2.24	$CDCl_3$
Me	Me	Me	2.33, 1.49, 1.24	7.79	8.00	7.79	$CDCl_3$
				7.98	8.36	8.31	C_6H_6
			1.61, 1.30, 1.02				HMPT
			1.88, 1.12, 0.77	7.40	7.74	7.68	TFA
Me	Br	Me	2.31, 1.44, 1.18	7.73		7.73	$CDCl_3$
			1.94, 1.36, 1.11	7.86		7.72	DMSO
Me	NO_2	Me	2.25, 1.35, 1.01	7.41		7.44	$CDCl_3$
				7.67		8.06	C_6H_6
Ph	Br	Me	2.29, 1.46, 1.18			7.66	$CDCl_3$
Me	Br	Ph	2.64, 1.66, 1.32	7.64			$CDCl_3$
Ph	Br	Ph	1.68, 1.30				$CDCl_3$
CO_2Et	Br	Me	2.20, 1.36, 1.04	8.58, 5.56		7.72	$CDCl_3$
CO_2Et	Br	Ph	2.39, 1.56, 1.24	8.56, 8.51			$CDCl_3$
Me	Br	Cl	2.16, 1.40, 1.10	7.68			$CDCl_3$
Cl	Cl	Me	2.26, 1.40, 1.12			7.72	$CDCl_3$
Cl	Br	Me	2.26, 1.40, 1.10			7.72	$CDCl_3$
Br	Br	Me	2.26, 1.38, 1.10			7.72	$CDCl_3$
						8.41	C_6H_6
			2.06, 1.20, 0.84			7.65	TFA
Br	Br	Ph	2.51, 1.60, 1.29				$CDCl_3$

TABLE 3.14G SPECTRAL DATA FOR 1-(2,4,6-TRINITROPHENYL)PYRAZOLES (122)

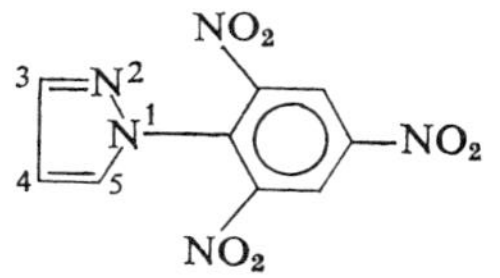

R3	R4	R5	τTNP	τ3	τ4	τ5	J_{34}	J_{35}	Solvent
H	H	H	2.22	2.70	3.98	2.95			Mesitylene
			1.08	2.15	3.40	2.26	1.9	2.6	$CDCl_3$
			0.81	2.18	3.36	1.91	1.9	2.5	Me_2CO
			0.76	2.08	3.34	1.74	1.9	2.5	DMSO
			0.44	2.15	3.39	1.48			HMPT
			0.65	1.54	2.95	1.76	2.6	2.6	TFA
Me	H	H	1.12	7.68	3.60	2.40		2.7	$CDCl_3$
				8.06					C_6H_6
			0.63	7.26	3.07	1.80		2.5	TFA
H	Me	H	1.12	2.34	7.84	2.56			$CDCl_3$
					8.43				C_6H_6
H	Br	H	1.05	2.21		2.21			$CDCl_3$
Me	Me	H	1.19	7.80	7.93	2.68		1.0	$CDCl_3$
				8.15	8.47				C_6H_6
Me	H	Me		8.00		8.22			C_6H_6
Me	Br	H	1.08	7.71		2.30			$CDCl_3$
Br	Me	H	1.07		7.87	2.54			$CDCl_3$
Me	Me	Me	1.04	7.82	8.03	7.95			$CDCl_3$
				8.06	8.44	8.27			C_6H_6
			0.60	7.40	7.74	7.69			TFA
Me	Br	Me	0.98	7.83		7.75			$CDCl_3$
Br	Br	Me	0.90			7.75			$CDCl_3$

TABLE 3.14H SPECTRAL DATA FOR MISCELLANEOUS *N*-SUBSTITUTED PYRAZOLES

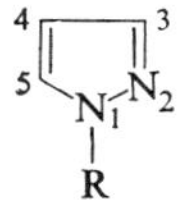

R1	R3	R4	R5	$\tau 1$	$\tau 3$	$\tau 4$	$\tau 5$	J_{34}	J_{35}	J_{45}	Solvent	Ref.
Et	H	H	H	8.53	2.50	3.77	2.62	2.0		2.3	$CDCl_3$	122
	CO_2Et	H	Me	5.84, 8.61	5.64, 8.60	3.48	7.70				$CDCl_3$	137
	Me	H	CO_2Et	5.50, 8.60	7.75	3.45	5.71, 8.66				$CDCl_3$	137
	CO_2Et	H	Ph	5.60, 8.63	5.77, 8.63	3.18	2.60				$CDCl_3$	139
	Ph	H	CO_2Et	5.38, 8.55	2.18, 2.67	2.88	5.70, 8.67				$CDCl_3$	139
	Ph	H	CO_2H	5.32, 8.50	2.60, 2.18	2.75	−1.5				$CDCl_3$	139
t-Bu	H	H	H	8.42	2.48	3.78	2.48				$CDCl_3$	122
				8.47	2.66	3.93	2.63				CCl_4	122
				8.69	2.57	4.02	2.94				Mesitylene	122
CO_2Et	H	H	H	5.46, 8.54	2.23	3.56	1.82				$CDCl_3$	139
	Me	H	Me	5.52, 8.56	7.73	4.01	7.47				$CDCl_3$	139
CH_2CO_2Et	H	H	H		2.53	3.69	2.46				$CDCl_3$	139
	Me	H	Me		7.80	4.16	7.76				$CDCl_3$	139
Tosyl	H	H	H		2.12	3.42	1.55	1.7	0.6	2.8	DMSO	122
					2.11	4.27	2.22	1.7	0.6	2.8	C_6D_6	122
	Me	H	Me		7.91	3.83	7.53		0.8		DMSO	122

TABLE 3.14H *(continued)*

R1	R3	R4	R5	$\tau 1$	$\tau 3$	$\tau 4$	$\tau 5$	J_{34}	J_{35}	J_{45}	Solvent	Ref.
$P(NMe_2)_2$	H	H	H	7.34	2.30	3.70	2.47	1.9		2.2	$CDCl_3$	149
				7.49	2.29	3.80	2.58	1.9		2.2	C_6D_6	149
	Me	H	H	7.39	7.76	3.92	2.53			2.4	Neat	149
				7.35	7.67	3.86	2.46			2.2	$CDCl_3$	149
				7.46	7.67	3.88	2.54			2.2	C_6D_6	149
				7.38	7.79	3.80	2.39			2.3	DMSO	149
	H	H	Me	7.39	2.53	4.03	7.76	2.0			Neat	149
				7.35	2.53	3.97	7.56	1.8			$CDCl_3$	149
				7.50	2.30	3.97	7.81				C_6D_6	149
				7.42	2.50	3.98	7.63				DMSO	149
	Me		Me	7.41	7.86	4.29	7.83				Neat	149
				7.38	7.81	4.22	7.73				$CDCl_3$	149
				7.38	7.73	4.22	7.80				DMSO	149
				7.42	7.88	4.19	7.83				Neat	149
$GeMe_3$	H	H	H	9.37	2.43	3.81	2.58	1.6		2.2	CCl_4	151
	Me	H	H	9.40	7.79	4.07	2.74			2.1	CCl_4	151
	Me	H	Me	9.37	7.89	4.34	7.80			0.6	CCl_4	151
$GeEt_3$	H	H	H	8.86	2.49	3.83	2.63	1.6		2.3	CCl_4	151
	Me	H	Me	8.85	7.88	4.35	7.80			0.6	CCl_4	151
$SiMe_3$	H	H	H	9.59	2.37	3.79	2.47	1.5		2.4	CCl_4	151
	Me	H	Me	9.58	7.87	4.28	7.78			0.8	CCl_4	151

TABLE 3.14I SPECTRAL DATA FOR 3-*H*-TYPE PYRAZOLE *N*-OXIDES (168)

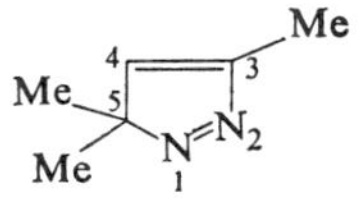

R1	R2	R4	$\tau 3$	$\tau 4$	$\tau 5$	Solvent
O		H	7.88	4.10	8.60	CCl_4
	O	H	7.92	2.90	8.62	CCl_4
O	O	H	7.75	3.65	8.38	$CDCl_3$
	O	OMe	7.78	5.88	8.70	CCl_4
	O	Cl	7.92		8.59	CCl_4
O	O	Br	7.78		8.42	CCl_4

N-substituent. In the case of 1-methylpyrazole in $CDCl_3$ and CCl_4 solutions, a combination of homo- and heteronuclear decoupling sharpened the spectrum enough to allow accurate determination of J_{34} and J_{45}, leading to the conclusion that H-5 absorbs at higher field than H-3. This assignment is opposite to that proposed from chemical shift data when 1,3- and 1,5-dimethylpyrazoles were used as models (133, 139). However, examination of the spectrum of the 5-deutero derivative **104** (140), prepared by decarboxylation of the corresponding deuterocarboxylic acid, showed unequivocally that the decoupling experiments gave the correct result.

The use of paramagnetic additions to solutions of these isomeric pyrazoles provides another method for identifying them (141). In the presence of small amounts of paramagnetic metal ions, exchange of the solute molecule between the first coordination sphere of the ion and the bulk medium is fast and the observed spectrum is the average of the paramagnetic and diamagnetic environments

$$\delta = \delta_p X_p + \delta_d X_d$$

where δ, δ_p, and δ_d are the shifts in the averaged spectrum, in the spectrum of the complex, and in the spectrum of the uncomplexed solute, respectively; X_p and X_d are the mole fractions of the two species. Under these conditions the line width is given by

$$\Delta\nu = \Delta\nu_p X_p + \Delta\nu_d X_d$$

If the electronic relaxation time of the metal ion is very short (e.g., Ni^{2+}) $\Delta\nu_p$ is given by

$$\Delta\nu_p = Ar^{-6} + B(\Delta\delta)^2$$

where r is the distance between the resonating proton and the paramagnetic ion, $\Delta\delta = \delta_p - \delta_d$ is the proton contact shift, and A and B are constants for the complex. Thus by estimating the distance r for the protons in a molecule one can obtain from the equations above an idea of the relative shifts and widths to be expected in dilute solutions of metal ions. This technique, using Ni^{2+} ions, was applied to the isomeric pair, 1,3- and 1,5-dimethylpyrazole, and proved to be a very simple method of distinguishing between them.

Spectra of a number of 1-carboxamidopyrazoles (**105**) (142) and other N–CO–X pyrazoles (X = Me or Ph, see Ref. 122) have been reported and in all cases in $CDCl_3$ or DMSO, the order of chemical shifts was H-4 > H-3 > H-5. Also, in these solvents as well as in benzene, the relative chemical shifts of methyl groups were Me-4 > Me-3 > Me-5. The resonance of the 5-substituent at lowest field in each case is expected because of the anisotropy of the adjacent C=O bond. The differences between J_{34} (~1.6 Hz) and J_{45}

(~2.9 Hz) in these compounds (see Table 3.14) are ascribed to a reduction of the aromatic character of the pyrazole ring by the *N*-substituent with associated localization of the double bonds.

Aryl-substituted pyrazoles are readily synthesized and the literature contains a considerable amount of data from their NMR spectra. It is interesting that many of the effects observed in the spectra of the phenyl substituents are independent of whether the group is attached to a nitrogen or a carbon atom. The spectral patterns from these phenyl substituents fall into two groups, those where the resonance is essentially a singlet, and those where a distinct multiplet pattern is observed (139, 143). In the latter case, protons ortho to the pyrazole ring are deshielded by about 0.4 ppm, compared with meta and para protons, which are not significantly different from those of benzene. These multiplet patterns are only observed for compounds where the phenyl and pyrazole rings can be coplanar. Since the effect is independent of whether *C*- or *N*-substitution of the pyrazole is involved, the major contributing factor is thought to be the magnetic anisotropy of the pyrazole ring, and not any of the possible electron-withdrawing mechanisms. The effect of substituents on the pyrazole ring α to the point of phenyl attachment is to reduce the coplanarity of the two rings, placing the ortho protons of the phenyl ring into the "zero shielding region" of the pyrazole ring current, thus producing the singlet-type spectra. Finar and Rackham (144), however, suggest the importance of the electric field associated with the lone pair of N-2 in the deshielding of the ortho protons, using 1-phenylpyrrole (**106**) as an example where, in spite of an obvious ring current in the

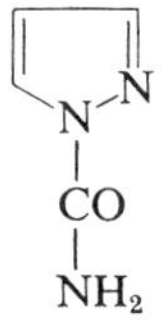

105 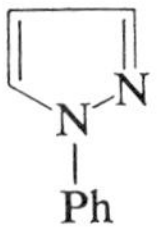 **106** 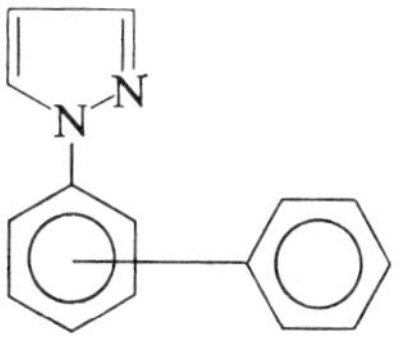 **107**

pyrrole ring, the phenyl resonance is essentially a singlet. Substitution at position 5 in 1-phenylpyrazole causes the phenyl signal to collapse to a singlet and greatly facilitates the identification of 1,5-diphenyl derivatives (145). Formation of a methiodide at N-2 causes the same simplification of the phenyl resonance. Spectra of the three (pyrazol-1-yl)biphenyls (**107**) showed similar effects. The monosubstituted phenyl rings of the meta and para isomers gave complex multiplets, whereas the ortho isomer showed a very simple signal, indicating considerable twist of the rings. A point of particular interest with these compounds was the chemical shift of the pyrazole 5-proton (τ 2.96) in the ortho isomer, an anomalously high value when compared with τ 2.08 and 2.09 for the meta and para isomers. In the spectrum of the 1-pentafluorophenylpyrazole, in addition to the expected couplings between the pyrazole protons, the signal from H-5 also contained long-range couplings to the *o*-fluoro substituents. With para-substituted phenyl groups the chemical shifts of the remaining phenyl protons vary according to the nature of the substituent (146) in the order expected from work on benzene. However, no such chemical shift parameters could be assigned to the 1-pyrazolyl substituent because its deshielding effect appeared to be tied to the nature of the para substituent, and was also influenced greatly by solvent changes. The steric effects of nitro groups in 4-nitro-, 2,4-dinitro-, and 2,4,6-trinitrophenyl derivatives have been studied in some detail (147, 122, 122a, b) and, in general, follow the trends expected from the above discussion.

An additivity relationship has been developed (139) which allows the calculation of the chemical shift of H-4 in 1,3,5-trisubstituted pyrazoles. It takes the form

$$\delta_4 = \delta_4(\mathrm{S}) + \alpha_1 + \alpha_3 + \alpha_5$$

where $\delta_4(\mathrm{S})$ is the chemical shift of H-4 of 1,3,5-trimethylpyrazole in solvent (S), and α_1, α_3, and α_5 are empirical constants representing the effect of replacing a methyl group by another group at the nominated position. The value of $\delta_4(\mathrm{S})$ for 5% solutions in $CDCl_3$ is 5.79 ppm and in CCl_4 is 5.66 ppm. The use of this equation gave calculated values of τ4 normally within 0.01 ppm of the observed value. The major effects on τ4 of the substituents studied in this work would be inductive in nature and this is demonstrated by a reasonable correlation between Hammett σ_p constants and the various α values. Temperature and concentration effects in the spectra of these compounds were shown to be small.

Nuclear magnetic resonance methods were also used to assign substitution patterns in a number of 3- or 5-amino-1-substituted pyrazoles (148a). The very extensive data available for substituted pyrazoles are collected in Tables 3.14A–I.

Spectra of pyrazolium ions, and indeed the protonation of pyrazoles, have received little attention. Protonation shifts between $CDCl_3$ and TFA solutions were used to assign substitution patterns (133), and protonation has normally been assumed to occur on nitrogen. 1-Phenylpyrazole, in concentrated sulfuric acid, gives a broad signal at τ −2.87, a position characteristic for NH protons, and signals from H-3, H-4, and H-5 were split by coupling to the N^+H proton. However, the methyl signal from *N*-methylpyrazole in the same solvent was only a singlet, indicating that protonation had occurred on N-2 to give the species **108**. A considerable amount of information on quaternary pyrazolium salts has been presented by Elguero *et al.* (148). Their assignments, based on model compounds, deuterium labeling, solvent effects, and internal consistency, are undoubtedly correct. Coupling over four and five bonds was observed between *N*-substituents and ring protons. If the results obtained for $CDCl_3$ solutions are ignored, chemical shifts of the ring protons move progressively to higher field as the polarity of the solvent increases. Solvent effects on methyl groups, however, were rather small and are of little use for structure determination. Peak positions in different solvents could be correlated by using mixed solvents to follow the proton shifts. Additivity relationships were also developed which allow the prediction of the chemical shifts of the ring protons in methyl- and phenyl-substituted pyrazolium ions. These relationships appear to be of little use to other workers because almost all possible combinations of these substituents occur in compounds included in this paper. Substituent effects were further discussed in terms of the mesomeric forms peculiar to each type of substitution and of the steric interaction of adjacent groups. The effect of the positive charge was evaluated by comparing the parent pyrazole with the corresponding pyrazolium ion to obtain a $\Delta\delta$ term for each group. In the 1,2-dimethyl series,

$$|\Delta\delta 3| > |\Delta\delta 5| > |\Delta\delta 4|$$

but for the 1-phenyl-2-methyl series,

$$|\Delta\delta 3| > |\Delta\delta 4| > |\Delta\delta 5|$$

Data for pyrazolium ions are collected in Table 3.15.

The spectra of a number of *N*-substituted pyrazoles have been measured where the substituent involves a direct bond between N-1 and a phosphorus, boron, germanium, or silicon atom. With the phosphorus compounds **109** (149) the spectra were similar to those of the *N*-alkylpyrazoles with the superposition of the ^{34}P couplings with protons (1.3 Hz) and methyl groups (0.9 Hz) on positions 5. The reaction between alkali metal borohydrides and pyrazole produced rather ill-defined products termed poly(1-pyrazolyl)-borates whose proton and ^{11}B spectra have been measured (150). In the case of the germanium and silicon compounds **110** (151) solvent shifts were used to identify isomers. Variations of the chemical shifts and coupling constants within the series N–C, N–Ge, and N–Si suggested a progressive increase in bond fixation. The chemical shift effects were opposite to those expected if inductive effects were dominant, and it was concluded that interaction between the N-1 lone pair and the outermost *d*-orbitals of silicon and germanium is important. Such interactions withdraw electrons from the pyrazole ring, reducing its aromatic character. Data for these compounds are included in Table 3.14.

TABLE 3.15 SPECTRAL DATA FOR PYRAZOLIUM IONS (148)[a]

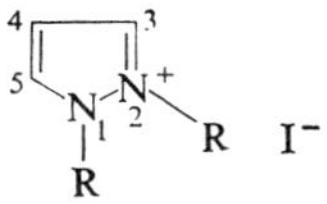

R1	R2	R3	R5	R4	$\tau 1$	$\tau 2$	$\tau 3$	$\tau 5$	$\tau 4$	Solvent
Me	Me	H	H	H	5.80	5.80	1.43	1.43	3.13	DMSO
CH_2Ph	Me	H	H	H	2.55, 3.98	5.82	1.26	1.26	2.98	DMSO
CH_2Ph	CH_2Ph	H	H	H	2.66, 4.06	2.66, 4.06	1.22	1.22	2.88	DMSO
Ph	Me	H	H	H	2.25	5.80	0.92	1.64	2.96	$CDCl_3$
						5.80	0.98	1.09	2.80	MMF
					2.18	6.01	1.05	1.12	2.82	DMSO
					2.20	5.88	1.29	1.55	2.94	$MeNO_2$
					2.22	5.97	1.31	1.40	2.97	MeOH
					2.40	6.00	1.50	1.55	2.94	D_2O
Ph	Et	H	H	H	2.25	5.66, 8.66	1.04	1.19	2.81	DMSO
Me	Me	Me	H	H	5.86	6.00	7.53	1.57	3.28	DMSO
Me	Me	Ph	H	H	5.73	5.93	2.38	1.13	2.93	DMSO
CH_2Ph	Me	Me	H	H	4.08, 2.62	6.08	7.51	1.37	3.16	DMSO
Ph	Me	Me	H	H	2.25	6.01	7.28	1.76	3.14	$CDCl_3$
					2.20	6.19	7.39	1.25	2.94	DMSO
Ph	Me	Ph	H	H	2.20	6.18	2.20	1.15	2.67	DMSO
Me	Me	H	H	Me	5.85	5.85	1.58	1.58	7.89	DMSO
Me	Me	H	H	Ph	5.78	5.78	0.92	0.92	2.5	DMSO
Ph	Me	H	H	Me	2.27	5.85	1.12	1.72	7.68	$CDCl_3$
					2.16	6.02	1.17	1.24	7.77	DMSO

Ph	Me	H	Me	H	2.28	6.23	1.24	7.74	3.01	DMSO
Ph	Me	H	Ph	H	2.3	6.09	0.99	2.61		DMSO
Me	Me	Me	H	Me	5.90	6.01	7.61	1.70	7.94	DMSO
Me	Me	Me	H	Ph	5.78	5.88	7.44	1.14	2.49	DMSO
Me	Me	Ph	H	Me	5.74	6.02	2.32	1.42	7.94	DMSO
Ph	Me	Me	H	Me	2.23	6.20	7.49	1.39	7.83	DMSO
Me	Me	Me	Me	H	5.99	5.99	7.55	7.55	3.36	DMSO
Et	Me	Me	Me	H	5.46, 8.68	5.97	7.50–7.53	7.50–7.53	3.32	DMSO
CH_2Ph	Me	Me	Me	H	2.7, 4.15	6.14	7.52–7.53	7.52–7.53	3.28	DMSO
Me	Me	Ph	Me	H	5.89	5.95	2.32	7.45	3.01	DMSO
Ph	Me	Me	Me	H	2.25	6.35	7.45	7.78	3.29	MeCN
					2.25	6.30	7.42	7.73	3.31	D_2O
					2.23	6.13	7.32	7.72	3.30	$CDCl_3$
					2.20	6.21	7.37	7.69	3.26	$MeNO_2$
					2.18	6.30	7.42	7.77	3.08	DMSO
CH_2Ph	CH_2Ph	Me	Me	H	2.7, 4.25	2.7, 4.25	7.48	7.48	3.02	DMSO
Me	Me	Ph	Ph	H	5.86	5.86	2.29	2.29	2.67	DMSO
Ph	Me	Me	Ph	H	2.31	6.27	7.36	2.64	2.72	DMSO
Ph	Me	Ph	Ph	H	2.1	6.19	2.1	2.52		DMSO
Me	Me	Me	Me	Me	6.04	6.04	7.63	7.63	7.99	DMSO
Me	Me	Me	Me	Et	6.06	6.06	7.62	7.62	7.51, 8.62	DMSO
Me	Me	Me	Me	CH_2Ph	6.02	6.02	7.63	7.63	2.74, 6.07	DMSO
Ph	Me	Me	Me	Me	2.29	6.38	7.53	7.87	7.87	MeCN
					2.25	6.30	7.48	7.83	7.83	D_2O
					2.29	6.17	7.42	7.82	7.87	$CDCl_3$
					2.27	6.23	7.45	7.75	7.82	$MeNO_2$
					2.28	6.35	7.51	7.84	7.87	DMSO

[a] Data for simple protonated pyrazoles are found in Table 3.14 when TFA is used as solvent.

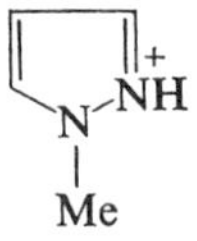

108

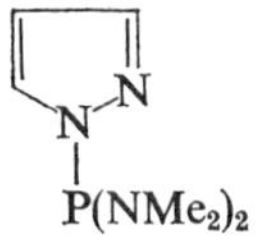

109

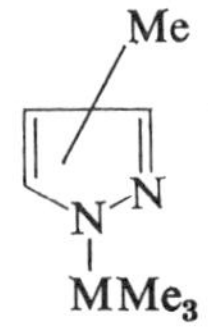

110, M = Ge or Si

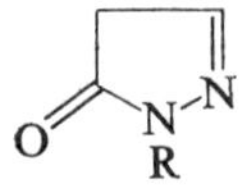

111

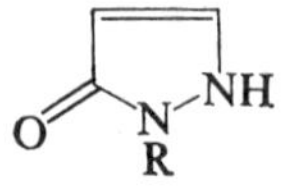

112

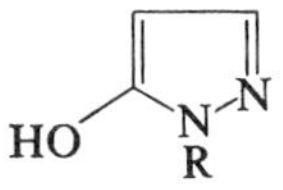

113

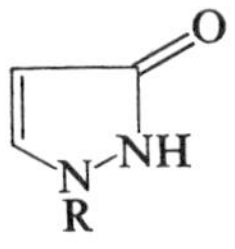

114

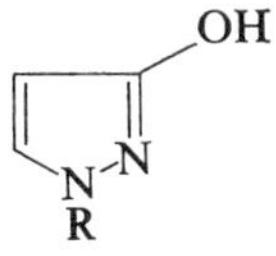

115

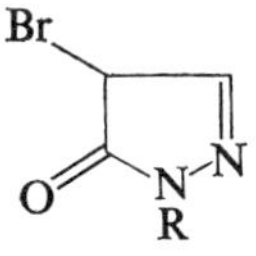

116

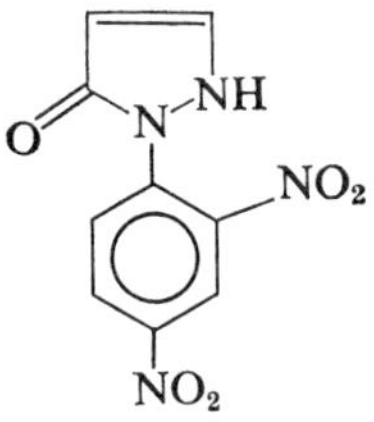

117

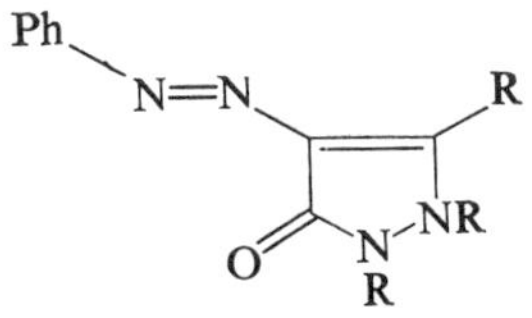

118

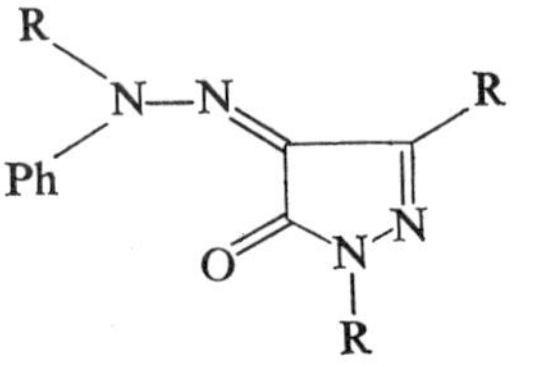

119

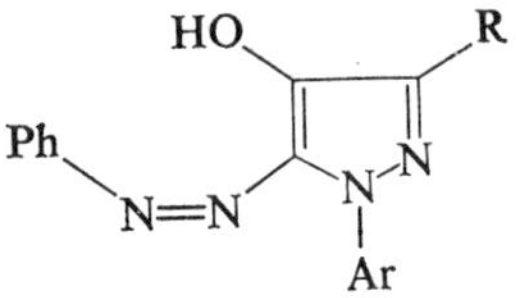

120

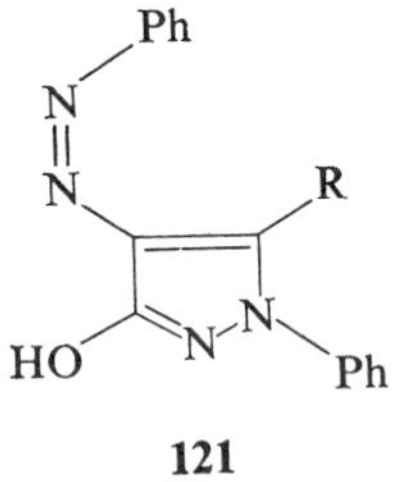

121

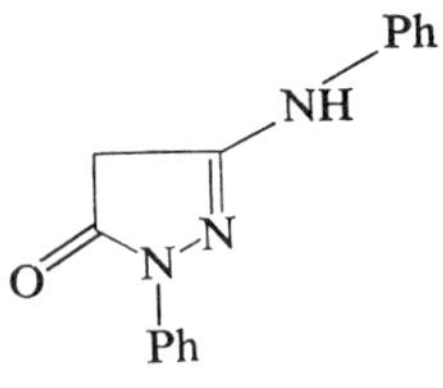

122

The pyrazole-3- and -5-ones can theoretically exist in a number of forms, the Δ^2-5-one **111**, the Δ^3-5-one **112**, and the 5-hydroxy form **113** for the 5-isomers and the 3-one **114** and 3-hydroxy form **115** for the 3-isomers. Proton magnetic resonance spectra of a number of these compounds have been used, in conjunction with other spectroscopic techniques, to determine the structures of the predominant tautomers in solution (152–154). The main use for NMR in these studies was to identify the Δ^2-5-one structures where the ring methylene protons could always be identified. Thus, in CCl_4 or $CDCl_3$ solutions, compounds without a 4-substituent have been shown to exist essentially in the Δ^2-5-one form (154a). However, in the presence of a 4-ethoxycarbonyl substituent the Δ^3-5-one form is favored. The sophistication of the approach has increased with successive publications and the valid point has been made in the most recent (154) that when a combination of physical methods is used, the data obtained from each technique are only comparable when solutions of the same concentration in the same solvent were used for each set of measurements. Unfortunately, NMR spectra alone were of little use with the majority of these compounds and the most important evidence was obtained by other techniques. Some data have been published for 4-bromopyrazolones (**116**) (155) and 1-(2,4-dinitrophenyl)-pyrazol-5-ones (**117**) (156) in papers on their synthesis. Spectra of phenylazo-pyrazolones (**118**) have been extensively studied (152, 154, 157, 158) and it was concluded that, in $CDCl_3$ solutions, the pyrazol-5-ones have the hydrazone structure (e.g., **119**) whereas the pyrazol-3- and -4-ones have the azohydroxy structures **120** and **121**. In the compounds studied, the hydrazone N*H* proton always absorbs 3–5 ppm to the low-field side of the O*H* proton from the azohydroxy structures.

An elegant piece of work involves the use of ^{15}N labeling to study the tautomerism of 3-anilino(^{15}N)-1-phenyl-2-pyrazolin-5-one (159). The only species present in $CDCl_3$, pyridine, or DMSO was the Δ^2-oxoamino isomer **122**, as shown by signals from the 4-methylene group and the presence of ^{15}N–H coupling in the ^{15}N-anilino derivative. Spectral data for many of these pyrazolones are collected in Table 3.16. Tautomerism in 5-aminopyrazolines has been extensively investigated by Elguero and his co-workers (154b, c).

2. *Reduced Pyrazoles*

The spectra of reduced pyrazoles, in particular Δ^2-pyrazolines (**123**), have been studied in great detail with the modern literature being again dominated by the Montpelier group.

The stereochemistry of the Δ^1-pyrazoline system **124** has received some attention in studies on the mechanisms of photochemical or thermal emission of nitrogen from these compounds to form cyclopropanes. A number of cis–trans pairs of 3,5-disubstituted Δ^1-pyrazolines (**125** and **126**) were studied

TABLE 3.16 SPECTRAL DATA FOR PYRAZOLONES AND HYDROXYPYRAZOLES

A. 3-HYDROXY- AND 3-ALKOXYPYRAZOLES

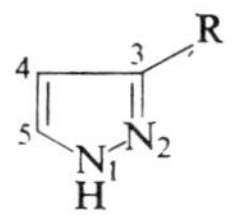

R1	R3	R4	R5	τ1	τ3	τ4	τ5	Solvent	Ref.
Ph	OH	H	Me	2.67	−1.1	4.46	7.84	$CHCl_3$	153
Ph	OMe	H	Me	2.63	6.13	4.34	7.90	Neat	153
Me	OH	H	Me	6.43	−1.14	4.59	7.83	$CHCl_3$	153
Me	OEt	H	Me	6.58	5.91, 8.75	4.65	7.98	Neat	153
Ph	OH	H	H		−0.24	4.11	1.75	DMSO	257

B. 5-Hydroxy- and 5-Alkoxypyrazoles

R1	R3	R4	R5	τ1	τ3	τ4	τ5	Solvent	Ref.
H	Me	H	OH		7.79	4.33		C_5H_5N	154
					7.88	4.65		C_5H_5N	256
				−0.94	7.88	4.69		DMSO	256
					7.55	4.11		TFA	256
					7.81	4.25		N HCl	256
					7.92			N NaOH	256
					7.79	4.61		SO_2	256
					7.83			D_2O	256
					7.80	4.67		DMF	256
H	H	Me	OH			7.82		C_5H_5N	154
H	Me	Me	OH		7.85	7.88		C_5H_5N	154
					7.62	7.98		TFA	256
Ac	Me	H	OH	7.42	7.42	4.23	1.10	$CDCl_3$	256
Ac	Me	H	OAc	7.39	7.43	3.89	7.71	$CDCl_3$	256
H	Me	H	OEt	1.16	7.44	4.50	5.81, 8.06	$CDCl_3$	256

TABLE 3.16 *(continued)*

C. Δ^3-PYRAZOL-5-ONES

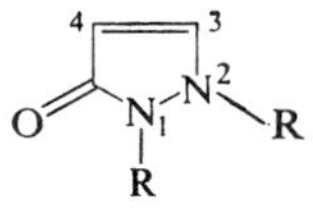

R1	R2	R3	R4	$\tau 1$	$\tau 2$	$\tau 3$	$\tau 4$	Solvent	Ref.
H	H	Me	H			7.82		D_2O	152
						8.5	4.9	C_5H_5N	152
H	H	Me	Me			8.52, 8.60	8.52, 8.60	C_5H_5N	152
Ph	H	Me	Me	2.1, 2.7	−0.6	8.10, 8.47	8.10, 8.47	$CDCl_3$	152
H	Me	H	Me		6.35	3.11	8.07	$CHCl_3$	154
					6.41		7.94	C_5H_5N	154
H	Me	Me	H		6.43	7.84	4.62	$CHCl_3$	154
					6.49	7.98	4.34	C_5H_5N	154
Me	Me	Me	H	6.71	6.79	7.87	4.91	Neat	153
H	Me	Me	Me		6.41	7.91	8.13	$CHCl_3$	154
					6.52	8.09	8.01	C_5H_5N	154
Me	Ph	Me	H	6.94	2.65	8.01	4.61	$CHCl_3$	153
Ph	Me	Me	H	2.70	6.97	7.80	4.63	$CHCl_3$	153
					6.49	7.55	4.01	TFA	255

D. Δ^2-PYRAZOL-5-ONES

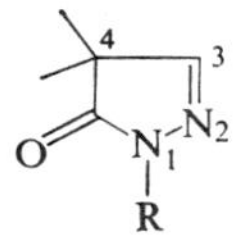

R1	R3	R4	R4′	τ1	τ3	τ4	τ4′	J_{34}	Solvent	Ref.
H	Me	Me	Me	0.87	8.02	8.90	8.90		$CDCl_3$	256
				0.28	8.07	8.88	8.88		$CDCl_3$	152
Me	Me	H	H	6.73	7.91	6.83	6.83		$CDCl_3$	256
Me	H	Me	H	6.69		8.65			$CDCl_3$	154
Me	Me	Me	Me	6.86	8.07	8.88	8.88		Neat	153
Me	Ph	H	H	6.59	2.7	6.40	6.40		$CHCl_3$	153
Ph	H	H	H		2.56	6.55	6.55	1.3	$CDCl_3$	154a
Ph	Me	H	H	2.2, 2.7	7.88	6.63	6.63		$CDCl_3$	152
Ph	Me	Me	Me	2.1, 2.8	8.04	8.82	8.82		$CDCl_3$	152
Ph	Ph	H	H	1.9–2.6	1.9–2.6	6.25	6.25		$CHCl_3$	153

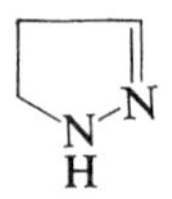

123 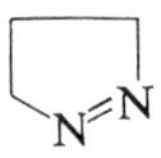 124 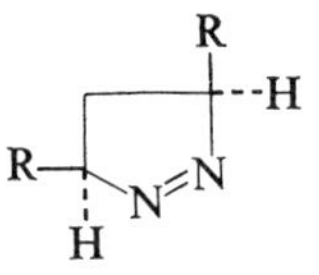 125 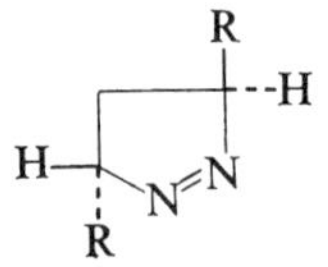 126 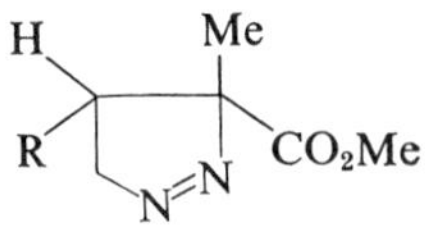 128

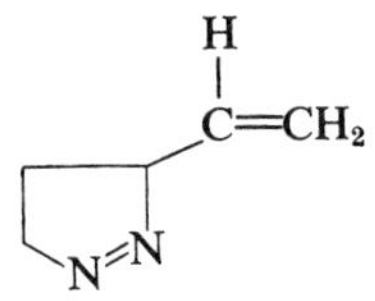

132 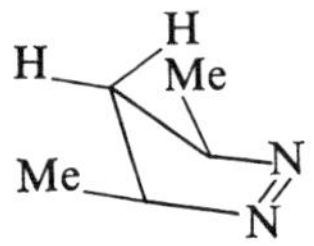 127 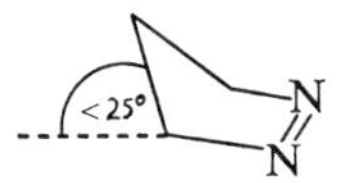 129

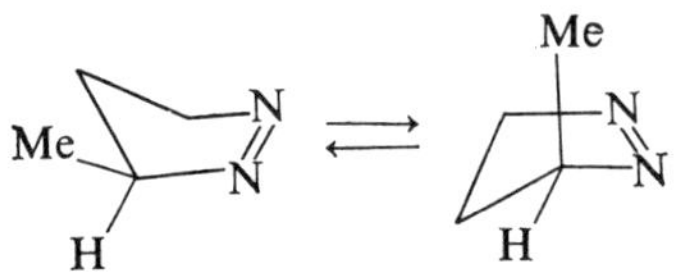

130 131 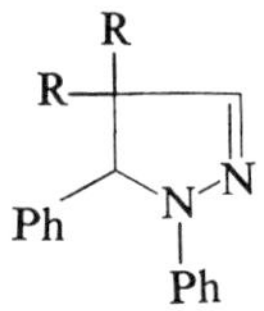 133

(160–163) and, in general, the trans isomers gave rise to symmetrical A_2X_2-type spectra, whereas in the cis compounds the protons on C-4 were not equivalent and a typical A_2MX pattern was obtained. In the spectra of these cis isomers, one proton from the 4-methylene group gives rise to peaks at extremely high field, and in the original work with diaryl substituents (163) this was thought to be due to aryl shielding. However, the effect was also observed with the dimethyl derivative (161) and was reasonably assigned to shielding effects of ring puckering. In such a conformation, stabilized by the large cis substituents, the pseudoaxial proton on C-4 is strongly shielded by the N=N bond **127**.

McGreer and his co-workers (164, 165) have studied the pucker of non-*N*-substituted Δ^1-pyrazolines of type **128**. The magnitudes of the vicinal couplings between protons on C-3 and C-4 suggest a puckered structure where the C-3, C-4, C-5 plane makes an angle of less than 25° with the C-2, N=N, C-5 plane **129**. Later chemical and NMR work (166) suggests that many of these compounds exist as a pair of rapidly interconverting puckered isomers **130** and **131**. Also, a complete analysis of the beautifully resolved spectrum at 100 MHz of 3-vinyl-Δ^1-pyrazoline (**132**) has been reported (167). The *N*-oxides of certain Δ^1-pyrazolines were prepared by Freeman in his study of the reactions of pernitrosomesityl oxide (168). Data for Δ^1-pyrazolines are collected in Table 3.17.

The Δ^2-pyrazolines contain an ethane-type fragment, C-4–C-5, in which the four substituents assume relatively fixed conformations, and hence the spectra produced are amenable to analysis. The results can be used to test theories concerned with the effects of geometry on bonding in such systems. Data for a large number of Δ^2-pyrazolines, most of them highly substituted, are collected in papers by several authors (169–173) and only those for more simple members of the series are included in Table 3.18.

The earliest work in the series (169) established the pattern of future papers by discussing the variations of coupling constants and chemical shifts between protons on C-4 and C-5 with known cis or trans stereochemistry. In general, J_{trans} was less than J_{cis}, but the actual values varied greatly with the nature of substituents on N-1 or on the ring carbons. In 1,5-diphenyl compounds of the type **133**, the C–C diamagnetic anisotropy from alkyl substituents on C-4 seems to be at least as important as the orientation of the 5-phenyl group in determining the chemical shift of H-5. According to McConnell (174), the contribution to the shielding of a proton ($\Delta\sigma$) due to neighboring C–C bonds can be calculated from the distance (r) of the proton from the midpoint of the C–C dipole, the acute angle (θ) that the vector of r makes with the symmetry axis of the group of electrons in the C–C bond, and the difference in the longitudinal and transverse magnetic susceptibilities of the electrons of the C–C bond ($\Delta X = X_L - X_T$)

$$\Delta\sigma = \frac{(1 - 3\cos^2\theta)\Delta X}{3_r^3} \tag{3.1}$$

The following approximations are inherent in the use of this equation. The major long-range shielding contribution by the C–C bonds is assumed to come from the β–γ bond. The C–C bond is considered to be axially symmetric and the geometric center of this bond is assumed to be the same as the electrical center of gravity of the C–C bond dipole. The magnetic anisotropy ΔX is an experimentally determined number based on approximations. Values of θ and r were calculated from the trigonometric relationships given by Bothner-By and Naar-Colin (175)

$$r = \left[\frac{19}{12}c^2 + b^2 + \frac{1}{9}bc(7 - 8\cos\psi)\right]^{1/2} \tag{3.2}$$

$$\cos\theta = \frac{15c + (2 - 16\cos\psi)b}{18r} \tag{3.3}$$

TABLE 3.17 SPECTRAL DATA FOR Δ^1-PYRAZOLINES

A. 3,4-SUBSTITUTED COMPOUNDS.[a] SOLVENT—CCl_4

R3	R3′	R4	R4′	τ3	τ3′	τ4	τ4′	τ5	J_{45}	J_{55}	Ref.
COMe	Me	H	Me	6.40	8.74	7.85	9.13	5.43, 6.05	9.0, 6.0	16.8	162
COMe	Me	Me	H	6.30	8.47	9.10	8 28	5.32, 6.02	8.1, 7.3	16.5	162
Me	CO_2Me	Et	H	8.74	6.29	8.77, 9.12	7.90	5.25, 6.04	7.9, 8.2	17.0	165
Me	CO_2Me	H	Et		6.33			5.19, 6.03	8.0, 8.0	17.0	165
Me	CO_2Me	H	H	8.50	6.29	7.95, 8.60	7.95, 8.60	5.43			165
CO_2Me	COMe	H	Me	6.24	7.67						259
COMe	CO_2Me	H	Me	7.48	6.24						259
CO_2Me	COMe	H	Ph	6.27	8.20			5.28			259
COMe	CO_2Me	H	Ph	7.50	6.82			4.98			259

B. 3,5-SUBSTITUTED COMPOUNDS

Substituents	$\tau 3$	$\tau 3'$	$\tau 4$	$\tau 4'$	$\tau 5$	$\tau 5'$	J_{34}	$J_{44'}$	Ref.
3,5′-Di-Me	8.71	5.43	8.73	8.73	5.43	8.71			161
3′,5′-Di-Me	5.80	8.46	9.52	7.92	5.80	8.46	8.5 av.	12.5	161
3,5′-Di-Ph	2.78	4.19	7.89	7.89	4.19	2.78	8.5 av.		163
3,5′-Di(4-ClPh)	2.83[b]	4.30	7.97	7.97	4.30	2.83[b]	8.5 av.		163
3,5′-Di(4-OMePh)	2.98[b]	4.25	7.95	7.95	4.25	2.98[b]	8.0 av.		163
3′,5′-Di(4-OMePh)	4.80	2.92	8.63[c]	7.66[c]	4.80	2.92			163

[a] For data on highly substituted derivatives see Ref. 259.
[b] Center of A_2B_2 multiplet.
[c] Assignments opposite to those published; see the text.

TABLE 3.18 SPECTRAL DATA FOR Δ^2 AND Δ^3-PYRAZOLINES

Δ^2-PYRAZOLINES

A. R = R′ = Ph. SOLVENT—$CDCl_3$ (171)

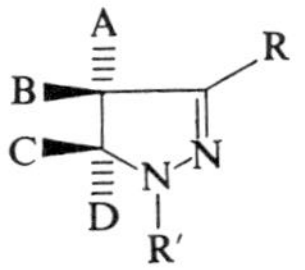

A	B	C	D	τA	τB	τC	τD	J_{AB} (gem)	J_{AC} (trans)	J_{BC} (cis)	J_{CD} (gem)
H	H	H	H	6.88	6.88	6.24	6.24	−16.0	8.7	12.0	−10.0
H	H	H	Me	7.19	6.63	5.56	8.74	−16.7	5.0	11.3	−6.3
H	H	Me	Me	6.86	6.86	8.51	8.51				
H	H	H	Bu	7.17	6.75	5.98		−16.6	5.8	10.5	
Me	H	H	Me	8.84	6.42	5.85	8.76	−7.0		10.0	−6.5
H	Me	H	Me	6.91	8.85	6.00	8.90	−7.0	2.3		−6.2
Me	H	Me	Me	8.86	6.91	8.53	8.89	−7.4			
H	H	H	Ph	6.94	6.25	4.81		−16.2	7.7	12.1	
H	Ph	H	H	5.48		6.25	5.98		6.0	9.9	−9.9
H	H	Ph	Ph	5.97	5.97						
H	Ph	H	Ph	5.52		4.91			5.2		
Ph	H	H	Ph		5.04	4.59				11.3	

H	Me	H	4-OMePh	6.67	8.72	5.28	6.45	−7.0	4.5		
H	4-OMePh	H	Me	5.9	6.45	5.9	8.85				−4.9
H	H	H	CO_2Me	6.62	6.80	5.38	6.40	−17.1	8.2	11.1	
H	H	Me	CO_2Me	6.43	6.85	8.43	6.37	−16.9			
H	Me	H	CO_2Me	6.27	8.66	5.57	6.39	−7.0	3.6		
H	CO_2Me	H	Me	6.08	6.44	5.29	8.81		3.6		−6.3
CO_2Me	H	Me	Me	6.43	6.01	8.49	8.74				
Me	Me	H	CO_2Me	8.49	8.58	5.63	6.49				
H	CO_2Me	H	CO_2Me	5.43	6.37	4.83	6.33		4.7		
CO_2Me	H	H	CO_2Me	6.38	5.22	5.03	6.29			13.3	
Me	CO_2Me	H	CO_2Me	8.33	6.39	5.27	6.32				
H	CO_2Me	Me	CO_2Me	5.70	6.39	8.33	6.39				
H	H	CH_2CO_2Me	CO_2Me	7.24	6.73	6.19	6.44	−16.2			
H	CO_2Me	*i*-Pr	CO_2Me	5.48	6.46		6.42				
H	CO_2Et	H	4-OMePh	5.83	5.85, 8.85	4.53	6.33		5.5		
H	CO_2Me	H	Ph	5.76	6.32	4.43			5.5		
H	Ph	H	CO_2Me	5.33		5.21	6.28		3.4		
H	4-OMePh	H	CO_2Et	5.41	6.33	5.86, 8.85			4.3		
H	4-ClPh	H	CO_2Me	5.40		5.26	6.41		4.1		
H	CO_2Me	H	4-ClPh	5.86	6.44	4.48			5.3		
H	H	H	CN	6.45	6.45	5.12					
H	4-OMePh	H	Ph	5.57	6.77	4.97			5.0		
H	Ph	H	4-OMePh	5.57		4.97	6.77		5.0		

TABLE 3.18 *(continued)*

B. R′ = H, R = CO_2Me. SOLVENT—$CDCl_3$ (179)

A	B	C	D	τR	τR′	τA	τB	τC	τD	J_{AB} (gem)	J_{AC} (trans)	J_{BC} (cis)
H	H	CO_2Me	H	6.24	4.10	6.75	6.75	6.19	5.49		8.5	8.5
H	H	4-OMePh	H	6.25	3.75	6.74	6.74	6.22	5.09	−16.9	10.5	11.2
4-ClPh	H	CO_2Me	H	6.24	3.36		5.80	6.10	5.24		3.9	
H	H	Ph	H	6.24	3.31	6.70	6.70		5.05	−17.1	9.8	11.5
Ph	H	Ph	H	6.39	3.27		5.77		5.16		6.5	
H	CO_2Me	Ph	H	6.24	3.12	5.74			4.69			13.4
Ph	H	CO_2Me	H	6.29	3.04		5.77		5.23		4.0	
H	H	Ph	Ph	6.24		6.45	6.45					
CO_2Me	H	Ph	Ph	6.24			5.12					

C. SOLVENT—$CDCl_3$ (172)

		B = Ph, D = CO_2Me			B = CO_2Me, D = Ph		
R	R′	τA	τC	J_{AC} (trans)	τA	τC	J_{AC} (trans)
Ph	4-OMePh	5.43	5.28	4.3	5.84	4.52	5.5
Ph	4-ClPh	5.40	5.26	4.1	5.86	4.48	5.3
Ph	2-ClPh	5.10	4.62	4.0	5.47	3.79	4.2
Ph	2,4-Di-BrPh	5.13	4.59	2.5	5.48	3.82	4.1
Ph	2,4,6-Tri-ClPh	5.15	4.77	7.3	5.33	4.12	7.2
Ac	Ph	5.22	5.05	4.5	5.82	4.28	6.8
CO_2Et	Ph	5.31	5.06	4.8		4.30	6.6

R	R′	A	B	C	D	R′ = CH_2 or Me_2								
						τ	τ'	J	τR	τA	τB	τC	τD	Solvent
Me	Et	Me	Me	Me	H	7.42	7.03	12.6	8.19		9.16	8.94	7.50	$CDCl_3$
Me	*i*-Pr	Me	Me	Me	H	9.05	8.95		8.18	8.97	9.15	8.70	7.25	$CDCl_3$
						9.09	9.02		8.32	9.20	9.28	8.54	7.35	C_6D_6
						9.11	8.98		8.25	8.98	9.22	8.76	7.28	$(CD_3)_2CO$
Me	CH_2Ph	Me	Me	Me	H	5.98	5.83	14.1	8.20	9.05	9.15	9.02	7.48	$CDCl_3$
						5.96	5.84	14.1	8.36	9.31	9.31	9.15	7.61	C_6D_6
						6.08	5.89	14.1	8.26	9.04	9.17	8.99	7.52	$(CD_3)_2CO$
Me	CH_2Ph	H	H	Ph	H	6.18	5.82	14.1	8.07				5.94	$CDCl_3$
						6.21	5.74	14.0	8.37				6.09	C_6D_6
						6.28	5.95	14.0	8.17				5.95	$(CD_3)_2CO$
						6.32	6.04	14.0	8.17				5.95	DMSO
						5.39	5.25	13.5	7.66				4.91	TFA
Me	CH_2Ph	Ph	H	H	H	5.85	5.71	13.5	8.27					$CDCl_3$
						5.79	5.69	13.5	8.36					C_6D_6
						5.82	5.82		8.33					$(CD_3)_2CO$
						5.85	5.85		8.31					DMSO
H	CH_2Ph	Me	H	H	H	6.02	5.79	13.0	3.55	8.96				$CDCl_3$
						5.97	5.68	13.3	3.62	9.27				C_6D_6
						6.01	5.82	13.0	3.44	8.98				$(CD_3)_2CO$
i-Pr	CH_2Ph	Me	Me	H	H	5.96	5.96		8.84	8.91	8.91	7.45	7.45	$CDCl_3$
						5.96	5.96		8.90	8.95	8.95	7.46	7.46	$(CD_3)_2CO$
						5.97	5.97		8.89	8.95	8.95	7.42	7.42	DMSO
Me	CH_2Ph	H	H	Me	Me	6.12	6.12		8.19	7.68	7.68	8.90	8.90	$CDCl_3$
H	Ph	H	H	Ph	H	2.72–3.12	2.72–3.12		3.24	6.54	7.22	2.73–3.12	5.00	$CDCl_3$
CHO	Ph	H	H	Ph	H	2.87, 3.05	2.87, 3.05		0.12	6.35	7.07	2.87–3.05	4.53	$CDCl_3$

TABLE 3.18 *(continued)*

E. C = Ph. SOLVENT—$CDCl_3$ OR CCl_4 (169)

R′	R	A	B	τD	J_{cis}	J_{trans}
Ph	Me	H	H	5.15	12	8
Ph	CH_2–CH_2–CH_2–CH_2		H	5.65		9
Ph	CH_2—CH_2—CH_2—CH_2			4.94	12	
Ph	CH_2—CH_2—CH_2		H	5.53		10
Ph	CH_2—CH_2—CH_2			5.00	12	
H	CO_2Me	CO_2Me	H	4.87		10
H	CO_2Me	H	CO_2Me	4.62	13	
Me	Ph	H	H	6.05	14	9.5

Δ^3-PYRAZOLINES

1,2-DIMETHYL-Δ^3-PYRAZOLINES. SOLVENT—$CDCl_3$ (186)

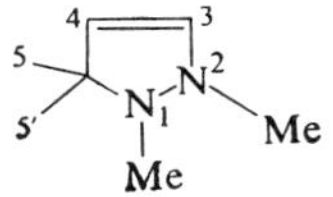

R3	R4	R5	τ1	τ2	τ3	τ4	τ5	τ5′	$J_{35'}$	$J_{45'}$	J_{34}
Ph	H	H	7.38	7.34	2.55	4.72	6.13	6.13		2.4	
Ph	Me	H	7.38	7.50	2.64	8.24	6.19	6.19		1.5	
Ph	Et	H	7.42	7.53	2.69	7.83, 9.50	6.18	6.18		1.8	
Ph	H	Me	7.41	7.33	2.64	4.88	8.76	6.47		2.7	
Me	Ph	H	7.38	7.30	8.09	2.77	5.79	5.79	1.8		
H	Ph	Me	7.41	7.24	3.74	2.75	8.67	6.07	1.5		
Me	H	Me	7.44	7.37	8.25	5.45	2.73	5.60	1.9	1.9	1.4

where b is the C–H bond length, c is the C–C bond length, ψ is the dihedral angle formed by the planes containing H-5–C-5–C-4 and C-5–C-4–alkyl, and r and θ are defined as above. These equations were used to calculate $\Delta\sigma$, the shift of H-5 on replacing C-4–H with C-4–C, but the results were not good.

McConnell's equation was later modified (176) by the addition of a term to correct for the fact that the bond dipoles were really not located at a single point, to give the rather complex expression

$$\Delta\sigma = \frac{\Delta\chi(1 - 3\cos^2\theta)}{3r^3} + \frac{s^2}{r^5}[-(\chi_L + 2\chi_T)/2 + 5(\chi_L\cos^2\theta + \chi_T\sin^2\theta) - \frac{35}{6}(\chi_L\cos^4\theta + \chi_T\sin^4\theta)] \quad (3.4)$$

Extension of the original pyrazoline work with more compounds and other values of $\Delta\chi^{cc}$ (172) and finally, use of Eq. (3.4) (177) produced more satisfactory results. All the data obtained are summarized in Table 3.18.

In spectra of Δ^2-pyrazolines with only one substituent on the C-4–C-5 bond **134**, the sum ($J_{cis} + J_{trans}$) was often readily available from a partial analysis of the system. In the case of the 1-substituted derivatives **135**, small variations in the sum reflect the electron-withdrawing or -donating power of the substituent (178). The spectra of a series of 3-carbomethoxy-Δ^2-pyrazolines (**136**) have been presented (179). Introduction of a phenyl group at position 5 has a large effect on the relative chemical shift of the two protons on C-4, although a carbomethoxyl group in the same position exerts very little effect. The cis and trans vicinal couplings in compounds with two protons on C-4 are nearly equal and differ from those found for 4-substituted derivatives. The presence of three substituents, in positions 4 and 5, appears to hold the ring rigid and the line width for the N–H proton varies with the number of phenyl groups on C-5. Restricted rotation of the carbonyl substituent in a series of 1-formyl- and 1-acylpyrazolines has also been inferred from the spectra of these compounds (179a).

Nonequivalence of ethyl, benzyl, and isopropyl groups on nitrogen atoms in 2- and 3-pyrazolines has been observed (173) and was shown to depend on asymmetry in the molecule, arising from an asymmetric ring carbon atom, a quaternary nitrogen atom, or a slowly inverting nitrogen atom. Such nitrogen inversion was temperature dependent and is affected by substituents. Spectra of a number of 1,3,5-triaryl-Δ^2-pyrazolines (**137**) have been reported by Huisgen and his co-workers (180, 181). Data for certain of these compounds are included in Table 3.18.

1,2-Dimethyl-Δ^2-pyrazolinium salts (**138**) have been synthesized and their spectra measured (182). It is remarkable how constant the difference between the signal from the N$^+$–Me and that from the adjacent N–Me group

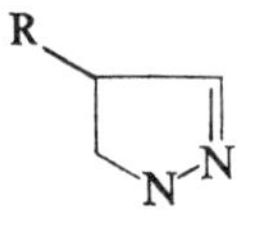

134 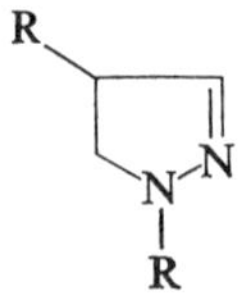 135 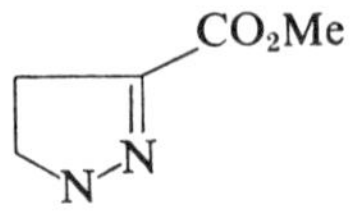 136 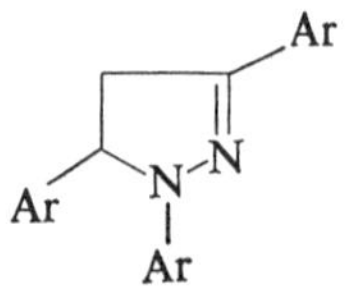 137

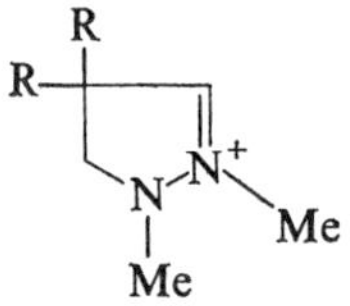

138
140, R = Me

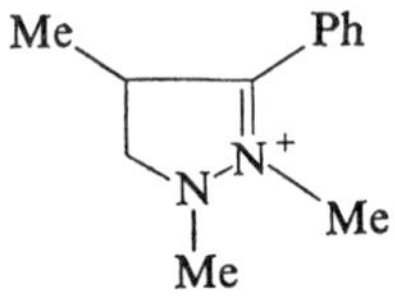

139 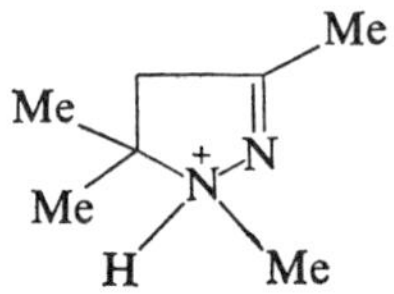 141 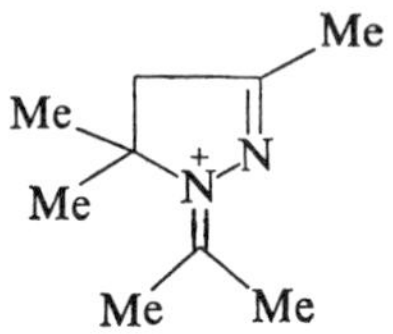 142

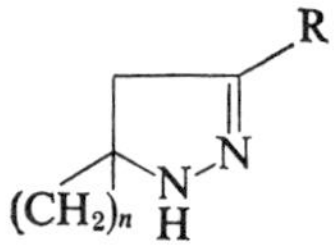

143 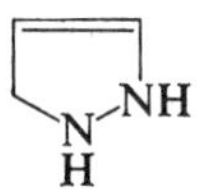 144

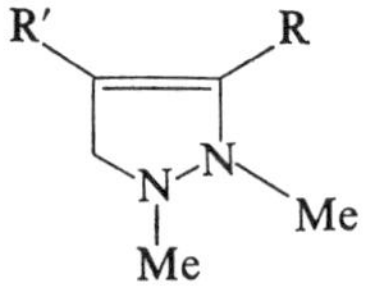

145
146, R = Ph, R′ = Me

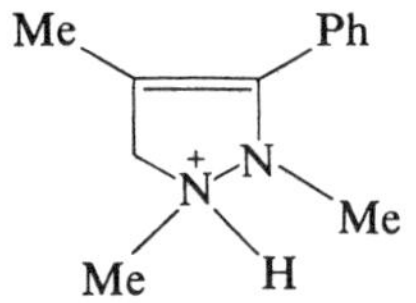

147

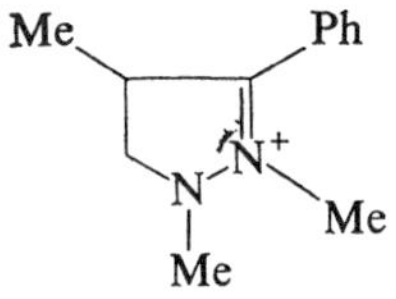

148 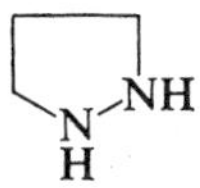 149 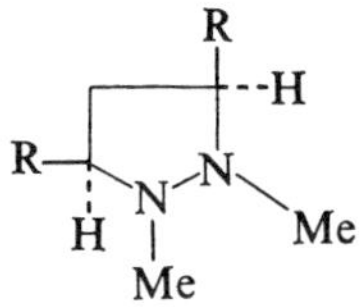 150 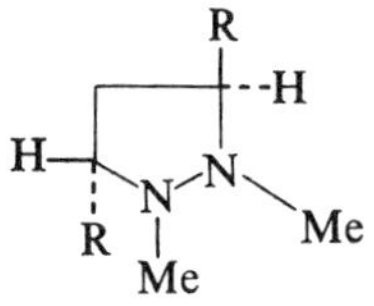 151

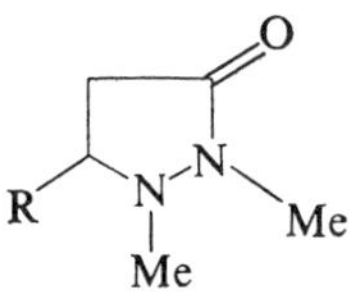

152

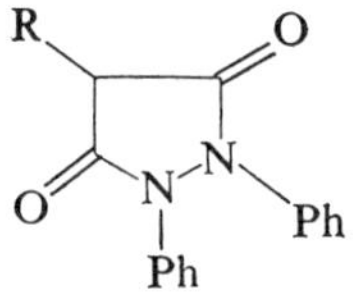

153
154, R = H

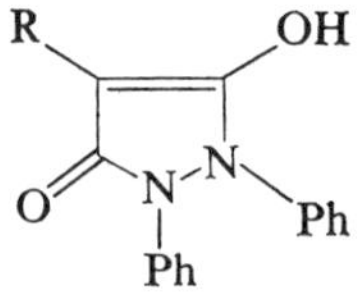

155 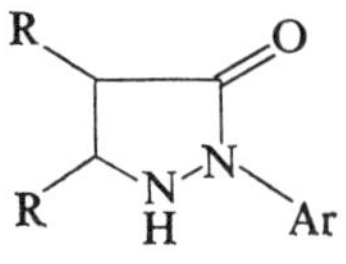 156

seems to be (~0.7 ppm), regardless of solvent. Solvent effects were unimportant; in particular the chemical shifts between $CDCl_3$ solutions and those in TFA are extremely small (~0.1 ppm), suggesting that the pyrazolium ion is not further protonated by the acidic solvent. Specific deuteration and double-resonance experiments made possible the analysis of the rather complex spin systems in many of these compounds. In compounds of the type **139** homoallylic couplings were observed between the 2-Me group and H-4 ($J = 1.7$ Hz), and in those of type **140**, allylic coupling between the 2-Me and H-3 ($J = 1.2$ Hz). Little detail was available from the spectra of 1,1-dimethyl-Δ^2-pyrazolinium ions.

In a study of the protonation of various Δ^2-pyrazolines (183), the spectrum of the methiodide (**141**) in $CDCl_3$ was measured at room temperature and at $-56°$. At room temperature the *gem*-dimethyl group gave a broad singlet, τ 8.36, but at $-56°$ this was replaced by a pair of singlets, τ 8.14, 8.58. This could only occur if N-1 was quaternary. Similar spectra were obtained with liquid SO_2 as solvent. Proof of 1-protonation of 3,5,5-trimethyl-Δ^2-pyrazoline was later obtained by condensation of the cation with acetone to give the adduct **142** (184). Representative spectra of these Δ^2-pyrazolinium ions are summarized in Table 3.19. Spectra of a number of 5,5-spiro-Δ^2-pyrazolines (**143**) have also been reported (185).

Little NMR information on Δ^3-pyrazolines (**144**) is available. Apart from a number of preparative papers, only two short communications (186, 187) have appeared which emphasize the NMR aspects of these compounds. In spectra of a number of *C*-substituted 1,2-dimethyl derivatives **145**, the signal from the 1-methyl group usually appears at higher field than that from the 2-methyl group. However, in compounds with a 3-phenyl and a 4-alkyl substituent this order is reversed, due to the shielding effect of the phenyl ring which is twisted considerably out of the plane of the pyrazoline system (186). With 3-phenyl-1,2,4-trimethyl-Δ^3-pyrazoline (**146**), protonation has been shown to occur first on N-1 to give the ion **147**, which slowly rearranges to the *C*-protonated species **148** (187).

Information on the spectra of pyrazolidines (**149**) is widely scattered and relatively few theoretical conclusions have been drawn from the data presented. An interesting study of conformational preferences in the cis **150** and trans **151** isomers of 3,5-disubstituted 1,2-dimethylpyrazolidines showed the presence of considerable steric interaction in the group C(R)–N(Me) but very little, if any, in the group N(Me)–N(Me) (188). This work was extended to the 1-benzyl derivatives where the benzyl methylene protons were shown to be nonequivalent (173). Evidence for double nitrogen inversion was obtained for these pyrazolidines, as well as for the inversion of the nonamide nitrogen atom in the similar pyrazolidones (**152**). Data for these two groups of compounds are collected in Table 3.20.

TABLE 3.19 SPECTRAL DATA FOR 1,2-DIMETHYLPYRAZOLINIUM IONS (182)

<table>
<tr><th>R3</th><th>R4</th><th>R4′</th><th>R5</th><th>R5′</th><th>τ1</th><th>τ2</th><th>τ3</th><th>τ4</th><th>τ4′</th><th>τ5</th><th>τ5′</th><th>J_{cis}</th><th>J_{trans}</th><th>J_{gem}</th><th>Solvent</th></tr>
<tr><td>Me</td><td>H</td><td>H</td><td>H</td><td>H</td><td>7.00</td><td>6.28</td><td>7.47</td><td></td><td></td><td></td><td></td><td></td><td></td><td></td><td>$CDCl_3$</td></tr>
<tr><td>Me</td><td>Me</td><td>H</td><td>H</td><td>H</td><td>6.99</td><td>6.26</td><td>7.49</td><td>8.58</td><td></td><td></td><td></td><td></td><td></td><td></td><td>$CDCl_3$</td></tr>
<tr><td></td><td></td><td></td><td></td><td></td><td>7.03</td><td>6.34</td><td>7.55</td><td>8.67</td><td></td><td></td><td></td><td></td><td></td><td></td><td>DMSO</td></tr>
<tr><td></td><td></td><td></td><td></td><td></td><td>6.98</td><td>6.33</td><td>7.55</td><td>8.58</td><td></td><td></td><td></td><td></td><td></td><td></td><td>TFA</td></tr>
<tr><td>Me</td><td>Me</td><td>Me</td><td>H</td><td>H</td><td>7.06</td><td>6.42</td><td>7.68</td><td colspan="2">8.73</td><td colspan="2">6.75</td><td></td><td></td><td></td><td>DMSO</td></tr>
<tr><td></td><td></td><td></td><td></td><td></td><td>6.95</td><td>6.31</td><td>7.49</td><td colspan="2">8.58</td><td colspan="2">6.63</td><td></td><td></td><td></td><td>TFA</td></tr>
<tr><td>Me</td><td>Ph</td><td>H</td><td>H</td><td>H</td><td>6.97</td><td>6.27</td><td>7.80</td><td>2.48</td><td>5.18</td><td>6.67</td><td>5.93</td><td>10.5</td><td>10.5</td><td>10.5</td><td>DMSO</td></tr>
<tr><td></td><td></td><td></td><td></td><td></td><td>6.87</td><td>6.23</td><td>7.70</td><td>2.55</td><td>5.34</td><td>6.40</td><td>5.94</td><td></td><td></td><td></td><td>TFA</td></tr>
<tr><td>Ph</td><td>H</td><td>H</td><td>H</td><td>H</td><td>6.85</td><td>6.08</td><td>2.38</td><td></td><td colspan="2">6.21</td><td></td><td></td><td></td><td></td><td>$CDCl_3$</td></tr>
<tr><td></td><td></td><td></td><td></td><td></td><td>6.95</td><td>6.27</td><td>2.26</td><td></td><td colspan="2">6.35</td><td></td><td></td><td></td><td></td><td>DMSO</td></tr>
<tr><td></td><td></td><td></td><td></td><td></td><td>6.84</td><td>6.17</td><td>2.35</td><td></td><td colspan="2">6.24</td><td></td><td></td><td></td><td></td><td>TFA</td></tr>
<tr><td>Ph</td><td>Me</td><td>H</td><td>H</td><td>H</td><td>6.86</td><td>6.14</td><td>2.37</td><td>8.65</td><td>5.80</td><td>6.86</td><td>5.80</td><td></td><td></td><td></td><td>$CDCl_3$</td></tr>
<tr><td></td><td></td><td></td><td></td><td></td><td>6.94</td><td>6.35</td><td>2.28</td><td>8.83</td><td>5.90</td><td>6.93</td><td>5.90</td><td></td><td></td><td>10.0</td><td>DMSO</td></tr>
<tr><td></td><td></td><td></td><td></td><td></td><td>6.87</td><td>6.19</td><td>2.33</td><td>8.62</td><td>5.80</td><td>6.87</td><td>5.80</td><td></td><td></td><td></td><td>TFA</td></tr>
<tr><td>H</td><td>Me</td><td>Me</td><td>H</td><td>H</td><td>7.07</td><td>6.29</td><td>1.70</td><td colspan="2">8.72</td><td colspan="2">6.65</td><td></td><td></td><td></td><td>DMSO</td></tr>
<tr><td></td><td></td><td></td><td></td><td></td><td>6.93</td><td>6.17</td><td>2.08</td><td colspan="2">8.55</td><td colspan="2">6.51</td><td></td><td></td><td></td><td>TFA</td></tr>
<tr><td>Me</td><td>H</td><td>H</td><td>Me</td><td>H</td><td>7.14</td><td>6.46</td><td>7.63</td><td></td><td></td><td>8.73</td><td></td><td></td><td></td><td></td><td>DMSO</td></tr>
<tr><td></td><td></td><td></td><td></td><td></td><td>7.04</td><td>6.38</td><td>7.55</td><td colspan="2">6.55</td><td>8.58</td><td>6.55</td><td></td><td></td><td></td><td>TFA</td></tr>
<tr><td>Me</td><td>H</td><td>H</td><td>Me</td><td>Me</td><td>7.08</td><td>6.22</td><td>7.43</td><td colspan="2">6.77</td><td colspan="2">8.65</td><td></td><td></td><td></td><td>$CDCl_3$</td></tr>
<tr><td></td><td></td><td></td><td></td><td></td><td>7.19</td><td>6.42</td><td>7.63</td><td colspan="2">6.88</td><td colspan="2">8.75</td><td></td><td></td><td></td><td>DMSO</td></tr>
<tr><td></td><td></td><td></td><td></td><td></td><td>7.07</td><td>6.23</td><td>7.53</td><td colspan="2">6.83</td><td colspan="2">8.62</td><td></td><td></td><td></td><td>TFA</td></tr>
<tr><td>Me</td><td>H</td><td>H</td><td>Ph</td><td>H</td><td>7.20</td><td>6.37</td><td>7.58</td><td colspan="2">6.37</td><td>2.58</td><td>5.55</td><td>13.0</td><td>10.0</td><td></td><td>DMSO</td></tr>
<tr><td></td><td></td><td></td><td></td><td></td><td>7.05</td><td>6.26</td><td>7.48</td><td>6.00–6.90</td><td>6.00–6.90</td><td>2.55</td><td>5.46</td><td>13.0</td><td>10.0</td><td></td><td>TFA</td></tr>
<tr><td>Ph</td><td>H</td><td>H</td><td>Me</td><td>H</td><td>6.96</td><td>6.26</td><td>2.22</td><td>6.34</td><td>8.61</td><td>6.34</td><td></td><td></td><td></td><td></td><td>DMSO</td></tr>
<tr><td></td><td></td><td></td><td></td><td></td><td>6.82</td><td>6.10</td><td>2.33</td><td>6.30</td><td>8.46</td><td>6.30</td><td></td><td></td><td></td><td></td><td>TFA</td></tr>
<tr><td></td><td></td><td></td><td></td><td></td><td>6.97</td><td>6.17</td><td></td><td>6.38</td><td>8.72</td><td>6.38</td><td></td><td></td><td></td><td></td><td>C_5H_5N</td></tr>
</table>

TABLE 3.20 SPECTRAL DATA FOR PYRAZOLIDINES. SOLVENT—$CDCl_3$

Substituents	Temperature	$\tau 1$	$\tau 2$	$\tau 3$	$\tau 5$	$\tau 4$	Ref.
1,2,3,5-Tetra-Me (cis)	32°	7.56	7.56	8.83	8.83		188
(trans)	32°	7.55	7.55	8.87	8.87		188
1,2-Di-Me,3,5-Di-Ph (cis)	32°	7.64	7.64			7.20, 7.59	188
(trans)	32°	7.63	7.63			7.73	188
2,3-Di-Me, 1-CH_2Ph	37°		7.75	8.83			173
2,5-Di-Me, 1-CH_2Ph	37°		7.62		8.94		173

Keto-enol tautomerism in 4-substituted 1,2-diphenyl-pyrazolidin-3,5-diones (**153**) in a number of solvents has been studied (189). The diketo form **154** prevails in the parent compound and its alkyl derivatives, but enolic structures **155** are predominant for compounds with electron-attracting substituents. Also, the spectra of some pyrazolidinones (**156**) have been reported (190).

C. The Imidazole System

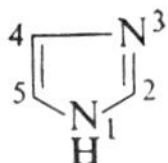

In contrast to the previous section on pyrazoles, comparatively little has been done on the spectra of imidazoles. The original work on imidazole and its methyl derivatives (191) established the relative shielding of the ring protons and the time-averaged equivalence of H-4 and -5. ^{13}C satellites were used to obtain a full analysis of the spectrum, but it is obvious that the values obtained are only approximate by more modern standards. The effect of methyl substitution was also discussed. This work was extended to the *N*-acyl derivatives **157** (192). In the spectrum of the 1:1 complex of *N*-acetyl-2-methylimidazole (**158**) and acetic acid, the concentration dependence of the N*H* signal was interpreted in terms of a salt structure **159**.

Joop and Zimmermann (193) have studied the spectra of imidazole and other azoles. Normally the NH protons give broadened signals, but fast exchange reactions were used to decouple the ^{14}N–H coupling, and hence, sharpen the spectrum. Imidazole and its 4-methyl derivative self-associate by an "*n*-donor" mechanism at sufficiently high concentrations in nonpolar solvents.

The most significant investigation of the spectra of imidazoles has been carried out by Mannschreck, Staab, and their co-workers (194–196) who have been particularly interested in the protonation of this ring system. Proton magnetic resonance data were obtained for a number of *C*- and *N*-substituted derivatives (194). In concentrated sulfuric acid, spectra of the *N*-protonated species were observed and all parameters for the system,

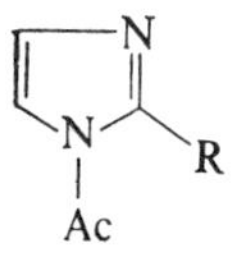

157, R = H **159** 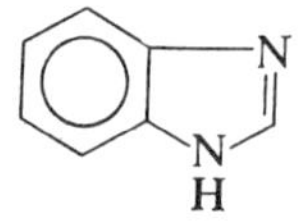 **160** 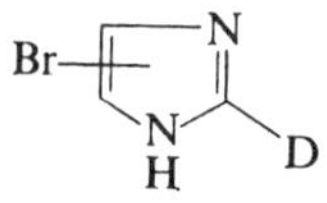 **161** 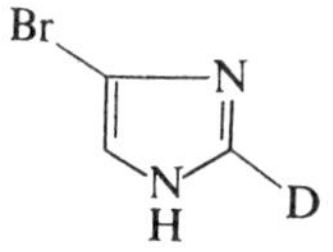 **162**
158, R = Me

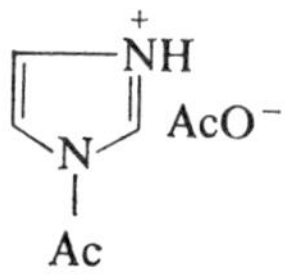

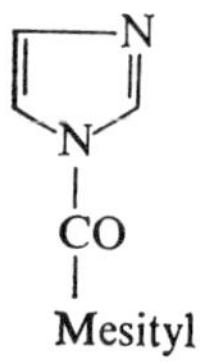

163 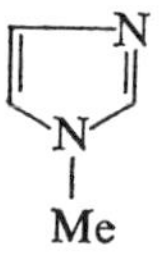 **164** 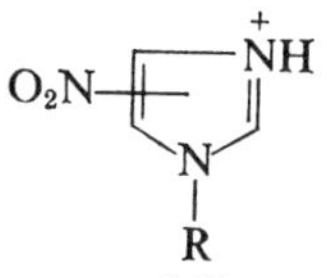 **165** 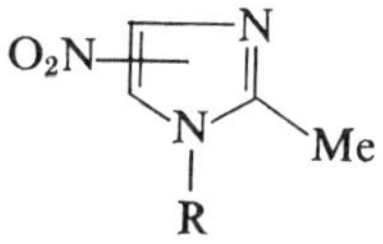 **166** 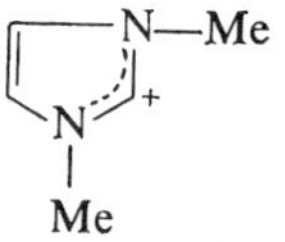 **167**

including couplings involving H-1, could be obtained. The results proved unambiguously that imidazole and benzimidazole (**160**) protonate on N-3. These workers also showed that 4- and 5-substituted imidazoles exist predominantly in one tautomeric form (195). This elegant piece of work involved *N*-protonation and/or *N*-deuteration of 4(or 5)-bromo-2-deutero-imidazole (**161**) and the results obtained were compatible only with protonation of the 4-bromo isomer **162**. In Ref. 196, a study was made of the rotational isomerism of *N*-mesitoylimidazoles (**163**). Protonation shifts, obtained by direct comparison of $CDCl_3$ and TFA solutions, for *N*-methylimidazoles have been used (197) to determine the site of protonation in a number of *N*-methyltriazoles. Thus for the parent 1-methylimidazole (**164**), which can protonate only on N-3, the signal from H-2 was shifted downfield by 1.26 ppm, and those from H-4 and H-5 moved only 0.52 and 0.64 ppm, respectively. This was not unexpected because H-2 is associated with the full positive charge whereas the other protons are adjacent to only part of this charge. The conjugate acids of 1-substituted -4- and 5-nitroimidazoles (**165**) in fluorosulfonic acid give broad signals which are not sharpened by ^{14}N decoupling (198). However, the chemical shift of the N–H proton was shown to depend on the relative position of the nitro group; N–H α to a nitro group, τ -2.6 to -1.8; and β to a nitro group, τ -1.7 to -1.3.

Protonation has also been used to differentiate between 1-substituted 2-methyl-4- or -5-nitroimidazoles (**166**) (199). The effect of the nitro group on the electron density around N-1 is pronounced and the N–CH_2CH_2–X protons give broader peaks (~0.8 ppm) for the 5-nitro than for the 4-nitro (~0.4 ppm) derivatives. The structure of 1-methylimidazole methiodide has been a subject of controversy, but NMR evidence shows that it has the 1,3-dialkylimidazolium structure **167** to the exclusion of any other suggested forms (200).

The available information on imidazole solvent effects is summarized by Wang and Li (201) in the preamble to their work on the zinc–imidazole complex. Imidazole undergoes extensive self-association through –NH· · ·N– bonds in solvents which do not compete effectively for hydrogen bonding. Infrared studies on carbon tetrachloride solutions showed the presence of linear oligomers (202) and this agrees with NMR data obtained for $CDCl_3$ solutions (194). However, there appears to be much less association in dimethyl sulfoxide solutions.

The relative chemical shifts of H-4 and H-5 in *N*-substituted imidazoles have been determined (138, 202). For the *N*-alkyl series in $CDCl_3$, H-4 normally absorbs at lower field than H-5, but as the bulk of the alkyl substituent increases the chemical shift difference becomes smaller. In the case of the *N*-arylimidazoles, aryl shielding ensures that H-5 resonates at a lower field than H-4. Proton magnetic resonance spectra have been used by a number of groups of workers to determine the orientation of substituents in the products of imidazole halogenations (203–205). During work on cephalosporidine (**168**), the spectrum of 1-imidazylacetic acid (**169**) was studied and the observed solvent and concentration effects were interpreted in terms of a zwitterionic structure **170** (206). Photolysis of 2-chloro-2-nitrosobutane produced a compound shown to have the dinitrone structure **171** (207).

Imidazole and certain of its derivatives have been included in general papers on the correlation of π-electron densities with proton chemical shifts (208) and on ^{13}C–H couplings in aza-aromatic systems (90). Spectral data for imidazoles are collected in Table 3.21.

Only scattered references are found to the spectra of reduced imidazoles. The imidazolines (**172**) are mentioned in a study of geminal couplings in reduced heterocycles where J_{22} was shown to have a value between -3.5 and -4 Hz (209). Also, the ^{19}F spectrum of the fluorinated imidazoline (**173**) has been measured (210). Greenhalgh and Weinberger (211) have done interesting work on the di-*N*-acylimidazolid-2-ones and -thiones (**174**). Apparently the preferred conformation for these compounds has the three carbonyl groups in trans coplanar positions (**175**). The thione grouping has a

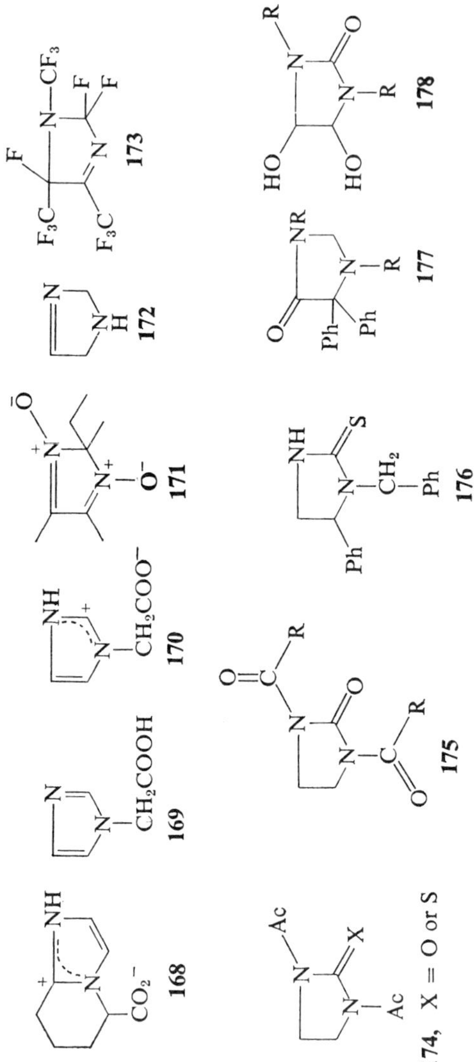
168
169
170
171
172
173
174, X = O or S
175
176
177
178

TABLE 3.21 SPECTRAL DATA FOR IMIDAZOLES

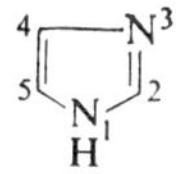

Substituents	Solvent	$\tau 1$	$\tau 2$	$\tau 4$	$\tau 5$	J_{24}	J_{25}	J_{45}	Ref.
None	$CDCl_3$		2.30	2.87	2.87				192
2-Me	$CDCl_3$		7.58	3.06	3.06				192
4-Me	$CDCl_3$		2.44	7.73	3.25				192
4,5-Di-Me	$CDCl_3$		2.41	7.81	7.81				192
4-Br	THF		2.54		2.98				194
2-Me, 4-Br	THF				3.13				194
1-Me	$CDCl_3$	6.40	2.63	3.02	3.22				197[a]
1-Me, 2-Br	$CDCl_3$	6.46		3.06	3.06				197
1-Me, 5-Br	$CDCl_3$	6.47	2.51	3.03					197
1-Me, 4-NO_2	$CDCl_3$	6.20	2.56		2.23				197
1-Me, 5-NO_2	$CDCl_3$	6.05	2.46	2.01					197
1-Ac	$CDCl_3$	7.40	1.85	2.92	2.54	0.8	1.5	1.6	192
1-Ac, 2-Me	$CDCl_3$	7.35	7.41	3.08	2.75			1.8	192
1-Ac, 4-Me	$CDCl_3$	7.41	1.95	7.74	2.83	0.4	1.2	1.2	192
1-Ac, 4,5-di-Me	$CDCl_3$	7.42	2.06	7.84	7.64	1.3			192

[a] Values in Ref. 197 have been corrected to compensate for a calibration error.

strong deshielding effect (with respect to the carbonyl group) on all protons. The strong deshielding of the thione grouping has been used in 1-benzyl-5-phenylimidazolid-2-thiones (**176**) to produce extremely large shifts (~2 ppm) between the benzyl methylene protons (212). Spectra of some 1- and 3-substituted 5,5-diphenylimidazolid-4-ones (**177**) have been reported (213). The cis and trans isomers of dihydroxyimidazolidinone (**178**) have also been studied (214). Data for imidazolidones and -thiones are collected in Table 3.22.

D. The Triazole Systems

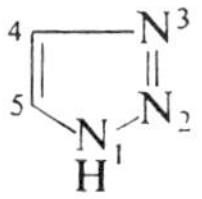

1,2,3 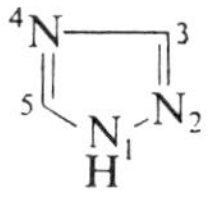 1,2,4

Spectra of the triazoles and their derivatives are not commonly encountered and the available information is contained in a small number of relatively modern papers. That of the parent 1,2,3-triazole (**179**) was first reported by Gold (215) who assigned the two peaks, observed for $CDCl_3$ solutions, to the 4,5-protons and the N*H* proton. Since this time, two other groups of workers (216, 197) have reported data for 1,2,3-triazole in a number of

TABLE 3.22 SPECTRAL DATA FOR IMIDAZOLIDINONES AND -THIONES

A. IMIDAZOLIDIN-2-ONES AND -2-THIONES (211)

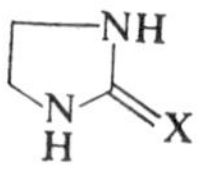

Substituents	X	Solvent	τ4,5	τNR
None	O	D_2O	6.50	
	S	D_2O	6.32	
1,3-Di-Ac	O	D_2O	6.26	7.62
		$CHCl_3$	6.17	7.47
	S	$CHCl_3$	6.00	7.18

B. 5,5-DIPHENYLIMIDAZOLIDIN-4-ONES (213). SOLVENT—$CDCl_3$

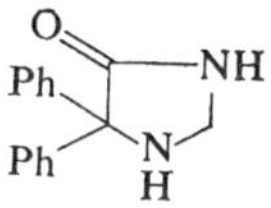

Substituents	τ1	τ2	τ3
None	7.24	5.60	2.04
1-Me	7.93	5.88	1.54
1-CH_2Ph	6.68, 2.60	6.03	1.70
1-CHO	2.00	4.95	2.07
1-COMe	8.48	4.85	1.75

solvents. Spectral parameters for a number of simple derivatives are also given in these papers and these are included in Table 3.23. In 1-methyl-1,2,3-triazole (**180**) in $CDCl_3$, H-4 and H-5 are not equivalent, and the assignment of specific signals to each of these protons depends on the negligible effect of a bromo substituent on the chemical shift of an adjacent hydrogen atom. Solvent effects for the parent triazole, as well as its simple derivatives, are reported in both of these papers, but they seem to be rather erratic and no sensible correlation pattern emerges. The difference between the N-1 and N-2 isomers of ω-aminoalkyl-1,2,3-triazoles (**181** and **182**) is evident from the AB-type spectrum (actually two broad singlets because of the weakness of the coupling) observed for the ring protons of the 1-isomer and the singlet observed for the symmetrical 2-isomer (215). The spectra of a

TABLE 3.23 SPECTRAL DATA FOR TRIAZOLES[a,b]

1,2,3-TRIAZOLES

Substituents		Solvent	τ4	τ5	J_{45}	τNR	Ref.
None		Neat	2.14	2.14		−2.	216
		$CDCl_3$	2.25	2.25		−2.05	216
		DMSO	2.09	2.09		−3.50	216
		C_6D_6	2.70	2.70		−3.40	216
	Cation	TFA	1.40	1.40			197
	Anion	NaOD	2.14	2.14			197
		C_5D_5N	2.00	2.00			197
		D_2O	2.00	2.00			197
		DMF	2.07	2.07			197
		$MeNO_2$	2.12	2.12			197
		MeOH	2.12	2.12			197
		$(Me)_2CO$	2.16	2.16			197
		MeCN	2.20	2.20			197
		Dioxan	2.28	2.28			197
		CH_2Cl_2	2.19	2.19			197
		$(n\text{-}Pr)_2O$	2.35	2.35			197
		EtOAc	2.22	2.22			197
4-CHO		$CDCl_3$		1.82			197
	Cation	TFA		1.00			197
4-COOH							
	Cation	TFA		1.07			197
1-Me		Neat	2.11	1.88		5.80	216
		$CDCl_3$	2.26	2.41		5.90	216
		DMSO	2.28	1.92		5.91	216
		C_6D_6	2.60	3.10		6.63	216
		C_5D_5N	2.22	2.17		6.05	216
		N_2H_4	2.05	1.75		5.77	216
		HMPT	2.37	1.63	1.0	5.85	216
	Cation	TFA	1.56	1.51		5.52	216
1-Me, 4-Br		$CDCl_3$		2.43		5.82	216
		DMSO		2.23		5.83	216
		C_6D_6		2.68		6.35	216
		HMPT		2.25		5.80	216
	Cation	TFA		1.61		5.52	197
1-Me, 5-Br		$CDCl_3$	2.12			6.02	197
	Cation	TFA	1.52			5.58	197
2-Me		$CDCl_3$	2.43	2.43		5.82	216
		DMSO	2.23	2.23		5.83	216
		C_6D_6	2.68	2.68		6.35	216
		HMPT	2.25	2.25		5.80	216
2-Me, 4-Br		$CDCl_3$		2.45		5.86	197
	Cation	TFA		2.20		5.60	197

TABLE 3.23 (*continued*)

Substituents		Solvent	τ4	τ5	J_{45}	τNR	Ref.
1-Me, 4-CHO		$CDCl_3$		1.89		5.80	197
	Cation	TFA		1.20		5.52	197
1-Me, 4-COOH		$CDCl_3$		1.75		5.82	197
	Cation	TFA		1.19		5.50	197
1-PNP		DMSO	1.93	0.97	1.2	1.76, 1.56	216
1-DNP		$CDCl_3$	2.05	2.10	1.2	1.13, 1.37, 2.08	216
		DMSO	1.90	1.12	1.2	1.04, 1.22, 1.77	216
1-TNP		DMSO	1.93	1.20	1.2	0.68	216
2-PNP		$CDCl_3$	2.12	2.12		1.70	216
		DMSO	1.79	1.79		1.80, 1.62	216
1-Tosyl, 5-Me		$CDCl_3$	2.80	7.70		2.5	229
1-Ph, 5-Me		$CDCl_3$	2.70	7.79		2.70	229

5-PROPYLOXY-4-METHYL-1-(*p*-NITROPHENYL)-Δ^2-1,2,3-TRIAZOLINE

	τ4	τ5	J_{45}
Cis-isomer	5.73	4.60	7.5
Trans-isomer	5.35	4.77	2.0

1,2,4-TRIAZOLES

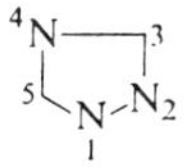

Substituents		Solvent	τ3	τ5	τNR	Temperature (°C)	Ref.
None		Neat	2.15	2.15	−3.9	125	224
		$CDCl_3$	1.77	1.77			222
		DMSO	1.74	1.74	−3.5		247
			1.75	1.75	−3.9	37	224
			1.67	1.67			222
		HMPT	1.83	1.83	−5.1	37	224
			1.15	2.08	−5.25	−34	224
	Cation	TFA	0.70	0.70			197
	Anion	NaOD	1.90	1.90			197
		C_5D_5N	1.35	1.35			197
		D_2O	1.52	1.52			197
		DMF	1.52	1.52			197
		$MeNO_2$	1.75	1.75			197
		MeOH	1.63	1.63			197

TABLE 3.23 (*continued*)

Substituents		Solvent	τ3	τ5	τNR	Temperature (°C)	Ref.
		$(Me)_2CO$	1.66	1.66			197
		MeCN	1.78	1.78			197
		Dioxan	1.82	1.82			197
		CH_2Cl_2	1.79	1.79			197
		$(n\text{-Pr})_2O$	1.97	1.97			197
		EtOAc	1.80	1.80			197
3-Me		$CDCl_3$	7.46	1.91			222
		DMSO	7.65	1.99			222
		$(Me)_2CO$	7.58	2.08			226
3,5-Di-Me		$CDCl_3$	7.55	7.55			222
		DMSO	7.75	7.75			222
1-Me		$CDCl_3$	2.06	1.91	6.07		222
		DMSO	2.05	1.53	6.12		222
	Cation	TFA	1.30	0.55	5.70		197
	Picrate	DMSO	1.20	0.44	5.97		222
4-Me		$CDCl_3$	1.99	1.99	6.37		222
		DMSO	1.52	1.52	6.34		222
1,3-Di-Me		$CDCl_3$	7.62	2.06	6.16		222
		DMSO	7.77	1.73	6.21		222
	Picrate	DMSO	7.53	0.39	6.01		222
1,5-Di-Me		$CDCl_3$	2.26	7.54	6.18		222
		DMSO	2.26	7.54	6.25		222
	Picrate	DMSO	1.16	7.32	6.03		222
3,4-Di-Me		$CDCl_3$	7.56	1.93	6.36		222
		DMSO	7.69	1.67	6.45		222
	Picrate	DMSO	7.38	0.71	6.20		222
1,3,5-Tri-Me		$CDCl_3$	7.73	7.67	6.32		222
		DMSO	7.90	7.73	6.39		222
	Picrate	DMSO	7.57	7.36	6.13		222
3,4,5-Tri-Me		$CDCl_3$	7.62	7.62	6.53		222
		DMSO	7.72	7.72	6.58		222
	Picrate	DMSO	7.42	7.42	6.33		222
3-Ph		$(Me)_2CO$	1.8–2.7	1.78			226
1,3-Di-Ph		MeCN		1.65			225
1,5-Di-Ph		MeCN	2.08				225
1-Ac		$(Me)_2CO$	1.97	1.05			225
3-NH_2		$(Me)_2CO$		2.53			226
3-NHMe		$(Me)_2CO$	7.12	2.48			226
3-N(Me)_2		$(Me)_2CO$	7.03	2.36			226
3-OMe		$(Me)_2CO$	6.15	2.03			226
3-SMe		$(Me)_2CO$	7.44	1.75			226
3-I		$(Me)_2CO$		1.69			226
3-Br		$(Me)_2CO$		1.59			226
3-Cl		$(Me)_2CO$		1.57			226
3-COOMe		$(Me)_2CO$	6.13	1.52			226

TABLE 3.23 *(continued)*

1-Me, 5-Br		$CDCl_3$	2.15		6.10	197
	Cation	TFA	1.30		5.80	197
4-Me, 5-Br		$CDCl_3$	1.76		6.35	197
	Cation	TFA	0.45		5.93	197
1-Me, 5-NH_2		$CDCl_3$	2.65		6.38	197
	Cation	TFA	1.81		6.08	197
4-Me, 5-NH_2						
	Cation	TFA	1.72		6.15	197
1-PNP		DMSO	1.64	0.45		223
4-PNP		DMSO	0.67	0.67		223
1-PNP, 3-Me		$CDCl_3$	7.49	1.39		223
		DMSO	7.60	0.64		223
1-PNP, 5-Me		DMSO	1.84	7.38		223
4-PNP, 3-Me		$CDCl_3$	7.49	1.67		223
		DMSO	7.54	1.14		223
1-PNP, 3,5-di-Me		$CDCl_3$	7.57	7.38		223
		DMSO	7.69	7.40		223
1-DNP		$CDCl_3$	1.82	1.47		223
		DMSO	1.64	0.66		223
1-DNP, 3,5-di-Me		$CDCl_3$	7.59	7.54		223
		DMSO	7.83	7.75		223

[a] Values reported for chemical shifts vary considerably from publication to publication. Those given for Ref. 216 have been corrected to compensate for a calibration error.
[b] The abbreviations PNP, DNP, and TNP refer to the *p*-nitro-, 2,4-dinitro-, and 2,4,6-trinitrophenyl groups.

number of 5-substituted 1-methyl-4-hydroxy-1,2,3-triazoles (**183**) have been reported (217) and, from the ^{13}C–H coupling constants, it was possible to distinguish between the signals from *O*- and *N*-methyl groups. Phenyloso-triazoles (e.g., **184**) are formed by many sugars and some data are available from their NMR spectra (see Ref. 218). Ring-chain tautomerism between compounds of the types **185a** and **186a** has also been studied by NMR techniques (219).

Spectra of a number of Δ^2-1,2,3-triazolines (**185**) have been reported by Huisgen and Szeimies in their study of 1,3-dipolar cycloaddition reactions (220, 221).

The only extensive study of 1,2,4-triazoles (**186**) was reported by Jacquier *et al.* (222, 223). The spectra of the parent 1,2,4-triazole and its 3,5-dimethyl derivative **187** in $CDCl_3$ or DMSO were singlets, the NH protons not being observed with normally purified solvents. Thus, the time-averaged picture of the molecule must be symmetrical. As in pyrazoles, the addition of an N-1 substituent removes the symmetry and raises the problem of assignment of peaks to protons or methyl groups in positions 3 and 5. However, substituent

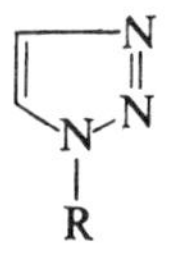

179, R = H
180, R = Me
181, R = $(CH_2)_nNH_2$

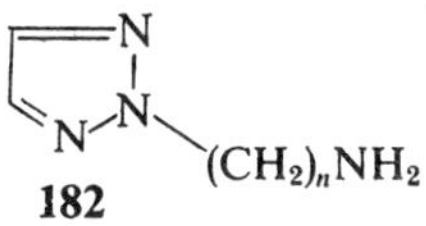

182

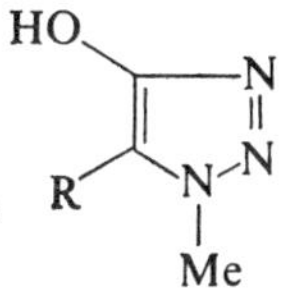

183

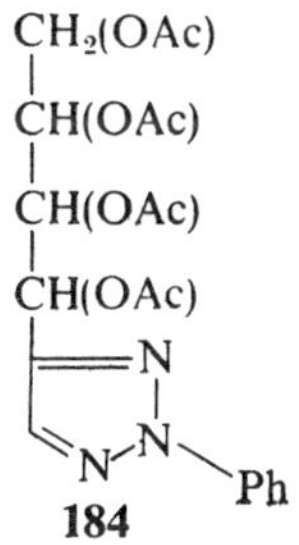

184

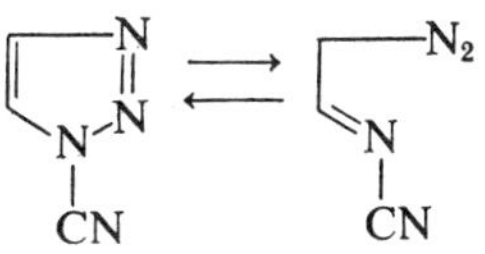

185a

186a

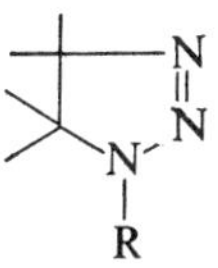

185

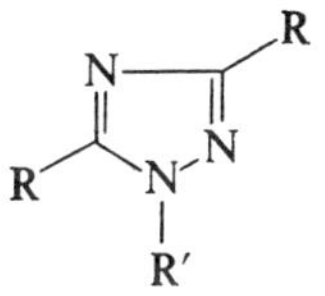

186, R = R′ = H
187, R = Me, R′ = H
188, R = H, R′ = Me
189, R = R′ = Me
190, R = Ar

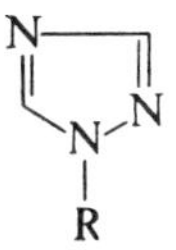

191, R = H
193, R = Ac

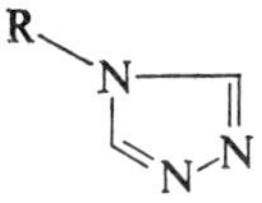

192, R = H
194, R = Ac

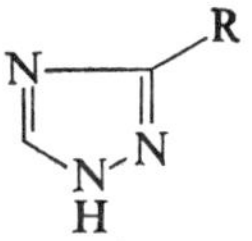

195

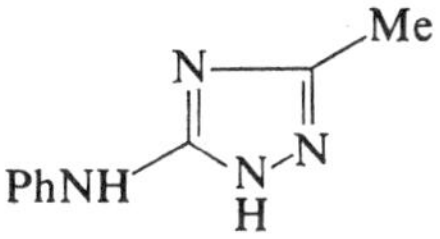

196

and solvent effects provide unambiguous assignments for the 1-methyl-1,2,4-triazole (**188**) and its 3,5-dimethyl derivative **189**. *C*-Alkyl substitution produces a general shielding of protons or methyl groups on C-3 or C-5, probably by inductive donation of electrons to the ring system. However, this effect for ring protons is much smaller with DMSO as solvent ($\Delta\tau$ = 0.14 ppm) than with $CDCl_3$ as solvent ($\Delta\tau$ = 0.32 ppm), but for methyl groups appears to be independent of solvent. Methyl substitution on nitrogen atoms produces a downfield shift (~ -0.24 ppm) of signals from α-protons but an upfield shift ($\sim +0.27$ ppm) of those from β-protons. Similarly the 5-methyl group of 1,3,5-trimethyl-1,2,4-triazole (**189**) absorbs at lower field than the 3-methyl group. Also, an interesting solvent effect was observed. Signals from protons α to an *N*-methyl group shift upfield by about 0.4 ppm when the solvent is changed from DMSO to $CDCl_3$, whereas signals from β-protons are not affected at all. In general, spectra of the corresponding picrates and *N*-aryl derivatives **190** (223) behaved in a similar manner.

In very dry solvents, the signal from the N–H proton of 1,2,4-triazole is sharp enough to be observed (224) and in carefully distilled HMPT, two distinct C–H peaks are observed at temperatures below 0°. In a 4.3% solution, a maximum separation of 56 Hz (at 60 MHz) is observed at −34°. Such a separation of peaks can only be interpreted in terms of the predominance of the 1-H tautomer **191** over the 4-H tautomer **192**.

The position of acylation of 1,2,4-triazole was determined from the lack of symmetry in the spectrum of the acyl derivative **193** (225). If acylation occurred in N-4 the symmetrical species **194** would have been formed. H-5 shifts have been measured for a series of 3-substituted 1,2,4-triazoles (**195**) (226) and the values obtained correlate well with the Hammett σp constants. Protonation of 1-methyl-1,2,4-triazole has been shown to occur on N-4 and the spectrum of the 1,2,4-triazole anion has been reported (197). Some data have been given for 5-anilino-3-methyl-1,2,4-triazole (**196**) and its phosphorylation products (227).

E. The Tetrazole System

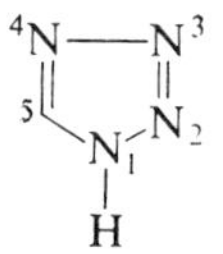

⁴N——N³
5 N₁ N₂
H

It is quite interesting to note that although tetrazole has only two protons, both missing in many of its derivatives, PMR has proved useful in the solution of quite a few structural problems involving this system (Table 3.24).

Data for tetrazole in a number of solvents have been reported (197), but no correlation could be drawn between the dielectric constant of the solvent and the H-5 chemical shift. The earliest study of tetrazole derivatives (229) compared N-1 (**197**) and N-2 (**198**) substituted tetrazoles as neat liquids. The extremely small substituent effects on H-5 led to the suggestion that the overriding factor in determining the chemical shift of this proton was the

TABLE 3.24 SPECTRAL DATA FOR TETRAZOLES[a]

Substituents		Solvent	τ5	τNR	Ref.
None		C_5D_5N	0.27		197
		D_2O	0.60		197
		DMF	0.37		197
		DMSO	0.57		197
		$MeNO_2$	0.91		197
		MeOH	0.66		197
		Me_2CO	0.61		197
		MeCN	0.85		197
		Dioxan	1.08		197
		EtOAc	0.97		197
	Cation	TFA	0.15		197
	Anion	NaOD	1.37		197
1-Me		$CDCl_3$	1.12	5.80	197
	Cation	TFA	0.22	5.50	197
2-Me		$CDCl_3$	1.50	5.60	197
	Cation	TFA	0.98	5.50	197
2,5-Di-Me		$CDCl_3$	7.47	5.70	231
1,5-Di-Me		$CDCl_3$	7.42	5.98	234
1-Me, 5-Br		$CDCl_3$		5.90	197
	Cation	TFA		5.80	197
2-Me, 5-Br		$CDCl_3$		5.65	197
1-Me, 5-NH_2		D_2O		6.18	234
1-Me, 5-NHAc		$CDCl_3$		5.79	234
1-Me, 5-NHNMe		$CDCl_3$	6.85	6.24	234
2-Me, 5-NH_2		$CDCl_3$		5.84	234

[a] Data from Ref. 197 have been corrected for a calibration error.

bond structure. From a comparison of the shift obtained for H-5 of tetrazole in DMF with those for the *N*-substituted derivatives it was concluded that the parent compound existed predominantly as the 1-tautomer **199**. An attempt was made to extend this approach to determine the predominant tautomer present in solutions of 5-methyltetrazole (230). Unfortunately, the *C*-methyl resonances of the isomeric 1,5- and 2,5-dimethyltetrazoles (**200** and **201**) were insufficiently different for conclusions to be drawn. From temperature- and concentration-dependence studies of the N–H resonance of 5-methyltetrazole it was concluded that, in liquid SO_2, the compound exists as one or more dimeric hydrogen-bonded species, for example, **202**.

Spectra of 5-phenyl- and 5-(*p*-nitrophenyl)tetrazole (**203**) and the corresponding 1- and 2-methyl derivatives **204** and **205** have been reported (231). The ortho protons of the phenyl ring in these compounds are deshielded by the coplanar tetrazole ring except in the case of the 1-methyl derivative.

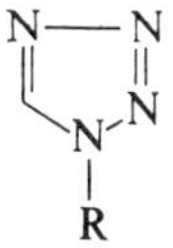

197
199, R = H

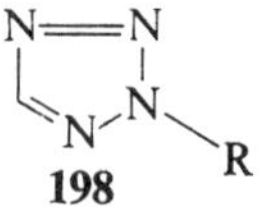

198

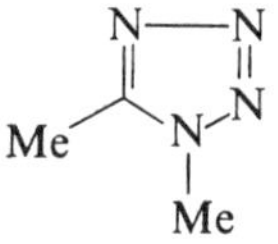

200

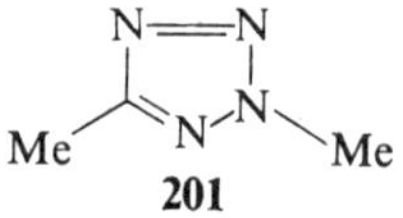

201

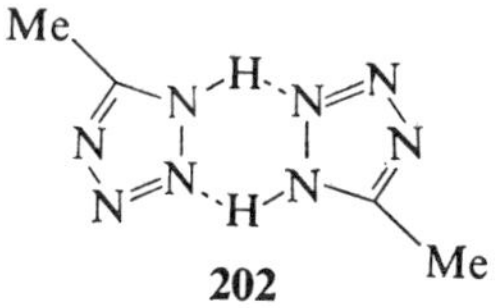

202

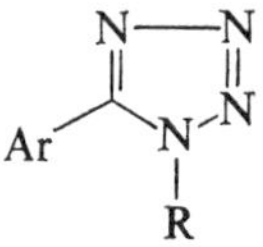

203, R = H
204, R = Me

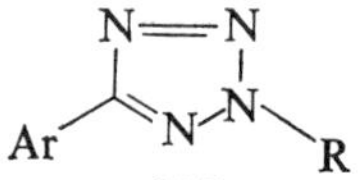

205

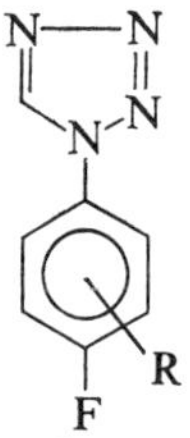

206

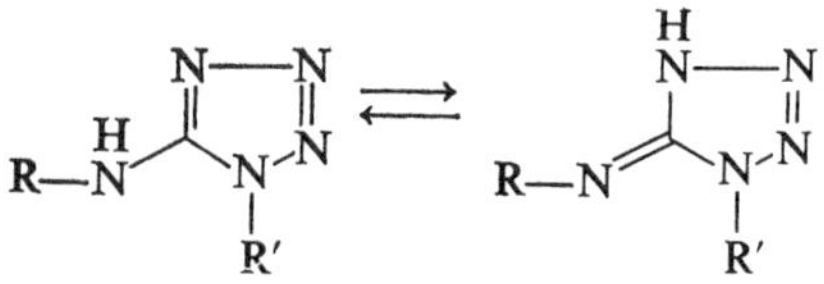

207 ⇄ **208**

209, R = N:CHAr

Here the methyl group causes twisting of the bond between the two rings and, as seen for the arylpyrroles and -pyrazoles, the phenyl resonance becomes a singlet. It was concluded that, while coplanarity of the two rings is necessary for deshielding, the chemical shifts of the phenyl protons are determined in part by the electron-donating interaction of the tetrazole ring. Two ^{15}N-labeled tetrazoles were also examined. ^{19}F shifts have been measured for an extensive series of 1-(fluorophenyl)tetrazoles (**206**) (232).

Tautomerism between the amino form **207** and the imino form **208** of 5-aminotetrazoles has often been suggested without any firm evidence. However, peaks from both isomers can be seen in the spectrum of 1-methyl 5-methylaminotetrazole (233). Proton magnetic resonance spectra have been reported for a series of 5-aminotetrazoles substituted on the exocyclic nitrogen atom (234). With tetrazol-5-ylhydrazones (**209**) in $CDCl_3$, the C–Me signals is found at $\tau \approx 5.75$ and is almost independent of the nature of the hydrazone grouping.

IV. SEVEN-MEMBERED RINGS

A. The Azepine System

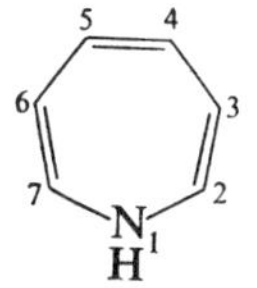

Over the last decade a number of papers have appeared on the preparation of azepines, mainly by ring-expansion reactions from appropriately substituted pyridines. Unfortunately, NMR techniques have been used in this work only to determine structures and for this reason, the data are sparse and scattered. The azepines can exist in four distinct tautomeric forms, with the sp^3-hybridized atom at position 1, 2, 3, or 4 to give the 1*H*-, 2*H*-, 3*H*-, and 4*H*-isomers. Nuclear magnetic resonance provides an excellent method for studying this isomerism because the extended ethylenic systems in each isomer give unique spectral patterns which are often spread widely enough for complete analysis.

Spectra of the three monomethyl isomers of 1-methoxycarbonylazepine (**210**) have been reported (235). Also, Johnson and his co-workers (236, 237), in their general study of the preparation and reactions of azepines, measured

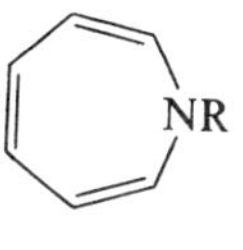

210, R = CO_2Me
211, R = Me

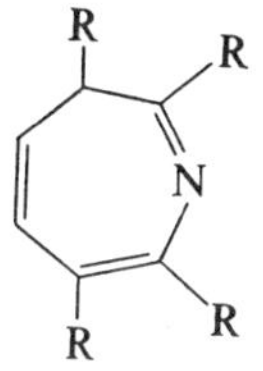

212 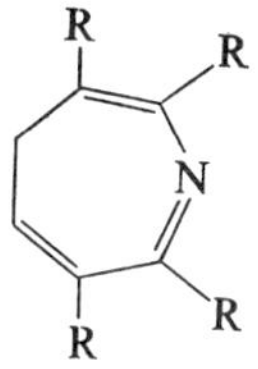 **213** 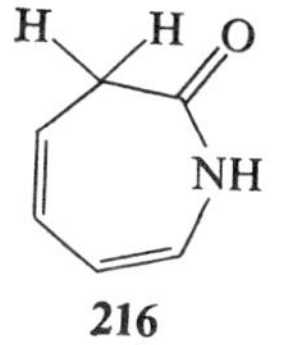 **216** 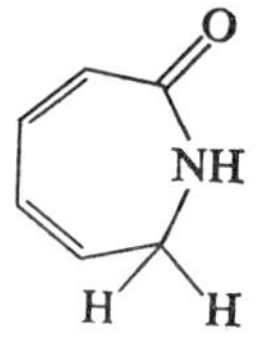 **217** 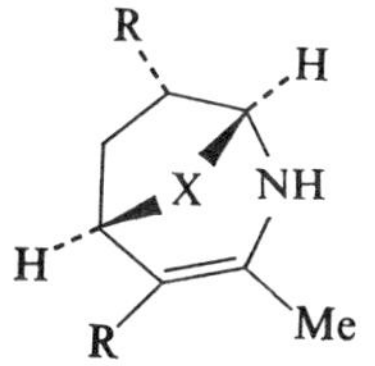 **218** 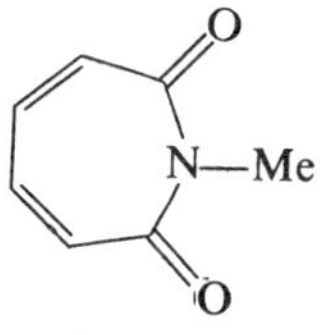 **219** 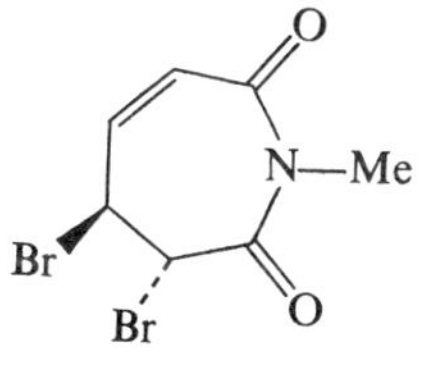 **220** 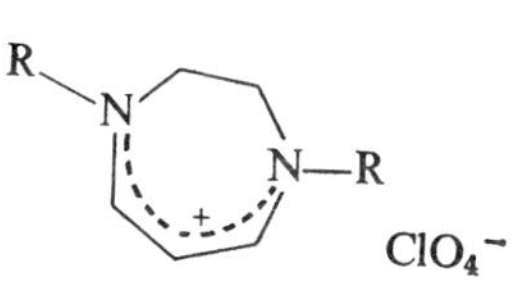 **221** 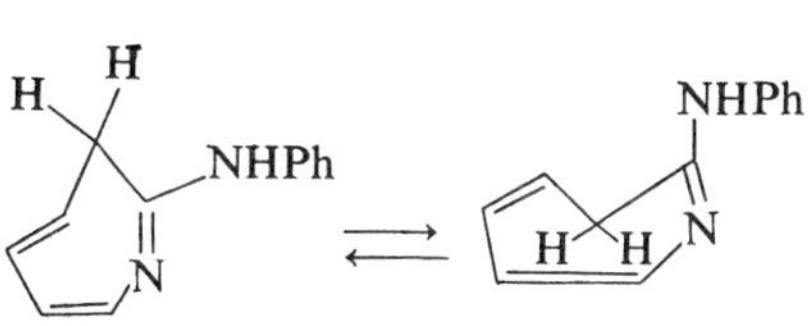 **214** **215**

TABLE 3.25 SPECTRAL DATA FOR AZEPINES

A. 1*H*-AZEPINES

Substituents	Solvent	$\tau 1$	$\tau 2$	$\tau 3$	$\tau 4$	$\tau 5$	$\tau 6$	$\tau 7$	Ref.
1-Me	CCl_4	7.55	5.0–5.5	5.0–5.5	5.0–5.5	5.0–5.5	5.0–5.5	5.0–5.5	260
1-CO_2Me,2Me	CCl_4	6.35	7.95	3.73–4.38	3.73–4.38	3.73–4.38	3.73–4.38	3.73–4.38	235
1-CO_2Me,3Me	CCl_4	6.31	4.23–4.67	8.33	3.85–4.19	3.85–4.19	3.85–4.19	4.23–4.67	235
1-CO_2Me,4Me	CCl_4	6.28	5.13–5.92	5.13–5.92	8.23	5.13–5.92	5.13–5.92	5.13–5.92	235
1-SO_2Me	$CDCl_3$	7.11	3.80	4.24	4.24	4.24	4.24	3.80	235
1-SO_2Ph	$CDCl_3$	2.38		4.24	4.24	4.24	4.24		235
1-P(Ph)$_2$	$CDCl_3$		3.80	4.52	4.52	4.52	4.52	3.80	235
1,2,7-Tri-Me	CS_2	7.42	8.27	4.83	4.28	4.28	4.83	8.27	235
1,2,7-Tri-Me,3,6-Di-CO_2Me	$CDCl_3$	7.18	7.79	6.36	3.66	3.66	6.36	7.79	237
1-H,4,5-Di-hydro, 2,7-Di-Me, 3,6-Di-CO_2Me,4-CN	$CHCl_3$	4.41	7.56		5.44	6.20, 6.43		7.59	236[a]

B. 3H-AZEPINES

Substituents	Solvent	$\tau 2$	$\tau 3$	$\tau 4$	$\tau 5$	$\tau 6$	$\tau 7$	Ref.
2-NEt_2	CCl_4		7.47	5.02	3.80	4.40	2.98	239[b]
2-NHPh	$(CD_3)_2CO$							
	(−75°)		8.26, 6.36	4.7				241[c]
	(+34°)		7.23					241
2-OMe, 3-COPh, 5-Cl_3	C_6D_6	6.88	6.60	3.65		4.03	3.12	238[d]
2-OMe, 3-COMe	C_6D_6	6.60	7.02, 8.15	3.7–4.4	3.7–4.4	3.7–4.4	3.00	238[e]
2-OMe, 3-CHO, 6-Cl	C_6D_6	6.72	7.32, 0.83	4.68	3.90		2.83	238[f]

1,2-DIHYDRO-2-OXO

		$\tau 1$	$\tau 3$	$\tau 4$	$\tau 5$	$\tau 6$	$\tau 7$	J_{34}	J_{46}	J_{67}	J_{17}	Ref.
None	$CDCl_3$	0.2	7.2	3.8–4.3	3.8–4.3	3.8–4.3	3.8–4.3					240
3-COPh, 5-Cl	C_6D_6	−0.4	5.37	3.87		4.17	3.57	6.5	1.0	9.0	5	238

TABLE 3.25 *(continued)*

C. BRIDGED AZEPINES (242)

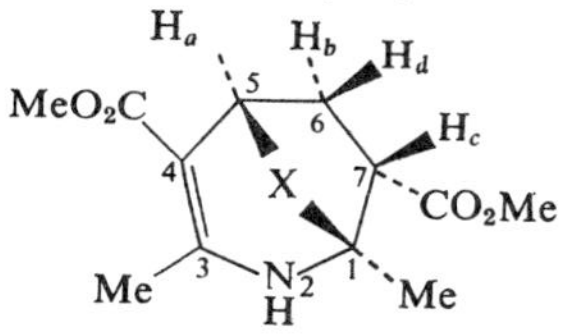

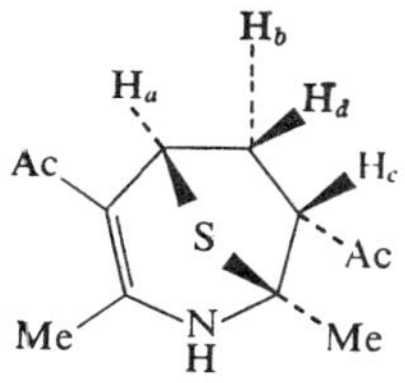

Solvent:	CDCl$_3$		C$_6$H$_6$		CDCl$_3$
X	S	Se	NMe	NCH$_2$Ph	
τa	5.29	4.82	5.82	5.88	5.24
τb	6.92	6.78	7.3	7.13	7.1
τc	7.24	7.00	7.3	7.43	7.1
τd	7.58	7.50	7.82	7.95	7.1
$\tau 1$	8.03	7.86	8.81	8.76	8.04
$\tau 3$	7.86	7.74	7.94	7.88	7.89
$\tau 4$	6.26	6.30	6.48	6.59	7.74
$\tau 7$	6.27	6.34	6.68	6.71	7.81
τNH	5.37	5.00	6.31	6.20	4.60
τNMe			7.98		
τNCH$_2$-				6.45, 6.70	
J_{ad}	5.5	5.5	6.0		5.0
J_{bd}	11.5	11.0	11.0		
J_{bc}	4.0	4.0	4.0		
J_{cd}	12.0	12.0	12.0		

Coupling constants:

[a] J_{45} 6.2, 1.5; J_{55} 15.1.

[b] J_{34} 1.10; J_{45} 8.80; J_{56} 5.65; J_{67} 7.80.

[c] J_{34} 6.8; J_{33} 11.0.

[d] J_{34} 6.0; J_{36} 0.4; J_{46} 1.4; J_{47} 0.6; J_{67} 8.5.

[e] J_{34} 6.0; J_{67} 7.5.

[f] J_{45} 9.4; J_{47} 0.7.

the spectra of a few 1-methylazepines (**211**). In the earlier paper (237), data were given for some 2,3,6,7-tetrasubstituted 3*H*- and 4*H*-azepines, for example, **212** and **213**. The isomer for which most data are available is the 3*H*-azepine system and detailed analyses have been made of the spectra of a number of simple derivatives formed in photorearrangement reactions (238–241) (see Table 3.25). These 3*H*-azepines are known to be puckered, and low-temperature NMR studies allowed a rough estimate to be obtained of the thermodynamic parameters for the inversion **214** to **215** (241). Also, associated with the 3*H*-azepines, the corresponding 3*H*-azepin-2-ones have been examined (238, 240). In the parent compound **216**, the position of the ring methylene group was obvious from the chemical shift of its two protons (τ 7.2). The only alternative structure which would fit other spectral data was the 7*H*-isomer **217** but in this case the methylene protons, adjacent to an amide nitrogen atom, would absorb at much lower field.

The bridged compounds **218** give well-resolved spectra, and data are available for compounds with a sulfur, selenium, or substituted nitrogen atom in the bridge position (242). These compounds are really bicyclic but for completeness are included here.

Spectra of 1-methylazepin-2,7-diones (**219**) have been reported recently (243). Selective solvent shifts were an important aid in assignment of the stereochemistry of these compounds. Thus, the spectrum of the parent in $CDCl_3$ showed two singlets, one for the *N*-methyl protons and the other for the ring protons, while, with benzene solutions, the low-field singlet was split into an AA′BB′ system with $\delta AB \sim 36$ Hz. Also, in the 3,4-dibromo-3,4-dihydro derivative **220**, benzene causes a preferential shielding of H-4 and H-5.

B. The Diazepine Systems

Little information is available for monocyclic diazepines. Barnett (244) measured the spectra of a number of the perchlorate salts of 2,3-dihydro-1,4-diazepenium ions **221** in DMSO. Signals from NH protons exchanged rapidly with added D_2O and a slow exchange of H-6 was observed. Data are summarized in Table 3.26.

A complete analysis of the spectrum of 5,7-dimethylhexahydro-1,4-diazepin-2-one (**222**) has been presented (245).

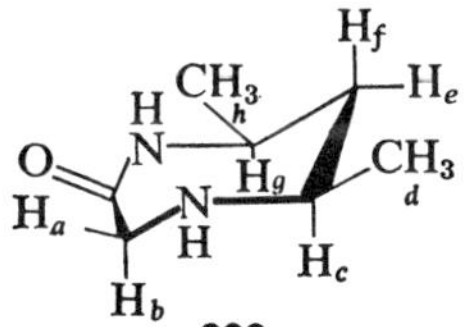

222

TABLE 3.26 SPECTRAL DATA FOR 2,3-DIHYDRO-1,4-DIAZEPINIUM PERCHLORATES (244). SOLVENT—DMSO, τNH FROM MeCN SOLUTIONS

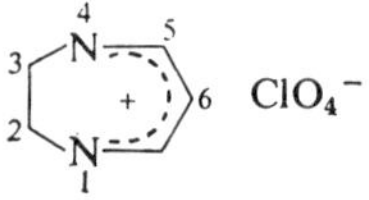

Substituents	τ2(3)	τ5(7)	τ6	τNH
None	6.38	2.5	4.9	2.05
6-Me	6.40	2.4	7.9	1.95
5,7-Di-Me	6.45	7.7	4.9	2.25
5,6-Di-Me	6.4	2.6, 7.75	7.95	2.15
1,4,6-Tri-Me	6.35	2.65	7.95	6.65
1,4,5,7-Tetra-Me	6.45	7.7	4.9	6.65
6-Br; 5,7-Di-Me	6.40	7.75		1.4
6-NO_2; 5,7-Di-Me	6.40	7.75		2.8
6-NO_2; 1,4,5,7-Tetra-Me	6.40	7.50		6.65

Solvent D_2O

τ*a*	6.35 or 6.72	J_{ab}	16.0
τ*b*	6.72 or 6.35	J_{cd}	6.5
τ*c*	7.00	J_{ce}	8.0
τ*d*	8.92 or 8.78	J_{cf}	10.5
τ*e*	8.20	J_{ef}	14.2
τ*f*	8.75	J_{eg}	2.0
τ*g*	6.75	J_{fg}	10.2
τ*h*	8.78 or 8.92	J_{gh}	6.5

REFERENCES

1. Harvey and Ratts, *J. Org. Chem.*, **31**, 3907 (1966).
2. Isomura, Okada, and Taniguchi, *Tetrahedron Lett.*, **1969**, 4073.
3. Isomura, Kobayashi, and Taniguchi, *Tetrahedron Lett.*, **1968**, 3499.
4. Mortimer, *J. Mol. Spectrosc.*, **5**, 199 (1960).
5. Brois, *J. Org. Chem.*, **27**, 3532 (1962).
6. Manatt, Elleman, and Brois, *J. Am. Chem. Soc.*, **87**, 2220 (1965).
7. Brois and Beardsley, *Tetrahedron Lett.*, **1966**, 5113.
8. Hassner and Heathcock, *Tetrahedron Lett.*, **1964**, 1125.
9. Ohtsuru and Tori, *J. Mol. Spectrosc.*, **27**, 296 (1968).
10. Yonezawa and Morishima, *J. Mol. Spectrosc.*, **27**, 210 (1968).
11. Brois, *J. Am. Chem. Soc.*, **89**, 4242 (1967).

11a. Ohtsuru and Tori, *Tetrahedron Lett.*, **1970**, 4043.

12. Deyrup and Greenwald, *J. Am. Chem. Soc.*, **87,** 4538 (1965).
12a. Brois, *Tetrahedron*, **26,** 227 (1970).
12b. Alvernhe and Laurent, *Bull. Soc. Chim. Fr.*, **1970,** 3003.
13. Bottini and Roberts, *J. Am. Chem. Soc.*, **78,** 5126 (1956); **80,** 5203 (1958).
14. Gutowsky, *Ann. N.Y. Acad. Sci.*, **70,** 786 (1958).
15. Loewenstein, Neumer, and Roberts, *J. Am. Chem. Soc.*, **82,** 3599 (1960).
16. Logothetis, *J. Org. Chem.*, **29,** 3049 (1964).
17. Bardos, Szantay, and Navada, *J. Am. Chem. Soc.*, **87,** 5796 (1965).
18. Bystrov, Kostyanovski, Panshin, and Iuzhakova, *Opt. Spectrosc.* (*USSR*), **19,** 122 (1965); Kostyanovski, Tchervin, Fomichov, and Samojlova, *Tetrahedron Lett.*, **1969,** 4021; Kostyanovski, Samojlova, and Tchervin, *ibid.*, **1968,** 3025.
19. Anet and Osyany, *J. Am. Chem. Soc.*, **89,** 352 (1967); Anet, Trepka, and Cram, *ibid.*, **89,** 357 (1967).
20. Boggs and Gerig, *J. Org. Chem.*, **34,** 1484 (1969).
21. Atkinson, *Chem. Commun.*, **1968,** 676.
22. Lehn and Wagner, *Chem. Commun.*, **1968,** 148; **1968,** 1298.
23. Felix and Eschenmoser, *Angew. Chem.*, **79,** 197 (1968).
24. Brois, *J. Am. Chem. Soc.*, **90,** 506 (1968); **90,** 508 (1968); Stogryn and Brois, *ibid.*, **89,** 605 (1967).
25. Turner, Heine, Irving, and Bush, *J. Am. Chem. Soc.*, **87,** 1050 (1965).
25a. Andose, Lehn, Mislow, and Wagner, *J. Am. Chem. Soc.*, **92,** 4051 (1970).
25b. Paulsen and Greve, *Chem. Ber.*, **103,** 486 (1970).
26. Jautelat and Roberts, *J. Am. Chem. Soc.*, **91,** 642 (1969).
27. Boykin, Turner, and Lutz, *Tetrahedron Lett.*, **1967,** 817.
28. Saito, Nukada, Kobayashi, and Morita, *J. Am. Chem. Soc.*, **89,** 6605 (1967).
29. Tori, Kitahonoki, Takano, Tanida, and Tsuji, *Tetrahedron Lett.*, **1965,** 869, and references cited therein.
30. Tori, Aono, Kitahonoki, Muneyuki, Takano, Tanida, and Tsuji, *Tetrahedron Lett.*, **1966,** 2921, and references cited therein.
31. Williams, *Acta Chem. Scand.*, **23,** 149 (1969).
31a. Graham, *J. Am. Chem. Soc.*, **88,** 4677 (1966); Stevens and Graham, *ibid.*, **89,** 182 (1967).
32. Uebel and Martin, *J. Am. Chem. Soc.*, **86,** 4618 (1964).
33. Mannschreck, Radeglia, Gründemann, and Ohme, *Chem. Ber.*, **100,** 1778 (1967); see also Radeglia, *Spectrochim. Acta*, **23A,** 1677 (1967).
34. Greene, Stowell, and Bergmark, *J. Org. Chem.*, **34,** 2254 (1969).
34a. Quast and Schmitt, *Chem. Ber.*, **103,** 1234 (1970).
35. Vigevani and Gallo, *J. Hetercycl. Chem.*, **4,** 583 (1967), and references cited therein.
36. Doomes and Cromwell, *J. Org. Chem.*, **34,** 310 (1969), and references cited therein.
37. Barrow and Spotswood, *Tetrahedron Lett.*, **1965,** 3325.
38. Smith and Cox, *J. Chem. Phys.*, **45,** 2848 (1966), Cox and Smith, *J. Phys. Chem.*, **71,** 1809 (1967).
39. Kagan, Basselier, and Luche, *Tetrahedron Lett.*, **1964,** 941; Decazes, Luche, and Kagan, *ibid.*, **1970,** 3661.
40. Fahr, Fischer, Jung, and Sauer, *Tetrahedron Lett.*, **1967,** 161.
40a. Fahr, Rohlfing, Thiedmann, Mannschreck, Rissmann, and Seitz, *Tetrahedron Lett.*, **1970,** 3605.
40b. Firl and Sommer, *Tetrahedron Lett.*, **1970,** 1925, 1929.
41. Reeves, *Can. J. Chem.*, **35,** 1351 (1957).
42. Abraham and Bernstein, *Can. J. Chem.*, **37,** 1056 (1959); **39,** 905 (1961).

43. Freymann and Freymann, *Compt. Rend.*, **248,** 677 (1959); Dischler, *Z. Naturforsch.* **20a,** 888 (1965).
44. Schaefer and Schneider, *J. Chem. Phys.*, **32,** 1224 (1960).
45. Reddy and Goldstein, *J. Am. Chem. Soc.*, **83,** 5020 (1961).
46. Katekar and Moritz, *Aust. J. Chem.*, **22,** 1199 (1969).
47. Gil, *Mol. Phys.*, **9,** 443 (1965).
48. Fukui, Shimokawa, and Sohma, *Mol. Phys.*, **18,** 217 (1970).
48a. Rahkamaa, *Z. Naturforsch.*, **24a,** 2004 (1969).
48b. Rahkamaa, *Mol. Phys.*, **19,** 727 (1970).
49. Abraham, Sheppard, Thomas, and Turner, *Chem. Commun.*, **1965,** 43.
50. Elvidge, *Chem. Commun.*, **1965,** 160.
51. Page, Alger, and Grant, *J. Am. Chem. Soc.*, **87,** 5333 (1965).
52. Black, Brown, and Heffernan, *Aust. J. Chem.*, **20,** 1325 (1967).
53. Happe, *J. Phys. Chem.*, **65,** 72 (1961).
54. Chenon and Lumbroso-Bader, *Compt. Rend.*, *Series C*, **266,** 293 (1968).
55. Strohbusch and Zimmermann, *Ber. Bunsenges. Phys. Chem.*, **71,** 567 (1967).
56. Chalaye, *Compt. Rend.*, *Series B*, **263,** 1227 (1966).
57. Porter and Brey, *J. Phys. Chem.*, **72,** 650 (1968).
58. Ronayne and Williams, *J. Chem. Soc.*, *B*, **1967,** 805.
58a. Perkampus, Krüger, and Krüger, *Z. Naturforsch.*, **24b,** 1365 (1969).
59. Connolly and McCrindle, *J. Chem. Soc.*, *C*, **1966,** 1613.
60. Bahtnagar and Nakajima, *Chim. Anal.*, **49,** 206 (1967).
61. Abraham, Bullock, and Mitra, *Can. J. Chem.*, **37,** 1859 (1959).
62. Whipple, Chiang, and Hinman, *J. Am. Chem. Soc.*, **85,** 26 (1963).
63. Chiang, Hinman, Theodoropulos, and Whipple, *Tetrahedron*, **23,** 745 (1967).
64. Hinman and Theodoropolus, *J. Org. Chem.*, **28,** 3052 (1963).
65. Skell and Bean, *J. Am. Chem. Soc.*, **84,** 4655 (1962).
66. Gronowitz, Hornfeldt, Gestblom, and Hoffman, *Ark. Kemi*, **18,** 133 (1961).
67. Gronowitz, Hörnfeldt, Gestblom, and Hoffman, *Ark. Kemi*, **18,** 151 (1961); *J. Org. Chem.*, **26,** 2615 (1961).
68. Karabatsos and Vane, *J. Am. Chem. Soc.*, **85,** 3886 (1963).
69. Jones and Wright, *Tetrahedron Lett.*, **1968,** 5495.
69a. Arlinger, Dahlqvist, and Forsen, *Acta Chem. Scand.*, **24,** 662 (1970).
70. Khan, Rodmar, and Hoffman, *Acta Chem. Scand.*, **21,** 63 (1967).
71. Wasserman, McKeon, Smith, and Forgione, *Tetrahedron*, Suppl. No. 8, Part ii, 647 (1966).
72. Hayes, Jackson, Judge, and Kenner, *J. Chem. Soc.*, **1965,** 4385.
73. Corwin, Chivvis, and Storm, *J. Org. Chem.*, **29,** 3702 (1964).
74. Brügel, Ankel, and Krückeberg, *Z. Electrochem.*, **64,** 1121 (1960).
75. Jones, Spotswood, and Cheuychit, *Tetrahedron*, **23,** 4469 (1967).
76. Anderson and Griffiths, *Can. J. Chem.*, **45,** 2227 (1967).
77. Birch, Hodge, Rickards, Takeda, and Watson, *J. Chem. Soc.*, **1964,** 2641.
78. Hodge and Rickards, *J. Chem. Soc.*, **1965,** 459.
79. Binns and Brettle, *J. Chem. Soc.*, *C*, **1966,** 341.
80. Anderson and Hopkins, *Can. J. Chem.*, **44,** 1831 (1966).
81. Rapoport and Castagnoli, *J. Am. Chem. Soc.*, **84,** 2178 (1962).
82. Rapoport, Castagnoli, and Holden, *J. Org. Chem.*, **29,** 883 (1964).
83. Rapoport and Bordner, *J. Org. Chem.*, **29,** 2727 (1964).
84. Grigg, Johnson, and Wasley, *J. Chem. Soc.*, **1963,** 359.
85. Grigg and Johnson, *J. Chem. Soc.*, **1964,** 3315.

86. Smith and Jensen, *J. Org. Chem.*, **32,** 3330 (1967).
87. Bocchi, Chierci, and Gardini, *Tetrahedron*, **23,** 737 (1967).
88. Hoft, Katritzky, and Nesbit, *Tetrahedron Lett.*, **1967,** 3041.
89. Markovac, Kulkarni, Shaw, and MacDonald, *Can. J. Chem.*, **44,** 2329 (1966).
90. Tori and Nakagawa, *J. Phys. Chem.*, **68,** 3163 (1964).
91. Weigert and Roberts, *J. Am. Chem. Soc.*, **90,** 3543 (1968).
92. Pugmire and Grant, *J. Am. Chem. Soc.*, **90,** 4232 (1968).
93. Rahkamaa, *J. Chem. Phys.*, **48,** 531 (1968).
93a. Gagnaire, Ramasseul, and Rassat, *Bull. Soc. Chim. Fr.*, **1970,** 415.
94. Cavalla, Katritzky, Sewell, and Bedford, *J. Chem. Soc.*, **1965,** 4546.
95. Bordner and Rapoport, *J. Org. Chem.*, **30,** 3824 (1965).
96. Atkinson, Atkinson, and Johnson, *J. Chem. Soc.*, **1964,** 5999.
97. Queen and Reipas, *J. Chem. Soc., C*, **1967,** 245.
98. Johnson, Robertson, Simpson, and Witkop, *Aust. J. Chem.*, **19,** 115 (1966).
99. Batterham, Riggs, Robertson, and Simpson, *Aust. J. Chem.*, **22,** 725 (1969).
100. Cox, Robertson, and Simpson, *Aust. J. Chem.*, **20,** 1539 (1967).
101. Abraham and McLauchlan, *Mol. Phys.*, **5,** 195 (1962).
102. Abraham and McLauchlan, *Mol. Phys.*, **5,** 513 (1962).
103. Abraham and Thomas, *J. Chem. Soc.*, **1964,** 3739.
104. Abraham, McLauchlan, Dalby, Kenner, Sheppard, and Burroughs, *Nature*, **192,** 1150 (1961).
105. Donohue and Trueblood, *Acta Crystallogr.*, **5,** 419 (1952).
106. Kollonitch, Scott, and Doldouras, *J. Am. Chem. Soc.*, **88,** 3624 (1966).
107. Mauger, Irreverre, and Witkop, *J. Am. Chem. Soc.*, **88,** 2019 (1966).
108. Blake, Wilson, and Rapoport, *J. Am. Chem. Soc.*, **86,** 5293 (1964).
109. Andreatta, Nair, Robertson, and Simpson, *Aust. J. Chem.*, **20,** 1493 (1967).
110. Andreatta, Nair, and Robertson, *Aust. J. Chem.*, **20,** 2701 (1967).
111. Garner and Watkins, *Chem. Commun.*, **1969,** 386.
112. Kwok and Pranc, *J. Org. Chem.*, **32,** 738 (1967).
113. Crabtree and Bertelli, *J. Am. Chem. Soc.*, **89,** 5384 (1967).
114. Flitsch and Peters, *Tetrahedron Lett.*, **1968,** 1475.
115. Duquette and Johnson, *Tetrahedron*, **23,** 4539 (1967).
116. Franklin, Mohrle, and Kilian, *Tetrahedron*, **25,** 437 (1969).
117. Morel and Foucaud, *Compt. Rend., Series C*, **262,** 373 (1966).
118. Morel and Foucaud, *Compt. Rend., Series C*, **265,** 1193 (1967).
119. Foucaud and Plusquellec, *Bull. Soc. Chim. Fr.*, **1968,** 3813.
120. Bryce-Smith and Hems, *J. Chem. Soc., B*, **1968,** 812.
121. Matsuo, *Can. J. Chem.*, **45,** 1829 (1967).
122. Elguero, Jacquier, and Tien Duc, *Bull. Soc. Chim. Fr.*, **1966,** 3727.
122a. Elguero, Jacquier, and Tarrago, *Bull. Soc. Chim. Fr.*, **1970,** 1345.
122b. Elguero, Gélin, Gélin, and Tarrago, *Bull. Soc. Chim. Fr.*, **1970,** 231.
123. Fowden, Noe, Ridd, and White, *Proc. Chem. Soc.*, **1959,** 131.
124. Williams, *J. Org. Chem.*, **29,** 1377 (1964).
125. Moore and Habraken, *J. Am. Chem. Soc.*, **86,** 1456 (1964).
126. Habraken and Moore, *J. Org. Chem.*, **30,** 1892 (1965).
127. Cola and Perotti, *Gazz. Chim. Ital.*, **94,** 1268 (1964).
128. Bystrov, Grandberg, and Sharova, *Opt. Spectrosc.*, **17,** 31 (1965).
129. Anderson, Duncan, and Rossotti, *J. Chem. Soc.*, **1961,** 140.
130. Anderson, Duncan, and Rossotti, *J. Chem. Soc.*, **1961,** 4201.
131. Vinogradov and Kilpatrick, *J. Phys. Chem.*, **68,** 181 (1964).

132. Perotti and Cola, in *Nuclear Magnetic Resonance in Chemistry* (Pesce, Ed.), Academic Press, New York, 1965, p. 249.
133. Finar and Mooney, *Spectrochim. Acta*, **20,** 1269 (1964).
134. Roumestant, Viallefont, Elguero, Jacquier, and Arnal, *Tetrahedron Lett.*, **1969,** 495.
135. Bystrov, Grandberg, and Sharova, *Zh. Obshch. Khim.*, **35,** 293 (1965).
136. Albright and Goldman, *J. Org. Chem.*, **31,** 273 (1966).
137. Habraken, Munter, and Westgeest, *Rec. Trav. Chim.*, **86,** 56 (1967).
138. Elguero and Jacquier, *J. Chim. Phys.*, **63,** 1242 (1966); Elguero, Imbach, and Jacquier, *ibid.*, **62,** 643 (1965).
139. Tensmeyer and Ainsworth, *J. Org. Chem.*, **31,** 1878 (1966).
140. Batterham and Bigum, *Org. Magnetic Resonance*, **1,** 431 (1969).
141. Zaev, Voronov, Shvartsberg, Vasilevsky, Molin, and Kotljarevsky, *Tetrahedron Lett.*, **1968,** 617.
142. Elguero and Jacquier, *Compt. Rend.*, **260,** 606 (1965).
143. Cohen-Fernandes and Habraken, *Rec. Trav. Chim.*, **86,** 1249 (1967).
144. Finar and Rackham, *J. Chem. Soc., B*, **1968,** 211.
145. Finar and Rackham, *J. Chem. Soc., C*, **1967,** 2650.
146. Lynch and Hung, *Can. J. Chem.*, **42,** 1605 (1964).
147. Lynch, *Can. J. Chem.*, **41,** 2380 (1963).
148. Elguero, Jacquier, and Tizane, *Bull. Soc. Chim. Fr.*, **1969,** 1687.
148a. Elguero, Jacquier, and Mignonac-Mondon, *Bull. Soc. Chim. Fr.*, **1970,** 4436.
149. Elguero and Wolf, *Compt. Rend., Series C*, **265,** 1507 (1967).
150. Trofimenko, *J. Am. Chem. Soc.*, **89,** 3170 (1967).
151. Elguero, Riviere-Baudet, and Satge, *Compt. Rend., Series C*, **266,** 44 (1968).
152. Jones, Ryan, Sternhell, and Wright, *Tetrahedron*, **19,** 1497 (1963).
153. Katritzky and Maine, *Tetrahedron*, **20,** 299, 315 (1964).
154. Elguero, Jacquier, and Tarrago, *Bull. Soc. Chim. Fr.*, **1967,** 3772, 3780.
154a. Newman and Pauwels, *Tetrahedron*, **25,** 4605 (1969).
154b. Baraseut, Elguero, and Jacquier, *Bull. Soc. Chim. Fr.*, **1970,** 1571.
154c. Elguero, Jacquier, and Mondon, *Bull. Soc. Chim. Fr.*, **1970,** 1576.
155. Elguero, Guiraud, Jacquier, and Tien Duc, *Bull. Soc. Chim. Fr.*, **1967,** 328.
156. Elguero, Guiraud, Jacquier, and Tarrago, *Bull. Soc. Chim. Fr.*, **1968,** 5019.
157. Snavelly and Yoder, *J. Org. Chem.*, **33,** 513 (1968).
158. Elguero, Jacquier, and Tarrago, *Bull. Soc. Chim. Fr.*, **1966,** 2990.
159. Lestina, Happ, Maier, and Regan, *J. Org. Chem.*, **33,** 3336 (1968).
160. Crawford, Dummel, and Mishra, *J. Am. Chem. Soc.*, **87,** 3023 (1965).
161. Crawford and Mishra, *J. Am. Chem. Soc.*, **87,** 3768 (1965).
162. van Auken and Rinehart, *J. Am. Chem. Soc.*, **84,** 3736 (1962).
163. Overberger, Weinshenker, and Anselme, *J. Am. Chem. Soc.*, **87,** 4119 (1965).
164. McGreer, Chiu, Vinji, and Wong, *Can. J. Chem.*, **43,** 1407 (1965).
165. McGreer and Wu, *Can. J. Chem.*, **45,** 461 (1967).
166. Crawford, Mishra, and Dummel, *J. Am. Chem. Soc.*, **88,** 3959 (1966); Crawford and Mishra, *ibid.*, **88,** 3963 (1966).
167. Crawford and Cameron, *Can. J. Chem.*, **45,** 691 (1967).
168. Freeman, *J. Org. Chem.*, **27,** 1309 (1962).
169. Hassner and Michelson, *J. Org. Chem.*, **27,** 3974 (1962).
170. Jacquier and Maury, *Bull. Soc. Chim. Fr.*, **1967,** 306.
171. Sustmann, Huisgen, and Huber, *Chem. Ber.*, **100,** 1802 (1967).
172. Huisgen, Sustmann, and Wallbillich, *Chem. Ber.*, **100,** 1786 (1967).
173. Elguero, Marzin, and Tizane, *Org. Magnetic Resonance*, **1,** 249 (1969).

174. McConnell, *J. Chem. Phys.*, **27,** 226 (1957).
175. Bothner-By and Naar-Colin, *Ann. N.Y. Acad. Sci.*, **70,** 833 (1958).
176. ApSimon, Craig, Demarco, Mathieson, Saunders, and Whalley, *Tetrahedron*, **23,** 2339 (1967).
177. Elguero, Fruchier, and Gil, *Bull. Soc. Chim. Fr.*, **1968,** 4403.
178. Aubagnac, Bouchet, Elguero, Jacquier, and Marzin, *J. Chim. Phys.*, **64,** 1649 (1967).
179. Brey and Valencia, *Can. J. Chem.*, **46,** 810 (1968).
179a. Elguero and Marzin, *Bull. Soc. Chim. Fr.*, **1970,** 3466.
180. Clovis, Eckell, Huisgen, Sustmann, Wallbillich, and Weberndörfer, *Chem. Ber.*, **100,** 1593 (1967).
181. Huisgen, Knupfer, Sustmann, Wallbillich, and Weberndörfer, *Chem. Ber.*, **100,** 1580 (1967).
182. Aubagnac, Elguero, and Jacquier, *Bull. Soc. Chim. Fr.*, **1967,** 3516.
183. Elguero and Jacquier, *Tetrahedron Lett.*, **1965,** 1175.
184. Elguero and Jacquier, *Bull. Soc. Chim. Fr.*, **1965,** 2961.
185. Elguero, Jacquier, and Muratelle, *Bull. Soc. Chim. Fr.*, **1968,** 2506.
186. Aubagnac, Elguero, Jacquier, and Tizané, *Tetrahedron Lett.*, **1967,** 3705.
187. Aubagnac, Elguero, Jacquier, and Tizané, *Tetrahedron Lett.*, **1967,** 3709.
188. Elguero, Marzin, and Tizané, *Tetrahedron Lett.*, **1969,** 513.
189. Mondelli and Merlini, *Gazz. Chim. Ital.*, **95,** 1371 (1965).
190. Willems, Vandenberghe, Poot, Roosen, Janssen, and Ruysschaert, *Tetrahedron*, **20,** 2723 (1964).
191. Reddy, Hobgood, and Goldstein, *J. Am. Chem. Soc.*, **84,** 336 (1962).
192. Reddy, Mandell, and Goldstein, *J. Chem. Soc.*, **1963,** 1414.
193. Joop and Zimmermann, *Z. Elektrochem.*, **66,** 440, 541 (1962).
194. Mannschreck, Seitz, and Staab, *Ber. Bunsenges. Phys. Chem.*, **67,** 470 (1963).
195. Staab and Mannschreck, *Angew. Chem.*, **75,** 300 (1963); Staab and Mannschreck, *Tetrahedron Lett.*, **1962,** 913.
196. Mannschreck, Staab, and Wurmb-Gerlich, *Tetrahedron Lett.*, **1963,** 2003.
197. Barlin and Batterham, *J. Chem. Soc., B*, **1967,** 516.
198. Cox, Fitzmaurice, Katritzky, and Tiddy, *J. Chem. Soc. B*, **1967,** 1251.
199. Sunjic, Kajfez, Slamnik, and Kolbah, *Bull. Sci. Cons. Acad. RSF Yougosl.*, **12,** 59 (1967); through *Chem. Abstr.*, **68,** 2483*n* (1967).
200. Overberger, Salamone, and Yaroslavsky, *J. Org. Chem.*, **29,** 3580 (1965); Caesar and Overberger, *J. Org. Chem.*, **33,** 2971 (1968).
201. Wang and Li, *J. Am. Chem. Soc.*, **88,** 4592 (1966).
202. Anderson, Duncan, and Rossotti, *J. Chem. Soc.*, **1961,** 2165.
202a. Imbach and Jacquier, *Compt. Rend.*, **257,** 2683 (1963).
203. Naidu and Bensusan, *J. Org. Chem.*, **33,** 1307 (1968), and references cited therein.
204. Lutz and DeLorenzo, *J. Heterocycl. Chem.*, **4,** 399 (1967).
205. Imbach, Jacquier, and Romane, *J. Heterocycl. Chem.*, **4,** 451 (1967).
206. Bishop and Richards, *Biochem. J.*, **86,** 277 (1963).
207. Baldwin and Rogers, *Chem. Commun.*, **1965,** 524.
208. Lynch and Dou, *Tetrahedron Lett.*, **1966,** 2627.
209. Cookson and Crabb, *Tetrahedron*, **24,** 2385 (1968).
210. Ogden and Mitsch, *J. Am. Chem. Soc.*, **89,** 5007 (1967).
211. Greenhalgh and Weinberger, *Can. J. Chem.*, **43,** 3340 (1965).
212. Southwick, Fitzgerald, and Milliman, *Tetrahedron Lett.*, **1965,** 1247.
213. Edward and Lantos, *Can. J. Chem.*, **45,** 1925 (1967).
214. Vail, Barker, and Mennitt, *J. Org. Chem.*, **30,** 2179 (1965).

215. Gold, *Ann. Chem.*, **688,** 205 (1965).
216. Elguero, Gonzalez, and Jacquier, *Bull. Soc. Chim. Fr.*, **1967,** 2998.
217. Begtrup and Pederson, *Acta Chem. Scand.*, **23,** 1091 (1969).
218. Lyle and Piazza, *J. Org. Chem.*, **33,** 2478 (1968).
219. Hermes and Marsh, *J. Am. Chem. Soc.*, **89,** 4760 (1967).
220. Huisgen and Szeimies, *Chem. Ber.*, **98,** 1153 (1965).
221. Szeimies and Huisgen, *Chem. Ber.*, **99,** 475, 491 (1966).
222. Jacquier, Roumestant, and Viallefont, *Bull. Soc. Chim. Fr.*, **1967,** 2630.
223. Jacquier, Roumestant, and Viallefont, *Bull. Soc. Chim. Fr.*, **1967,** 2634.
224. Creagh and Truitt, *J. Org. Chem.*, **33,** 2956 (1968).
225. Potts and Crawford, *J. Org. Chem.*, **27,** 2631 (1962).
226. Freiberg, Kröger, and Radeglia, *Tetrahedron Lett.*, **1967,** 2109.
227. van den Bos, Schipperheyn, and van Deursen, *Rec. Trav. Chim.*, **85,** 429 (1966).
228. Harvey, *J. Org. Chem.*, **31,** 1587 (1966).
229. Moore and Whittaker, *J. Am. Chem. Soc.*, **82,** 5007 (1960).
230. Markgraf, Backmann, and Hollis, *J. Org. Chem.*, **30,** 3472 (1965).
231. Fraser and Haque, *Can. J. Chem.*, **46,** 2855 (1968).
232. Kauer and Sheppard, *J. Org. Chem.*, **32,** 3580 (1967).
233. Butler, *Chem. Commun.*, **1969,** 405.
234. Scott, Butler, and Feeney, *J. Chem. Soc., B*, **1967,** 919.
235. Paquette and Kuhla, *Tetrahedron Lett.*, **1967,** 4517.
236. Anderson and Johnson, *J. Chem. Soc.*, **1965,** 2411.
237. Childs and Johnson, *J. Chem. Soc., C*, **1966,** 1950.
238. Ogata, Kano, and Matsumoto, *Chem. Commun.*, **1968,** 397.
239. von Doering and Odum, *Tetrahedron*, **22,** 81 (1966).
240. Vogel, Erb, Lenz, and Bothner-By, *Ann. Chem.*, **682,** 1 (1965).
241. Mannschreck, Rissmann, Vögtle, and Wild, *Chem. Ber.*, **100,** 335 (1967).
242. Ashby, Cort, Elvidge, and Eisner, *J. Chem. Soc., C*, **1968,** 2311.
243. Shapiro and Nesnow, *J. Org. Chem.*, **34,** 1695 (1969).
244. Barnett, *J. Chem. Soc., C*, **1967,** 2436.
245. McDougall and Malik, *J. Chem. Soc., C*, **1969,** 2044.
246. Bhacca, Hollis, Johnson, and Pier, *NMR Spectra Catalog*, Varian Associates, Palo Alto, California, **2,** 379 and 441 (1963).
247. Wilshire, *Aust. J. Chem.* **19,** 1935 (1966).
248. Sheehan and Lengyel, *J. Am. Chem. Soc.*, **86,** 1356 (1964).
249. Reinecke, Johnson, and Sebastian, *J. Am. Chem. Soc.*, **85,** 2859 (1963).
250. Morgan and Morrey, *Tetrahedron*, **22,** 57 (1966).
251. Anderson and Huang, *Can. J. Chem.*, **45,** 897 (1967).
252. Streith and Sigwalt, *Tetrahedron Lett.*, **1966,** 1347.
253. Papesch and Dodson, *J. Org. Chem.*, **30,** 199 (1965).
254. Elguero, Jacquier, and Tien Duc, *Compt. Rend., Series C*, **263,** 1456 (1966).
255. Guiraud and Jacquier, *Bull. Soc. Chim. Fr.*, **1963,** 22.
256. Evans, Whelan, and Johns, *Tetrahedron*, 21, 3351 (1965).
257. O'Brien and Gates, *J. Org. Chem.*, **31,** 1538 (1966).
258. Bouchet, Elguero, and Jacquier, *Tetrahedron*, **22,** 2461 (1966).
259. Danion-Bougot and Carrié, *Bull. Soc. Chim. Fr.*, **1968,** 2526.
260. Hafner and Mondt, *Angew. Chem.*, **78,** 822 (1966).
261. Shimokawa, Fukui, and Sohma, *Mol. Phys.*, **19,** 695 (1970).

4 BICYCLIC NITROGEN HETEROCYCLES

I. THE 3,3 SYSTEMS

A. 1-Azabicyclo[1,1,0]butanes

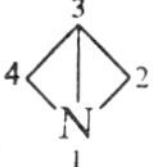

The spectra of a few members of this recently synthesized system have been reported. In bicyclobutanes the two three-membered rings are not coplanar and the geometry is as represented in **1**. Data have been reported for the

R
H_X $H_{X'}$ exo
N
H_A $H_{A'}$ endo

1

parent compound and its 3-alkyl derivatives (1) and for the 3-phenyl derivative (2). In the latter case, a complete analysis of the AA′XX′ system was obtained, using ^{13}C side bands. Previous work on bicyclobutanes (see Ref. 1) had shown that strong coupling occurred between exo ring protons in these systems. Since the coupling constants $J_{XX'}$ and $J_{AA'}$ have values of 6.25 and 0.65 Hz, the former value is assigned to the exo protons, and this in turn shows that the endo protons absorb at higher field than the exo protons. Data for these systems are collected in Table 4.1.

II. THE 4,3 SYSTEMS

A. 1-Azabicyclo[2,1,0]pentanes

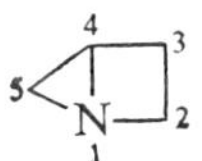

Data from the only compound of this series whose spectrum has been reported are given in Table 4.2.

TABLE 4.1 SPECTRAL DATA FOR THE 1-AZABICYCLO[1,1,0]BUTANES. SOLVENT—CCl_4

R	τexo	τendo	τR	$J_{XX'}$	J_{AX}	$J_{AA'}$	$J_{AX'}$	Ref.
H	7.87	9.07	7.67					1
Me	7.97	9.04	8.41					1
Et	7.92	9.07	8.08, 9.03					1
Ph[a]	7.36	8.67	2.5–2.9	6.25	2.75	0.65	~0	2

[a] $J_{^{13}CX} = 166$ Hz; $J_{^{13}CA} = 175$ Hz.

III. THE 5,3 SYSTEMS

A. The 1-Azabicyclo[3,1,0]hexane System

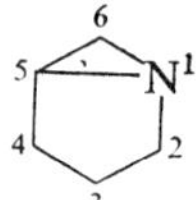

Data for the parent compound and its 3-methyl derivative have been reported (4). The parent gave a continuous multiplet τ 6.1–8.97 ppm, the methyl derivative τMe 9.2, τ*CH*Me 7.0, remainder τ 7.9–9.05 ppm.

B. The 1,3-Diazabicyclo[3,1,0]hex-3-ene System

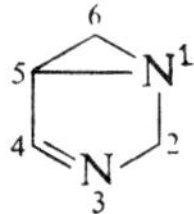

Nuclear magnetic resonance spectra were used to prove the structure of a number of these compounds (5). Data are collected in Table 4.3.

TABLE 4.2 SPECTRAL DATA FOR 4-PHENYL-3-PHENYLIMINO-1-AZABICYCLO-[2,1,0]PENTANE (3). SOLVENT—CCl_4

τa	τb	τc	τd	J_{ab}	J_{cd}	τPh
7.2	8.33	4.9	5.2	~2	16	2.4, 2.8

TABLE 4.3 SPECTRAL DATA FOR DIAZABICYCLO[3,1,0]-HEXANES. SOLVENT—$CDCl_3$

A. 1,3-DIAZABICYCLO[3,1,0]HEX-3-ENES (5)

Ar	R	R′	Ar′	τ2	τ5	τ6	J_{25}	J_{26}	J_{56}
p-NO_2Ph	H	Me	Ph	4.30	6.42	7.38	1.2	1.0	1.6
p-NO_2Ph	H	n-Pr	Ph	4.45	6.40	7.35	1.4	1.0	1.4
p-NO_2Ph	H	Ph	Ph	3.20	6.25	7.51	1.4	0.8	2.0
Ph	H	Ph	p-NO_2Ph	3.16	6.16	7.52	1.6	0.9	2.2
p-NO_2Ph	Me	Me	Ph		6.37	7.37			1.8

B. 3,6-DIAZABICYCLO[3,1,0]HEXANES (5a)

R1	R2	τA	τB	τC	J_{AB}
H	CH_2Ph	6.93	7.69	7.61	9.5
COMe	CH_2Ph	6.77	7.72	6.99	10.5

C. The 3,6-Diazabicyclo[3,1,0]hexane System

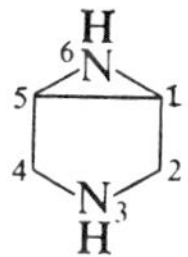

Data for a number of these 3,4-epiminopyrrolidines have been reported (5a) and are included in Table 4.3.

IV. THE 5,4 SYSTEMS

A. The 2-Azabicyclo[3,2,0]heptane System

Various unsaturated members of this ring system exist in tautomeric equilibrium with azepines, and data are spread throughout the extensive modern literature on the seven-membered ring system. The particular derivatives found in this work are the 3,6-dienes (**2**) (6) and the 3-one-6-enes (**3**) (7, 8).

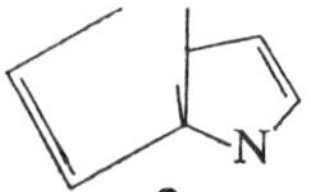

N **2** N O **3**

HO Me Ph N N Ac **4**

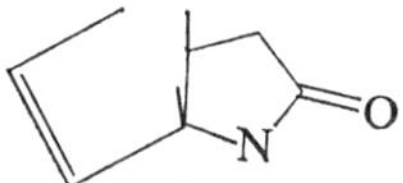

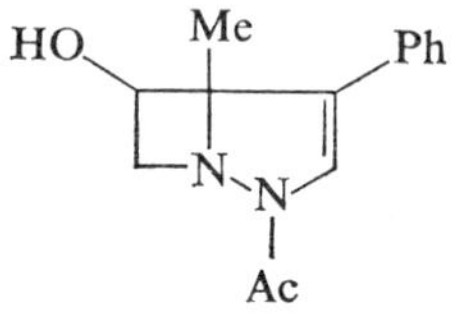

Spectral data for these compounds are collected in Table 4.4.

B. The Diazabicyclo[3,2,0]heptane System

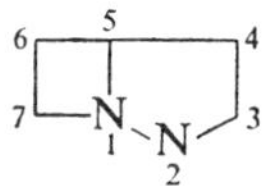

6 5 4 7 N 1 N 2 3

The available data for this system (Table 4.5) are found in two papers (9, 10), which describe a number of these compounds in various stages of oxidation. Protonation shifts observed for **4** suggested that protonation occurred on N-1.

V. THE 5,5 SYSTEMS

A. The Pyrrolizine System (1-Azabicyclo[3,3,0]octane)

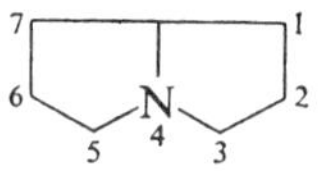

7 1 6 N 2 5 4 3

A detailed analysis of the spectra of pyrrolizine and some of its derivatives has shown them to exist as the 3*H*-tautomers **5** rather than the 1*H*-tautomers **6** (11). Data are collected in Table 4.6.

Chemical shifts were reported for 2,3-dihydro-7-methyl-1*H*-pyrrolizin-1-one (**7**), the major component of the hairpencil secretion of a male butterfly (13); τ2 7.16, τ3 5.87, τ5 3.31, τ6 3.91 ppm; $J_{56} = 2.5$ Hz.

TABLE 4.4 SPECTRAL DATA ON 2-AZABICYCLO[3,2,0]-HEPTENES. NEAT LIQUIDS

A. 3,6-DIENES (6)

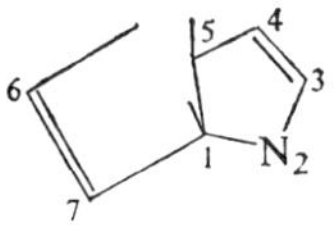

Substituents	τ1	τ3	τ4	τ5	τ6	τ7	Other data
1-COMe	5.12	3.44	4.82	6.12	3.58	3.96	
1-COMe,3-Me	5.25	7.96	5.20	6.49	3.81	4.11	
1-COMe,1-Me	8.40	3.76	5.13	6.74	3.85	3.95	
1-COMe,4-Me	5.21	3.88	8.35	6.36	3.60	4.01	
1-COMe,7-Me	5.29	3.60	4.92	6.40	4.01	8.32	
1-COMe,5-Me	5.62	3.65	5.03	8.70	3.65	4.11	
1-COMe,6-Me	5.31	3.52	4.88	6.34	8.28	4.28	
1,2,3-Tri-Me,4,7-di-CO_2Me	7.8	7.57	6.38	5.47	5.60	6.38	τ2 6.89

B. 3-ONE-6-ENES

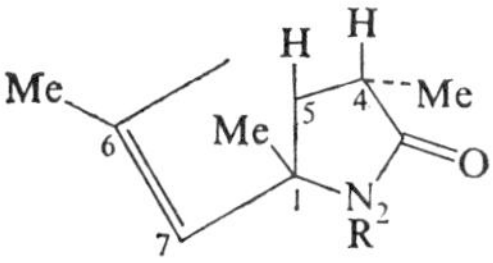

R	Solvent	τ1	τ2	τ4		τ5	τ6	τ7	J_{45}	Ref.
				H	Me					
H	CCl_4	8.61	1.22	7.42	8.87	7.15	8.24	3.85		8
H	$CDCl_3$	8.66		7.32	8.79	6.92	8.07	3.9		7
Me[a]	CCl_4	8.65	7.36	7.4	8.94	7.10	8.26	3.84	9.8	8
Me[a]	CCl_4	8.61	7.34	7.84	8.85	7.55	8.29	3.92	3	8

[a] Isomers around the 4,5-bond.

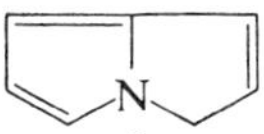

5 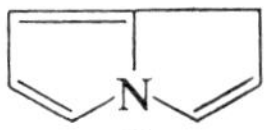 **6** 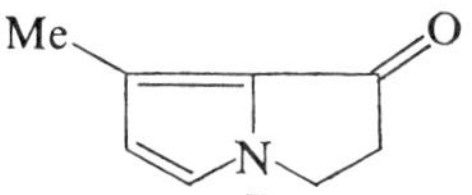 **7** 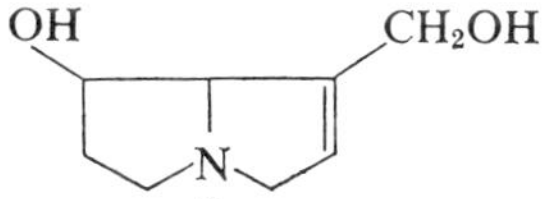 **8**

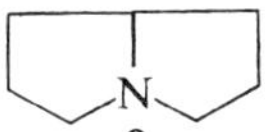

9 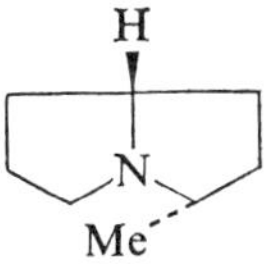 **10** 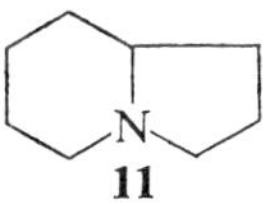 **11** 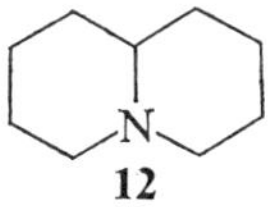 **12**

TABLE 4.5 SPECTRAL DATA FOR 1,2-DIAZABICYCLO-[3,2,0]HEPTANE SYSTEM. SOLVENT—$CDCl_3$

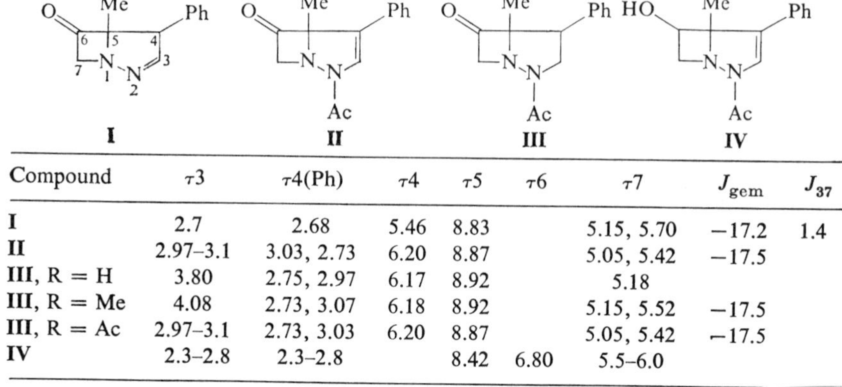

Compound	τ3	τ4(Ph)	τ4	τ5	τ6	τ7	J_{gem}	J_{37}
I	2.7	2.68	5.46	8.83		5.15, 5.70	−17.2	1.4
II	2.97–3.1	3.03, 2.73	6.20	8.87		5.05, 5.42	−17.5	
III, R = H	3.80	2.75, 2.97	6.17	8.92		5.18		
III, R = Me	4.08	2.73, 3.07	6.18	8.92		5.15, 5.52	−17.5	
III, R = Ac	2.97–3.1	2.73, 3.03	6.20	8.87		5.05, 5.42	−17.5	
IV	2.3–2.8	2.3–2.8		8.42	6.80	5.5–6.0		

TABLE 4.6 SPECTRAL DATA FOR THE PYRROLIZINES

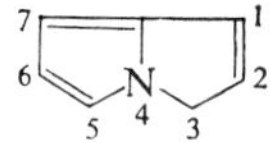

Substituents	Solvent	Chemical shifts τ1	τ2	τ3	τ5	τ6	τ7	Ref.
None		3.80	4.37	6.25	3.46	3.92	4.32	11
		3.60	4.20	6.18	3.60	3.60	4.20	12
	CS_2	3.56	4.28	5.73	3.30	3.91–4.1	3.91–4.1	11
1-Me		8.10	4.50	6.10	3.28	3.75	4.07	11
	CS_2	8.10	4.50	6.00	3.45	4.05	4.38	11
1,5,7-Tri-Me,6-CO_2Et	C_6H_6	7.90	4.35	6.32	7.22	8.66, 5.54	7.50	11
	CS_2	7.91	4.29	5.83	7.63	8.70, 5.83	7.73	11
2-COPh	CS_2	2.98		5.26	2.88	3.75	3.75	11
6-COPh	CS_2	3.45	3.78	5.57		2.15–2.7	3.73	11
6-CO_2Me	CS_2	3.47	3.82	5.56	2.67	6.32	3.83	11

Coupling constants (11)

Substituents	J_{12}	J_{13}	J_{15}	J_{23}	J_{26}	J_{35}	J_{37}	J_{56}	J_{57}	J_{67}
None	6.2	2.2	0.6	2.2	1.0	0	0.3	2.7	1.1	3.5
1-Me	2.2	1.5	0	2.2	1.0	0.7		2.5	1.0	3.2
1,5,7-Tri-Me,6-CO_2Et	1.5	1.5								
2-COPh		2.0				1				
6-COPh	6.0	2.0		2.0		1				
6-CO_2Me	6.0	2.0	0	2.0		1	0			

TABLE 4.7 SPECTRAL DATA FOR 1,2-DEHYDROPYRROLIZIDINES (15)

Compound	Solvent	$\tau 2$	$\tau 3$	$\tau 5$	$\tau 6$	$\tau 7$	$\tau 8$	$\tau 9$	Other data
R = α-OH,	$CDCl_3$	4.42	6.7, 6.1	7.25, 6.7	8.1	5.9	6.0	5.68	OH 4.63
R′ = H	C_5H_5N	4.26	6.64, 6.02	7.32, 6.59	8.06	5.62	5.62	5.68	OH 3.56
	D_2O	4.44	6.76, 6.34	7.38, 7.00	8.37, 8.14	5.90	6.25	5.87	
R = β-OH,	$CDCl_3$	4.3	6.6, 6.2	7.25, 6.8	8.05	5.7	5.8	5.75	OH 4.3
R′ = H	C_5H_5N	4.20	6.5, 6.0	7.1, 6.7	7.95	5.5	5.5	5.5	OH 4.1
	D_2O	4.22	6.6, 6.2	7.3, 6.9	8.05	5.7	5.85	5.77	
R = R′ = H	$CDCl_3$	4.53	6.73, 6.15	7.5, 6.9				5.86	OH 4.28

An extremely complete study has been made of the spectra of the pyrrolizidine alkaloids (14, 15). These compounds, most of which are based on the 7-hydroxy epimers of 1-hydroxymethyl-1,2-dehydropyrrolizidine (**8**), have reasonably rigid ring systems of known absolute stereochemistry, and it was possible to obtain values for all spectral parameters. Data for the more simple members of the series are collected in Table 4.7 but the reader is referred to the original papers for the wealth of information which they contain.

The PMR spectra of hexahydropyrrolizine (pyrrolizidine) (**9**), its salts, its 3-methyl derivative, and its methiodide have been examined in an attempt to determine the predominant conformations of these substances (16). Over the range −70 to +90°, hexahydropyrrolizine appears to exist mainly, if not entirely, in the cis-fused form, that is, with the bridgehead proton and the nitrogen lone pair cis to each other. The 3-*endo*-methyl derivative (**10**), however, shows rapid nitrogen-inversion at room temperature and only below −60° does it exist predominantly in the cis-fused form. The relative shielding of quaternary methyl groups in the cis and trans forms of the methiodides of pyrrolizidine, indolizidine (**11**), and quinolizidine (**12**) are explained in terms of single-bond anisotropies in the preferred conformations (see also Ref. 17). Some information on pyrrolizidine alkaloids containing the fully reduced ring system is included in Ref. 15. Data for these pyrrolizidines are collected in Table 4.8.

TABLE 4.8 SPECTRAL DATA FOR HEXAHYDRO-PYRROLIZINES (16)

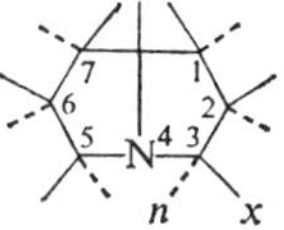

Substituents[a]	Solvent	Temperature (°C)	τ8x	τ3x	τ5x	τ5n	τ3n	τ1n,7n	τ4x	$J_{3x,3n}$
None	Neat	35	6.67	7.10	7.10	7.52	7.52	8.66		
	CCl_4	35	6.55	6.97	6.97	7.40	7.40			9.75
4-H+	TFA	33	5.51	6.21	6.21	6.78	6.78			
4-Me+ I−	D_2O	33	5.93	6.49	6.49	6.49	6.49		6.86	
3x-Me	C_5H_{10}	35	6.55	8.99	7.20	7.48	7.48			6.20
	C_6H_6	33	6.48	8.84			7.48			5.82
3x-Me,4-H+	TFA	33	5.46	8.41			6.48			6.46
3n-Me	C_5H_{10}	−85	6.69	7.15	7.40	7.60				6.60
		30	6.78	7.10	7.48	7.48				
	C_6H_6	33	6.56	7.05			8.88			6.90
3n-Me,4-H+	TFA	33	5.47	6.01			8.45			6.90

[a] x = exo, n = endo.

B. The Pyrazolo[1,2-a]pyrazole System (Diazapentalene)

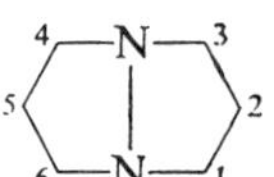

The available data for this system (17) are collected in Table 4.9.

TABLE 4.9 SPECTRAL DATA FOR THE PYRAZOLO[1,2-a]-PYRAZOLES

Compound	Solvent	τ1	τ2	τ3	τ4	τ5	τ6	J_{45}	J_{25}	Ref.
1	D_2O	2.99	3.35	2.99	2.99	3.35	2.99	2.5		17a
1	DMSO	2.95	3.52	2.95	2.95	3.52	2.95	2.5		17b
1 1,3-Di-Ac	D_2O	7.57	2.60	7.57	1.66	3.26	1.66	2.8	1.1	17a
1 1,3-Di-COPh	D_2O	2.15–2.58			1.36	3.12	1.36	2.9	1.1	17a
2 2-Br	D_2O	5.32	4.52	5.32	1.65	3.02	1.65	2.9		17a
	DMSO	4.96	4.38	4.96	1.33	2.98	1.33			17b
2 2,5-Di-Br	D_2O	4.65	4.3	4.65	1.16		1.16			17a
3	DMSO	2.4–3.3								17c

C. The Pyrrolo[1,2-*c*]-*v*-triazole System

The spectra of 5-methyl- and 5,5-dimethyl-1,6,7,7*a*-tetrahydro-5*H*-pyrrolo-[1,2-*c*]*v*-triazoles (**13** and **14**) have been reported (4). For neat samples of **13**, τMe 8.9, τCH_2 8.5, τNH 6.0 ppm, and **14**, τMe 8.84, 8.48, τCH_2 8.50, τNH 6.0 ppm.

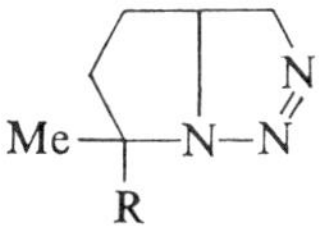

13, R = H
14, R = Me

VI. THE 6,3 SYSTEMS

A. The 7-Azabicyclo[4,1,0]heptane System

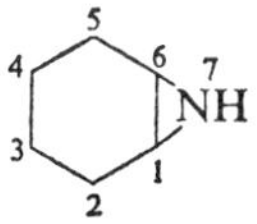

The two 7-chloro derivatives of the fully reduced system have been obtained pure and the spectra have been reproduced (18).

B. The Diazanorcaradiene Systems (Diazabicyclo[4,1,0]heptanes)

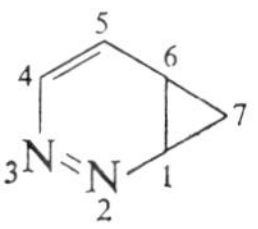

2,3- 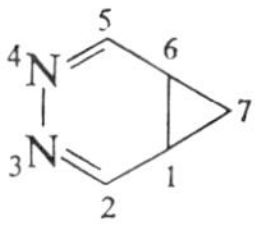 3,4-

Data for a number of substituted 3,4-diazanorcaradienes have been published (19). The cyclopropane ring was shown to be twisted well out of the plane of the six-membered ring, with one of the 7-substituents held rigidly in the shielding cone of the double bond system. (Data for these diazanorcaradienes are collected in Table 4.52.)

VII. THE 6,5 SYSTEMS

A. One Nitrogen Atom

1. *The Indole System*

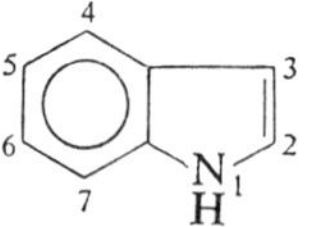

The literature on the spectra of indole derivatives is relatively extensive, and this reflects the considerable interest in biologically active compounds related to tryptophan (**15**). Since the original report of the indole spectrum (20) in which H-2 and H-3 were assigned, and the characteristic couplings to H-1 were observed, there have been a series of papers giving chemical shifts of these two protons for solutions in a wide range of solvents (21–23). Finally, Black and Heffernan (24) have produced a complete analysis of the spectrum of indole dissolved in acetone. Correlations between these results and π-electron densities were not good (25). Data for indole are summarized in Table 4.10.

From the earliest paper (20) it was stated that the difference between the chemical shifts of α- and β-protons was a useful method of determining the position of substituents on the hetero-ring. While this works well for alkyl

TABLE 4.10 SPECTRAL DATA FOR INDOLE

Solvent	$\tau 1$	$\tau 2$	$\tau 3$	$\tau 4$	$\tau 5$	$\tau 6$	$\tau 7$	Ref.
CCl_4		3.32	3.62		3.10	3.10		20
		3.46	3.66					21
		3.25	3.58					22
		3.48	3.71					23
$CDCl_3$		3.26	3.59					21
DMSO		2.65	3.53					21
Dioxan		2.84	3.52					21
CS_2		3.45	3.71					21
Et_2N		2.80	3.57					21
Me_2CO		2.72	3.52					21
Me_2CO[a]	−0.12	2.73	3.55	2.45	3.00	2.92	2.60	24

[a] Coupling constants: J_{12} 2.5, J_{13} 2.0, J_{14} 0.8, J_{23} 3.1, J_{37} 0.7, J_{45} 7.8, J_{46} 1.2, J_{47} 0.9, J_{56} 7.0, J_{57} 1.2, J_{67} 8.0.

substituents it could not be expected to work where the substituent is highly anisotropic or electron withdrawing, or both. However, as can be seen from Table 4.11, there are marked solvent shifts between polar and nonpolar solvents for the α-proton, but almost no shift for the β-proton. This property has been shown to extend to a range of carbonyl-substituted indoles and provides an excellent method for determining substitution site (21) (see Table 4.11). A word of warning has been published on this method (26). At least in polar solvents the α- and β-protons of the indole ring experience large specific concentration effects which could conceivably cancel the solvent shifts. This concentration dependence might aid in distinguishing between H-2 and H-3 in certain indoles but, in any case, must be taken into account when structural interpretations based on NMR spectra are presented.

Most data on substituted indoles come from papers dealing with the preparation of the compounds, and NMR is often used to check structural assignments. Thus, information is available on alkylindoles (20, 23), 3-vinylindole (**16**) (27), benzylindoles (22), chloroindoles (28), hydroxyindoles (29), and the iodoaminochromes and similar compounds (**17**) (30), carbonyl-substituted indoles (21), nitroindoles (31), fluoroindoles (32), and others. The spectrum of tryptophan (**15**) at 220 MHz has been reported (33) and, at lower field strengths, those of tryptamine (**18**) (20) and certain of its derivatives (34) have been used in structural studies. Data for substituted indoles are collected in Table 4.11.

The tautomeric equilibrium between 2-methyl- or 2-phenylindolenine *N*-oxide (**19**) and the corresponding *N*-hydroxyindole (**20**) has been extensively investigated using PMR methods (35, 36). Variation of the basicity of the solvent allowed complete control over the composition of the equilibrium mixture. Data are collected in Table 4.12.

The spectra of a number of 2,3,3-trisubstituted 3-*H*-indoles (indolenines) (**21**) have been reported (37) and the nonequivalence of various $-CX_2R$ groups (X = H or Me) on C-3 was used to assign the most favorable conformations.

Protonation of indoles has been shown by NMR methods to occur in concentrated solutions of mineral acids predominantly on C-3 (38) to give the ion **22** and similar results were obtained for ring protonation of tryptamine derivatives (39).

The structure of the indole Grignard reagent has been determined by NMR methods (40). Spectra of the Na, Li, Mg, and K salts of the indolyl anion **23** have been measured (41). As with other aromatic systems, the strong deshielding of ortho protons in these compounds is thought to be due mainly to paramagnetic influences of the metal ion rather than to electron-density effects.

TABLE 4.11 SPECTRAL DATA FOR SUBSTITUTED INDOLES

Compound	Solvent	τ1	τ2	τ3	τ4	τ5	τ6	τ7	Other Data				Ref.
1-Me	$CDCl_3$	6.63	3.18	3.52		2.83							20
2-Me	$CDCl_3$		7.80	3.87		2.92							20
	$CDCl_3$			3.95									21
	CCl_4			4.06									21
	DMSO			3.88									21
	Me_2CO			3.86									21
	Dioxan			3.88									21
	CS_2			4.10									21
	Et_2N			3.93									21
3-Me	$CDCl_3$		3.20	7.70		2.82							20
	$CDCl_3$		3.39										21
	CCl_4		3.65										21
	Me_2CO		3.0										21
	DMSO		2.94										21
	Dioxan		3.13										21
	CS_2		3.63										21
	Et_2N		3.25										21
									J_{12}	J_{13}	J_{23}	J_{37}	
4-Me	CCl_4		3.35	3.60					2.4	2.1	3.3	0.8	23
5-Me	CCl_4		3.42	3.76					2.4	2.0	3.2	0.6	23
6-Me	CCl_4		3.33	3.66					2.3	2.1	3.3	1.0	23
7-Me	CCl_4		3.33	3.69					2.3	2.15	3.3		23

2,3-Di-Me	$CDCl_3$		7.95	7.88	———	——2.95——	———	———	20
1-CH_2Ph	CCl_4	4.83[a]	3.07	3.57					22
2-CH_2Ph	CCl_4		6.02[a]	3.79					22
3-CH_2Ph	CCl_4		3.30	5.94[a]					22
1,3-Di-CH_2Ph	CCl_4	4.88[a]	3.30	5.94[a]					22
2,3-Di-CH_2Ph	CCl_4		6.02[a]	5.90[a]					22
3-Ph	$CDCl_3$		2.97		———	——2.83——	———	———	20
1-Me, 2-Ph	$CDCl_3$	6.43		3.47	———	——2.83——	———	———	20
3-$CHCH_2$	CCl_4	2.28	0.7		———	——3.0——	———	———	27
3-COMe	Me_2CO		1.82						21
	DMSO		1.66						21
2-COPh	$CDCl_3$			2.54					21
	Me_2CO			2.30					21
3-COPh	DMSO		2.0						21
2-CO_2H	Me_2CO			2.79					21
	DMSO			2.81					21
	Dioxan			2.74					21
3-CO_2H	DMSO		1.84						21
2-CO_2Me	$CDCl_3$	0.80	6.04	2.78	2.26	2.5	2.5	2.9	57
2-CO_2Et	$CDCl_3$			2.77					21
	Me_2CO			2.78					21
	DMSO			2.80					21
	Dioxan			2.80					21
	Et_2NH			2.80					21
3-CO_2Me	$CDCl_3$	1.20	2.10	6.07	1.79	2.5	2.5	2.8	57
3-CO_2Et	$CDCl_3$		2.18						21
	Me_2CO		1.99						21
	DMSO		1.88						21
	Dioxan		2.10						21
	Et_2NH		2.22						21

TABLE 4.11 *(continued)*

Compound	Solvent	τ1	τ2	τ3	τ4	τ5	τ6	τ7	Other Data	Ref.
1-Me, 4-CO_2Me	$CDCl_3$	6.28	2.87	2.87	6.03	2.05	2.77	2.48		57
5-CO_2Me	$CDCl_3$	1.15	2.77	3.37	1.53	6.08	2.08	2.63		57
1-Me, 6-CO_2Me	$CDCl_3$	6.21	2.81	3.46	2.35	2.13	6.07	1.87		57
2-Cl	CCl_4			3.56						28
3-Cl	CCl_4		2.87							28
1-CH_2Ph, 2-Cl	CCl_4			3.4						28
2-Me, 3-Cl	CCl_4		7.78							28
3-Br	$CDCl_3$		2.98							21
	Me_2CO		2.58							21
	DMSO		2.42							21
	Dioxan		2.77							21
	CS_2		3.13							21
	Et_2NH		2.95							21
4,5,6,7-Tetra-F	CCl_4	1.70	2.90	3.40						32
	Me_2CO	−1.0	2.60	3.40						32
4,5,6,7-Tetra-F, 2-CO_2H	Me_2CO	−1.6	−0.2	2.80						32
4-OH	$CDCl_3$		3.05	3.50		3.60	3.15	3.15		29
5-OH	$CDCl_3$		2.89	3.74	3.05		3.32	2.80		29
6-OH	$CDCl_3$		3.19	3.70	2.65	3.38		3.25		29
7-OH	$CDCl_3$		2.92	3.77	3.32	3.02	3.54			29
2-CO_2Et, 7-NO_2	Me_2CO	−0.46	5.58 8.60	2.71	1.85	2.73	8.15			31

2-CO_2Et, 3,7-di-NO_2	Me_2CO	−2.26	5.49, 8.56		1.57	2.42	1.57			31
3,7-Di-NO_2	Me_2CO		1.75		1.43	2.39	1.43			31
3,5,7-Tri-NO_2	Me_2CO		1.18		0.93		0.63			31
5,6-Di-OMe	$CDCl_3$	1.84	2.93	3.56	2.89	6.08, 6.12		3.15		30
5,6-Di-OMe, 7-I	$CDCl_3$	1.80	2.85	3.40	2.91	6.10	6.10			30
5,6-Di-OAc	$CDCl_3$	1.56	3.12	3.71	2.74	7.60	7.60	3.03		30
5,6-Di-OAc, 7-I	$CDCl_3$	1.54	2.89	3.43	2.66	7.63, 7.71				30
5,6-Di-OMe, 2-Me	$CDCl_3$	2.25	7.63	3.86	2.98	6.08, 6.15		3.25		30
5,6-Di-OMe, 2-Me, 4-I	$CDCl_3$	2.42	7.64	4.15		6.15, 6.19		3.15		30
5,6-Di-OMe, 2-Me, 7-I	$CDCl_3$	2.19	7.58	3.73	3.03	6.09, 6.12				30
5,6-Di-OH, 2-Me	Me_2CO		7.75	4.08	3.09			3.20		30
5,6-Di-OH, 2-Me, 7-I	Me_2CO		7.74	3.92	3.15					30
5,6-Di-OAc, 2-Me	$CDCl_3$	1.98	7.64	3.91	2.82	7.72	7.72	2.94		30
5,6-Di-OAc, 2-Me, 7-I	$CDCl_3$	2.10	7.65	3.94	2.86	7.71	7.71			30
5,6-Di-OH, 1-Me	Me_2CO	6.47	3.18	3.82	3.02			3.23		30
5,6-Di-OH, 1-Me, 7-I	Me_2CO	5.93	3.14	3.83	3.07					30
Tryptophan	D_2O (p*D* 7.5)		2.69		2.27[b]	2.80, 2.71		2.45[b]		33
Tryptamine derivatives										
Tryptamine	$CDCl_3$	1.33	3.08						N*b*-H 8.72	20
N*a*-Me	$CDCl_3$	6.36	3.20						N*b*-H 8.63	20
N*b*-Me	$CDCl_3$	0.70	3.12						N*b*-H 8.73, N*b*-Me 7.58	20
N*ab*-di-Me	$CDCl_3$	6.35	3.18						N*b*-Me 7.62	20
N*bb*-di-Me	$CDCl_3$	0.90	3.13						N*b*-Me 7.65	20
5-OH, N*bb*-di-Me	CD_3OD		3.02		3.05		3.30	2.82	N*b*-Me 7.70	34

[a] CH_2 groups.
[b] Assignment doubtful.

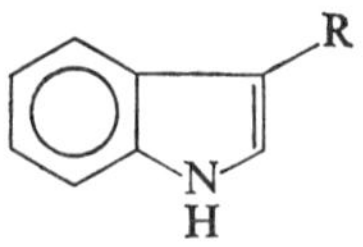

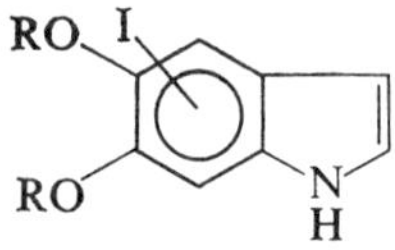

15, R = $CH_2CH(NH_2)CO_2H$
16, R = $CHCH_2$
18, R = $CH_2CH_2NH_2$

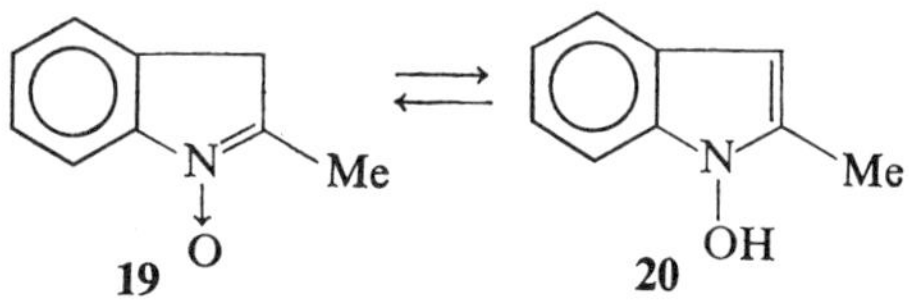

17

19 ⇌ **20**

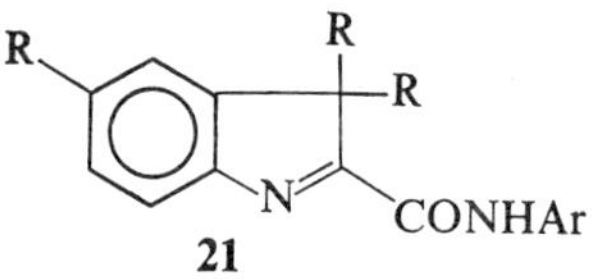

21 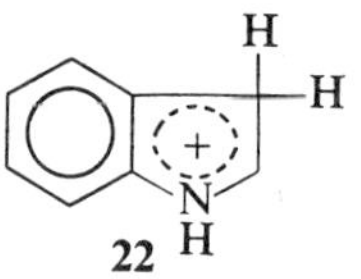 **22** 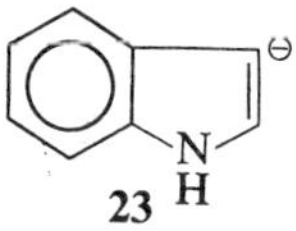 **23**

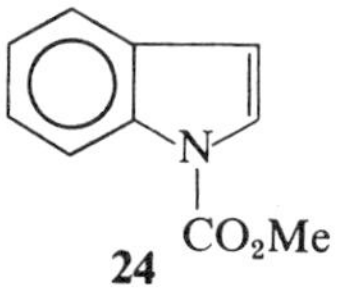

24 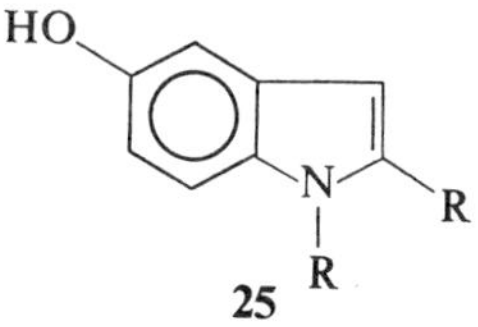 **25**

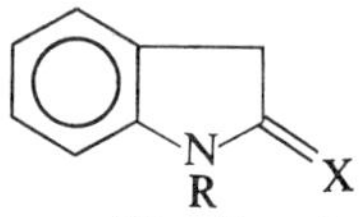

26, X = 0
27, X = S

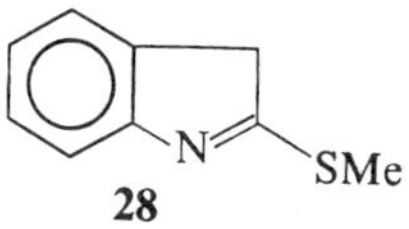

28 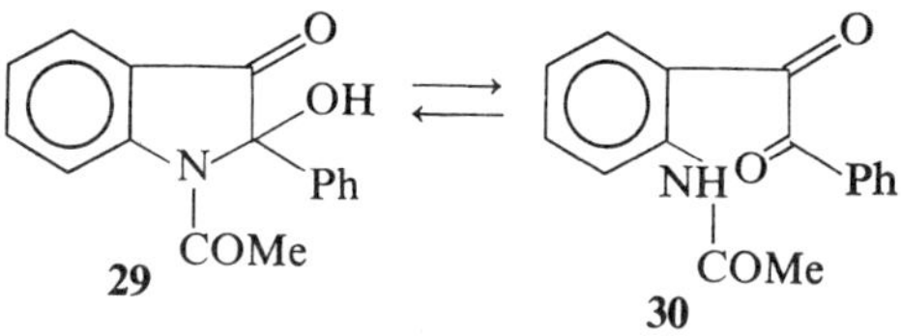 **29** ⇌ **30**

Apart from data in the papers discussed above, certain information is available on 1-substituted indoles. Thus the spectrum of 1-ethoxycarbonylindole (**24**) has been reported (42) and a large amount of data has been collected for a series of 1,2-disubstituted 5-hydroxyindoles (**25**) (43). These data are included in Table 4.11.

The oxindoles (**26**) form the basis of a large group of alkaloids, but very little NMR work has been done on the parent ring system. With simple

TABLE 4.12 SPECTRAL DATA FOR OXINDOLES AND THE CORRESPONDING THIOINDOLES

X	R_1	R_2	R_3	Solvent	τR_1	τR_2	τR_3	τarom	J_{gem}	Ref.
O	H	H	H	$CDCl_3$	0.43	6.49	6.49			45
	Me	H	H	$CDCl_3$	6.80	6.52	6.52		7.7	45
	H	Me	H	$CDCl_3$	0.41	8.51	6.53			45
				CCl_4	−0.56	8.51	6.68	2.97		47
	H	Me	Me	$CDCl_3$	0.40	8.60	8.60			45
	NH_2	H	H		5.7	6.55	6.55	2.53–3.18		48
	H	Me	Br	$CDCl_3$	1.24	7.93		3.13		47
	H	Me	NC_5H_{10}	$CDCl_3$		8.45		2.85		47
	H	Me	SPh	$CDCl_3$		8.26		2.76		47
	H	Me	Cl	CCl_4	−0.7	8.14				28
	H	Me	OMe	CCl_4	−0.42	8.55	7.04			28
S	H	H	H	$CDCl_3$	−1.22	5.94	5.94			45
	Me	H	H	$CDCl_3$	6.40	5.94	5.94			45
	H	Me	H	$CDCl_3$	−1.25	8.39	6.20			45
	H	Me	Me	$CDCl_3$	−1.16	8.55	8.55			45

N-substituted derivatives, the 3-methylene group gives clearly resolved peaks which are quite diagnostic for the system. The chemical shift values for this methylene group have been shown to correlate linearly with the appropriate Hammett σ constants for substituents in the benzene ring (44). A comparative study of a series of oxindoles (2-indolinones) and the corresponding thiones (**27**) has been reported (45). The C=S group, as expected, deshields the 3-position much more strongly than the C=O group. Also in this paper, data are reported for a number of indolenines or 3-*H*-indoles (**28**), although there seems to be some doubt in the literature about the actual tautomeric structure of these substances; see Ref. 46. In other reports of these compounds, the NMR information is of secondary importance and the relevant data are simply collected in Table 4.12.

Nuclear magnetic resonance methods have been used to demonstrate the ring-chain tautomerism existing between **29** and **30** (49).

A certain amount of information is available on the 2,3-dihydroindole or indoline system **31**. A study has been made of the spectra of indoline-2-carboxylic acid (**32**) (50) and its derivatives. Vicinal coupling constants of the ABX system formed by the 3-methylene group and the single 2-proton varied with substitution over a very wide range, making it hazardous to assign cis

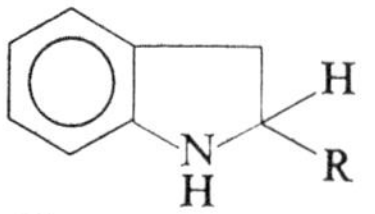

31
32, R = CO_2H

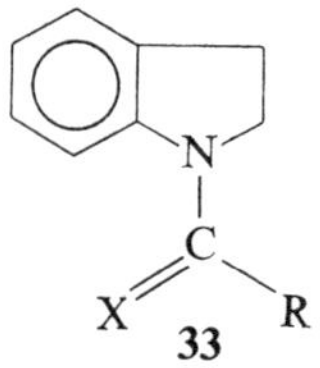

33

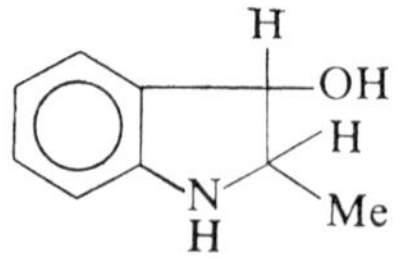

34

and trans stereochemistry from these values. Nagarajan and Nair (51, 52) have studied the orientation of the substituent group in a series of *N*-acyl- or thioacylindolines (**33**). Restricted rotation was observed about the amide bond and where X = O, R = Me or Ph, the C=O bond was very predominantly oriented toward the benzene ring as in **33**. However, when the substituent was a formyl or thioacetyl group the compounds exist as mixtures of rotamers. The solvent and temperature dependence of these mixtures has been studied (52). Also, spectra of *N*-acylindolines were reported quite early in the history of NMR (53). Parameters for the cis and trans isomers of 2-methyl-3-hydroxyindoline (**34**) have been reported (54) and for some other incomplete data on these compounds see Ref. 55. Data for these 2,3-dihydroindoles are collected in Table 4.13.

Finally, for the indoles, spectral parameters have been published for a number of compounds reduced in the benzene ring (56, 57), and these are collected in Table 4.14.

2. *The Isoindole System*

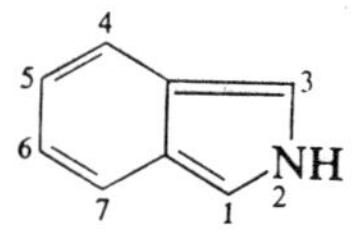

The bond structure of these molecules is normally assumed to be ortho quinonoid in type **35**, but in the case of the 1-phenyl derivative **36**, NMR shows an equilibrium between this form and the aromatic isoindolenine (**37**) (58). Also, Fletcher (59) reports data for some 1,3,4,7-substituted isoindoles and isoindolenines. NMR techniques have been used to study the equilibrium in aqueous solutions between compound **38** and its hydrate **39** (60).

The cis and trans isomers of 1,3,4,7-tetramethylisoindolines (**40**) have been prepared (61) and where the nitrogen carries a benzyl or neopentyl substituent, the isomers can be distinguished by the multiplicity of the N–CH_2 signal (62). In the cis case, the compound is symmetrical and the expected singlet is obtained, but with the asymmetric trans isomer, an AB quartet was observed. At low temperatures the spectrum of the methylene groups of *N*-acylisoindolines (**41**) resolves into a normal A_2B_2 pattern which shows

TABLE 4.13 SPECTRAL DATA FOR 2,3-DIHYDROINDOLES

Compound	Solvent	$\tau 1$	$\tau 2$	$\tau 3$	$\tau 3'$	$\tau 4,5,6$	$\tau 7$	J_{23}	$J_{23'}$	$J_{33'}$	Ref.
Parent	$CDCl_3$	6.47	6.73	7.22		2.8–4.0	3.55				51
1-Ac	$CDCl_3$	7.87	6.05	6.93		2.5–3.25	1.78				51
1-CHO	$CDCl_3$	1.11, 1.53	5.98	6.93		2.5–3.23	1.95				51
1-CSMe	$CDCl_3$	7.05, 7.23	5.45, 5.77	6.92		2.50–3.10	0.50				51
1-Ac, 3,3-di-Me	$CDCl_3$	7.80	6.23	8.68		2.65–3.08	1.82				51
5-Ac	$CDCl_3$	5.58	6.31	6.97			3.50				51
2-CO_2H	D_2O		5.78	6.63	6.82	2.7–3.3	2.7–3.3	9.8	7.2	16.5	50
2-$CONH_2$	DMSO	4.15	5.78	6.67	7.02	2.8–3.5	2.8–3.5	10.0	8.4	16.0	50
1-Tosyl, 2-CO_2Me	$CDCl_3$		5.18	6.81	6.92	2.2–3.1	2.2–3.1	10.1	5.3	16.4	50
1-Tosyl, 2-$CONH_2$	$CDCl_3$		5.35	6.75	7.23	2.2–3.1	2.2–3.1	2.6	10.5	16.7	50
1-Tosyl, 2-CN	$CDCl_3$		4.88	6.76		2.2–3.0	2.2–3.0	6.9	6.9		50

TABLE 4.14 DATA FOR INDOLES REDUCED IN THE BENZENE RING. SOLVENT—$CDCl_3$

I II III

Compound	τ1	τ2	τ3	τ3a	τ4[a]	τ5	τ6	τ7	τ7a	Ref.
I R = H	6.5	2.78	6.3		5.42, 6.3	6.81	4.05	4.08		57
I R = CO_2Me	6.35	6.17	6.26		5.49, 6.35	6.80	4.06	4.09		57
I R = CO_2Me, 6,7-di-H	6.33	6.16	6.22		6.16, 6.33	8.05	8.05	7.46		57
II	7.20	6.21	6.24	6.37	6.10	2.94	3.68	4.01	5.17	57
II 6,7-Di-H	7.20	6.24	6.28	6.39	6.12	2.99	7.78	8.06	5.99	57
III R = H		4.00	3.32				5.21			56
III R = Me		4.09	3.45		6.77		5.25	6.77		56

[a] The first value is for H, the second for CO_2Me.

35 36 37

38 39 40

41 42

long-range couplings between protons on C-1 and those on C-2. The largest of these is assigned to the trans pairs of protons (63). Throughout this work the author assumes that no nitrogen inversion is occurring and that rotation of the N–Ac bond is slow. The spectra of the bridged compounds **42** have also been reported (64). Data for the isoindole system are collected in Table 4.15.

TABLE 4.15 SPECTRAL DATA FOR ISOINDOLES

Compound	Solvent	τ2	τ3	τ4,5,6	Ref.
1-Ph,2*H*	$CDCl_3$	0.5	1.9–3.2	1.9–3.2	58
1-Ph, 3*H*	$CDCl_3$		5.13		58
1-(4-OMePh),2*H*	$CDCl_3$	0.0	1.9–3.2	1.9–3.2	58
1-(4-OMePh),3*H*	$CDCl_3$		5.19	1.9–3.2	58

1,3-DIHYDROISOINDOLES

Substituents	Solvent	Temperature (°C)	τ1,3	Δ1,3[a]	τ2	τ4,7	τ5,6	J_{11}	J_{13cis}	$J_{13trans}$	Ref.
1,3,4,7-Tetra-Me	CCl_4		5.65, 8.62		7.95	7.80	3.21				61
2-CO_2Me	$CDCl_3$	−25		7.6				12.1	0.9	2.6	63
2-COMe	$CDCl_3$	36		8.4				12.7	1.1	1.7	63
2-COPh	$CDCl_3$	38		25.5					1.2	1.2	63
2-$COCF_3$	$CDCl_3$	36		11.4					0.8	0.8	63

[a] Δ1,3 is the difference (in hertz) between the geminal protons on C-1 and C-3 when nonequivalence is observed.

3. *The Indolizine System*

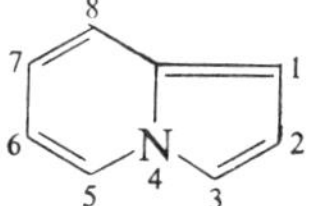

The only NMR study of simple indolizines gives a complete analysis of the spectra of the parent ring system and of a number of methyl derivatives (65). Apart from this paper, information on the unreduced indolizine system is rarely encountered and generally is found as an addition to preparative work (66, 67). A large amount of data for highly substituted indolizine can be found in papers by Acheson and his co-workers, for example, Refs. 68–70. These are not included in this treatment because the observed spectra are essentially those of the substituents. Data for indolizines are collected in Table 4.16.

Protonation of indolizines has been thoroughly studied and the NMR technique gave unique and clear-cut information which was not available from other spectral techniques. A tabulation of the spectra of a number of indolizinium salts (71a) was followed by two papers which appeared almost simultaneously and which fully investigated the protonation of indolizine and its alkyl derivatives (71b, 72). Protonation of the indolizine ring system

TABLE 4.16 SPECTRAL DATA FOR INDOLIZINES. SOLVENT—CCl_4, EXCEPT WHERE SPECIFIED OTHERWISE

	Parent	1-Me	2-Me	2,3-Di-Me	2,5-Di-Me	2,6-Di-Me	2,7-Di-Me	1-CO_2Et[a], 2-CO_2H
$\tau 1$	3.72	7.65	3.91	3.90	3.81	3.98	4.08	5.6, 8.7
$\tau 2$	3.36	3.53	7.73	7.77	7.70	7.76	7.76	
$\tau 3$	2.86	2.96	3.06	7.69	3.19	3.14	3.14	2.4
$\tau 5$	2.24	2.30	2.33	2.52	7.59	2.54	2.44	1.85–3.5
$\tau 6$	3.69	3.75	3.78	3.68	3.99	7.86	3.94	1.85–3.5
$\tau 7$	3.50	3.57	3.58	3.54	3.54	3.69	7.80	1.85–3.5
$\tau 8$	2.75	2.84	2.88	2.88	2.83	2.96	3.14	1.85–3.5
J_{12}	3.90							
J_{13}	1.2		0.5					
J_{15}	1							
J_{23}	2.74	3						
J_{26}	0.5							
J_{38}	0.5							
J_{56}	6.82	6–7	6.5	6				
J_{57}	1.02	1	1	2		2		
J_{58}	1.15	1	1	1–2				
J_{67}	6.39	6–7	6	7	7			
J_{68}	1.02	1	1.5	2	2		2	
J_{78}	8.98	9	8.5	7–8	9	8		
Ref.	65	65	65	65	65	65	65	67

[a] In N NaOH.

occurs preferentially at C-3 to give ions of the type **43**. When position 3 was

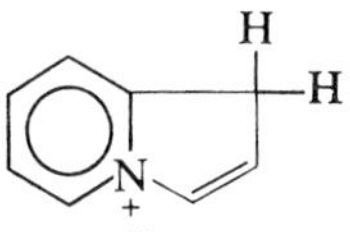

43 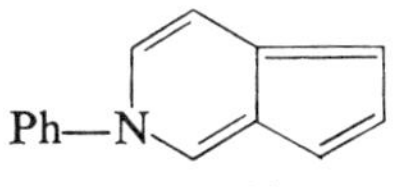 **44** 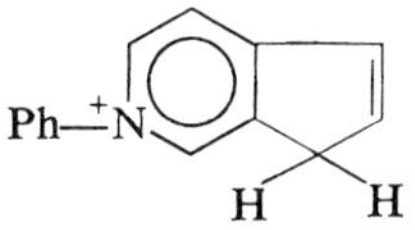 **45**

unsubstituted the spectrum observed was entirely that of the 3*H*-isomer. However, when 3-substituents were present protonation occurred both on C-3 and on C-1. Protonation of indolizine at C-3 agrees with HMO predictions in terms of change of frontier electron densities, but not of electrophilic localization energies. The extensive data on indolizinium cations are summarized in Table 4.17.

Very little is known of the reduced indolizine systems. The difference in chemical shift between the 4-methyl groups of the cis and trans isomers of indolizidine was noted (16) and tentatively explained in terms of bond anisotropies (16a). The preferred conformations of the epimeric pairs of 7- and 8-hydroxyindolizidines were investigated using a similar approach to that used for the piperidinols (73) (see Chapter 2, Section I.F).

4. *The 2-Pyrindine System*

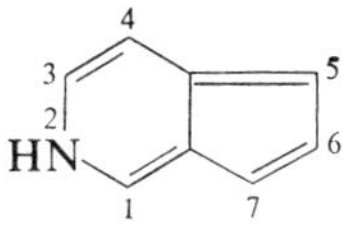

The PMR spectrum of 2-phenyl-2-pyrindine (**44**) has been reported, but the phenyl group obscured most of the spectrum (74). However, in 98% H_2SO_4 it was shown to protonate on C-7 to give the ion **45** (75).

B. Two Nitrogen Atoms

1. *The Indazole System*

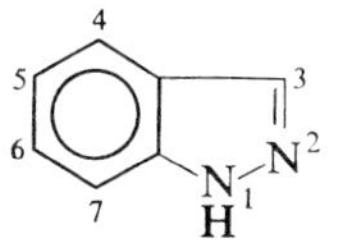

The spectrum of indazole in acetone at 100 MHz consists of a series of well-separated multiplets which proved simple to analyze (76). One of the main problems in indazole chemistry has been a lack of understanding of the indazole (**46**)–isoindazole (**47**) tautomeric equilibrium. While alkylation reactions show the presence of the equilibrium, the results above, together

TABLE 4.17 SPECTRAL DATA FOR INDOLIZINIUM IONS

Compound	Anion	Solvent	τ1	τ2	τ3	τ5	τ6	τ7	τ8	Ref.
Parent[a]	ClO_4^-	TFA	2.81	2.49	4.45	1.02	2.16	1.53	1.96	71
	TFA	TFA	2.77	2.45	4.43					72
	Cl^-	HCl	2.63	2.31	4.29					72
2-Me[a]	ClO_4^-	TFA	3.17	7.59	4.60	1.16	2.33	1.63	2.14	71
2-*t*-Bu	ClO_4^-	TFA	3.14	8.52	4.43					72
	Cl^-	HCl	2.95	8.52	4.29					72
3-Me	ClO_4^-	TFA	5.78	3.20	7.39					72
	TFA	TFA[b]	5.72	3.13	7.33					72
		[c]	2.77	2.44	4.30, 8.06					72
5-Me	ClO_4^-	TFA	2.72	2.37	4.62	7.04				72
7-Me	ClO_4^-	TFA	2.82	2.47	4.40			7.25		72
1,2-Di-Me[a]	ClO_4^-	TFA	7.75	7.68	4.68	1.15	2.32	1.57	2.16	71
2,3-Di-Me	ClO_4^-	TFA[b]	5.79							72
		[c]	4.13	8.09	4.57					72
2,5-Di-Me	ClO_4^-	TFA	3.07	7.38	4.75	7.07				72
2,6-Di-Me	ClO_4^-	TFA	3.17	7.57	4.59	1.27	7.41	1.74	2.17	71
2,7-Di-Me	TFA	TFA	3.16	7.55	4.63			7.31		72
2,8-Di-Me	ClO_4^-	TFA	3.05	7.54	4.54	1.25	2.36	1.77	7.33	71
3,5-Di-Me	ClO_4^-	TFA	2.86	2.54	4.17, 8.11	6.91				72
	Cl^-	HCl	2.79	2.46	4.05, 8.19	6.92				72
3,7-Di-Me	ClO_4^-	TFA	5.79	3.29	7.38			7.16		72
1,2,3-Tri-Me	ClO_4^-	TFA	7.73	7.73	4.71, 8.19	1.18	2.26	1.58	2.15	71
	Cl^-	HCl	7.69	7.69	8.15					72
2,3,6-Tri-Me	ClO_4^-	TFA	5.78	7.64	7.45		7.26			72
	ClO_4^-	HCl[b]	5.84							72
		[c]	3.09		8.27					72
2,3,7-Tri-Me	TFA	TFA[b]	5.87							72
		[c]	3.18		4.65					72
2-Ph, 3-Me	ClO_4^-	TFA[b]	5.44		7.30					71
		[c]	2.69		8.09					71

[a] Coupling constants: parent—J_{12} 6, J_{56} 6.5, J_{78} 7.5; 2-Me and 1,2-di-Me—J_{56} 6.5, J_{78} 7.5.
[b] Protonation on C-1.
[c] Protonation on C-3.

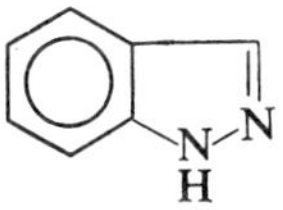

46 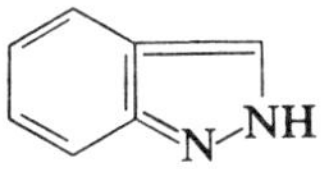 47 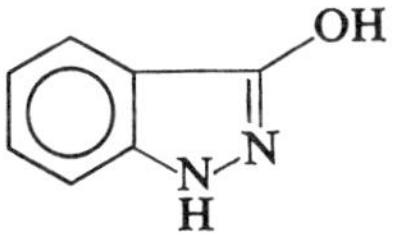 48

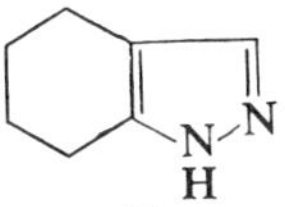

49 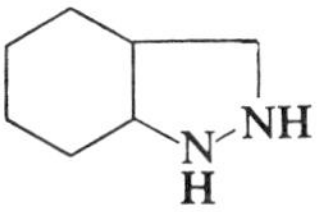 50

with the observation of an NH signal when DMSO is used as solvent (77), confirm the predominance of the indazole isomer. Nuclear magnetic resonance methods have been used in a general approach to this problem in which spectra of 1- and 2-aryl derivatives were studied and alkylation reactions were investigated (78–80). To complement this work, data for an extensive series of 1*H*- and 1-arylindazoles have been published (81). Also, a number of indazoles substituted on N-1 with germanium- (82) or phosphorus-containing groups (83) have been synthesized and their spectra reported. Spectra of a number of 3-hydroxyindazoles show them to exist almost exclusively in the enol tautomer **48** (84).

Little is known of the spectra of reduced indazoles but some data for 4,5,6,7-tetrahydro-1*H*- (**49**) (85) and 3,3*a*,4,5,6,7-hexahydro-2*H*-indazoles (**50**) (86) are available. Information on indazoles is collected in Table 4.18.

2. *The Benzimidazole System*

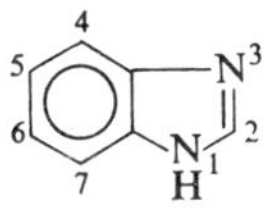

Data for simple benzimidazoles are extremely scarce and I can find no satisfactory analysis of the parent ring system. However, the chemical shift of H-2, together with that for H-2 in its 1-acetyl derivative, has been reported (87), and data for the benzenoid protons were obtained by Black and Heffernan (see footnote to Table 4.19). These values and a few others (88–91) are collected in Table 4.19.

The structure of *N*-acyl-*N*,*N*′-dialkyl-*o*-phenylenediamines has been subject to some debate, but their NMR spectra show conclusively that they can exist both as the open-chain amide **51** and as the closed carbinol **52** (92). It is interesting to note that H-2 of 3-methoxy-1-methylbenzimidazolium iodide (**53**) resonates at extremely low field, τ -1.0 (93).

Nuclear magnetic resonance methods were used, together with other

TABLE 4.18 SPECTRAL DATA FOR INDAZOLES[a]

Compound	Solvent	τ1	τ3	τ4	τ5	τ6	τ7	Other Data	Ref.
Indazole	Me_2CO		1.94	2.23	2.89	2.66	2.41	Couplings[b]	76
		−2.4	1.88						77
	$CDCl_3$		1.80	2.15	2.80	2.66	2.40		81
3-Me	$CDCl_3$		7.35	2.33	2.95	2.70	2.50		81
5-Me	$CDCl_3$		1.95	2.52	7.59	2.83	2.63		79
	Me_2CO		2.07	2.53	7.62	2.85	2.56		79
	DMSO		2.00	2.48	7.61	2.82	2.50		79
7-Me	$CDCl_3$		1.84	2.42	2.86	2.86	7.40		79
	Me_2CO		1.92	2.40	2.90	2.90	7.43		79
	DMSO		1.90	2.42	2.92	2.92	7.45		79
3,4-Di-Me	$CDCl_3$		7.25	7.30	3.15	2.78	2.72		81
3,6-Di-Me	$CDCl_3$		7.40	2.43	3.03	7.53	2.80		81
3,5,6-Tri-Me	$CDCl_3$		7.42	2.57	7.65	7.65	2.78		81
3-Me, 4-Cl	$CDCl_3$		7.20		2.67	2.87	2.67		81
3-Me,5,6-di-OMe	$CDCl_3$		7.33	2.85	5.95	5.98	3.05		81
1-Ph	$CDCl_3$		1.85						79
	Me_2CO		1.78						79
	DMSO		1.60						79
2-Ph	$CDCl_3$		1.67						79
	Me_2CO		1.25						79
	DMSO		0.91						79
1-P$(NMe_2)_2$	C_6D_6		1.87		2.27–3.10			$J_{37} = 0.8$	83
	DMSO		1.72		2.09–2.98				83
1-$GeMe_2$	CCl_4		1.94		2.32–3.20			$J_{37} = 0.95$	82
1-Me,4,5,6,7-tetrahydro	$CDCl_3$	6.30	2.75						85
2-Me,4,5,6,7-tetrahydro	$CDCl_3$	6.20	2.97						85

[a] A large amount of data on 1-arylindazoles can be found in Refs. 79 and 81.
[b] $J_{37} = 0.8$, $J_{45} = 7.8$, $J_{46} = 1.2$, $J_{47} = 1.0$, $J_{56} = 7.0$, $J_{57} = 1.2$, $J_{67} = 7.9$ Hz.

TABLE 4.19 SPECTRAL DATA FOR BENZIMIDAZOLES. SOLVENT—$CDCl_3$

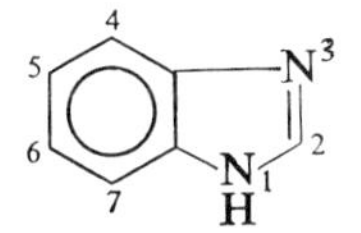

Compound	τ1	τ2	τ4	τ5–7	Ref.
Parent[a]		1.92			87
1-Ac		1.45			87
2-Me		7.35			87
2-Me, 1-Ac	7.26	7.19			87
1,2-Di-Me	6.67	7.63	2.35	2.6–3.0	88
1-Et, 3-Me	6.10, 8.79	7.54	2.32	2.7–2.9	88
2-Alkyl	−2.13	4.82, 6.22	2.58	2.58	89
2-Propenyl	−1.47	3.19, 8.13	2.64	2.64	89

[a] τ4,7 2.30, τ5,6 2.74, J_{45} 8.2, J_{46} 1.4, J_{47} 0.7, J_{56} 7.1 (solvent—acetone) (111).

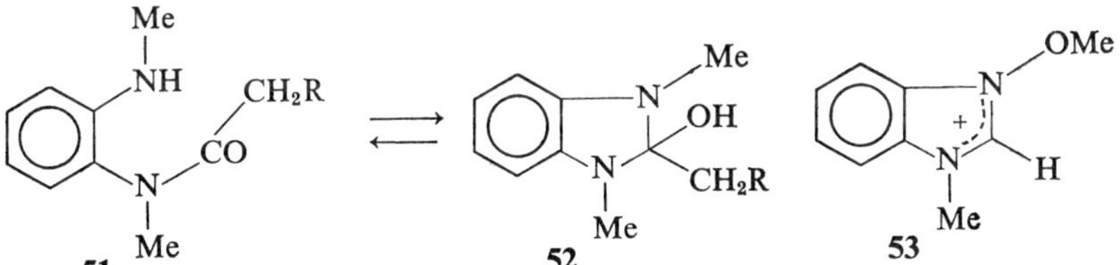

51 ⇌ **52** **53**

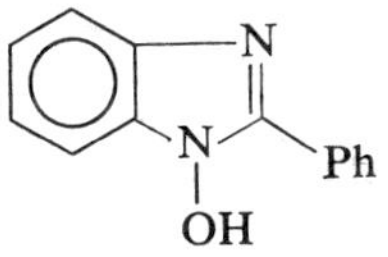

54

spectral techniques, to show that **54** is the predominant tautomer of 2-phenylbenzimidazole-1-oxide (94).

3. *The Imidazo[1,2-a]pyridine System*

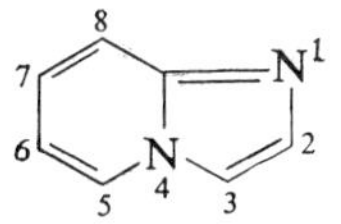

This system is probably the most thoroughly studied of all the polyazaindenes although quite a lot of the effort has been overlapping.

Paudler and his co-workers (95–99), as part of their investigation of aromaticity in heterocycles with ten π-electrons, have prepared a large number of imidazo[1,2-*a*]pyridines and have provided an approximate analysis of the spectra of all of them. A full analysis of the parent compound has also been reported (65). The fact that the 5-, 6-, 7-, and 8-protons resonate at fields somewhat lower than those observed for model pyridine compounds has led two sets of authors to conclude that the system must be reasonably flat and "aromatic" in the sense that it supports a considerable ring current (96, 100, 101). A rough "qualitative" correlation was obtained between π-electron densities on the carbon atoms and the chemical shifts of the attached protons (96). This was later extended using a semiempirical method utilizing charge densities obtained from HMO calculations to obtain quite acceptable results (98). The success of the method was ascribed to the choice of indolizine as reference compound. The assignment of specific signals to H-2 and H-3 in the case of the parent compound was possible because of long-range couplings between H-3 and H-5. Data for the neutral molecules of imidazo[1,2-*a*]pyridines are given in Table 4.20.

Chemical shifts were also used to show that protonation and methylation occur predominantly at N-1 to give ions of the type **55**, which are also thought to be flat and "aromatic" (96, 102). Electron-density calculations on the

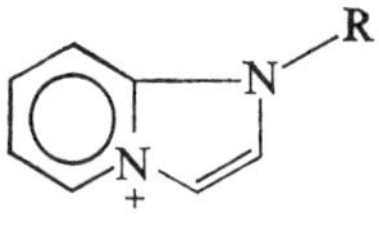

55 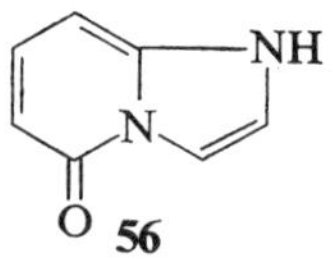 **56** 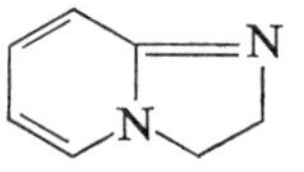 **57** 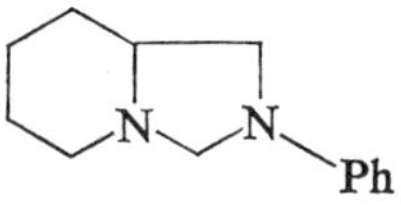 **58**

cation ascribe a higher charge to N-1, in agreement with the preferred ionic structure. Spectral parameters for these cations are collected in Table 4.21.

The preferred tautomeric structure for the 5-hydroxy derivative of this system was shown to be the (1*H*)-5-pyridone (**56**), although the 5-amino compound exists predominantly as such (101). Data for the neutral molecule and cation 7-methyl-2,3-dihydroimidazo[1,2-*a*]pyridine (**57**) are also included in the tables (96).

4. *The Imidazo[1,5-a]pyridine System*

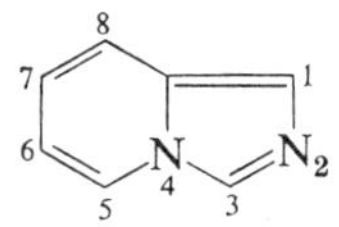

The spectrum of the parent heterocycle in CCl_4 has been reported (65) and the chemical shifts have been compared with those calculated from charge densities (98):

$\tau 1$ 2.73, $\tau 3$ 2.03, $\tau 5$ 2.12, $\tau 6$ 3.59, $\tau 7$ 3.42, $\tau 8$ 2.66;

J_{56} 7.13, J_{57} 0.88, J_{58} 1.11, J_{67} 6.38, J_{68} 1.05, J_{78} 9.21.

TABLE 4.20 SPECTRAL DATA FOR IMIDAZO[1,2-*a*]PYRIDINES

Substituents	Solvent	$\tau 2$	$\tau 3$	$\tau 5$	$\tau 6$	$\tau 7$	$\tau 8$	J_{23}	J_{35}	J_{56}	J_{57}	J_{58}	J_{67}	J_{68}	J_{78}	Ref.
None	CCl_4	2.33	2.40	1.95	3.45	3.03	2.31	1.2	0.7	6.8	1.3	1.3	6.6	1.3	9.0	94
	CCl_4	2.52	2.52	1.91	3.35	2.97	2.49			6.89	1.19	1.02	6.77	0.95	9.29	65
	D_2O			1.80	3.20	2.79	2.44									95
2-Me	$CDCl_3$	7.60	2.82	2.12	3.53	3.08	2.58		0.7	6.8	1.4		6.6	1.3	9.0	94
3-Me	$CDCl_3$	2.63	7.73	2.32	3.38	2.98	2.45									96
5-Me	$CDCl_3$	2.37	2.68	7.77	3.65	3.05	2.52	1.2					7.0	1.2	9.0	94
6-Me	$CDCl_3$	2.35	2.48	2.10	7.78	3.01	2.40	1.2	0.5		1.2				9.2	95
7-Me	$CDCl_3$	2.48	2.52	2.04	3.45	7.65	2.64	1.2	0.5				7.2	1.4		94
	D_2O	2.43	2.48	2.08	3.54	7.80	2.85									95
8-Me	$CDCl_3$	2.32	2.40	1.96	3.32	3.02	7.41	1.2	0.7	6.5	1.4		6.7			94
2,3-Di-Me	$CDCl_3$	7.65	7.83	2.32	3.34	3.00	2.45			7.2	1.3		6.6	1.4	9.4	94
5,7-Di-Me	$CDCl_3$	2.31	2.55	7.51	3.50	7.68	2.59	1.2								95
3-Br	$CDCl_3$	2.38		2.10	3.25	2.90	2.43									96
3-Br, 2-Me	$CDCl_3$	7.60		2.24	3.40	3.01	2.42									96
3-Br, 5-Me	$CDCl_3$	2.57		7.20	3.75	3.15	2.65									96
3-Br, 5,7-di-Me	$CDCl_3$	2.63		7.17	3.80	7.78	2.82									96
3-CH_2NMe_2	$CDCl_3$	2.48		1.70	3.23	2.86	2.27									99
	D_2O	2.31		1.70	2.90	2.52	2.27									99
2-Me, 3-CO_2Et	$CDCl_3$	7.30		0.74	3.04	2.61	2.27									99
2-Me, 3-CH_2NMe_2	$CDCl_3$	7.55		1.78	3.25	2.87	2.45									99
7-Me, 2,3-di-H	$CDCl_3$	6.10	6.10	3.12	3.59	8.00	2.92									95

TABLE 4.21 SPECTRAL DATA FOR IMIDAZO-[1,2-*a*]PYRIDINIUM IONS

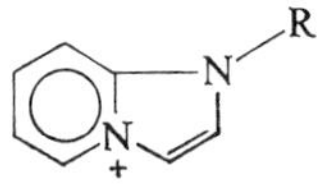

Substituents	Anion	Solvent	τ2	τ3	τ5	τ6	τ7	τ8	τNR	Ref.
None	Br^-	$CDCl_3$	1.20	1.94	0.60	2.42		1.6	−3.84	99
		D_2O	1.44	1.64	0.86					99
2-Me	Cl^-	$CDCl_3$	7.34	1.52	0.66	2.54	2.08	1.08	−5.16	99
		D_2O	7.14		1.08	2.26				99
6-Me	Cl^-	D_2O	1.89	1.78	1.38	7.44	2.13	2.13		95
1,6-Di-Me	I^-	D_2O	2.00	1.85	1.44	7.31	2.07	2.07	5.85	95
7-Me	Cl^-	D_2O	2.10	1.93	1.42	2.67	7.38	2.32		95
	I^-	D_2O	2.04	1.88	1.38	2.62	7.35	2.24		95
1,7-Di-Me	I^-	D_2O	2.04	1.84	1.38	2.60	7.32	2.16	5.91	95
8-Me	Cl^-	D_2O	1.89	1.68	1.27	2.47	2.10	7.28		95
1,8-Di-Me	I^-	D_2O	1.98	1.76	1.33	2.54	2.17	7.04	5.58	95
5,7-Di-Me	Cl^-	D_2O	2.14	2.14	7.62	3.02	7.44	2.74		95
7-Me, 2,3-di-H	Cl^-	D_2O	5.99	5.29	2.08	3.14	7.85	3.07		95

Nuclear magnetic resonance has been used to help determine the conformational preferences in a number of 2-phenyloctahydroimidazo[1,5-*a*]-pyridines (**58**) (103), also named 8-phenyl-1,8-diazabicyclo[4,3,0]nonanes. Information on these compounds is summarized in Table 4.22.

5. *The Pyrazolo[1,5-a]pyridine System*

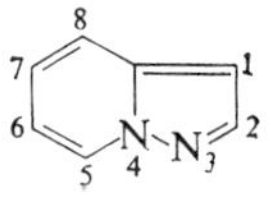

Analysis of the spectra of the parent heterocycle (65) and certain derivatives (104) have been reported, and the results are given in Table 4.23.

6. *The Pyrrolo[1,2-b]pyridazine System*

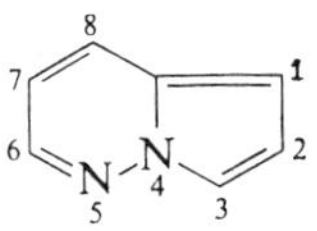

The spectra of a number of members of this ring system have been reported in a publication dealing with the synthesis and reactions of this type of compound (105). Data are collected in Table 4.24.

TABLE 4.22 SPECTRAL DATA FOR 2-PHENYLOCTAHYDROIMIDAZO[1,5-*a*]PYRIDINES (103)

A. CIS FUSED

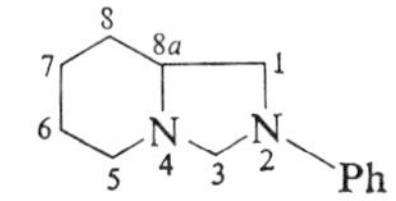

Substituents	Stereochemistry	$\tau 3$	$\tau 3'$	$J_{33'}$
6-Me	*cis*-6,8*a*-H	5.78	6.12	−6.0
7-Me	*trans*-7,8*a*-H	5.84	6.21	−6.0
8-Me	*cis*-8,8*a*-H	5.78	6.15	−6.0

B. TRANS FUSED

Substituents	Stereochemistry	$\tau 1$	$\tau 1'$	$\tau 3$	$\tau 3'$	$J_{11'}$	$J_{33'}$	J_{18a}	$J_{1'8a}$
None		6.67	6.93	5.67	6.39	−7.8	−3.8	6.1	9.3
5-Me	*cis*-5,8*a*-H	6.67	6.93	5.54	6.43	−7.9	−3.7	6.5	9.6
6-Me	*trans*-6,8*a*-H	6.62	6.94	5.66	6.43	−7.9	−3.6	6.7	8.7
7-Me	*cis*-7,8*a*-H	6.61	6.93	5.63	6.41	−7.9	−3.7	6.3	9.5
8-Me	*trans*-8,8*a*-H	6.56	6.88	5.64	6.41	−7.9	−3.6	6.0	8.4
1-Me	*cis*-1,8*a*-H	6.32		5.60	6.62		−3.7	6.4	
1-Me	*trans*-1,8*a*-H		6.63	5.77	6.32		−4.3		8.2

7. *The Pyrrolo[1,2-c]pyrimidine System*

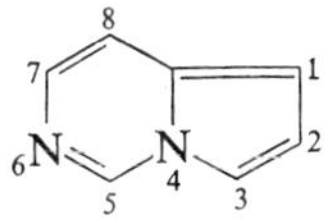

The spectrum of the parent compound in $CDCl_3$ has been reported (106): $\tau 5$ 1.16, $\tau 7$ 3.06, $\tau 8$ 3.50, $\tau 1,2,3$ 2.63 ppm.

8. *The Pyrrolo[1,2-a]pyrazine System*

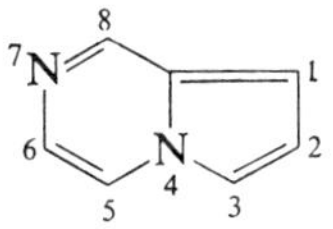

An analysis of the spectra of the parent heterocycle and the 1,3-dibromo derivative has been reported (104) and the data are given in Table 4.25.

TABLE 4.23 SPECTRAL DATA FOR PYRAZOLO(1,5-*a*)PYRIDINES (65, 104) SOLVENT—$CDCl_3$

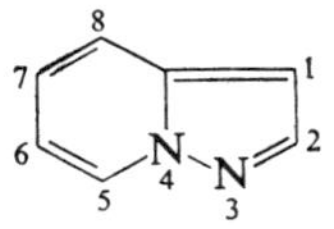

Substituents	$\tau 1$	$\tau 2$	$\tau 5$	$\tau 6$	$\tau 7$	$\tau 8$	J_{12}	J_{15}	J_{27}	J_{56}	J_{57}	J_{58}	J_{67}	J_{68}	J_{78}
None	3.62	2.20	1.61	3.38	3.03	2.56	2.18	0.9	0.5	6.93	1.00	1.02	6.97	1.17	8.94
1-Br		2.08	1.62	2.88	2.72	2.52			0.5	7.0	1.2	1.2	6.8	1.6	8.3
1,2-Di-Me	7.85	7.63	1.74	3.48	3.08	2.74				6.93	1.00	1.02	6.79	1.17	8.90
1-Ac, 2-Me	7.47	7.34	1.60	3.14	2.59	1.82				6.79	1.00	1.54	7.10	1.52	9.12
1,2,5-Tri-Me	7.80	7.57	7.33	3.57	3.07	2.75							6.95	1.20	8.72
1-Ac, 2,5-di-Me	7.45	7.28	7.28	3.24	2.67	1.87							7.01	1.51	9.02

TABLE 4.24 SPECTRAL DATA FOR PYRROLO[1,2-*b*]PYRIDAZINES (105). SOLVENT—CS_3

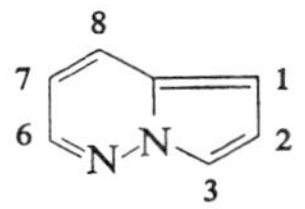

Substituents	$\tau 1$	$\tau 2$	$\tau 3$	$\tau 6$	$\tau 7$	$\tau 8$	J_{12}	J_{13}	J_{23}	J_{67}	J_{68}	J_{78}
None	3.69	3.33	2.48	2.21	3.68	2.49	4	1	2.5	4.5	2	9
6,8-Di-Me	3.76	3.46	2.60	7.68	3.94	7.68	4	1.5	2.5			
6,8-Di-Ph	3.48	3.24	2.30	2.0–2.7	3.08	2.0–2.7	4.5	1.5	2.5			
6-Me, 8-Ph	3.61	3.38	2.48	7.62	3.72	2.3–2.8	4.4	1.6	2.8			
6-H, 8-Ph	3.48	3.24	2.34	2.07	3.58	2.3–2.7	4.5	1.5	2.5	4.5		
6-Ph, 8-H	3.62	3.26	2.1–2.7	2.1–2.7	3.12	2.33	5	1.5	3			9

TABLE 4.25 SPECTRAL DATA FOR PYRROLO[1,2-*a*]PYRAZINES (104). SOLVENT—$CDCl_3$

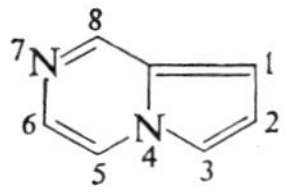

Substituents	$\tau 1$	$\tau 2$	$\tau 3$	$\tau 5$	$\tau 6$	$\tau 8$	J_{12}	J_{13}	J_{15}	J_{23}	J_{38}	J_{56}	J_{58}
None	3.15	3.03	2.54	2.11	2.42	1.09	4.50	1.45	0.85	2.55	1.00	5.5	1.55
1,3-Di-Br		3.10		2.19	2.33	1.29						5.0	1.55

9. *The Pyrrolo[1,2-a]pyrimidine System*

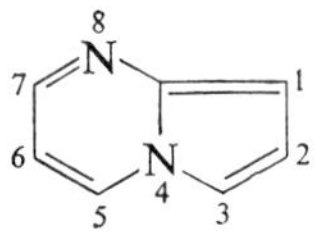

An extensive study has been made of a series of derivatives of this ring system which are reduced in the five-membered ring and most of which contain oxygen functions in positions 5 and/or 7 (107–109). Data for the many types of compounds included in this work are given in Table 4.26.

TABLE 4.26 SPECTRAL DATA FOR PYRROLO[1,2-*a*]-PYRIMIDINE DERIVATIVES (107–109)

A.

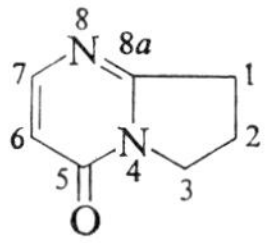

Substituents	Solvent	τ1	τ2	τ3	τ6	τ7	τ8	τ8*a*
7-Me	$CDCl_3$	6.88	7.70	5.88	3.90	7.72		
7-Ph	$CDCl_3$	6.90	7.80	5.92	3.44	2.20, 2.65		
7-OH	$CDCl_3$	6.82	7.67	5.84	3.67			
7-OMe	$CDCl_3$	6.90	7.74	5.89	4.47	6.15		
6-Et, 7-OH	$CDCl_3$	6.82	7.55	5.84	7.55, 8.90	−1.50		
6-Ph, 7-OH	$CDCl_3$	6.89	7.80	5.89	2.62			
6,7-Di-H	$CDCl_3$	7.34	8.00	6.27	7.64	6.40		
6,7-Di-H, 7-Me	$CDCl_3$	7.50	8.00	6.33	7.50	6.30, 8.74		
6,7-Di-H, 7-Ph	$CDCl_3$	7.30	8.00	6.25	7.30	5.25, 2.65		
6,7,8,8*a*-Tetra-H, 7-Me	Neat	8.30	8.30	6.90	8.30	7.0, 9.08	7.88	5.95
	D_2O	8.10	8.10	6.55	7.70	6.90, 8.80		5.70

B.

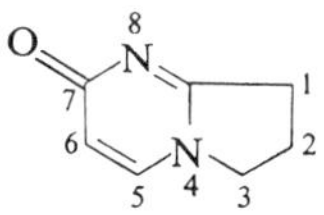

Substituents	Solvent	τ1	τ2	τ3	τ5	τ6
5-Me	$CDCl_3$	7.10	7.70	6.00	7.83	4.30
5-Ph	$CDCl_3$	6.95	7.75	6.10	2.67	4.10
5,6-Di-H, 5-Me	HCl	6.90	7.65	5.85	5.85, 8.85	6.90
5,6-Di-H, 5-Ph	$CDCl_3$	7.20	7.90	6.65	5.40, 2.82	7.20
5,6-Di-H	HCl	6.80	7.65	6.0	6.0	6.70
5,6-Di-H, 6-Me	$CDCl_3$	7.25	7.80	6.50	6.50	7.25, 8.82

TABLE 4.26 (*continued*)

C.

Substituents	Solvent	τ1	τ2	τ3	τ5	τ6	τ7
5-Me	$CDCl_3$	7.5	8.0	6.7	5.7, 8.75	5.3	3.75
6-Me	$CDCl_3$	7.6	8.0	6.8	6.1	8.5	3.95
7-Me	$CDCl_3$	7.5	8.05	6.9	6.0	5.5	8.3

D.

Substituents	Solvent	τ1	τ2	τ3	τ5	τ6	τ7
5-Me	$CDCl_3$	7.6	8.1	6.7	6.7, 8.8	8.2	5.1
6-Me	$CDCl_3$	7.6	8.1	6.7	6.7	8.1, 9.0	5.4
7-Me	$CDCl_3$	7.6	8.1	6.7	6.7	8.1	8.6

E.

Substituents	Solvent	τ1	τ2	τ3	τ5	τ6	τ7	τ8	τOH
5-Me	$CDCl_3$	6.9	7.8	6.2	6.2, 8.6	7.8	4.9	6.6	4.7
6-Me	$CDCl_3$	7.0	7.8	6.25	6.8	7.8, 9.0	5.3	6.8	
7-Me	$CDCl_3$	6.9	7.75	6.2	6.5	7.75	8.6	6.85	4.82

10. The Pyrrolo[2,3-b]pyridine System

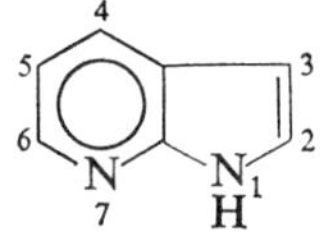

The spectra of the parent compound in Me_2CO and in CCl_4 have been reported (24):

Me_2CO: τ2 2.53, τ3 3.52, τ4 2.04, τ5 2.95, τ6 1.73; J_{23} 3.4, J_{45} 7.7, J_{46} 1.5, J_{56} 4.7.

CCl_4: τ2 2.64, τ3 3.59, τ4 2.11, τ5 2.99, τ6 1.71; J_{23} 3.3, J_{45} 7.7, J_{46} 1.5, J_{56} 4.8.

C. Three Nitrogen Atoms

1. *The Benztriazole System*

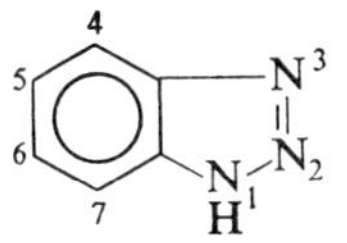

Spectra of bentriazole and its 1- and 2-methyl derivatives were originally used to distinguish between the symmetric 2-methyl compound **59** and the asymmetric 1-methyl compound **60** (110). A full analysis of the spectrum of

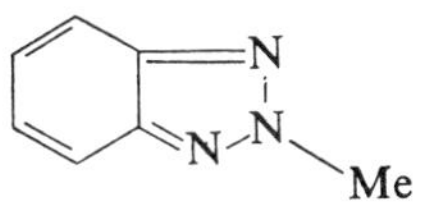

59 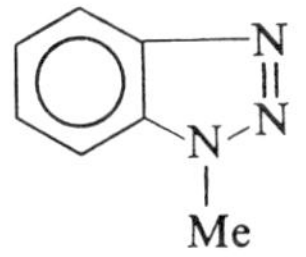 **60**

the time-averaged symmetrical parent compound was carried out by Black and Heffernan (111), and recently a thorough study has been made of this spectrum and that of the *N*-methyl derivatives (112, 113). In this latter work iterations were carried out using the two computer programs LAOCOON and NMRIT. The statement is made that the results obtained by the two methods were very close, but some of the tabulated results seem to be rather unsatisfactory, for example, for one case the values 6.89 and 6.98 were obtained for the same coupling constant. The differences between the two analyses are worrying, especially when one realizes that the work of Black and Heffernan on various heteroaromatic systems has been taken as standard by theoretical chemists wishing to calculate chemical shifts. Concentration effects were very small indeed, with H-4 and H-7 being most affected.

Coupling constants for the benzene ring of 2-methylbenztriazole have been used in a study of bond fixation in such systems (114). The parameters obtained from these analyses together with those for another substituted benztriazole (115) are given in Table 4.27.

2. *The s-Triazolo[4,3-a]pyridine System*

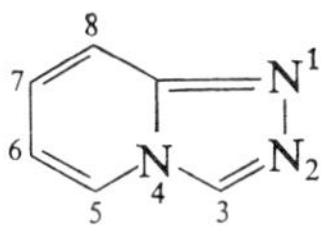

The spectra of a series of simple derivatives of this system have been studied (116–118) mainly in an attempt to determine the "aromaticity" of the system. After extensive discussion of chemical shifts and the effects of

TABLE 4.27 SPECTRAL DATA FOR BENZTRIAZOLES

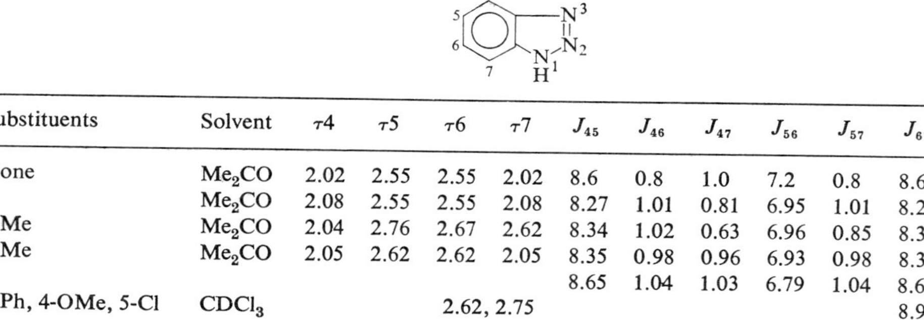

Substituents	Solvent	$\tau4$	$\tau5$	$\tau6$	$\tau7$	J_{45}	J_{46}	J_{47}	J_{56}	J_{57}	J_{67}	Ref.
None	Me_2CO	2.02	2.55	2.55	2.02	8.6	0.8	1.0	7.2	0.8	8.6	111
	Me_2CO	2.08	2.55	2.55	2.08	8.27	1.01	0.81	6.95	1.01	8.27	112
1-Me	Me_2CO	2.04	2.76	2.67	2.62	8.34	1.02	0.63	6.96	0.85	8.36	113
2-Me	Me_2CO	2.05	2.62	2.62	2.05	8.35	0.98	0.96	6.93	0.98	8.35	113
						8.65	1.04	1.03	6.79	1.04	8.65	114
2-Ph, 4-OMe, 5-Cl	$CDCl_3$			2.62, 2.75							8.9	115

TABLE 4.28 SPECTRAL DATA FOR *s*-TRIAZOLO [4,3-*a*] PYRIDINES (117). SOLVENT—CDCL$_3$ (INFINITE DILUTION)

Substituents	$\tau3$	$\tau5$	$\tau6$	$\tau7$	$\tau8$	J_{56}	J_{57}	J_{58}	J_{67}	J_{68}	J_{78}
None	1.14	1.79	3.12	2.72	2.21	6.7	1.2	1.2	6.7	1.2	9.0
3-Me	7.26	2.12	3.15	2.78	2.27	6.8	1.2	1.2	6.8	1.3	8.9
5-Me	1.27	7.31	3.33	2.76	2.30				6.9		9.2
6-Me	1.28	2.03	7.65	2.89	2.32		1.5				9.4
7-Me	1.26	1.94	3.29	7.54	2.52	7.2				1.4	
8-Me	1.20	1.95	3.23	2.96	7.33	6.6			6.6		
3-Ph	2.21, 2.40	1.71	3.13	2.71	2.16	6.9	1.2	1.2	6.9	1.2	9.5
3-Cl		2.01	2.97	2.62	2.20	6.4	1.1	1.1	6.4	1.4	9.1
3-Br		1.94	3.02	2.66	2.25	6.8	1.1	1.1	6.8	1.4	9.4
3,5-Di-Me	6.98	7.16	3.53	2.96	2.46				6.8		9.3
3,6-Di-Me	7.28	2.44	7.63	2.93	2.40		1.8				9.3
3,7-Di-Me	7.29	2.25	3.32	7.57	2.54	7.0					
3,8-Di-Me	7.28	2.25	3.24	2.97	7.36	6.5			6.5		
3-Ph, 5-Me	2.48	7.79	3.40	2.79	2.29				6.5		9.2
3-Ph, 6-Me	2.20, 2.41	1.95	7.65	2.86	2.27		1.5				9.7
3-Ph, 7-Me	2.14, 2.37	1.82	3.27	7.50	2.59	7.1				1.6	
3-Ph, 8-Me	2.24, 2.46	1.88	3.20	2.96	7.26	6.7			6.7		
5,7-Di-Me	1.34	7.34	3.53	7.58	2.46						
3,5,7-Tri-Me	7.02	7.22	3.69	7.67	2.73					1.0	

methyl substitution the conclusion was reached that there was considerable double-bond character present in the 5,6- and 7,8-bonds although the presence of some ring current was necessary to rationalize certain of the substituent effects. Long-range couplings were observed and were a useful aid to assignment. Data for representative members of this class of compounds are given in Table 4.28.

The extremely low field resonance of H-3 in spectra of 2-methyl-*s*-triazolo[4,3-*a*]pyridinium iodides has been successfully used to distinguish between the 1- and 2-methiodides, **61** and **62**, respectively, of this ring system (119). Data are collected in Table 4.29.

61 **62**

TABLE 4.29 SPECTRAL DATA FOR *s*-TRIAZOLO[4,3-*a*] PYRIDINIUM METHIODIDES (119)

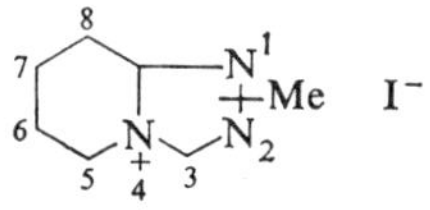

Substituents	NMe	Solvent	τ3	τ5	τ6	τ7	τ8	τNMe
None	1	D_2O	0.53	1.20	2.42	1.87	1.82	5.67
		DMSO	0.13	0.92	2.45	1.96	1.58	5.67
	2	D_2O	−0.43	1.33	2.53	2.10	2.10	5.58
		DMSO	−0.78	1.00	2.25	1.96	1.58	5.58
3-Me	1	D_2O	7.08	1.37	2.20	1.82	1.82	5.70
		DMSO	7.08	1.00	2.33	2.03	1.66	5.70
	2	D_2O	6.87	1.50	2.20	1.82	1.82	5.63
		DMSO	6.83	1.12	2.33	2.03	1.66	5.67
5-Me	1	D_2O	0.53	7.08	2.60	2.18	1.92	5.68
		DMSO	0.04	7.09	2.50	2.07	1.75	5.67
	2	D_2O	−0.23	7.18	2.62	2.18	1.92	5.56
		DMSO	−1.03	7.18	2.52	2.07	1.75	5.58
7-Me	1	D_2O	0.63	1.28	2.50	7.30	2.04	5.71
		DMSO	0.24	1.03	2.42	7.32	1.67	5.73
	2	D_2O	−0.33	1.40	2.62	7.42	2.28	5.58
		DMSO	−0.68	1.13	2.60	7.46	2.13	5.60
8-Me	1	D_2O	0.60	1.37	2.52	2.20	7.08	5.43
		DMSO	0.16	1.06	2.46	1.94	7.08	5.46
	2	D_2O	−0.40	1.37	2.52	2.20	7.33	5.52
		DMSO	−0.80	1.16	2.46	2.12	7.42	5.52
3,8-Di-Me	1	D_2O	7.08	1.59	2.53	2.20	7.08	5.43
		DMSO	7.11	1.29	2.63	2.30	7.08	5.52
	2	D_2O	6.87	1.59	2.53	2.20	7.36	5.62
		DMSO	6.85	1.29	2.63	2.30	7.42	5.68

3. *The s-Triazolo[1,5-a]pyridine System*

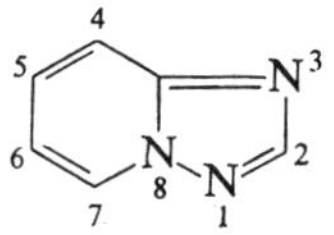

These compounds were dealt with in conjunction with the *s*-triazolo-[4,3-*a*]pyridines (116, 117) and the comments made in the previous section (VI.C.2) about neutral molecules hold equally well for both systems. Data for the *s*-triazolo [1,5-*a*] pyridines are collected in Table 4.30.

4. *The Imidazo[1,2-b]pyridazine System*

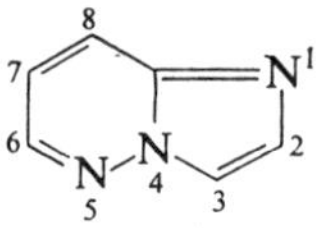

Spectra of imidazo[1,2-*b*]pyridazines are included in a number of papers on the synthesis of this ring system and were used to show that electrophilic

TABLE 4.30 SPECTRAL DATA FOR *s*-TRIAZOLO[1,5-*a*]-PYRIDINES (116, 117). SOLVENT—$CDCl_3$

Substituents	τ2	τ4	τ5	τ6	τ7	J_{45}	J_{46}	J_{47}	J_{56}	J_{57}	J_{67}
None	1.65	2.18	2.48	2.96	1.38	8.8	1.2	1.2	6.6	1.3	6.6
2-Me	7.26	2.36	2.78	3.13	1.97	9.2	1.2	1.2	6.5	1.3	6.5
2-Ph	1.70	2.20	2.50	3.00	1.37	8.7	1.3	1.3	6.9	1.5	6.9
	2.52										
2,4-Di-Me	7.39	7.37	2.73	3.14	1.66				6.9		6.9
2,5-Di-Me	7.42	2.59	7.53	3.23	1.67						7.1
2,6-Di-Me	7.43	2.43	2.70	7.61	1.72	9.2				2.2	
2,7-Di-Me	7.40	2.39	2.53	3.10	7.24	7.8			6.3		
2-Ph, 4-Me	1.67	7.27	2.74	3.12	1.60				7.1		7.1
	2.53										
2-Ph, 5-Me	1.71	2.54	7.54	3.22	1.55						7.2
	2.54										
2-Ph, 6-Me	1.75	2.37	2.56	7.61	1.59	8.6					
	2.50										
2-Ph, 7-Me	1.65	2.30	2.57	3.12	7.17	8.0			6.8		
	2.42										

substitution occurred at the 3-position (120, 121). Data are collected in Table 4.31.

63

Spectra of the hydrochlorides and methiodides (**63**) have been reported (122) and the parameters obtained are summarized in Table 4.32.

5. *The Imidazo[1,2-c]pyrimidine System*

Nuclear magnetic resonance has been used to establish the positions of substituents in a number of derivatives of this ring system (123, 124). The data obtained are given in Table 4.33.

TABLE 4.31 SPECTRAL DATA FOR IMIDAZO[1,2-*b*]PYRIDAZINES

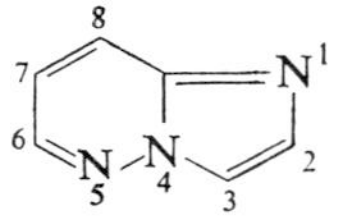

Substituents	Solvent	τ2	τ3	τ6	τ7	τ8	J_{23}	J_{38}	J_{67}	J_{68}	J_{78}	Ref.
None	$CDCl_3$	2.21	2.01	1.70	3.00	2.05	1.0	0.8	4.5	2.0	10.0	120
2-Me	$CDCl_3$	7.67	2.78	2.32	3.54	2.70		0.8	4.5	2.0	10.0	120
3-Br	$CDCl_3$	2.21		1.53	2.90	2.03			4.5	2.0	10.0	120
3-NO_2	$CDCl_3$	1.34		1.22	2.56	1.80			4.5	2.0	10.0	120
6-OMe	$CDCl_3$	2.24, 2.36		6.03	3.32	2.19						121
6-Cl	$(CD_3)_2CO$	2.04	1.72		2.68	1.75	1.0	0.8			10.0	120
2-Me, 6-OMe	$CDCl_3$	7.56	2.47	6.06	3.40	2.33						121
2-Me, 6-Cl	$(CD_3)_2CO$	7.56	1.97		2.70	1.91		0.8			10.0	120
3-Br, 6-Cl	DMSO	1.85			2.40	1.56					10.0	120
2-CO_2Et, 6-OMe, HBr	$CDCl_3$		1.64	5.83	2.52	1.14					10.0	121
2-Me, 3-Br, 6-Cl	TFA	7.90			2.78	2.10					10.0	120

TABLE 4.32 SPECTRAL DATA FOR CATIONS OF IMIDAZO-[1,2-*b*]PYRIDAZINES (122). SOLVENT—D_2O

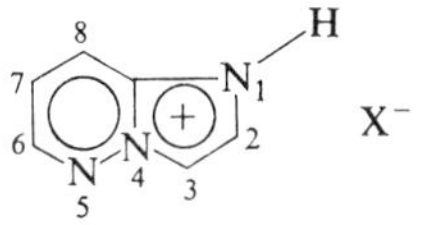

Substituents	Anion	τ1	τ2	τ3	τ6	τ7	τ8	J_{23}	J_{38}	J_{67}	J_{68}	J_{78}
1-Me	I^-	5.77	1.80	1.61	1.00	2.05	1.40	2.0	1.0	4.5	1.5	9.5
1,6-Di-Me	I^-	5.81	1.90	1.67	7.25	2.21	1.56	2.0	1.0			9.5
2-Me	Cl^-			1.82	1.12	2.22	1.62		1.0	4.5	1.5	9.5
7,8-Di-Me	Cl^-		1.80	1.56	1.16	7.32	7.24	2.0				
1,7,8-Tri-Me	I^-	5.62	2.02	1.72	1.17	7.42	7.17	2.0				
6-Cl, 1,8-di-Me	I^-	5.61	1.90	1.60		2.24	7.05	2.0				
6-Cl, 7,8-di-Me	Cl^-			1.56		7.32	7.17	2.0				

6. *The Imidazo[1,2-a]pyrimidine System*

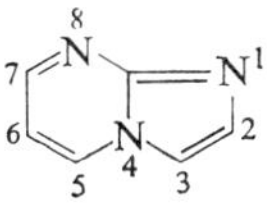

This ring system was included in the study of aromaticity in the poly-azaindene series (125). Using the corresponding imidazo[1,2-*a*]pyridines (**64**) as reference compounds, the calculation of chemical shifts from charge densities gave excellent results for all protons except H-7. This position would be strongly influenced by the anisotropy of N-8, which was not included in the calculations. Protonation and methylation were shown to

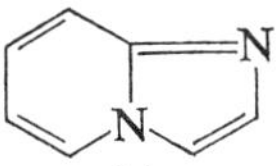

64 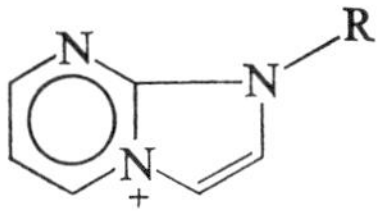 **65** 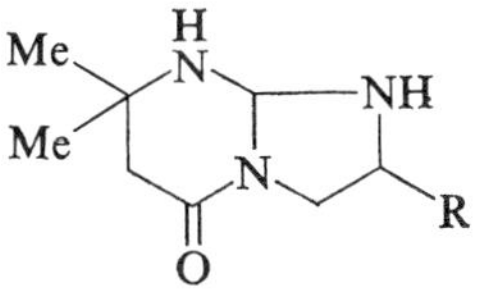 **66**

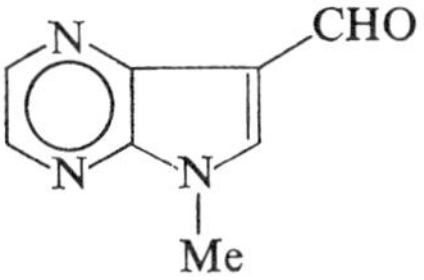

67

occur at N-1 to give ions of the type **65**. Data for this system are collected in Table 4.34.

TABLE 4.33 SPECTRAL DATA FOR IMIDAZO[1,2-*c*]PYRIMIDINES

Substituent	Solvent	τ2	τ3	τ8	τ5	τ7	τNH	J_{78}	Ref.
2,5,7-Tri-Cl	CCl_4		2.4, 2.6						122
2,7-Di-Cl, 5-SMe	CCl_4		2.2, 2.3		7.2				122
5-Oxo, 6-H, 7-SMe	DMSO	3.6	3.1			7.4	−7.1, −2.1		122
5-Oxo, 2,6-di-Me, 3-OAc	DMSO	7.90	7.65	3.54		2.64		7	123
2,7-Dioxo, 1,3-di-H, 5-SMe	DMSO[a]		5.7	4.9	7.45		2.8		122
2,5,7-Trioxo, 1,3,6-tri-H	DMSO		5.75	5.15			−0.5, 2.6		122

[a] Measured at 100°C.

TABLE 4.34 SPECTRAL DATA FOR IMIDAZO[1,2-*a*]PYRIMIDINES (124). SOLVENT—$CDCl_3$

A. NEUTRAL MOLECULES

Substituents	τ2	τ3	τ5	τ6	τ7	J_{23}	J_{56}	J_{57}	J_{67}
None	2.23	2.34	1.28	3.09	1.40	1.4	6.9	2.0	4.1
7-Me	2.30	2.41	1.40	3.20	7.42	1.4	6.9		
5,7-Di-Me	2.21	2.58	7.40	3.35	7.42	1.4			
2,3,7-Tri-Me	7.62	7.68	1.98	3.30	7.45		6.9		
2-Ph, 7-Me		2.35	1.78	3.40	7.42		6.9		
3-Br	2.16		1.66	3.08	1.38		6.9	2.0	4.2
3-Br, 5,7-di-Me	2.51		7.08	3.58	7.58				

B. CATIONS

Substituents	Anion	Solvent	τ1	τ2	τ3	τ5	τ6	τ7	J_{23}	J_{56}	J_{57}	J_{67}
None	TFA^-	TFA		1.80, 1.87		0.70	2.22	0.80				
	Cl^-	D_2O		1.69, 1.72		0.62	2.22	0.80	2.1	6.9	1.8	4.5
1-Me	I^-	D_2O	5.83	1.78, 1.85		0.74	2.27	0.87	2.1	6.9	1.8	4.5
7-Me	TFA^-	TFA		1.92, 2.02		1.02	2.40	7.12				
5,7-Di-Me	TFA^-	TFA		1.95, 2.04		7.07	2.55	7.18				
1,5,7-Tri-Me	I^-	D_2O	5.86	1.83	1.83	7.13	2.39	7.21				
2,3,7-Tri-Me	TFA^-	TFA		7.42	7.42	1.20	2.40	7.14				
1,2,3,7-Tetra-Me	I^-	D_2O	6.00	7.40	7.40	1.09	2.40	7.13		6.9		

TABLE 4.35 SPECTRAL DATA FOR IMIDAZO[4,5-*b*]PYRIDINES (127). SOLVENT—DMSO

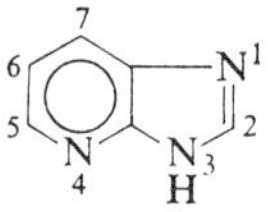

Substituents	$\tau 2$	$\tau 5$	$\tau 6$	J_{56}	τNH	τNH_2
7-Cl	1.37	1.60	2.58	5		
7-NH_2	1.67	1.90	3.42		−2.0	3.42
5,7-Di-Cl	1.38		2.48			

It is interesting to note that the hexahydro derivative **66** occurs naturally as an alkaloid (126), and its spectrum played a major part in determining its structure.

7. *The Imidazo[4,5-b]pyridine System*

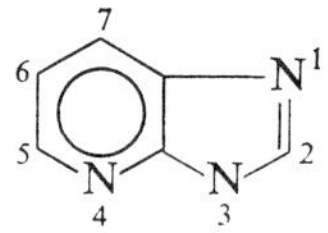

Data for a number of substituted members of this class of compound have been published (127) and are summarized in Table 4.35.

8. *The Imidazo[4,5-c]pyridine System*

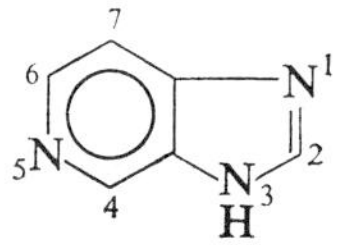

Data for a number of compounds from this system (127, 128) are collected in Table 4.36.

9. *The Pyrrolo[3,4-d] pyrimidine System*

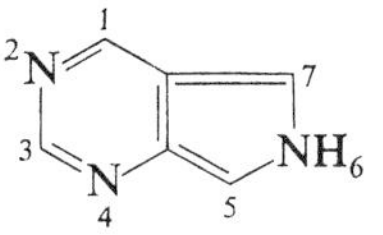

Data for a number of 5,7-dihydro derivatives (129) are summarized in Table 4.37.

10. *The Pyrrolo[2,3-d]pyrimidine System*

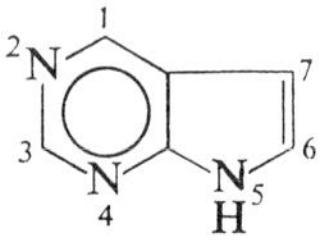

Spectra of a number of 4-substituted or 4,5-disubstituted derivatives of this heterocycle have been reproduced (130), but few accurate numerical data are available.

TABLE 4.36 SPECTRAL DATA FOR IMIDAZO[4,5-*c*]-PYRIDINES

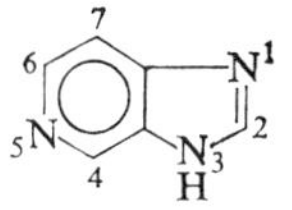

Substituents	Solvent	$\tau 2$	$\tau 4$	$\tau 6$	$\tau 7$	J_{67}	Ref.
4-Cl	DMSO	1.43		1.78	2.33	5.5	127
4-NH_2	DMSO	1.75		2.40	3.12	6.0	127
6-NH_2	DMSO	1.87	1.50	3.35			127
5-NO_2, 6-NHAc	DMSO	1.58	1.20	7.80	1.58		127
4-Ar, 4,5,6,7-tetra-H	$CDCl_3$	2.95	5.15	6.98	7.42		128
	TFA	1.28	3.97	6.06, 6.54			128

11. The Pyrrolo[2,3-b]pyrazine System

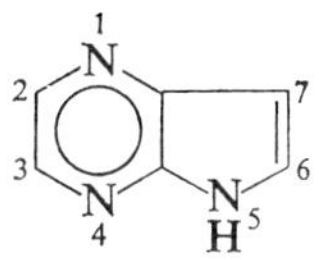

The NMR spectrum was used to prove the structure of 7-formyl-5-methylpyrrolo[2,3-*b*]pyrazine (**67**) (131): $\tau 2,3$ 1.50, $\tau 6$ 1.72, τNMe 5.98, τCHO -0.23 (solvent not given).

12. The Pyrrolo[3,2-d]pyrimidine System

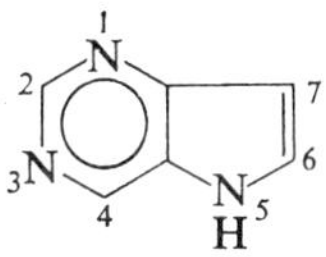

Although no information is available on the unsubstituted heterocycle, quite a lot of work has been done on derivatives which can be regarded as

TABLE 4.37 SPECTRAL DATA FOR 5,7-DIHYDROPYRROLO[3,4-*d*]-PYRIMIDINES (129). SOLVENT—TFA

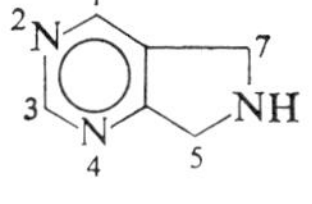

Substituents	$\tau 2$	$\tau 5, 7$	τNR
1-OH, 6-Ph	0.93	4.41, 4.56	2.18–2.40
1-OH, 6-*p*-ClPh	0.95	4.48, 4.61	2.18–2.40
1-NH_2, 6-Ph	0.92	4.35, 4.45	2.38
1-NH_2, 6-*p*-ClPh	1.05	4.49, 4.58	2.20–2.60
1-OH, 3-NH_2, 6-Ph		4.58, 4.78	2.38
1,3-Di-NH_2, 6-Ph		4.59	2.27

deaza analogs of naturally occurring purines (132, 133). Thus, NMR showed clearly that, as in the purine series, an amino group on C-2 or C-4 exists as such but an hydroxyl group takes the 2- or 4-oxo form. The data obtained from these studies are collected in Table 4.38.

D. Four Nitrogen Atoms

1. *The s-Triazolo[4,3-b]pyridazine System*

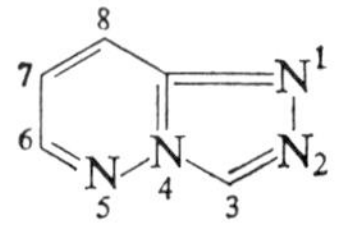

Little information is available on the neutral molecules of this series but spectra have been reported for the 3-methyl or 3-phenyl-5-oxides (**68**) (134).

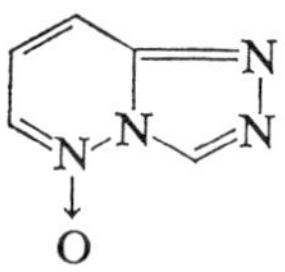

68 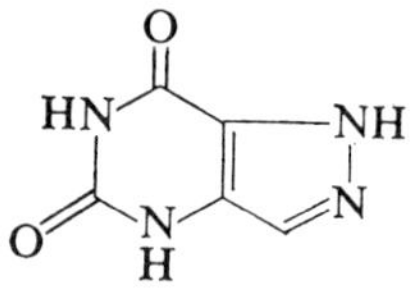 **69**

However, *N*-methylation of *C*-methyl derivatives has been studied (122) and NMR provides a simple method of distinguishing between products methylated on positions 1 or 2. Thus $\tau 3$, whether for H or Me, is always of lower value for the 2-methyl isomers than for the 1-methyl isomers. Data for these neutral molecules and cations are included in Table 4.39.

2. *The s-Triazolo[4,3-a]pyrimidine System*

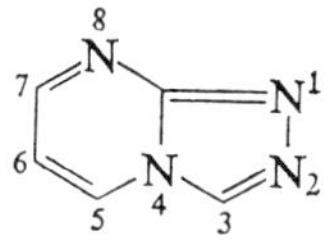

Data for the parent heterocycle and some of its methyl derivatives have been reported (135) and are collected in Table 4.40.

3. *The s-Triazolo[1,5-a]pyrimidine System*

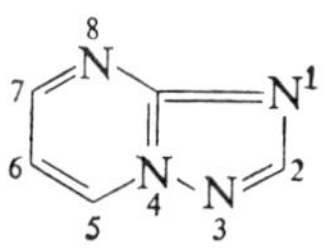

The spectra of a number of *s*-triazolo[1,5-*a*]pyrimidine derivatives have been recorded (136) in order to obtain information about their reactivities

TABLE 4.38 SPECTRAL DATA FOR PYRROLO[3,2-*d*]PYRIMIDINES[a]

Substituent	Solvent	$\tau 2$	$\tau 4$	$\tau 6$	$\tau 7$	τNH	J_{67}	J_{56}	J_{57}	Ref.
None	D_2O	1.60	1.60	2.43	3.70		3			132
	N NaOD	1.26, 1.46	1.26, 1.46	1.96	3.43		2			132
	N D_2SO_4	0.78, 0.96	0.78, 0.96	1.55	3.03		3			132
	DMSO	1.08, 1.18	1.08, 1.18	2.10	3.38	−1.8	3	3		132
6-Me	N D_2SO_4	1.06	1.06	7.26	3.25					
4-NH_2	DMSO	1.83	3.36	2.45	3.60	−1.0	4			133
	N D_2SO_4	1.63		2.30	3.46		3			132
4-OH	DMSO	2.20		2.63	3.61	−2.0	3	3	3	132
	N NaOD	1.86		2.58	3.56		3			132
2,6-Di-Me	DMSO		1.40		3.80	−1.4				132
	N D_2SO_4	7.18	1.19	7.34	3.45					132
6-Me, 4-NH_2	DMSO	1.97	3.63		3.93					132
	N D_2SO_4	1.72		7.58	3.84					132
6-Me, 4-OH[b]	DMSO	2.27			3.90	−1.7			2	132
	N NaOD	1.91		7.67	3.90					132
4-NH_2, 7-CO_2H	TFA	1.52		1.07						133
2,4-Di-NH_2	DMSO	1.72, 2.81	1.72, 2.81	2.55	3.86	−2.3	3			132
2-NH_2, 4-OH[b]	DMSO	4.05		2.90	4.05	−1.4	3	3		132
2,4-Di-OH[b]	DMSO			2.93	4.13	−1.7, −1.4, −1.6	3	3		132
	N NaOD			2.77	3.95		3			132

[a] Data for some highly substituted compounds are included in these references.
[b] Neutral molecules take the oxo form.

TABLE 4.39 SPECTRAL DATA FOR *s*-TRIAZOLO [4,3-*b*]-PYRIDAZINES

A. SOLVENT—DMSO (134)

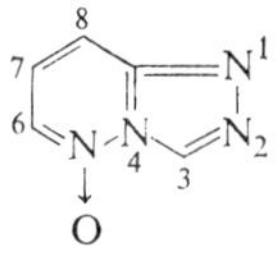

Substituents	τ3	τ6	τ7	τ8	J_{67}	J_{78}
3-Me	7.03	1.86	2.70	2.12	6.0	9.0
3-Ph	2.50	1.77	2.62	2.00	6.0	9.0

B. SOLVENT—D_2O (122)

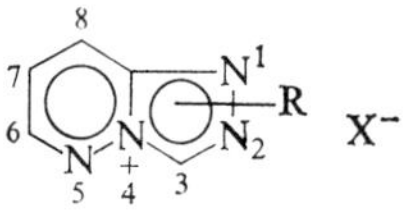

Substituents	Anion	τ3	τ6	τ7	τ8	τNMe	J_{67}	J_{78}	J_{68}	J_{38}
1-H	Cl^-	0.25	0.84	2.12	1.38		4.5	9.5	2.0	
1-H, 3-Me	Cl^-	7.00	0.91	2.04	1.40		4.5	9.5	2.0	
1-H, 6-Me	Cl^-	0.27	7.17	2.05	1.45			9.5		1.0
1-H, 7-Me	Cl^-	0.24	0.95	7.25	1.55				2.0	1.0
1-H, 8-Me	Cl^-	0.28	1.18	2.28	7.20		4.5			
1-Me	I^-	0.31	0.95	2.00	1.36	5.62	4.5	9.5	2.0	0.5
1,7-Di-Me	I^-	0.17	1.05	7.32	1.55	5.70			2.0	0.8
1,6-Di-Me	I^-	0.55	7.19	2.00	1.55	5.64		9.5		
1.3-Di-Me	I^-	7.08	0.80	1.89	1.25	5.71	4.5	9.5	2.0	
2,3-Di-Me	I^-	6.90	0.80	1.89	1.25	5.64	4.5	9.5	2.0	
1,8-Di-Me	I^-	0.64	1.11	2.11	7.00	5.42	4.5			
2,8-Di-Me	I^-	0.33	1.21	2.35	7.32	5.51	4.5			
1,3,8-Tri-Me	I^-	7.08	1.08	2.15	7.05	5.51	4.5			
2,3,8-Tri-Me	I^-	6.89	1.17	2.38	7.28	5.63	4.5			

TABLE 4.40 SPECTRAL DATA FOR *s*-TRIAZOLO[4,3-*a*]-PYRIMIDINES (135)

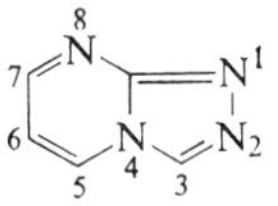

Substituents	Solvent	τ3	τ5	τ6	τ7	J_{56}	J_{67}	J_{57}	$J_{5Me,6}$
None	DMSO	0.72	0.98	2.85	1.23	6.8	4.0	2.0	
5,7-Di-Me	$CDCl_3$	1.30	7.28	3.34	7.38				1.0
	DMSO	0.77	7.32	3.07	7.47				
3,5,7-Tri-Me	$CDCl_3$	7.70	7.47	3.35	7.70				0.4
	DMSO	7.72	7.83	3.20	7.72				

and electronic structures. The concept of "total methyl shift" introduced by Reddy *et al.* (137) was explored thoroughly for this ring system. Although the quantitative nature of this effect was shown not to apply to these compounds, the authors were able to conclude that considerable localization of the double bonds occurred. Charge densities determined from the proton chemical shifts showed a remarkably good correspondence with those calculated by simple HMO methods, although this may simply reflect the intelligent choice of the many arbitrary variables used in these calculations. Data for these compounds are given in Table 4.41.

TABLE 4.41 SPECTRAL DATA FOR *s*-TRIAZOLO[1,5-*a*] PYRIMIDINES

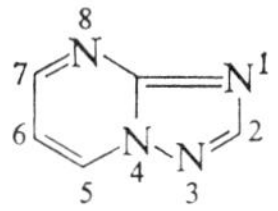

Substituents	Solvent	τ2	τ5	τ6	τ7	J_{56}	J_{57}	J_{67}	Ref.
None	$CDCl_3$	1.48	1.00	2.81	1.13	6.5	2.0	4.4	135
		1.44	1.00	2.77	1.13	6.7	2.0	4.4	136
	DMSO	1.24	0.48	2.54	1.02				135
2-Me	$CDCl_3$	7.37	1.15	2.89	1.24	6.4	2.0	4.4	136
5-Me	$CDCl_3$	1.49	7.11	2.96	1.28			4.4	136
6-Me	$CDCl_3$	1.53	1.29	7.50	1.29				136
7-Me	$CDCl_3$	1.57	1.23	2.96	7.27	7.0			136
2,5-Di-Me	$CDCl_3$	7.36	7.17	3.09	1.38			4.5	136
2,7-Di-Me	$CDCl_3$	7.41	1.36	3.07	7.32	6.8			136
5,7-Di-Me	$CDCl_3$	1.59	7.19	3.14	7.34				136
		1.62	7.18	3.15	7.32				135
	DMSO	1.41	7.23	2.78	7.39				135
6,7-Di-Me	$CDCl_3$	1.64	1.44	7.60	7.35				136
2,5,7-Tri-Me	$CDCl_3$	7.40	7.25	3.26	7.39				136
		7.40	7.23	3.19	7.39				135
	DMSO	7.51	7.30	2.88	7.44				135
5,6,7-Tri-Me	$CDCl_3$	1.66	7.19	7.66	7.37				136
2,5,6,7-Tetra-Me	$CDCl_3$	7.42	7.24	7.69	7.40				136
5-Cl	$CDCl_3$	1.38		2.68	1.19				136
6-Cl	$CDCl_3$	1.46	1.01		1.19				136
7-Cl	$CDCl_3$	1.47	1.15	2.79					136
5,7-Di-Cl	$CDCl_3$	1.43		2.67					136
5-OMe	$CDCl_3$	1.50	5.70	3.45	1.23				136
7-OMe	$CDCl_3$	1.69	1.40	3.38	5.87				136
5,7-Di-OMe	$CDCl_3$	1.70	5.78	4.07	5.88				136

4. *The Pyrazolo[4,3-d]pyrimidine System*

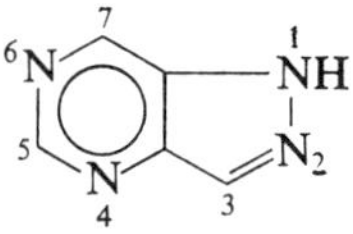

The only examples of this ring system for which spectra are available are all derivatives of the 4,6-di-H-5,7-dioxo compounds **69**, the pyrazolouracils (138–140). The useful data are all contained in one of these publications (138) and are collected in Table 4.42.

TABLE 4.42 SPECTRAL DATA FOR PYRAZOLO[4,3-*d*]URACILS (138). SOLVENT—CD_3CO_2D

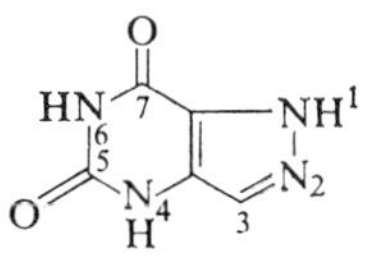

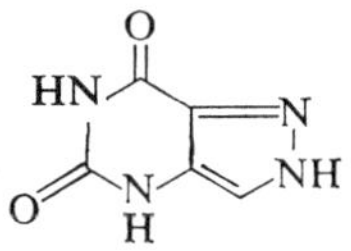

Substituents	$\tau 1$	$\tau 2$	$\tau 3$	$\tau 4, 6$
1-H, 4,6-di-Me			2.22	6.50, 6.59
1,4,6-Tri-Me	5.83		2.43	6.54, 6.66
1,4,6-Tri-Et			2.37	
2,4,6-Tri-Me		5.94	2.35	6.5, 6.61
2,4,6-Tri-Et		5.62, 8.46	2.24	

5. *The Pyrazolo[3,4-b]pyrazine System*

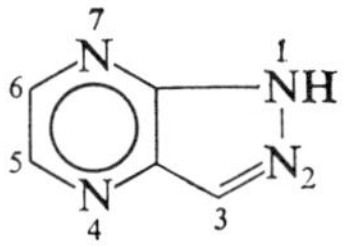

Spectra of the cations of the parent heterocycle and its 5,6-dimethyl derivative in DCl/D_2O have been described (141): parent—$\tau 5,6$ 0.84, 1.00, J_{56} 3, $\tau 3$ 1.28; 5,6-di-Me—$\tau 5,6$ 7.19, 7.22, $\tau 3$ 1.58.

6. *The Imidazo[4,5-d]pyrimidine or Purine System*

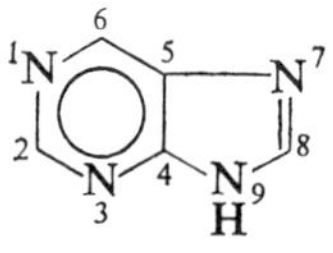

A large amount of experimental work has been done on this system, consistent with its importance in biological processes. Unfortunately, much of the data available, while of interest to the biochemist, has little to offer the serious student of NMR. The field is full of papers in which are reported spectra of solutions of purines in all sorts of aqueous solvents and with a wide variety of quite unsatisfactory standards. For this reason, much of the available information will be dealt with here as historical and the data will not be included in the tables.

The spectrum of purine itself has had a very varied history. Unfortunately, the spectrum consists of three sharp singlets, and the problem of assigning these to specific protons proved very difficult. Two different assignments (142, 143) based mainly on intuition proved to be wrong. The confusion was finally resolved by Matsuura and Goto (144, 145) who prepared specifically deuterated derivatives to show that neither of the original assignments was correct: H-6 was most deshielded, followed by H-2, with H-8 absorbing farthest upfield. This order was confirmed by two other groups of workers (146, 147). Prediction of the relative chemical shifts on the basis of electron-density calculations has also given considerable trouble (149–152).

Most of the work on the spectrum of purine has involved the association of the molecule either with itself or with added basic aromatic compounds (146, 152–154). Throughout this investigation the various groups of authors have spent much effort in trying to assess accurate three-dimensional structures for what must be loose agglomerations of molecules. I have, in other parts of this book, expressed my distrust of such calculations and, after careful perusal of the experimental techniques used in the present studies, I can see no reason to modify my views. Even so, the general conclusions of these workers appear justified. The spectrum of purine was found to be very dependent on concentration, solvent, pH, and temperature, and the rings are thought to be stacked in vertical columns with partial overlap. The proton signals of purine in aqueous solutions are shifted upfield on the addition of aniline. This is ascribed to the influence of neighboring aniline molecules in a weak association complex (155). The references cited for purine "stacking" do not include those dealing with nucleosides.

Throughout all of the preceding work, spectra of purine in solutions of varying pHs, some of which guarantee the presence of the purine anion **70**, and others the purine cation, have been reported. The spectrum of the anion is very similar to that of the neutral molecule, but that of the cation is shifted considerably downfield. Purines are thought to protonate predominantly on N-1 but the large shift of H-8 suggests that the positive charge is delocalized with major contributions from the two structures **71** and **72** (147). That protonation occurs mainly at N-1 is shown by the observation of coupling between H-2 and H-6 in the protonated species (156) (see Chapter 2). Representative data for the three species are given in Table 4.43.

Spectra of simple *C*-substituted purines are not commonly encountered. Chemical shifts obtained from a routine examination of potential anticancer agents are given in the only extensive accumulation of data (156, 157). A reasonable correlation was obtained between the chemical shifts of H-8 in

TABLE 4.43 DATA FOR THE IONIC SPECIES OF PURINE (176)

Species	Solvent	τ2	τ6	τ8
Neutral	D_2O	1.18	1.04	1.40
Anion[a]	N NaOD	1.26	1.07	1.47
Cation	N DCl	0.61	0.44	0.88

[a] These values are regarded as approximate. They were calculated from the data of Matsuura and Goto (144).

6-substituted and 2,6-disubstituted purines and the appropriate Brown's electrophilic substituent constants. An extensive NMR study has been made of the equilibrium between the 2- and 6-azidopurines (**73** and **74**) and the corresponding tetrazolopurines (**75** and **76**) (158). Data for purine and its *C*-substituted derivatives are included in Table 4.44.

Much more information is available for the *N*-substituted derivatives, particularly those related to adenine (**77**), hypoxanthine (**78**), and purine-6(1*H*)-thione (**79**). This information is contained mainly in papers dealing with other aspects of purine chemistry (159–162). The use of DMSO as standard solvent for this type of compound makes possible the observation of the various N–H or NH_2 groups in these molecules and hence, the very facile assignment of tautomeric structures. Protonation of adenine and its exocyclic *N*-methyl derivatives has normally been assumed to take place on the 6-amino group. However, an examination of the spectra of suitably substituted derivatives over a range of acid strengths has led to the suggestion that protonation occurs first on N-1 (163). In the case of the 9-substituted adenines (**80**), where R can be CO_2Et, CN, Ph, or α-pyridyl, H-2 and H-8 resonate between τ 1.65 and 2.01, NH_2 between 2.05 and 2.75, NCH_2 between 5.32 and 5.57, and CH_2R between 6.60 and 7.22 (164). Also, some data are available on the spectra of caffeine (**81**), theophylline (**82**), and theobromine (**83**) and their derivatives (165–167). The spectrum of caffeine shows three methyl peaks, and a study of solvent and pH effects made possible the assignment of each of these to specific groups. Confirmation was obtained from the spectra of theophylline and theobromine. The addition of sodium benzoate to an aqueous solution of caffeine produces large chemical shifts of the three methyl signals. These effects were thought to be due to two different association phenomena (168).

Nuclear magnetic resonance was used together with ultraviolet spectroscopy to study the structures of mercury and chloromercurypurines (169). A comparison of the spectra of the purines, their sodium salts, and the

TABLE 4.44 SPECTRAL DATA FOR *C*-SUBSTITUTED PURINES

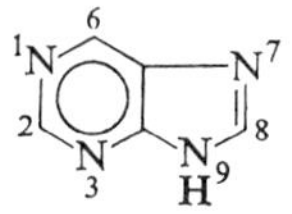

Substituents	Solvent	$\tau 2$	$\tau 6$	$\tau 8$	τNH	Ref.
None	DMSO	1.01	0.81	1.32	−1	156
2-N_3	DMSO		0.97	1.42		158
	TFA		0.45	0.65		158
	AcOH		0.90	1.42		158
6-Me	DMSO	1.24		1.47		157
6-CN	DMSO	0.92		1.09		157
6-NH_2	DMSO	1.89		1.86		157
6-NMe_2	DMSO	1.81		1.93		157
6-N_3	TFA	0.87		0.87		158
	AcOH	1.26		1.41		158
6-OMe	DMSO	1.49		1.62		157
6-SMe	DMSO	1.27		1.53		157
6-Cl	DMSO	1.24		1.29		157
6-I	DMSO	1.40		1.37		157
2-Et, 6-Cl				1.42		156
2-NH_2, 6-Me				2.02		156
2-Cl, 6-Me				1.36		156
2,6-Di-NH_2				2.24		156
2-NMe_2, 6-NH_2				2.28		156
2-NH_2, 6-SMe				2.07		156
2-SMe, 6-NMe_2				2.03		156
28SMe, 6-NH_2				1.98		156
2-F, 6-NH_2				1.84		156
2-F, 6-NMe_2				1.97		156
2-Cl, 6-NMe_2				1.87		156
2,6-Di-SMe				1.68		156
2-Cl, 6-OMe				1.55		156
2,6-Di-Cl				1.23		156
2,6-Di-Br				1.30		156
2-CF_3, 6-Cl				1.03		156

mercury compounds showed the latter to be predominantly covalent. The chloromercury groups were shown to be attached to N-7 of 3-benzylhypoxanthine (**84**) and to N-9 of 1-benzylhypoxanthine (**85**), 1-benzylpurine-6(1*H*)-thione (**86**), and 6-dimethylaminopurine (**87**). Data for *N*-substituted purines and the parent oxo-, thio-, and aminopurines are collected in Table 4.45.

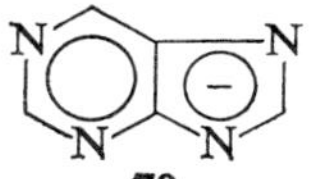

70

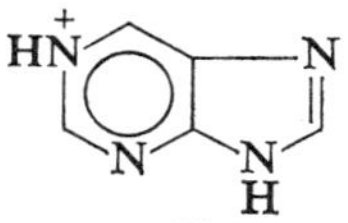

71

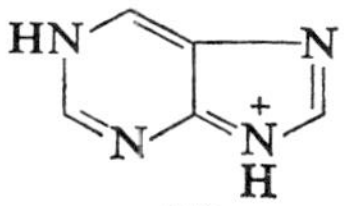

72

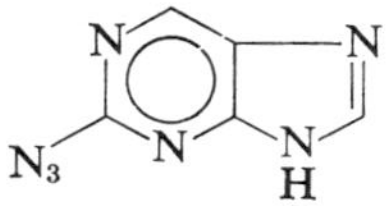

73

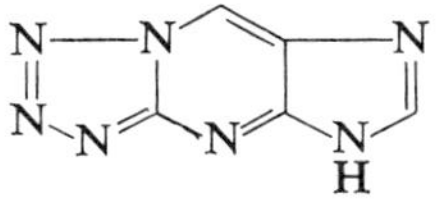

74

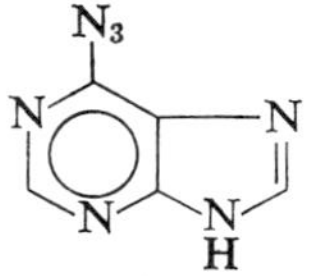

75

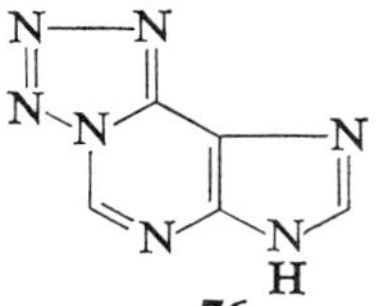

76

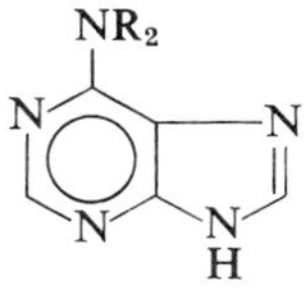

77, R = H
87, R = Me

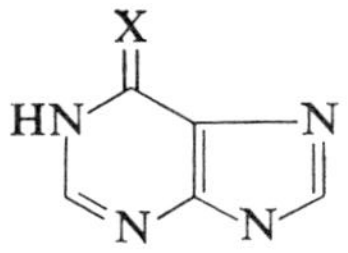

78, X = O
79, X = S

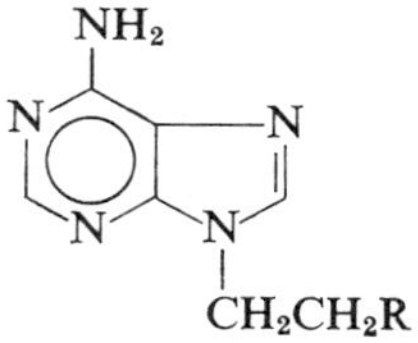

80

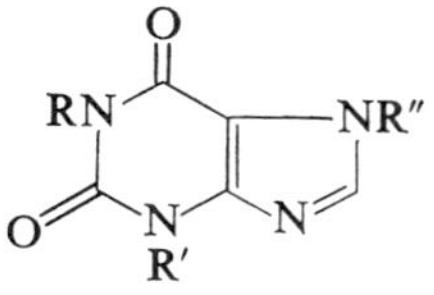

81, R = R′ = R″ = Me
82, R = R′ = Me, R″ = H
83, R = H, R′ = R″ = Me

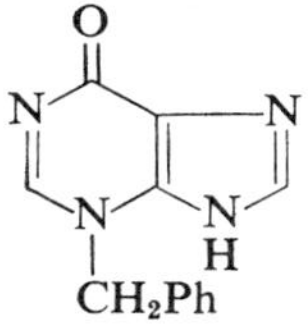

84

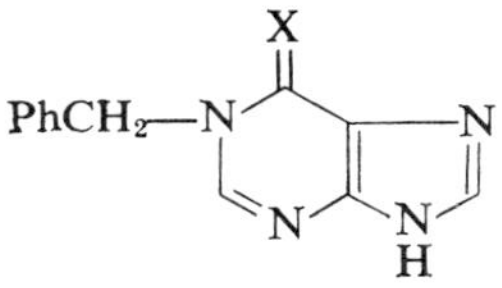

85, X = O
86, X = S

TABLE 4.45 SPECTRAL DATA FOR *N*-SUBSTITUTED PURINES

A.

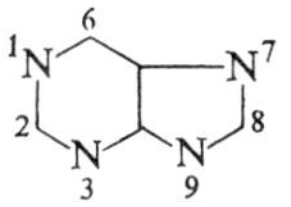

Substituents	Solvent	τ6	τ2	τ8	τNMe	Ref.
3-Me	DMSO	1.0, 0.9		1.55		159
3-CH_2CHCMe_2, 6-NH_2	SO_2	1.81, 1.95				162
2-SMe, 9-Me	$CDCl_3$	0.96	7.28	1.97	6.26	161
	DMSO	1.07	7.41	1.61	6.23	161
2-SOMe, 9-Me	$CDCl_3$	0.82	6.61	1.73	6.00	161
2-SO_2Me, 9-Me	$CDCl_3$	0.62	6.53	1.52	5.90	161
6-SMe, 9-Me	$CDCl_3$	7.29	1.26	2.08	6.13	161
	DMSO	7.29	1.08	1.41	6.11	161
6-SOMe, 9-Me	$CDCl_3$	6.86	0.91	1.85	6.05	161
6-SO_2Me, 9-Me	$CDCl_3$	6.52	0.91	1.65	6.00	161
6-Cl, 9-Me	$CDCl_3$		1.29	1.91	6.08	161
	DMSO		1.25	1.36	6.12	161
8-SMe, 9-Me	$CDCl_3$	1.10	1.17	7.21	6.29	161
	DMSO	1.09	1.21	7.22	6.33	161
8-SOMe, 9-Me	$CDCl_3$	0.84	0.97	6.74	5.82	161
8-SO_2Me, 9-Me	$CDCl_3$	0.79	0.91	6.44	5.84	161

B. SOLVENT—DMSO (160, 169)
X = O Hypoxanthine Derivatives

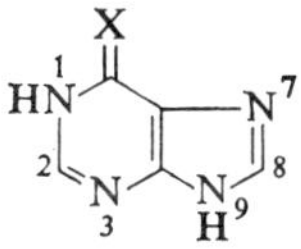

Substituents	τ2, 8	τNCH_2	τPh
None	1.88, 2.03		
1-CH_2Ph	1.49, 1.81	4.73	2.69
1-CH_2Ph, Na salt	1.95, 2.47	4.82	2.74
1-CH_2Ph, chloromercury	1.53, 1.72	4.77	2.71
3-CH_2Ph	1.40, 1.79	4.53	2.62
3-CH_2Ph, Na salt	1.81, 2.55	4.63	2.67
3-CH_2Ph, chloromercury	1.37, 2.05	4.52	2.69
7-CH_2Ph	1.59, 1.98	4.39	2.65
9-CH_2Ph	1.79, 1.95	4.61	2.68
1,3-Di-CII_2Ph·IIBr	−0.80, 1.50	4.25, 4.56	2.59
1,7-Di-CH_2Ph	1.49, 1.57	4.77, 4.40	2.70
1,9-Di-CH_2Ph	1.43, 1.77	4.77, 4.62	2.71
3,7-Di-CH_2Ph	1.45, 1.63	4.59, 4.38	2.58
7,9-Di-CH_2Ph	0.58, 1.87	4.50, 1.17	2.62
7,9-Di-CH_2Ph·HBr	−0.23, 1.56	4.37, 4.20	2.57

TABLE 4.45 (*continued*)

X = S PURINE-6(1*H*)-THIONE DERIVATIVES

Substituents	τ2, 8	τNCH_2	τPh
1-CH_2Ph	1.12, 1.37	4.13	2.68
1-CH_2Ph, chloromercury	0.93, 1.67	4.19	2.67
1,7-Di-CH_2Ph	1.08, 1.37	4.14, 3.85	2.69
1,9-Di-CH_2Ph	1.10, 1.57	4.10, 4.57	2.67

C.

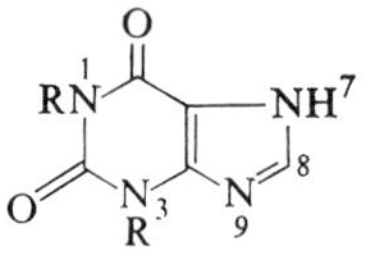

Substituents	Solvent	τ1Me	τ3Me	τ7Me	Ref.
1,3,7-Tri-Me (caffeine)	TFA	6.41	6.20	5.67	166
	$PhNO_2$	6.77	6.62	6.18	166
1,3-Di-Me (Theophylline)	TFA	6.35	6.14		166
	$PhNO_2$	6.73	6.55		166
3,7-Di-Me (theobromine)	TFA		6.25	5.72	166
	$PhNO_2$		6.65	6.19	166
1,3-Di-Me, 7-OMe, 8-Ph	$CDCl_3$	6.65	6.45	5.92	167

The ^{13}C satellite spectra of purine in solutions of pH 0–8 have been reported (170). All parameters demonstrate similar responses to changes in pH in a way that suggests a simple acid–base equilibrium. The results suggest that purine protonates on N-1, N-3, and N-7, with N-1 protonation predominating. Although this conclusion contradicts the commonly held view that protonation occurs exclusively on N-1, I do not feel that it is incompatible with the results from proton spectra. Using deuterated compounds of known structure, the ^{13}C chemical shifts of C-2, C-6, and C-8 were assigned unequivocally (171). The two remaining peaks, found at lowest and highest field, were assigned to C-4 and C-5, respectively, on the strength of theoretical π-electron densities. In an increasing magnetic field the ^{13}C pattern differs from the ^{1}H pattern, suggesting that chemical shift values should be used with care as a means of predicting charge densities. Nevertheless, a gross correlation of ^{13}C chemical shift data and theoretical estimates of charge distribution in purine do exist. ^{15}N data for purine are given in Table 4.46. Data for reduced purines are difficult to find but some of those available (172) are collected in Table 4.47.

TABLE 4.46 ^{13}C DATA FOR PURINE. SOLVENT—D_2O

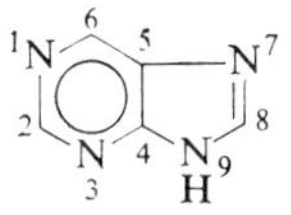

CHEMICAL SHIFTS (ppm RELATIVE TO BENZENE) (171)

C-2	C-4	C-5	C-6	C-8
-23.1 ± 0.1	-25.9 ± 0.2	0.4 ± 0.2	-15.9 ± 0.1	-19.0 ± 0.1

^{13}C—H COUPLING CONSTANTS (Hz) (170)

	$J_{^{13}C-H}$		
p*D*	2	6	8
0.15	219.1	196.6	218.5
1.09	218.8	196.1	217.3
2.72	211.5	191.4	215.2
3.22	208.6	188.8	213.8
5.63	207.0	187.4	213.5

TABLE 4.47 SPECTRAL DATA FOR REDUCED PURINES (172). SOLVENT—$CDCl_3$

A. 1,6-DIHYDRO

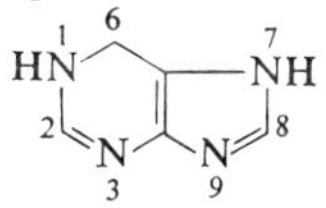

Substituents	τ2, 8	τ6	τCOMe
1-COMe	0.96, 1.44	4.58	7.88
1,7-Di-COMe	1.75, 1.96	4.34	7.26, 7.75

B.

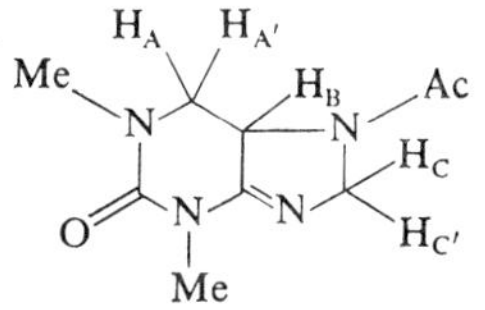

τA	τA′	τB	τC, C′	τNMe
6.8	5.83	5.4	4.75	7.88

7. *The Imidazo[4,5-d]pyridazine System*

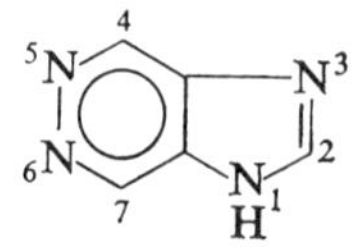

Data for the parent heterocycle and various 4-substituted derivatives (173, 174) are collected in Table 4.48.

TABLE 4.48 SPECTRAL DATA FOR IMIDAZO[4,5-*d*]-PYRIDAZINES (174)

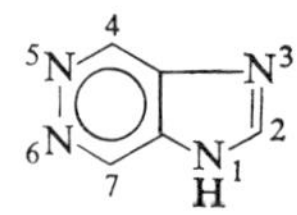

Substituents	Solvent	$\tau 2$	$\tau 4$	$\tau 7$
None	D_2O	1.48	0.49	0.49
4-O^-	NaOD	1.94		1.26
4-S^-	NaOD	1.83		0.98
4-SMe	NaOD	1.68	7.20	0.78
4-$NHNH_2$	NaOD	1.81		

8. *The Imidazo[4,5-c]pyridazine System*

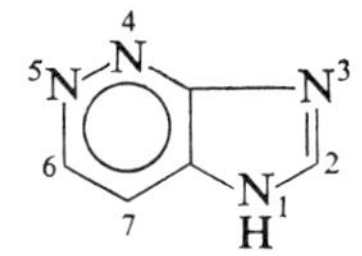

Some incomplete data for members of this system are found in Ref. 174.

E. Five Nitrogen Atoms

1. *The Tetrazolo[1,5-a]pyrazine System*

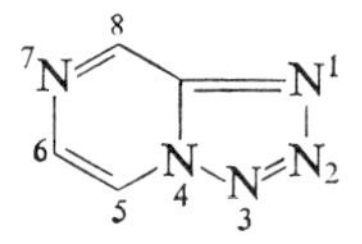

The spectrum of the parent compound in $CDCl_3$ has been reported (175): $\tau 5$ 1.17, $\tau 6$ 1.64, $\tau 8$ 0.36; J_{56} 4.7, J_{58} 1.6, $J_{68} \sim 0.5$.

2. *The v-Triazolo[4,5-d]pyrimidine or 8-Azapurine System*

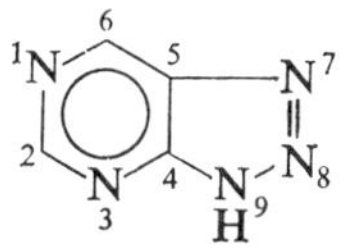

Spectra of this system are found almost exclusively in papers dealing with the addition of water to the 1,6-bond to give the 1,6-dihydro compounds **88**

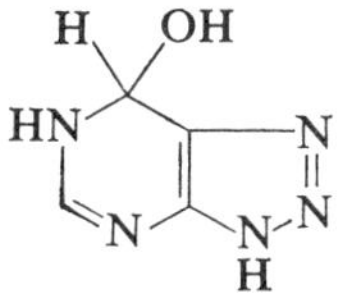

88

known as "covalent hydrates" (176, 177). Spectra for the neutral molecules, cations, and various "hydrated" species are summarized in Table 4.49. The

TABLE 4.49 SPECTRAL DATA FOR THE *v*-TRIAZOLO[4,5-*d*]PYRIDINES (8-AZAPURINES)

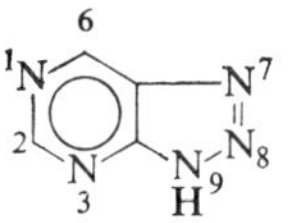

Substituents	Species[a]	Solvent	τ2	τ6	τNMe	Ref.
None	NM	D_2O	0.80	0.32		176
	C	TFA	0.40	−0.31		176
	HC	DCl	1.46	3.19		176
2-Me	HC	DCl	7.38	3.29		176
6-Me	C	TFA	0.54	6.62		176
7-Me	NM	D_2O			5.38	177
8-Me	NM	D_2O	0.82	0.31	5.4	177
	HNM	D_2O	2.56	3.57		177
	HC	DCl	1.56	3.35	5.81	177
9-Me	NM	D_2O			5.52	177
$2-NH_2$	C	TFA		0.32		176
	HC	DCl		3.51		176

[a] NM—neutral molecule, HNM—hydrated neutral molecule, C—cation, HC—hydrated cation.

assignment of peaks to H-2 and H-6 in the spectrum of the parent neutral molecule was based on protonation shifts.

3. *The v-Triazolo[4,5-d]pyridazine System*

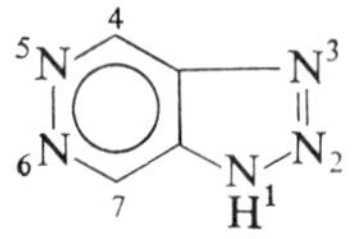

Data for a number of 4-substituted derivatives are given in Table 4.50 (174).

TABLE 4.50 SPECTRAL DATA FOR *v*-TRIAZOLO[4,5-*d*]-PYRIDAZINES (174). SOLVENT—NaOD

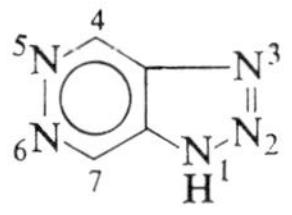

Substituents	τ7	τSMe
4-O^-	0.93	
4-S^-	0.63	
4-SMe	0.61	7.25
4-NH_2	0.83	
4-$NHNH_2$	1.05	

4. *The s-Triazolo-as-triazine Systems*

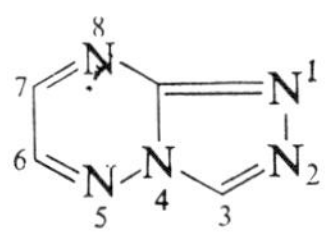

[3,4-*b*] 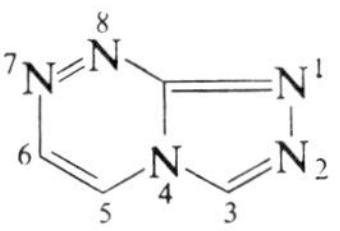 [3,4-*c*] 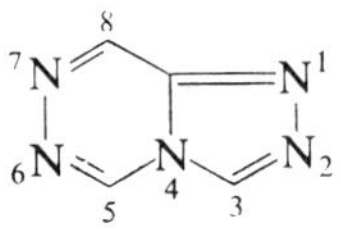 [3,4-*d*]

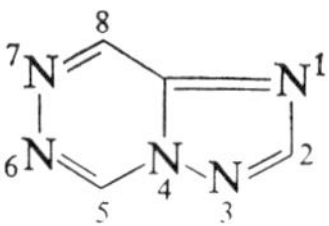

[1,5-*d*] 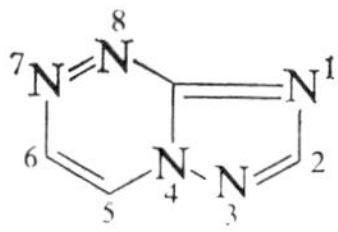 [1,5-*c*]

Data for a number of hydroxy compounds are listed in Table 4.51 (178). Information is also given in this reference for some dihydro-oxo derivatives.

TABLE 4.51 SPECTRAL DATA FOR THE *s*-TRIAZOLO-*as*-TRIAZINES (178). SOLVENT—DMSO

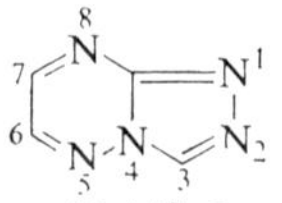

[3,4-*b*] I 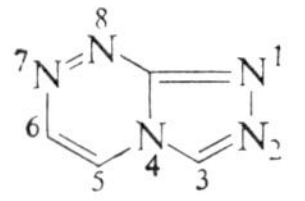 [3,4-*c*] II 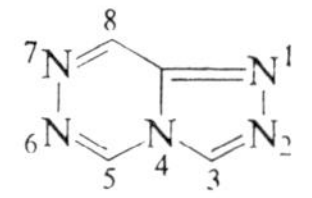 [3,4-*d*] III

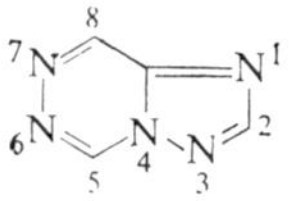

[1,5-*d*] IV 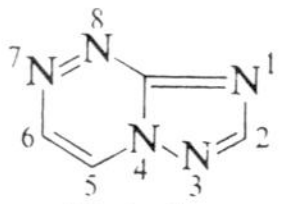 [1,5-*c*] V

Isomer	Substituents	τ2	τ3	τ5	τ6	τ7	τ8
I	7-OH		0.93		2.10		
I	3-Me, 7-OH		7.52		2.15		
I	6-Me, 7-OH		1.02		7.68		
I	6-Me, 7-SH		0.85		7.53		
I	6-Me, 7-SMe		0.63		7.52	7.32	
II	5-OH		0.79		2.34		
II	6-Me, 5-OH		0.87		7.72		
II	3,6-Di-Me, 5-OH		7.53		7.75		
III	8-Me, 5-OH		0.55				7.48
III	8-Me, 5-SH		0.47				7.42
III	8-Me, 5-SMe		0.40	7.17			7.22
IV	8-Me, 5-OH	1.33					7.50
IV	8-Me, 5-SH	1.25					7.44
V	5-OH	1.60			2.15		
V	6-Me, 5-OH	1.60			7.63		
V	6-Me, 5-SH	1.47			7.48		
V	2-Me, 5-OH	7.58			2.23		
V	2,6-Di-Me, 5-OH	7.62			7.70		

VIII. THE 6,6 SYSTEMS

A. The Quinoline System

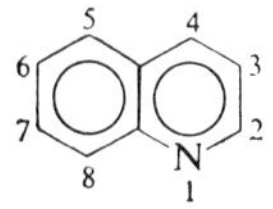

Spectra of the parent quinoline were first reported by Schaefer and Schneider (179) who studied the solvent shifts between hexane, acetone, and benzene solutions; H-3 and H-4 showed the largest benzene shifts (upfield), H-2 behaved in a similar manner to the α-proton of pyridine, but H-8 was almost unaffected. This work was later extended to a series of dimethylquinolines

TABLE 4.52 SPECTRAL DATA FOR THE DIAZANORCARADIENES

A. 3,4-DIAZA. SOLVENT—$CDCl_3$ (X over ring)

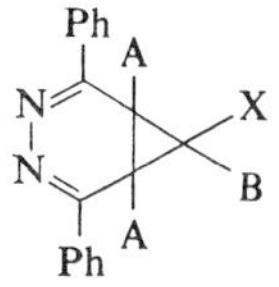

Substituents	τA	τB	τX	J_{AB}	J_{AX}	J_{BX}	Ref.
None	7.38	7.91	9.77	9.0	4.8	3.5	19a
B = Me	7.58	8.63	9.12		6.0		19a
X = Me	7.20	7.80	9.40	9.0			19a
B = Ph	6.99		8.15		4.5		19b

B. 2,3-DIAZA. SOLVENT—C_2Cl_6 AT 120° (19d)

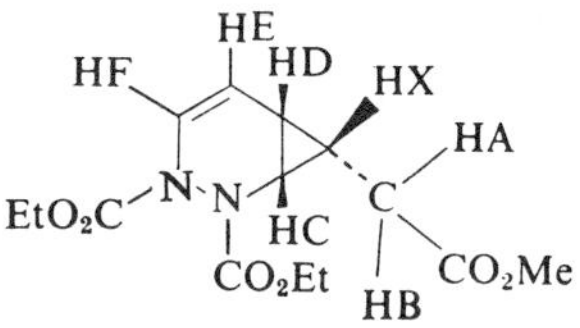

τA, B	τX τD	τC	τE	τF	τOMe	τOEt
7.68, 8.02	8.3–8.7	6.69	4.83	3.22	6.48	5.90, 8.85

J_{AB}	J_{AX}	J_{BX}	J_{EF}	J_{DF}
17	9	5	7.5	1.5

(180) where similar trends were noted. The large variation in the benzene shifts from proton to proton suggests highly specific solute–solvent interactions. More recently, this interaction has been studied in a rather unsatisfactory manner by Ronayne and Williams (181). Dilution studies of the hetero-ring protons of 8-hydroxyquinoline (**89**) and 8-methylquinoline (**90**) in CCl_4, benzene, and acetone showed them to behave in a similar manner (182). The spectra of six methyl- and dimethylquinolines have been carefully analyzed (183), and a reasonable correlation was obtained between the proton chemical shifts and the charge densities on the respective carbon atoms, with the exception of the 8-position where the chemical shift of the proton is affected strongly by the lone pair of the peri nitrogen atom. Also, Seiffert (184) obtained a reasonable linear correlation between out-of-plane C–H vibrations in the infrared and the chemical shifts of the ring protons for 15 quinolines.

Black and Heffernan (185) have fully analyzed the spectra of quinoline and its nitro derivatives. Again, solvent shifts suggest that quinoline–quinoline interactions in the neat liquid are similar to the benzene–quinoline interactions. The substituent effects of nitro groups were largely in accord with those obtained for nitrobenzofuroxans (186) and dinitronaphthalenes (187). There is a general, presumably inductive, shift of 0.1–0.2 ppm to low field in all positions around the ring. Protons ortho and para to the nitro groups are subject to a larger low-field shift attributable to mesomeric effects; ortho protons also experience a large proximity effect resulting in further deshielding. Similar deshielding effects were observed for the addition of ring nitrogen atoms to the naphthalene system. The heteroatom also exerts a large effect on coupling constants involving adjacent nuclei. Thus J_{23} is about 2.6 Hz less and J_{24} 0.3 Hz greater than comparable couplings in similar carbocyclic systems. The effect of a nitro group on coupling constants in the ring containing it is essentially opposite to and weaker than the effect of the heteroatom. Thus, when effects presumably dominated by the π-electron structure of the molecule are considered, the ring nitrogen and a nitro substituent exert a qualitatively similar influence. Short-range influences on coupling constants are expected to be due predominantly to variations in σ-electron structure and are perhaps connected with small changes in bond angles. Here the ring nitrogen and a nitro group exert quite dissimilar effects. The assignments for quinoline itself have been checked by the preparation of all of the monodeuteroquinolines (188).

In an extremely detailed study (189), the spectral behavior of 5- or 7-halogenoquinoline (**91**) has been investigated. The influence of the halogens on chemical shifts parallels that found in substituted benzenes. Steric effects appear to be important although the small effect on protons peri to the halogen seems anomalous. The easily confused protons 4 and 8 can be distinguished by their different dependence on concentration. Diehl's additive substituent theory of solvent effects (190) was extended and used to interpret solvent and concentration effects. This paper also contains a discussion of partly degenerate ABX spectra.

The problem of the tautomeric structure of 4-aminoquinolines has been studied by NMR, with two sets of authors reaching mutually contradicting conclusions (191, 192); J_{23} was found to be about 5 Hz in model compounds with fixed amino structure, and about 8 Hz for the imino counterparts. An examination of the spectra of 4-aminoquinoline led to the conclusion that the imino form **92** predominates in aqueous solutions at neutral pH (a conclusion not supported by any other form of spectroscopy) while the amino form **93** is dominant in the presence of alkali amide. These results prompted a very thorough investigation of the situation (192) in both aprotic and aqueous solvents in which it was shown conclusively that the amino tautomer was the only species which could be detected in any of the solutions studied. The suggestion was made that all simple 4-aminoquinolines have the amino structure.

The signs of coupling constants in the quinoline system have been studied by a variety of methods. Paterson and Bigam (193) first showed the proton–proton couplings of the hetero-ring to have the same sign. ^{15}N–H couplings have been measured for the ^{15}N isomers of quinoline, *N*-ethylquinolinium iodide (**94**), and quinoline *N*-oxide (**95**) (194). The sign of these couplings in quinoline was shown to be negative (opposite J_{HH}) and to increase algebraically (i.e., to decrease numerically) as the acidity of the solvent or the *s*-character of the nitrogen atom increased. This assignment was later confirmed (195). The reduced coupling constants (196, 197)

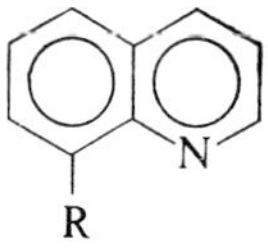

89, R = OH
90, R = Me

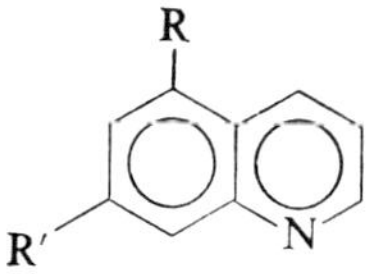

91, R or R′ = Hal

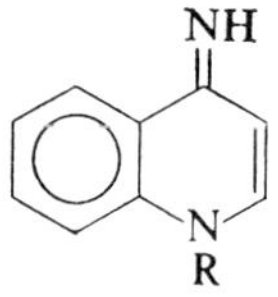

92

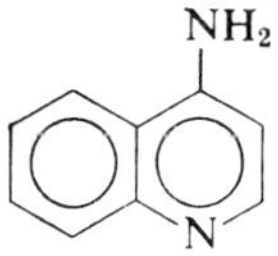

93

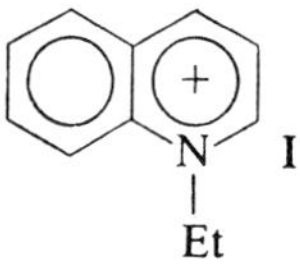

94

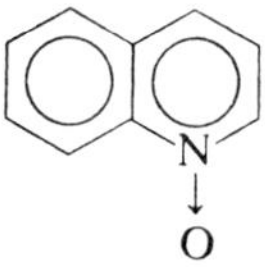

95

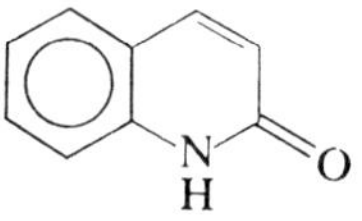

96

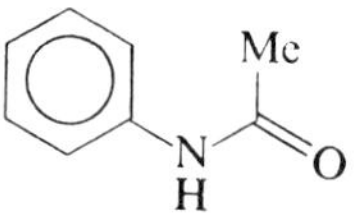

97

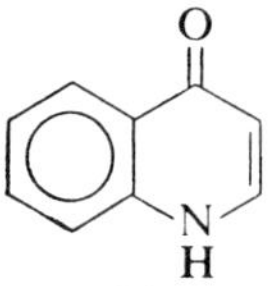

98

$$^nK_{kl} = (2\pi/h\gamma_k\gamma_l)^nJ_{kl}\ \text{cm}^{-3}$$

where γ_k is the gyromagnetic ratio of nucleus k, however, are of reversed sign, and hence $^2K_{NH\alpha}$ and $^3K_{NH\beta}$ both have positive absolute signs in quinoline.

In general, spectral data for substituted quinolines are difficult to find. Those for compounds in the previous discussion, together with a number from papers on other aspects of quinoline chemistry (198–200) are collected in Table 4.53.

TABLE 4.53 SPECTRAL DATA FOR QUINOLINES

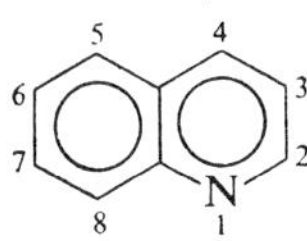

Substituents	Solvent	Chemical shifts $\tau 2$	$\tau 3$	$\tau 4$	$\tau 5$	$\tau 6$	$\tau 7$	$\tau 8$	Ref.
None	Neat	0.47	2.90	2.17	2.41	2.65	2.40	1.56	185
	CCl_4	1.19	2.73	2.00	2.31	2.56	2.38	1.94	185
	Me_2CO	1.10	2.53	1.71	2.08	2.46	2.26	1.93	185
2-CD_3	Neat		2.94	2.10					186
6-Me	CCl_4	1.31	2.87	2.15	2.63	7.58	2.63	2.03	183
7-Me	CCl_4	1.25	2.88	2.00	2.45	2.73	7.51	2.17	183
8-Me	CCl_4	1.28	2.88	2.18				7.23	198
2,4-Di-Me	CCl_4	7.47	3.21	7.60	2.40			2.08	183
2,6-Di-Me	CCl_4	7.41	3.09	2.40	2.81	7.65	2.73	2.20	183
2,7-Di-Me	CCl_4	7.41	3.10	2.60			7.55	2.33	183
5,7-Di-Me	CCl_4	1.38	2.95	2.09	7.65	3.08	7.57	2.43	198
3-NO_2	Me_2CO	0.43		0.76	1.71	2.18	1.98	1.79	185
5-NO_2	Me_2CO	0.93	2.22	1.10		1.57	2.04	1.57	185
6-NO_2	Me_2CO	0.87	2.27	1.33	1.03		1.51	1.74	185
7-NO_2	Me_2CO	0.90	2.28	1.50	1.79	1.67		1.16	185
8-NO_2	Me_2CO	0.98	2.30	1.49	1.77	2.24	1.87		185
2-Me, 4-NH_2	Me_2CO	7.60	3.61	3.90		2.0–3.1			192
	DMSO	7.54	3.50	3.30		1.8–2.8			192
2-Me, 4-NMe_2	$CDCl_3$	7.38	3.44	7.13		1.9–2.9			192
	DMSO	7.40	3.22	7.04		1.9–3.0			192
2-Me, 4-NHMe[a]	DMSO	7.50	3.66	7.10		1.8–2.8			192
	Me_2CO	7.42	3.65	6.99		1.9–2.8			192
5-F, 8-OH	DMSO[b]	1.40	2.74	1.92		3.10	3.32		199
5-Cl, 8-OH	DMSO[b]	1.32	2.63	1.88		2.73	3.18		199
5-Br, 8-OH	DMSO[b]	1.32	2.58	1.87		2.50	3.15		199
5-I, 8-OH	DMSO[b]	1.42	2.70	1.95		2.45	3.07		199
5-SCN, 8-OH	DMSO[b]	1.32	2.52	1.68		2.32	3.10		199
2-Me, 5-Cl, 8-OH	DMSO[b]	7.57	2.85	2.06		2.80	3.21		199
2-Me, 5-I, 8-OH	DMSO[b]	7.46	2.74	1.98		2.44	3.01		199
2-Me, 5-SCN, 8-OH	DMSO[b]	7.47	2.60	1.75		2.33	3.07		199
2-SMe	$CDCl_3$	7.27	2.75	2.05				1.95	200
	DCl	7.06	2.3	1.38	2.05	2.05	2.05		200
2-SO_2Me	$CDCl_3$	6.58	1.77	1.43		2.1	2.1	1.66	200
	DCl	6.13		0.18					200
4-SMe	$CDCl_3$	1.12	2.78	7.40	1.75	2.15	2.3	1.75	200
	DCl	1.39	2.55	7.26	2.2	2.2	2.2	2.2	200
4-SO_2Me	$CDCl_3$	0.68	1.75	6.71	1.15	2.05	2.05	1.57	200
	DCl	0.27	1.12	6.32		1.5	1.5		200
5,7-Di-Cl	CCl_4	1.18	2.66	1.61		2.50		2.05	198

TABLE 4.53 (*continued*)

	Coupling constants (references as above)									
	J_{23}	J_{24}	J_{34}	J_{56}	J_{57}	J_{58}	J_{67}	J_{68}	J_{78}	J_{48}
None (neat)	4.2	1.7	8.3	8.1	1.6	0.3	6.8	1.1	8.2	0.9
(CCl_4)	4.1	1.7	8.2	8.2	1.4	0.5	6.9	1.1	8.6	0.9
(Me_2CO)	4.2	1.7	8.3	8.0	1.4	0.4	6.8	1.1	8.4	0.8
6-Me	4.0	2.0	9.0						9.0	
7-Me	4.1	1.9	9.0							
8-Me	4.1	1.8	7.9							
2,4-Di-Me					2.0				8.0	
2,6-Di-Me			8.5		1.9				9.0	
2,7-Di-Me			8.0							
5,7-Di-Me	3.9	1.7	8.6							0.8
3-NO_2		2.7		8.2	1.3	0.5	6.8	1.0	8.6	0.8
5-NO_2	4.0	1.4	8.6					1.2		0.8
6-NO_2	4.1	1.7	8.2		2.5	0.5			9.2	0.8
7-NO_2	4.2	1.7	8.4	8.9		0.5		2.3		
8-NO_2	4.3	1.8	8.3	8.4	1.3		7.4			
5,7-Di-Cl	4.2	1.6	8.5					2.0		0.8

[a] $J_{NH,NMe} = 5.2$ Hz.
[b] Reference—external TMS.

Protonation of quinolines, and in fact any other process which removes the nitrogen lone pair and leads to the development of a positive charge on the nitrogen atom, causes an increase in J_{23} (1.5 ± 0.2 Hz) but has little effect on the other coupling constants (201).

^{19}F spectra of some fluoroquinolines have been reported (202, 203). In the spectrum of 2-fluoroquinoline in CCl_4, the 2-fluoro signal, at normal temperature, showed as a broad hump which could be sharpened by cooling the solution. Similar effects have been observed by myself in the proton spectra of quinoline. In compounds with peri F–H atoms the peri couplings are not large (<1.5 Hz), but with two fluorine atoms in these positions extremely large couplings (45–50 Hz) are observed. ^{19}F spectra are summarized in Table 4.54.

The ^{13}C–H coupling constants for quinoline and its methyl derivatives have been reported (204). Malinowski's additivity rule (205), developed for monocyclic systems, was shown to hold for these compounds. The coupling constants obtained are collected in Table 4.55.

^{14}N chemical shifts in a number of 8-hydroxyquinolines (**89**) and -quinol-2-ones (**96**) have been used to confirm the oxo structure of the latter (206). The 8-hydroxy compounds and quinoline itself absorb at about 90 ppm (from NH_4NO_3 in 3 *N* aq. HCl) whereas the oxo compounds absorb at about

TABLE 4.54 ^{19}F SPECTRA OF FLUOROQUINOLINES AND FLUOROISOQUINOLINES

A. QUINOLINES

Substituents	Solvent	Standard[a]	Chemical shifts[b]	Ref.
2-F	CCl_4	1	−17.51 (2)	203
5-Cl, 8-F	CCl_4	1	44.36	203
5,8-Di-F	CCl_4	1	50.12	203
6,8-Di-F	CCl_4	1	32.80, 40.63	203
5,6,7,8-Tetra-F	CCl_4	1	76.03, 76.97, 80.96, 83.76,	203
2-H, hexa-F	CH_2Cl_2	2	133.4 (4), 146.0 (5), 148.2, 151.3, 154.3	202
2-OMe, hexa-F	CH_2Cl_2	2	132.4 (4), 147.0 (5), 150.2, 154.0, 159.5, 160.5	202
2-NH_2, hexa-F	Me_2CO	2	137.1 (4), 147.7 (5), 152.5, 155.7, 159.8, 164.6	202
4-OMe, hexa-F	CH_2Cl_2	2	76.0 (2), 143.2, 149.6, 153.7, 158.0, 160.1	202
4-NH_2, hexa-F	Me_2CO	2	81.9 (2), 147.1, 149.8, 155.4, 161.8, 168.0	202
Hepta-F	CH_2Cl_2	2	77.2 (2), 126.0 (4), 145.7 (5), 148.3, 150.7, 154.4, 160.6	202

TABLE 4.54 *(continued)*

B. ISOQUINOLINES—STANDARD 2,[a] SOLVENT—Me_2CO (202)

Substituents	Chemical shifts (ppm)
1,6-Di-OMe, penta-F	100.0 (3), 138.85 (8), 139.7 (5), 150.9 (7), 165.2 (4)
1-H, hexa-F	97.2 (3), 145.6, 146.4, 148.1, 150.9, 154.6
1-OMe, hexa-F	97.6 (3), 136.2 (8), 146.9 (5), 148.7 (7), 156.5 (6), 164.7 (4)
1-NH_2, hexa-F	94.4 (3), 140.0, 146.6, 147.3, 150.4, 159.9
6-OMe, hexa-F	62.7 (1), 98.9, 141.5, 142.5, 147.4, 155.9
Hepta-F	61.0 (1), 96.5 (3), 138.9, 144.5, 145.2, 152.4, 154.6
	Coupling constants (Hz)
1,6-Di-OMe, penta-F	J_{34} 16.3, J_{35} 2.0, J_{37} 4.0, J_{38} 4.3, J_{45} 50.7, J_{47} 3.8, J_{48} 1.5, J_{57} 2.0, J_{58} 14.3, J_{78} 14.3
Hepta-F	J_{34} 17.7, J_{35} 0.7, J_{37} 7.9, J_{38} 4.7, J_{45} 49.7, J_{46} 3.7, J_{47} 3.7, J_{48} 1.5, J_{56} 17.3, J_{57} 4.0, J_{58} 15.8, J_{67} 19.0, J_{68} 8.0, J_{78} 17.8

[a] Standard—1, external TFA; 2, $CFCl_3$.

[b] Chemical shifts of multiplets in order of frequency. Assignment, if known, is given in parentheses.

TABLE 4.55 ^{13}C–H COUPLING CONSTANTS FOR QUINOLINES AND ISOQUINOLINE (204). NEAT LIQUIDS

	Coupling constants (Hz)					
Compound	C-1	C-2	C-3	C-4	C-8	Me
Quinoline		177.5	163.5			
2-Me			160.8	156.8	160.0	126.8
3-Me		174.8				127.3
4-Me		175.0	162.0		160.0	127.4
6-Me		176.6	164.0		160.0	126.8
7-Me		175.6			159.8	126.5
8-Me		176.8	162.4	162.0		127.0
Isoquinoline	176.0		176.0	163.0		

240 ppm, a position very similar to that observed for acetanilide (**97**) (243.3 ppm). In the case of the 4-quinolone (**98**), a coupling of 1.5 Hz was observed between the NH proton and H-3 (207).

Nuclear magnetic resonance has been used to determine the structure of the "acetone anils" of various toluidines, and for this reason quite a lot of data have accumulated on these 2,2′,4-trimethyl-1,2-dihydroquinolines (**99**) (208, 209), their dimers (210), and other 2,2′-dimethyl-1,2-dihydroquinolines (211, 212). Also, spectra have been reported for pseudobases (**100**) and anhydrobases (**101**) obtained from various quinolinium salts (213, 214). Base-catalyzed ring-chain tautomerism was observed between the pseudobase **100** and the ring-open form **102** (215). The size of the olefinic coupling constant has been used to distinguish between 1,2- and 1,4-dihydroquinolines formed by the action of nucleophiles on 1-substituted quinolinium ions (214): 1,2-dihydro, $J \approx 10$; 1,4-dihydro, $J \approx 8$ Hz.

Spectra of a number of 2,3-dihydro-4-quinolones (**103**) have been reported (216, 217), and those for a number of 3-substituted derivatives have been analyzed (218). The magnitudes of the vicinal couplings were interpreted as proof that a 3-halogen occupies a pseudoaxial position but a 3-methyl group adopts a quasiequatorial orientation.

Booth (219) has made an extensive study of the NMR spectra of 1,2,3,4-tetrahydroquinolines (**104**). The parent and its 1-methyl derivative appear to invert rapidly between the two conformations of the reduced ring, with nitrogen inversion taking place at an even more rapid rate. With the introduction of a 2-methyl group, inversion is still rapid but populations favor the invertomer with an equatorial methyl group. In the case of the 1,1-dimethyltetrahydroquinolinium iodide (**105**), nitrogen inversion is prevented,

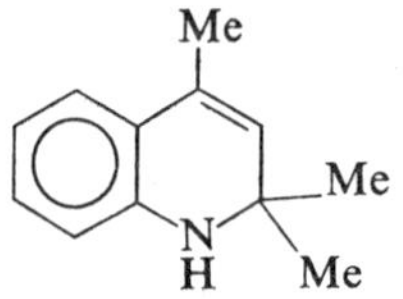

99 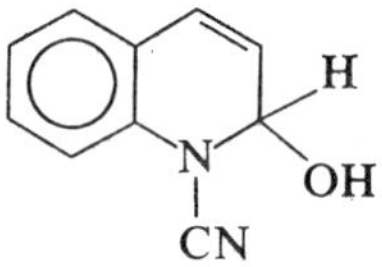 **100** 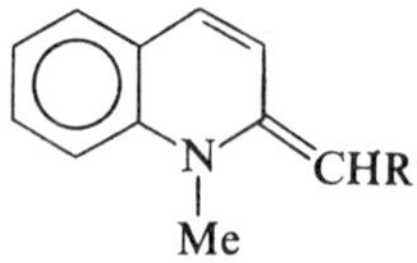 **101**

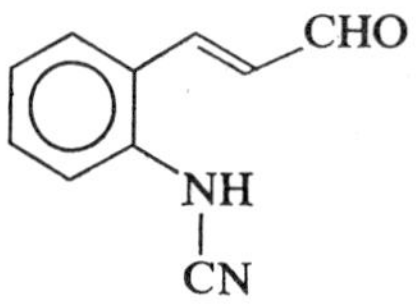

102 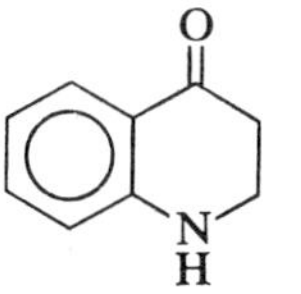 **103** 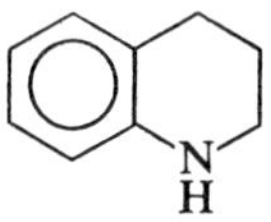 **104**

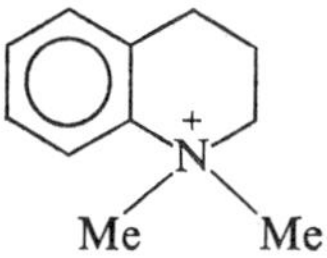

105 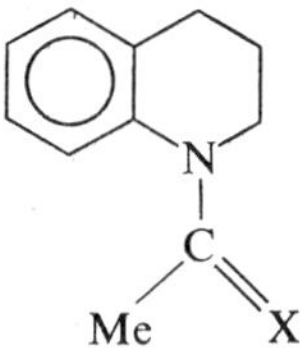 **106**, X = O
108, X = S **109**

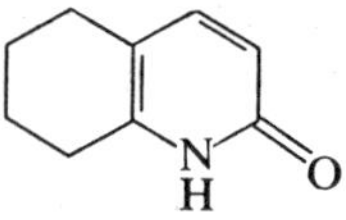

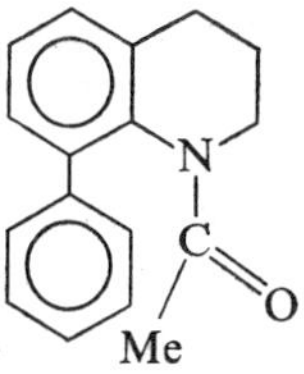

107 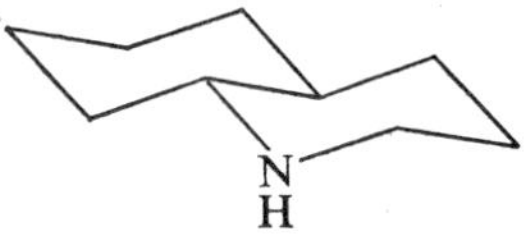 **110** 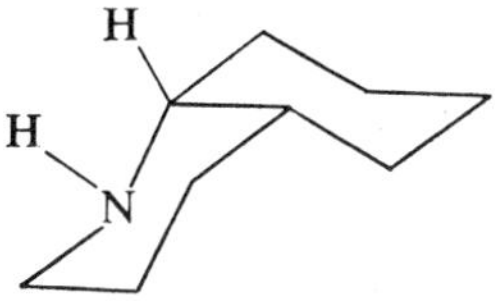 **111**

but ring inversion is still fast. However, introduction of a 2-phenyl substituent fixes the conformation, and in other compounds with heavier substitution, a deformed half-chair conformation is postulated for the hetero-ring. Quaternization of the nitrogen atom deshields all protons on the reduced ring.

Another interesting piece of work involves restricted rotation around the amide bond in *N*-acyl-1,2,3,4-tetrahydroquinolines (**106**) (220). The chemical shift of H-8 with respect to the rest of the aromatic protons was used to show that the preferred conformation of the acyl group had the carbonyl group twisted away from the aromatic ring as in **106** (the exo conformation). In the case of the 8-phenyl derivative **107** the very strong shielding (τ 8.57) experienced by the methyl group of the acetyl indicates that the two phenyl rings are highly twisted, with the methyl group sitting right in the shielding cone of the 8-phenyl ring. In the thioacyl series **108**, the C=S group has a larger effect on H-2 than the corresponding C=O group. More recent work (221), using variable-temperature measurements, suggests that the *N*-formyl compounds exist, as above, in the exo conformation, whereas the *N*-acetyl derivatives populate both exo and endo conformations to an appreciable extent. Some details are available on the spectra of 5,6,7,8-tetrahydroquinolones (**109**) (222).

Spectra of *cis*- and *trans*-decahydroquinoline have been reported (223). In that of the trans isomer, which exists predominantly in the chair–chair form **110**, signals from the geminal protons on C-2 are separated from the broad absorption band of the other protons in the molecule. In the case of the cis isomers, as well as these two protons showing up separately, the signal from the bridgehead proton adjacent to the nitrogen atom is also resolved. The authors conclude that the cis isomer exists mainly in a single conformation **111** and that a hydrogen substituent is appreciably more space demanding than the nitrogen lone pair. Spectral data for reduced quinolines are collected in Table 4.56.

B. The Isoquinoline System

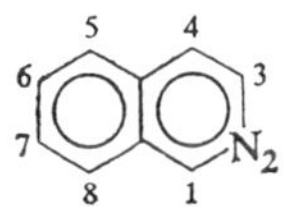

The complex and closely coupled spectrum of isoquinoline resisted analysis for some time but a complete set of parameters was produced by Black and Heffernan (224). These assignments have been, in part, checked by the preparation of 1- and 5-deuteroisoquinolines (188). Data for three nitroisoquinolines have been reported (225), together with those for other azanaphthalenes and their nitro derivatives. Shielding effects of ring nitrogen atoms and of nitro groups in these systems were calculated and used to predict the spectrum of isoquinoline. A reasonable agreement between the predicted and observed spectrum was obtained.

The remaining references to the spectra of aromatic isoquinolines are in papers dealing with the synthesis or reactions of this ring system, for example, Refs. 200 and 226. It is interesting to note the difference in chemical shift between the two NH protons of the 3-NH_2 group of 1,3-diamino-4-nitrosoisoquinoline (**112**) (226). One is hydrogen bonded to the adjacent NO

TABLE 4.56 SPECTRAL DATA FOR REDUCED QUINOLINES

A. 1,2-DIHYDROQUINOLINES

Substituents	Solvent	τ2	τ3	τ4	τ5	τ6	τ7	τ8	τNR	J_{23}	J_{24}	J_{34}	Ref.
2,2′,4-Tri-Me	CCl_4	8.81	4.78	8.06					6.40				209
2,2′,6-Tri-Me	CCl_4	8.80	4.60	3.77	3.30	7.81	3.22	3.74	6.61				208
	$CDCl_3$	8.71	4.45	3.66	3.17	7.77	3.10	3.57	6.51				208
.HCl	?	8.30	4.19	3.49	2.98	7.65	2.89	2.22					211
2,2′,4,7-Tetra-Me	CCl_4	8.81	4.80	8.07	3.05	3.58	7.70	3.83	6.65				208
2,2′,4,6-Tetra-Me	CCl_4	8.80	4.71	8.03	3.10	7.66	3.19	3.70	6.70				208
1,2-Di-Me	CCl_4	5.57, 8.82	3.91			2.10–3.30			6.93				213
1,2,2′-Tri-Me	CCl_4	8.56	4.16			2.20–3.35			6.95				213
1,2-Di-Me, 3-Br	$CDCl_3$	5.82, 8.80				2.75–3.6			7.16				213
1-Me, 2-Ph	CCl_4	4.94	4.45	3.76					7.39	4.7	0.5	9.9	214
1-CN, 2-OH	THF/D_2O	3.94	4.16	3.20						4.5	−0.1	9.9	215
1-CN, 2-O$^-$	THF/D_2O	3.6	3.90	3.6								10.2	215
1-CN, 2-OMe	CCl_4	3.98	4.36	3.22						4.6	−0.5	9.8	214

B. 1,4-DIHYDROQUINOLINES (214)

Substituents	Solvent	$\tau 1$	$\tau 2$	$\tau 3$	$\tau 4$	J_{23}	J_{24}	J_{34}
1-Me, 4-CN	CCl_4	6.98	3.92	5.17	5.36	8.2	0.9	3.6
1-Me, 4-PhC(CN)Me	$CDCl_3$	7.04	3.75	5.33	5.97	7.6	−0.1	5.5

C. 1,2,3,4-TETRAHYDROQUINOLINES

Substituents	Solvent	$\tau 1$	$\tau 2$	$\tau 3$	$\tau 4$	τ5-8	Ref.
None	C_6H_6	6.76	7.16	8.42	7.46		219
	$CDCl_3$	6.47	6.93	8.23	7.37	2.90–3.83	220
2-Me	C_6H_6	6.86	7.02, 9.16	8.25–8.85	7.30–7.55		219
1-Me	C_6H_6	7.41	7.12	8.25	7.41		219
1-CHO	$CDCl_3$	1.32	6.33	8.20	7.30	2.80–3.33	220
1-COMe	$CDCl_3$	7.80	6.20	8.07	7.28	2.50–3.10	220
1-COPh	$CHCl_3$		6.09	7.99	7.17		219
1-CSMe	$CDCl_3$	7.27	5.62	7.95	7.27	2.77	220
1-COMe, 2-Me	$CDCl_3$	7.87	5.22, 8.9	8.33–8.90	7.70–7.67	2.85	220
1-CSMe, 2-Me	$CDCl_3$	7.37	4.20, 8.83			2.7–2.9	220

TABLE 4.57 SPECTRAL DATA FOR ISOQUINOLINES

Substituents	Solvent	τ1	τ3	τ4	τ5	τ6	τ7	τ8	Ref.
		Chemical shifts							
None	Neat	0.55	1.23	2.51	2.42	2.54	2.63	2.26	224
	CCl_4	0.85	1.55	2.50	2.29	2.43	2.50	2.13	224
	Me_2CO	0.71	1.49	2.26	2.08	2.26	2.34	1.93	224
5-NO_2	$CDCl_3$	0.61	1.25	1.54		1.41	2.28	1.64	225
7-NO_2	$CDCl_3$	0.38	1.12	2.17	1.87	1.43		0.93	225
8-NO_2	$CDCl_3$	−0.40	1.26	2.20	1.85	2.23	1.65		225
1,3-Di-NH_2	Dioxan	4.30	5.37	4.08	2.3–3.3	2.3–3.3	2.3–3.3	2.3–3.3	226
	DMSO	3.58	4.77	4.10	2.6–3.2	2.6–3.2	3.03	2.08	226
1-NH_2, 3-NHOH	DMSO	3.36	2.33	3.68	2.55–3.05	2.55–3.05	2.55–3.05	1.96	226
1,3-Di-NH_2, 4-NO	DMSO	1.45	1.82, −2.38		1.23	2.23, 2.53		1.75	226
1-SMe	$CDCl_3$	7.25	1.53	2.55	2.25	2.25	2.25	1.66	200
	DCl	7.07	0.95	2.22	2	2	2	2	200
1-SO_2Me	$CDCl_3$	6.45	1.42	2.08	2.05	2.05	2.05	0.91	200
	DCl	6.08	0.9	0.9	1.4	1.4	1.4	0.8	200

Coupling constants (references as above)

		J_{13}	J_{14}	J_{15}	J_{34}	J_{45}	J_{48}	J_{56}	J_{57}	J_{58}	J_{67}	J_{68}	J_{78}
None	CCl_4	0		0.5	6.0		0.8	8.6	1.0	0.9	6.9	1.2	8.2
5-NO_2	$CDCl_3$				5.8		0.8				8.5	1.8	8.5
7-NO_2	$CDCl_3$		0.8	0.7	5.8		0.9	8.8		0.8		2.4	
8-NO_2	$CDCl_3$		0.8		5.7	0.7		8.3	1.5		7.4		

group and resonates over 4 ppm downfield from the other. Available data for simple members of this class of compounds are collected in Table 4.57. The ^{13}C–H coupling constants for isoquinoline have been measured and are included in Table 4.55.

Data for isocarbostyrils or 1-isoquinolones (**113**) are also extremely scant (227) and one reference (228) is found to the spectra of isoquinoline pseudobases (**114**). Incomplete information on some dihydroisoquinolines has been reported in work on the reduction of the isoquinoline nucleus (229).

Spectra of a number of 1,2,3,4-tetrahydroisoquinolines (**115**) have been reported (230, 231). In the case of the 1,2,3,4-tetrahydro-2-acetyl-6,7-

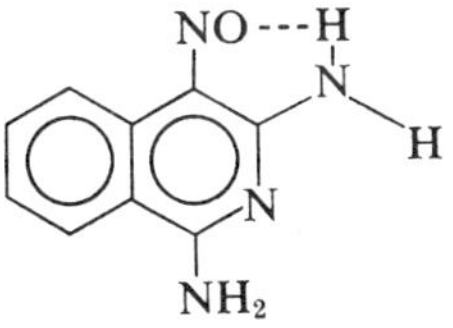

NO---H N H N NH_2
112

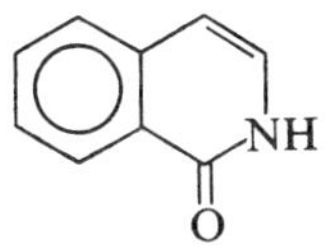

NH O
113

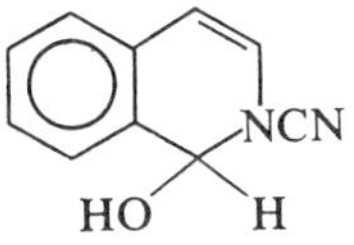

NCN HO H
114

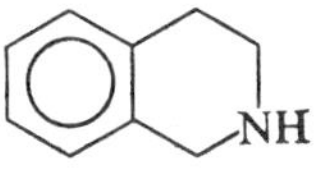

NH
115

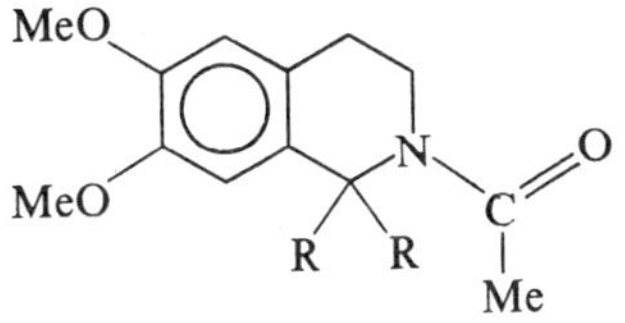

MeO MeO N C O Me R R
116

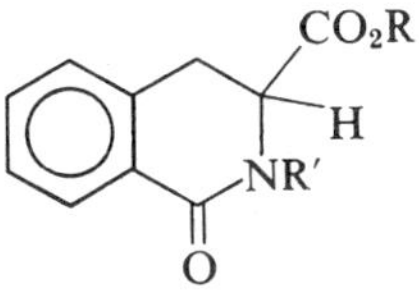

CO_2R H NR′ O
117

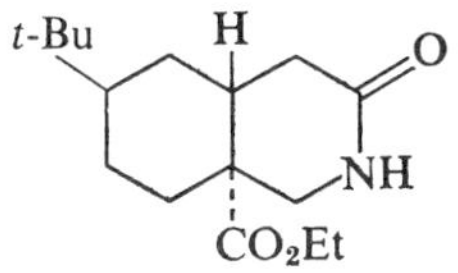

t-Bu H O NH CO_2Et
118

dimethoxyisoquinoline (**116**) and its C-1 derivatives, hindered rotation of the acetyl group was observed (232). In the light of the recent criticisms of methods used to determine thermodynamic parameters from NMR data, the authors have corrected their earlier interpretation of these spectra (233). Nuclear magnetic resonance techniques have also been used to determine the predominant conformations in some 3-carbomethoxy-3,4-dihydroisocarbostyrils (**117**) (234).

A series of papers on the conformations of chloro- or hydroxy-substituted decahydroisoquinolines have appeared (235–240). The techniques used are the same as those used for the piperidinols (see Chapter 2). A first-order approach to the spectrum of 6-*t*-butyl-9-carbethoxy-3-oxodecahydroisoquinoline (**118**) (240) indicates that this compound exists in a form with a twist-boat cyclohexane ring and a half-chair pyridone ring. Data for reduced isoquinolines are collected in Table 4.58.

TABLE 4.58 SPECTRAL DATA FOR REDUCED ISOQUINOLINES

A. ISOCARBOSTYRILS (227)—SOLVENT $CDCl_3$

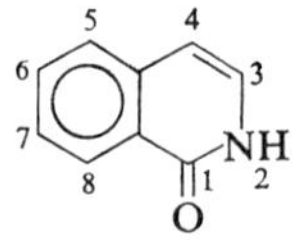

Substituents	τ2	τ4	τ3	τ5, 6, 7	τ8
3-Et	1.8	3.65	7.30, 8.62	2.0–3.0	2.0–3.0
3-Ph	1.8	2.0–3.0	2.0–3.0	2.0–3.0	2.0–3.0
2-OH, 3-Me		3.60	7.47	2.53	1.68

B. 1,2,3,4-TETRAHYDROISOQUINOLINES (230)[a]

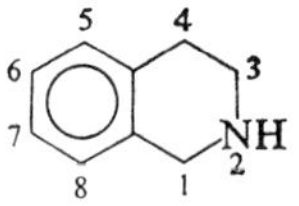

R_1	R_2	R_3	R_4	τ4	τ5	τ8	τ3	$J_{44'}$
H	H	H	Me	6.76, 7.04	3.28	3.28	8.41	17.0
H	Me	H	Me	6.74, 7.00	3.29	3.29	8.32	17.0
H	Ph	H	Me	6.61, 6.85	3.24	3.73	8.24	17.0
H	CH_2Ph	H	Me		3.27	3.42	8.36	
H	CH_2Ph	H	H		3.23, 3.34			
H	H	Me	Me	6.76, 7.03	3.31		8.42	17.0
H	Me	Me	Me	6.70, 6.96	3.28, 3.26		8.32	17.0

[a] A large amount of data on highly substituted derivatives can be found in Ref. 219.

C. The Quinolizine System

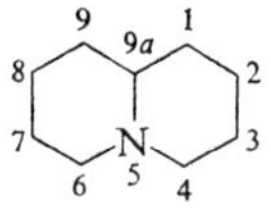

The main contributions to the study of quinolizine spectra have come from Acheson and his co-workers (241–243), and their results for 1,2,3,4-tetramethoxycarbonyl derivatives have been collected and summarized (244). Two tautomeric forms of quinolizines are commonly found, the 4*H*-isomer **119** and the 9a*H*-isomer **120**, and these can be readily detected by their different coupling patterns. The 4*H*-quinolizines have been classed as vinylogous pyridones, and a comparison of respective protons led to the suggestion that the quinolizines possess a greater ring current than do the pyridones. Protonation of the tetraethoxycarbonyl-4*H*-quinolizines (**121**)

occurs normally on C-3 to give ions of the type **122**, but with a 9-methyl substituent present protonation occurred at C-1 to give the ion **123**. Spectra of the 9*aH*-isomers of these tetrasubstituted compounds showed the unsubstituted ring to be olefinic in character. Data for these quinolizines and quinolizinium ions and a number of others from preparative papers (245–247) are included in Table 4.59.

Very little information is available for partly reduced quinolizines (see Refs. 244 and 248) but spectra of the fully reduced derivatives, the quinolizidines, have been studied in some detail in attempts to determine the stereochemistry and preferred conformations of these compounds. Each of the 1-, 2-, and 3-hydroxyquinolizidine racemates was synthesized and NMR confirmed the presence of a trans ring junction in all cases (249). Infrared and NMR techniques were used to show that with quinolizidine itself and with both racemic forms of the 1-, 2-, 3-, or 4-methyl derivatives, all except one isomer exist predominantly in a trans-fused conformation (250). The stereochemistry of the cations seems to be very similar to that of the free bases. However, formation of a methiodide from a compound with a trans-fused ring junction and an axial methyl substituent on the same side of the molecule as the nitrogen lone pair leads to a salt with a cis-fused ring system. The suggestion was made that in CCl_4 the lone pair is "smaller" than hydrogen but that the solvated lone pair (in hydrogen-bonding solvents) could be much larger. The stereochemistry of the *cis*- and *trans*-1-hydroxy-1-phenylquinolizidines was confirmed by NMR (251). In the cis isomer, the phenyl group gave rise to one broad multiplet, whereas with the trans counterpart hindered rotation separated the phenyl absorption into two

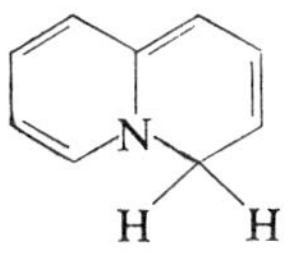

H H

119

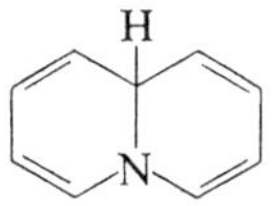

H

120

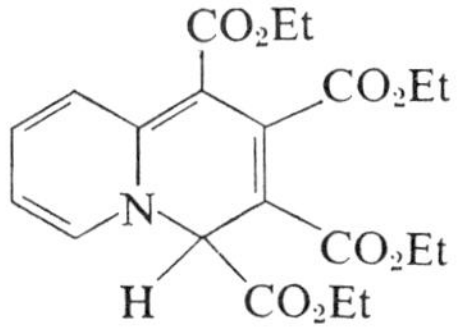

CO_2Et CO_2Et CO_2Et H CO_2Et

121

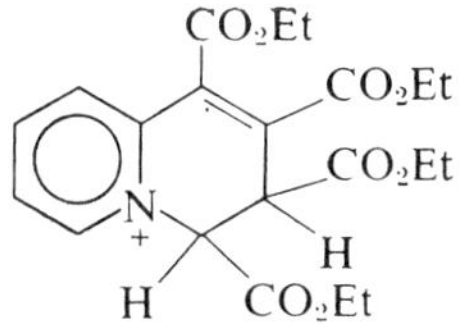

CO_2Et CO_2Et CO_2Et H H CO_2Et

122

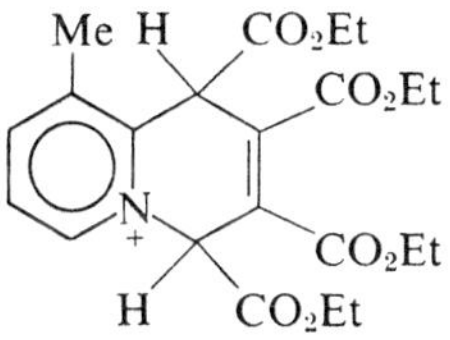

Me H CO_2Et CO_2Et CO_2Et H CO_2Et

123

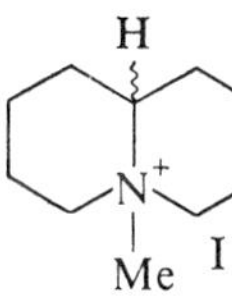

H Me I^-

124

O

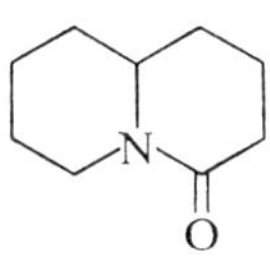

127

TABLE 4.59 SPECTRAL DATA FOR QUINOLIZINES AND QUINOLIZINIUM IONS

A. 4*H*-QUINOLIZINES (244). SOLVENT—$CDCl_3$

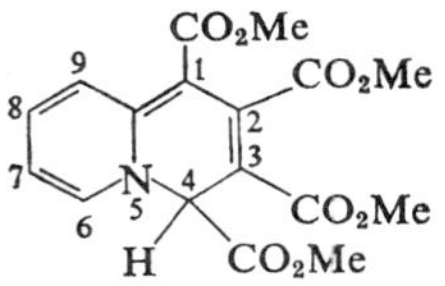

Substituents	τ4	τ6	τ7	τ8	τ9	J_{67}	J_{68}	J_{89}	τMe (ester)
None	3.94	2.3–2.6	3.15	2.3–2.6	1.32				6.07, 6.20, 6.24, 6.28
6-Me	3.44	7.42	3.22	2.56	1.42			9	6.08, 6.20, 6.27, 6.31
7-Me	3.98	2.62	7.71	2.52	1.38		1	9	6.05, 6.18, 6.23, 6.26
8-Me	3.99	2.53	3.28	7.61	1.47	7			6.10, 6.23, 6.28, 6.31
9-Me	3.89	2.48	3.07	2.33	7.72				

B. 9*aH*-QUINOLIZINES (244). SOLVENT—$CDCl_3$

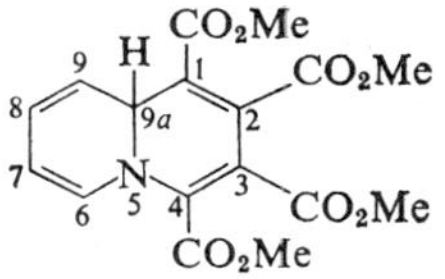

Substituents	τ6	τ7	τ8	τ9	τ9*a*	J_{67}	J_{89}	τMe (ester)
6-Me	8.04	3.9–4.2		4.35	5.03		9	6.12, 6.16, 6.24, 6.33
9-Me	3.78	4.05–4.22		8.19	5.01	7		6.09, 6.09, 6.24, 6.29
9*a*-Me	3.63	4.43	3.80	4.14	8.78		9	6.02, 6.19, 6.19, 6.28

C. 3*H*,4*H*-QUINOLIZINIUM IONS (244). SOLVENT—TFA

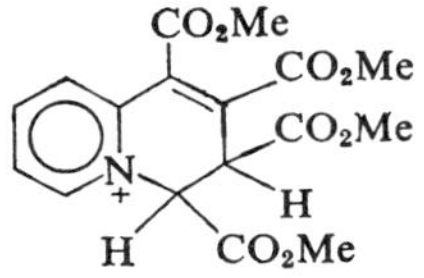

Substituents	τ3	τ4	τ6	τ7	τ8	τ9	J_{67}	J_{89}	τMe (ester)
None	4.72	3.54	0.89	1.68	1.14	1.62		8	5.80, 5.86, 6.01, 6.12
6-Me	4.67	3.31	6.86	1.86	1.35	1.86		8	5.82, 5.86, 6.01, 6.13
7-Me	4.80	3.67	1.13	7.29	1.40	1.87		8	5.85, 5.91, 6.05, 6.16
8-Me	4.79	3.66	1.14	1.93	7.17	1.87	7		5.85, 5.90, 6.05, 6.15

D. 1*H*,4*H*-QUINOLIZINIUM IONS (244). SOLVENT—TFA

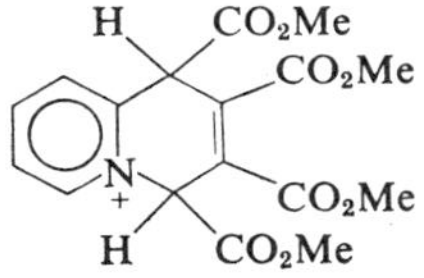

Substituents	τ1	τ4	τ6	τ7	τ8	τ9	τMe (ester)
S-Me	4.28	3.11	1.09	1.85	1.34	7.16	5.88, 5.88, 6.02, 6.08
7,9-Di-Me	4.34	3.22	1.32	7.33	1.55	7.22	5.89, 5.89, 6.02, 6.09

TABLE 4.59 *(continued)*

E. DEHYDROQUINOLIZINIUM IONS (243)

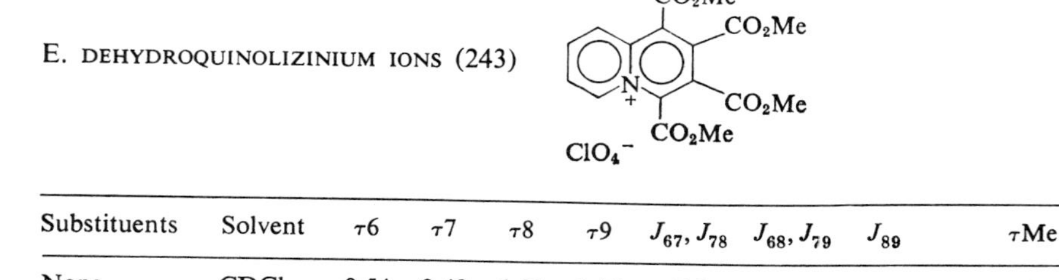

Substituents	Solvent	$\tau 6$	$\tau 7$	$\tau 8$	$\tau 9$	J_{67}, J_{78}	J_{68}, J_{79}	J_{89}	τMe (ester)
None	$CDCl_3$	0.54	2.48	1.95	1.17	7.5	2	9	6.03, 6.03, 6.10
	$MeNO_2$	0.53	2.27	1.74	1.28				

125 **126**

bands. Also, in the cis compound the OH is hydrogen bonded to the nitrogen and hence gives rise to a sharp singlet which contrasts with the broad OH peak from the trans isomer. Isomers of 2-hydroxy-4-phenylquinolizidines have also been identified by their NMR spectra (252). The half-height widths of the resonance from the *N*-methyl group of the cis- and trans-fused 5-methylquinolizidinium iodides (**124**) in D_2O have been used to determine the configuration of the ring junction (253). Also, the ring opening of the internal quaternary salt of tosyllupinine (**125**) to give the compound **126** was followed by NMR (253a). In an extensive study of the lupin alkaloid lactams, the 4-oxo group of quinolizidones (**127**) has been shown to deshield protons strongly in the 6-equatorial position (253b). Some representative data for these quinolizidines are given in Table 4.60.

TABLE 4.60 SPECTRAL DATA FOR QUINOLIZIDINES (250). SOLVENT—CCl_4

	Chemical shifts				
	Ring protons			Methyl groups	
Compound	τ4, 6	τ Others	$W_{1/2}$ (Hz)	Axial	Equatorial
Parent	7.23, 7.43	8.56	17		
cis-1-Me	7.25, 7.40	8.46	18	9.05	
trans-1-Me	7.24, 7.45	8.41	28		9.16
cis-2-Me	7.20, 7.44	8.59	22		9.09
trans-2-Me	7.82	8.68	26	9.01	
cis-3-Me	7.41, 7.63, 7.84	8.58	12	8.92	
trans-3-Me	7.27, 7.44	8.57	17		9.18
cis-4-Me	6.88, 7.75	8.57	11		9.00
trans-4-Me	7.7	8.56	11		9.07

TABLE 4.61 SPECTRAL DATA FOR CINNOLINES[a]

Substituents	Solvent	τ3	τ4	τ5	τ6	τ7	τ8	Ref.
None[b]	Me_2CO	0.67	1.91	1.98	2.16	2.06	1.51	254
	CCl_4	0.78	2.24 ± 0.05	2.24 ± 0.05	2.24 ± 0.05	2.24 ± 0.05		254
	CCl_4	0.90	2.27	2.43	2.43	2.43	1.70	255
	$CDCl_3$	0.72					1.47	257
4-Me	$CDCl_3$	1.04	7.47	2.32	2.32	2.32	1.70	255
	CCl_4	0.98	7.49					261
4-COOH	DMSO	0.15		1.50	2.07	2.07	1.37	255
4-Styryl	$CDCl_3$	0.70		2.03			1.68	255
3-NH_2	DMSO		3.00	2.47	2.47	2.47	1.83	255
4-NH_2	DMSO	1.27	2.63	2.35	2.35	1.83	1.83	255
5-NH_2	DMSO	0.87	1.75		3.18	2.48	2.48	255
8-NH_2	$CDCl_3$	0.97	3.15	2.45	3.15	2.70	4.62	255
3-NO_2	DMSO		0.77	1.70	1.93	1.93	1.40	255
5-NO_2	$CDCl_3$	0.60	1.20		1.17	2.10	1.38	255
8-NO_2	$CDCl_3$	0.60	2.05	2.80	2.05	2.05		255

[a] Data for some 4-substituted 6,7-dimethoxycinnolines can be found in Ref. 262.

[b] Coupling constants: J_{34} 5.7, J_{48} 0.8, J_{56} 7.8, J_{57} 1.5, J_{58} 0.8, J_{67} 6.9, J_{68} 1.3, J_{78} 8.6.

D. The Cinnoline System

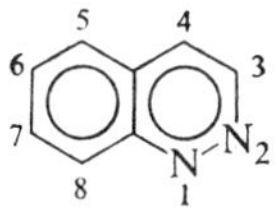

The spectrum of cinnoline in acetone was reasonably spread and amenable to analysis (254), whereas in CCl_4 protons 4, 5, 6, and 7 gave rise to a broad singlet. In spite of a reasonable number of references pertaining to the spectra of cinnolines, most of these describe the cations **128** or the 4-cinnolones (**129**) and very few are concerned with the neutral aromatic cinnolines. A

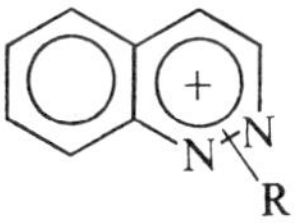

128 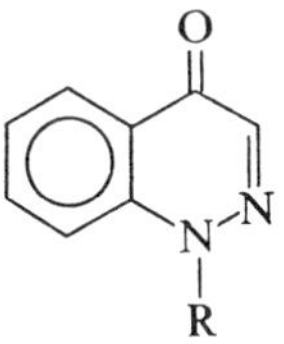 **129** 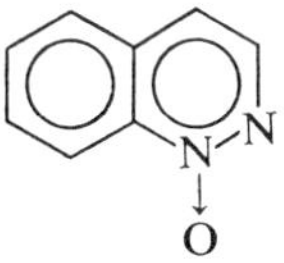 **130** 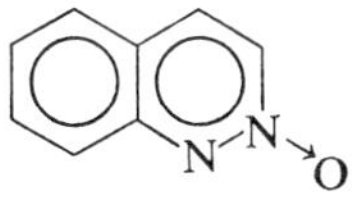 **131**

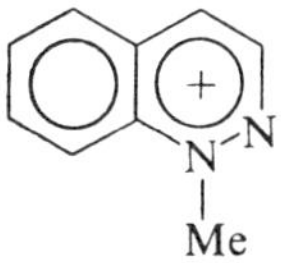

132 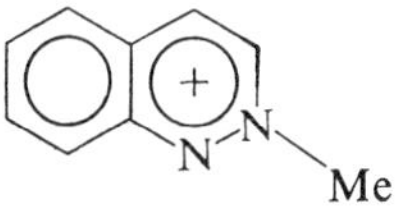 **133** 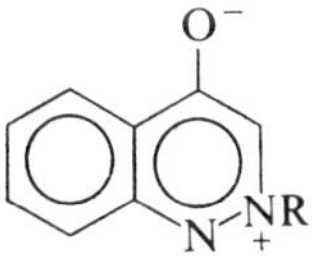 **134**

simple qualitative approach, involving deshielding of nearby protons by the nitrogen atoms and an evaluation of the rather distinctive coupling patterns in these molecules, was used to assign multiplets to specific protons in the spectra of a series of monosubstituted cinnolines (255). Most substituent effects were found to be quite normal, but anomalies were found in the spectra of 8-nitro- and 5-aminocinnolines. Other papers in this section contain a few data on cinnolines and these are collected in Table 4.61.

The spectra of cinnoline 1- and 2-oxides (**130** and **131**) have been discussed (256, 257). A strong deshielding effect of the 1-oxide grouping on the adjacent peri proton (H-8) was observed. Data for these compounds are included in Table 4.62.

The preferred site for the protonation of cinnoline has been the subject of much speculation. Palmer and Semple (201) applied NMR methods to this problem but their conclusions from chemical shift data were not consistent with those they obtained from coupling constants. It was later shown that

TABLE 4.62 SPECTRAL DATA FOR CINNOLINIUM IONS AND CINNOLINE *N*-OXIDES

A. CINNOLINIUM IONS

Substituents	Anion	Solvent	τ3	τ4	J_{34}	τNMe	Ref.
1-Me	ClO_4^-	D_2O	0.43	0.79	5.7		258
2-Me	ClO_4^-	D_2O	0.49	0.83	6.0		258
2-Me, 4-OEt	BF_4^-	DMSO	0.33	5.32, 8.60		5.18	259

B. CINNOLINE *N*-OXIDES. SOLVENT—$CDCl_3$

Substituent		τ3	τ4	τ8	J_{34}	J_{78}	J_{48}	Ref.
1-Oxide		1.75	2.61	1.41	6.0		0.9	257
		1.67	2.50	1.33	6.2		0.9	256
	4-Me	1.87	7.42	1.35	1.0			256
	3-Cl		2.48	1.48			0.9	256
	3-OMe	5.93	3.08	1.52			1.0	256
2-Oxide		1.85	2.05		7.0		1.0	257
		1.79	1.94		7.0			256
	4-Me	1.90	7.37		1.0			256
1,2-Dioxide		1.92	2.55	1.66	7.5	9.6	1.0	257

the spectra of 1- and 2-methylcinnolinium salts (**132** and **133**) were not sufficiently different for the isomers to be distinguished in this way, although ultraviolet spectra clearly showed that protonation occurred predominantly on N-2 (258). For data from these and other cinnolinium ions (259), see Table 4.62.

The biological activity of 4-cinnolones has led to the preparation of large numbers of simple derivatives which have made possible an extensive study of NMR correlations throughout this series of compounds. Unfortunately, the peak patterns from the aromatic ring are extremely complex and most authors have worked from only partial analyses. However, in one investigation of 4(1*H*)-cinnolones (**129**) (260), complete iterative analyses of the spectra of many derivatives were undertaken and the results obtained, when assisted by deuteration studies, led to unequivocal assignments of chemical shifts to specific protons. In all cases, H-3 and H-5 are easily identified, but assignments of peaks to H-8 in cinnolone (261) and a series of 1-alkylcinnolones (262, 263) must be viewed with caution. Tautomerism in a number of sulfonic acid derivatives has been studied (260) and a considerable amount of data is available on the so-called anhydrobases of the 2-alkyl-4-hydroxycinnolinium ions (**134**) (262, 263). Data for simple 4-cinnolones are summarized in Table 4.63.

Some information about the spectra of dihydro- (264–266), octahydro-, and decahydrocinnolines (267) is available.

E. The Quinazoline System

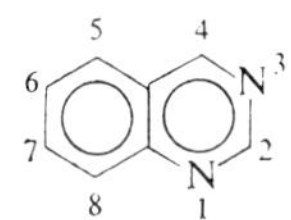

The spectra of quinazoline in acetone and CCl_4 have been fully analyzed (254). In these spectra, H-2 and H-4 give rise to a pair of singlets downfield from the multiplets from the benzenoid protons. It was assumed in these analyses that the broader of these two singlets would contain the cross-ring coupling J_{48} and hence could be assigned to H-4. An extensive study of substituent effects (268) produced evidence for the reversal of this assignment, and an examination of the spectrum of 4-deuteroquinazoline confirmed that the original assignments were indeed incorrect (225). In cyclohexane solutions these two peaks merge to a singlet, but in a range of other solvents (see Table 4.64) H-4 absorbs downfield from H-2.

The only investigation of simple substituted quinazolines is that mentioned above (268). In this work, the effects of substituents on the benzenoid ring were found to be qualitatively similar to those observed with benzenes but of different magnitude, although the influence of groups such as Cl was rather erratic. When the substituent occupied the 5- or 8-position, linear correlations were obtained between the Hammett σp constants and the substituent shift of the *p*-proton. Also of interest was the observation that meta coupling across a substituent is larger than across a proton. Data for these compounds and a number of others from the literature (269–272) are collected in Table 4.64.

The spectrum of quinazoline-3-oxide (**135**) has been investigated (271, 273). Peaks from H-2 and H-4 again appear at low field and close together. The assignment given in Table 4.64 (273) was presumably made intuitively and must be regarded with caution. Since the lone pair of N-3 has been removed by *N*-oxide formation, a normal meta-type coupling is observed between H-2 and H-4. More data on substituted 3-oxides are available from work on the cyclization of benzaldoximes (272).

Data for the cations of a number of *N*-methyldihydrooxo- or iminoquinazolines have been reported in a publication dealing with the Dimroth rearrangement in the quinazoline system (270). Also, the NMR spectrum of the condensation product **136** of *o*-aminoacetophenone and methyl

TABLE 4.63 SPECTRAL DATA FOR 4(1*H*)-CINNOLONES[a]

Substituents	Solvent	τNR	τ3	τ5	τ6	τ7	τ8	J_{56}	J_{57}	J_{58}	J_{67}	J_{68}	J_{78}	Ref.
None	DMSO		2.17	1.89	2.52	2.16	2.34	8.3	1.6	0.5	7.1	1.0	8.4	260
5-Cl	DMSO		2.33		2.63	2.32	2.50				7.6	1.0	8.6	260
6-Cl	DMSO		2.22	2.03		2.20	2.30		2.5	0.3			9.1	260
7-Cl	DMSO		2.23	1.99	2.59		2.40	8.8		0.5		1.9		260
8-Cl	DMSO		2.10	1.98	2.58	2.07		8.1	1.3		7.7			260
3-COOH	DMSO			1.80	2.32	2.01	2.15	8.2	1.4	0.5	7.0	1.0	8.6	260
5-OH	DMSO		2.19		3.32	2.33	2.98				7.9	0.9	8.4	260
8-OH	DMSO		2.24	2.74	2.50	2.93		8.1	1.1		7.7			260
1-Me	$CDCl_3$	5.92	2.23	1.70			2.60							263
	DMSO	5.70	2.21	1.80			2.46							263
	Me_2CO	5.95	2.36	1.78			2.59							263
	TFA	5.19	1.00	1.30			1.85							263
1-Et	$CDCl_3$		2.20	1.66			2.63							263
2-CH_2Ph	$CDCl_3$		2.06	1.68										263
1-Me, 6-Cl	$CDCl_3$	5.89	2.20	1.70		2.27	2.60		3.0				10.0	262
1-Me, 6-NO_2	$CDCl_3$	5.86	2.18	0.88		1.47	2.47		3.0				10.2	262

[a] A great deal of data on more highly substituted cinnolones is included in the references listed.

TABLE 4.64 SPECTRAL DATA FOR QUINAZOLINES

Substituents	Solvent	$\tau 2$	$\tau 4$	$\tau 5$	$\tau 6$	$\tau 7$	$\tau 8$	J_{24}	J_{56}	J_{57}	J_{58}	J_{67}	J_{68}	J_{78}	Ref.
None	CCl_4	0.77	0.71	2.16	2.42	2.17	1.99		7.9	1.2	0.8	6.9	1.2	8.5	254
	Me_2CO	0.73	0.48	1.89	2.25	1.99	1.97		8.2	1.3	0.6	7.0	0.8	8.5	254
	$CDCl_3$	0.62	0.55												225
	DMSO	0.59	0.29	1.74	2.12	1.86	1.86		8.40	2.02	0.21	6.89	1.63	8.62	268
2-Me	$CDCl_3$	6.98	0.45		1.7–2.5										269
4-Me	$CDCl_3$	0.66	7.13		1.7–2.5										269
5-OMe	DMSO	0.71	0.34		2.81	2.06	2.44					7.98	0.85	8.07	268
5-OH	DMSO	0.70	0.34		2.92	2.17	2.55					8.00	0.82	8.28	268
5-Me	DMSO	0.65	0.25												268
5-Cl	DMSO	0.53	0.22												268
5-NO_2	DMSO	0.48	−0.13		1.39	1.75	1.56					7.77	0.76	8.77	268
6-OMe	DMSO	0.78	0.48	2.54		2.40	2.12			2.97	0.22			9.42	268
6-OH	DMSO	0.88	0.55	2.67		2.39	2.88			2.63	0.20			9.17	268
6-Me	DMSO	0.73	0.46												268
6-Cl	DMSO	0.62	0.37	1.69		1.98	1.92			2.16	−0.16			8.40	268
6-NO_2	DMSO	0.48	0.06	0.80		1.33	1.78			2.70	0.77			9.73	268

TABLE 4.64 *(continued)*

7-OMe	DMSO	0.80	0.58	1.95	2.64		2.66		9.04		1.03		2.62		268
7-OH	DMSO	0.87	0.62	2.00	2.69		2.78		8.71		0.87		2.22		268
7-Me	DMSO	0.70	0.43	1.96	2.40		2.19		8.43		0.78		1.49		268
7-Cl	DMSO	0.63	0.33	1.78	2.20		1.93		8.75		0.53		2.02		268
7-NO_2	DMSO	0.48	0.12	1.51	1.43		1.21		8.30		1.90		0.10		268
8-OMe	DMSO	0.68	0.40	2.28	2.25	2.48			8.31	1.40		7.51			268
8-OH	DMSO	0.59	0.43	2.45	2.42	2.65			8.22	1.58		7.42			268
8-Me	DMSO	0.50	0.37	2.06	2.36	2.17			8.07	1.33		7.20			268
8-Cl	DMSO	0.48	0.22	1.75	2.16	1.72			8.20	1.35		7.75			268
8-NO_2	DMSO	0.45	0.10	1.46	2.01	1.37			8.42	1.29		7.68			268
2-Amino															
None	$CDCl_3$		0.60		1.9–3.0										270
5-OMe	$CDCl_3$	4.4	0.93	6.03		2.2–3.2									270
6-OMe	$CDCl_3$	4.7	0.87		6.06	2.4–2.9									270
7-OMe	$CDCl_3$	3.7	1.00	2.35	3.25	6.15	3.25								270
8-OMe	$CDCl_3$	4.4	1.10		2.8–3.4		6.05								270
3-Oxides															
None	D_2O	0.36	0.89		1.8–2.2			1.8							271
5-OMe	$CDCl_3$	0.90	1.08	5.97	3.02	2.55		1.6				6.0	3.0		272
6-OMe	$CDCl_3$	0.90	1.00	2.89	5.93	2.42	1.88	1.6		3.0				8.0	272
7-OMe	$CDCl_3$	1.05	1.12	2.70		6.02	2.4	1.6					3.0		272
8-OMe	$CDCl_3$	1.15, 1.20		3.02	2.08	2.60	6.05		8.0						272

TABLE 4.65 SPECTRAL DATA FOR DIHYDRO-OXO-, IMINO-, OR THIOQUINAZOLINES. SOLVENT—DCl/D_2O

A. 1,4-DIHYDRO (270)

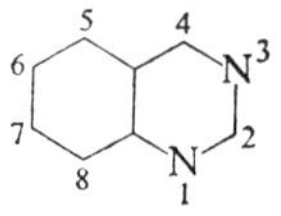

Substituents	τ1	τ2	τ3	τ5–8
1-Me, 4-NH	5.9	0.9		1.6–2.1
1-Me, 4-O	5.78	0.5		1.5–2.1
1,3-Di-Me, 4-$O^{+}I^{-}$	5.80	−0.40	6.23	1.3–2.0

B. 2,3-DIHYDRO (270)

Substituents	τ3	τ4	τ5–8	τOMe
3-Me, 2-NH	6.02	0.28	1.5–2.4	
3-Me, 2-O	6.18	0.02	1.7–2.4	
3-Me, 2-NH, 5-OMe	5.97	0.42		6.18
3-Me, 2-NH, 6-OMe	5.95	0.66		6.02

C. 3,4-DIHYDRO (270)

Substituents	τ2	τ3	τ5–8
3-Me, 4-O	0.37	6.10	1.5–2.2

D. 1,2,3,4-TETRAHYDRO (274)

Substituents	τ1	τ3	τ4 (= CH_2)	J_{CH_2}
3-Me, 2-O, 4-CH_2	6.75	6.52	5.10, 5.71	2.2
3-Me, 2-S, 4-CH_2	6.25	6.20	4.92, 5.45	2.5

isothiocyanate showed clearly the presence of the exocyclic methylene group (274). Data for these types of compound are summarized in Table 4.65.

The cations of quinazoline and its 3-oxide in aqueous solutions have been shown to add a molecule of water across the 3,4-bond to give the corresponding 4-hydroxy-3,4-dihydroquinazolines (**137** and **138**). While ultraviolet spectroscopy indicated the presence, in neutral and acidic solutions, of species with different chromophores, the NMR spectra showed them to be "hydrates" (271). The spectrum of the 4-deuteroquinazoline cation showed

135 136 137 138

139

conclusively that this addition occurred across the 3,4-bond. Information on the chemical shifts for these compounds is included in Table 4.66.

TABLE 4.66 SPECTRAL DATA FOR REDUCED QUINAZOLINES

A. 3,4-DIHYDRO ("HYDRATES")

Substituents	Solvent	τ2	τ4	τ5–8	Ref.
None	TFA/H_2O	1.65	3.58	2.3–2.8	269
3-OH	DCl	1.13	3.38	2.2–2.8	271

B. DECAHYDROQUINAZOLINE (275)[a]

Solvent	Ring fusion	τ2	τ4		τ8a	τ4a, 5–8 ($W_{1/2}$)	τNH
			Axial	Equatorial			
$CDCl_3$	Cis	5.98, 6.34	7.09	7.09	6.98	8.0–8.9 (15)	8.18
	Trans	6.03, 6.37	7.61	7.04	7.8	8.0–9.3 (54)	8.42
D_2O	Cis	6.12, 6.57	7.20	7.20	7.05	8.1–8.9 (12)	
	Trans	6.06, 6.52	7.67	7.09	7.8	7.9–9.2 (48)	

[a] Data for other compounds and for coupling constants can be found in Ref. 275.

The preferred conformations of *cis*- and *trans*-decahydroquinazolines (**139**) were deduced from their NMR spectra, using the normal coupling rules for reduced systems (275). Cis isomers can be distinguished easily from the corresponding trans isomers by the observation that the complex multiplet from H-3, H-4, H-5, and H-6 is much narrower in the cis case. Representative data for reduced quinazolines are collected in Table 4.66.

F. The Quinoxaline System

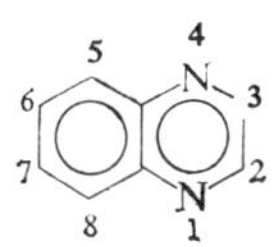

The spectrum of quinoxaline has been analyzed by Black and Heffernan in their general study of the diazanaphthalenes (254). Nuclear magnetic resonance parameters have been tabulated for the full analysis of the spectra of 14 monosubstituted quinoxalines in DMSO solutions (276). Ortho and meta substituent effects were calculated for nine or these substituents, and were compared with the corresponding values for monosubstituted benzenes. Generally, these effects were greater in quinoxalines for electron-donating substituents and in benzenes for electron acceptors. Spectra of a number of quinoxaline 2-*C*- or *O*-glycosides have been reported (277, 278).

Data for quinoxaline and its monosubstituted derivatives are included in Table 4.67. For spectra of a number of di- and trisubstituted quinoxalines, see Refs. 279 and 280. Also, the ^{19}F spectra of hexafluoroquinoxaline and 5-chloropentafluoroquinoxaline have been reported (281).

Blears and Danyluk (282) have studied the protonation of quinoxaline and a number of other heterocycles. A qualitative interpretation of the downfield shifts observed on going from CH_2Cl_2 to TFA solutions was given in terms of the magnetic anisotropy and electric field effects of neighboring nitrogen atoms and the π-electron charge densities in the neutral molecule and cation.

Two studies have been made of the spectra of 1,2,3,4-tetrahydroquinoxalines (283, 284). The shielding of protons on C-2 and C-3 in *cis*- and *trans*-2,3-dimethyl-1,2,3,4-tetrahydroquinoxalines (**140**) was shown to depend on their axial or equatorial orientation. This difference of about 0.5 ppm reflects the unequal effect of the diamagnetic anisotropy of the C–N single bond on the different C–H conformations and makes possible the assignment of relative stereochemistry to the pairs of isomers. The conclusion was reached that interconversion between conformers in both isomers was rapid down to $-87°$. The results obtained were used to assign stereochemistry to cis–trans pairs of reduced triazanaphthalenes, phenazines, and pteridines.

TABLE 4.67 SPECTRAL DATA FOR QUINOXALINES. SOLVENT—DMSO (UNLESS INDICATED OTHERWISE)

Substituents	$\tau 2$	$\tau 3$	$\tau 5$	$\tau 6$	$\tau 7$	$\tau 8$	J_{23}	J_{56}	J_{57}	J_{58}	J_{67}	J_{68}	J_{78}	Ref.
None (CCl_4)	1.26	1.26	1.93	2.32	2.32	1.93		8.4	1.5	0.5	6.8	1.5	8.4	254
(Me_2CO)	1.09	1.09	1.90	2.15	2.15	1.90		8.4	1.4	0.6	6.9	1.4	8.4	254
	0.90	0.90	1.77	2.03	2.03	1.77		8.6	1.4	0.8	7.1	1.4	8.6	276
2-SMe ($CDCl_3$)	7.29	1.23	1.9	2.2	2.2	1.9								200
(DCl)	6.92	0.57	1.65	1.65	1.65	1.65								200
2-SO_2Me ($CDCl_3$)	6.53	0.30	1.65	1.9	1.9	1.65								200
(DCl)	6.30	0.1	1.3	1.3	1.3	1.3								200
5-NH_2	1.03	1.19		2.95	2.37	2.70	1.7				7.9	1.2	8.4	276
5-OH	1.06	1.12		2.71	2.22	2.37	1.8				7.8	1.3	8.4	276
5-OEt	1.01	1.05		2.72	2.23	2.33	1.9				8.3	0.7	8.6	276
5-Me (neat)	1.17	1.25		2.29	2.25	2.06	1.7				7.0	1.0	8.4	276
5-OAc	0.96	0.99		2.30	2.08	1.92	1.9				7.7	1.2	8.5	276
5-NO_2	0.77	0.77		1.48	1.88	1.44					5.0	2.9	10.3	276
6-NH_2	1.25	1.40	2.87		2.57	2.09	1.9		2.7	0.3			9.2	276
6-OMe	1.04	1.13	2.48		2.44	1.93	1.7		2.7	0.7			9.3	276
6-OH	0.94	1.03	2.42		2.30	1.78	1.9		2.6	0.3			9.2	276
6-Me	0.98	0.98	2.00		2.23	1.90			2.3	0.3			8.8	276
6-OAc	1.02	1.02	2.11		2.28	1.82			2.6	0.4			9.1	276
6-CO_2Et	0.90	0.90	1.43		1.74	1.83			1.9	0.5			8.8	276
6-Cl	0.96	0.96	1.83		2.10	1.85			2.4	0.6			9.1	276
6-NO_2	0.83	0.83	1.07		1.40	1.62			2.6	0.4			9.3	276

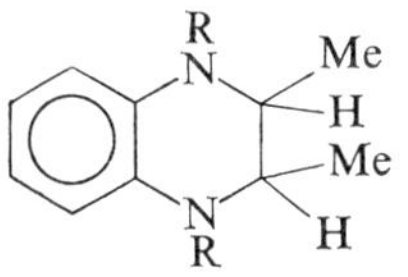

140

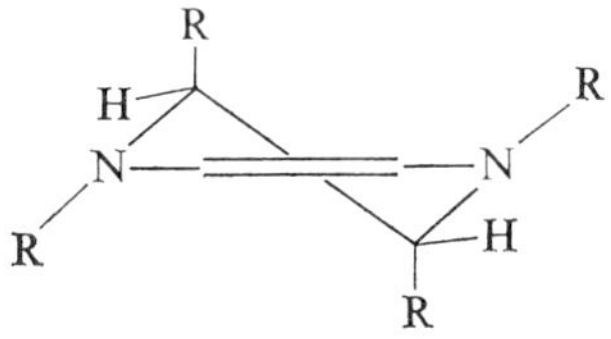

141

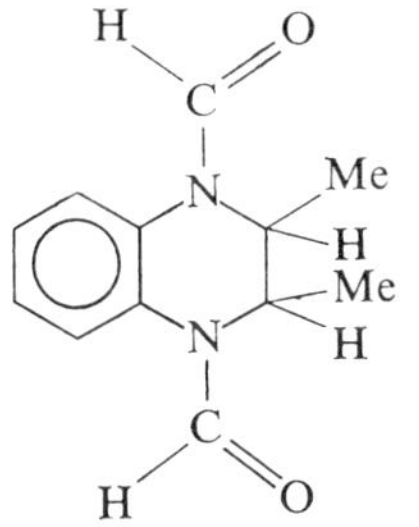

142

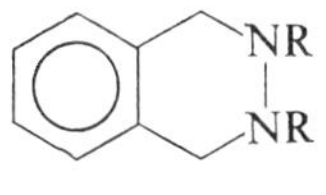

143

More recently, these tetrahydroquinoxalines were used as models in an attempt to elucidate the stereochemistry of tetrahydrofolic acid (284). A thorough investigation of coupling constants confirms the expected semichair structure **141** of the system, a structure which is somewhat flattened by the introduction of *N*-acyl substituents. The deshielding effects of the formyl group in some 1,4-diformyl derivatives led to the suggestion that these compounds exist predominantly in the cis(N-1)cis(N-4) conformation **142**. Representative data for reduced quinoxalines are given in Table 4.68.

G. The Phthalazine System

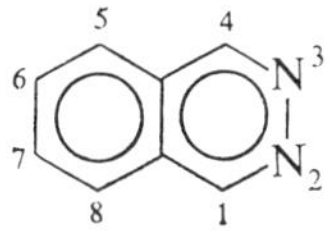

The spectra of phthalazine itself in acetone and CCl_4 have been analyzed (254).

CCl_4: τ1,4 0.56, τ5,8 2.07, τ6,7 2.15.

Me_2CO: τ1,4 0.40, τ5,8 1.87, τ6,7 1.99, $J_{56(78)}$ 8.1, $J_{57(68)}$ 1.2, J_{58} 0.5, J_{67} 6.7, J_{48} 0.4.

Ring inversion of a number of 1,2,3,4-tetrahydrophthalazines (**143**) has been studied by the usual coalescence temperature method (285).

TABLE 4.68 SPECTRAL DATA FOR 1,2,3,4-TETRAHYDROQUINOXALINES (283)

N-Substituent	*C*-Substituent	Solvent	τ2	τ3	τ2Me	τ3Me	τNR	τ5–8	J_{23}
H	None	$CDCl_3$	6.80	6.80			6.68	3.5	
	2,3-Di-Me (cis)	CCl_4	6.63	6.63		8.94	6.60	3.69	2.7 (cis)
	(trans)	CCl_4	7.05	7.05		8.99	6.92	3.67	7.0 (trans)
	2-Me	CCl_4	6.66	6.84 (eq.), 7.13 (ax.)	8.90		6.82	3.67	2.6 (cis) 8.2 (trans)
1,4-Di-COPh	None	$CDCl_3$	5.90	5.90			2.66	3.30	
	2,3-Di-Me (cis)	$CDCl_3$	4.79	4.79		8.83	2.67	3.33	8.4 (cis)
	(trans)	$CDCl_3$	5.57	5.57		8.75	2.59	3.00	1.9 (trans)
	2-Me	$CDCl_3$	5.00	5.50 (eq.), 6.55 (ax.)	8.73		2.68, 2.62	3.28	7.25 (cis) 5.80 (trans)
1,4-Di-CHO	None	$CDCl_3$	6.10	6.10			1.25	2.82	
	2,3-Di-Me (cis)	$CDCl_3$	5.30	5.30		8.75	1.47	2.75	5.6 (cis)
	2,3-Di-Me (trans)[a]	$CDCl_3$	5.19	5.19		8.80	1.05	2.75	1.9 (trans)
	[b]	$CDCl_3$	6.15	6.15			1.65	1.45	
	2-Me	$CDCl_3$	5.05	5.60, 6.80	8.90		1.12, 1.22	2.77	

[a] The cis(N-1)cis(N-4) conformation.
[b] The trans(N-1)trans(N-4) conformation.

H. The Azaquinolizine Systems

Data have been published for a number of pyrido[1,2-*a*]pyrimid-2-ones (**144**) (286), and these are listed in Table 4.69. Also, the spectrum of the reduced 6-one (**145**) (287) has been recorded.

TABLE 4.69 SPECTRAL DATA FOR THE PYRIDO[1,2-*a*] PYRIMID-2-ONES (286). SOLVENT—$CDCl_3$

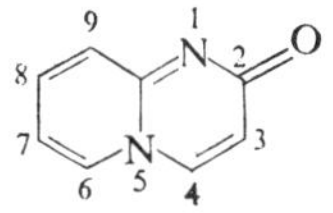

Substituents	τ3	τ4	τMe	J_{34}	J_{ArMe}
None	3.3	1.6		7.7	
6-Me	3.4	1.6	7.3	7.6	0.9
7-Me	3.3	1.7	7.6	7.1	0.9
8-Me	3.4	1.75	7.6	7.2	1.0
9-Me	3.35	1.7	7.75	7.6	1.0

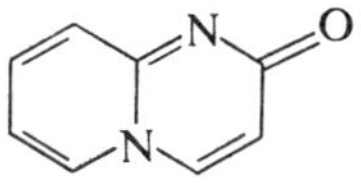

144 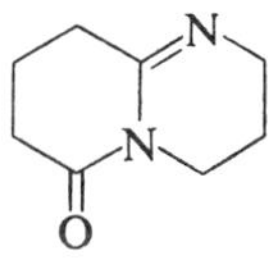 **145** 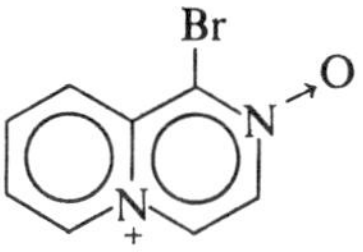 **146** 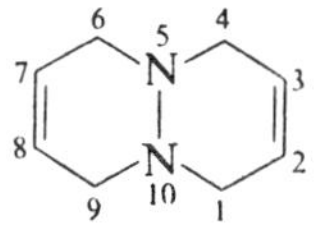 **147**

In the spectrum of the 2-oxide of the 1-bromopyrido[1,2-*a*]pyrazin-5-ium ion (**146**) in TFA, H-4 and H-6 absorb between τ 0.1 and 0.3, H-3 at τ 0.85, and the remaining ring protons between τ 1.1 and 1.7 (288).

Ring inversion of 1,4,6,9-tetrahydropyridazino[1,2-*a*]pyridazine (**147**) has been studied (285). In pyridine at 81° the methylene groups give a sharp singlet, τ 6.64. On cooling, the singlet collapses and finally, at $-20°$, resolves into two multiplets, τ 7.7 and 5.7.

I. The Naphthyridine Systems

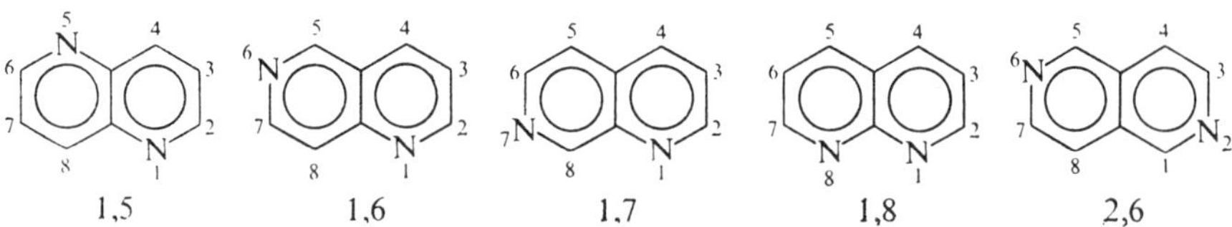

In recent years, a large amount of information on these systems has been collected, mainly by Paudler and his co-workers (289–293) in their investigations of the aromaticity of compounds containing ten π-electrons. This effort parallels the development of general synthetic methods for the various isomeric naphthyridines. Comparison of the chemical shifts of ring protons and substituents of naphthyridines with those of the corresponding naphthalenes led to the suggestion that the ring currents, and hence the "aromaticity," of the carbocyclic and heterocyclic systems were similar. Also, in many cases reasonable correlation was obtained between Brown's σp^+-substituent constants and chemical shifts of ring protons. Information on the 2,6-naphthyridine is included in a short communication dealing with its synthesis (294). Also, spectra of the 1-, 6-, and 1,6-di-*N*-oxides of 1,6-naphthyridine have been reported (295). Data for the naphthyridines and some of their simple derivatives are collected in Table 4.70.

Spectra of the cations of a number of disubstituted 1,6-naphthyridines have been reported (289). Protonation shifts have been used to show that 1,6-, 1,7-, and 1,8-naphthyridines, respectively, protonate and methylate on N-6, N-7, and N-1 (296). Data for the parent trifluoroacetates and methiodides are given in Table 4.71.

Comparison of the spectra of 2-methoxy-1,8-naphthyridine (**148**) and 1-methyl-2-oxo-1,2-dihydro-1,8-naphthyridine (**149**) with that of the 2-

148

149, R = Me
150, R = H

hydroxynaphthyridine shows conclusively that the latter exists predominantly in the oxo form **150** (297).

The cis and trans isomers of the decahydronaphthyridines can be distinguished by their NMR spectra (298). In these spectra, as with decahydroquinolines and -isoquinolines, the upfield band from ring CH_2 groups is much broader for the trans isomer than for its cis counterpart. In this work, the structures of a number of tetrahydronaphthyridines were determined with the aid of their NMR spectra.

TABLE 4.70 SPECTRAL DATA FOR NAPHTHYRIDINES. SOLVENT—$CDCl_3$

A. 1,5-NAPHTHYRIDINES

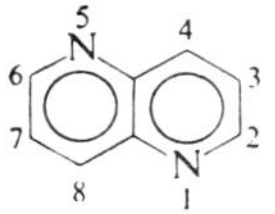

Substituents	$\tau 2$	$\tau 3$	$\tau 4$	$\tau 6$	$\tau 7$	$\tau 8$	J_{23}	J_{24}	J_{34}	J_{48}	J_{67}	J_{68}	J_{78}	Ref.
None	1.01	2.40	1.59	1.01	2.40	1.59	4.2	1.8	8.5	0.8	4.2	1.8	8.5	225[a]
	1.03	2.42	1.60	1.03	2.42	1.60	4.1	1.8	8.0		4.1	1.8	8.0	289
2-Me	7.30	2.52	1.77	1.17	2.63	1.80			8.8	0.8	4.2	1.6	8.4	291
3-Me	1.18	7.44	1.84	1.08	2.45	1.63		2.0	1.0	0.8	4.2	1.8	8.2	291
4-Me	1.03	2.44	7.14	1.08	2.28	1.47	4.4		0.9		4.2	1.8	8.7	291
2-NH_2		2.83	1.54	1.02	2.04	1.58			7.0	0.5	4.0	1.5	9.0	292
3-NO_2	0.02		0.55	0.61	2.08	1.25		2.7		0.8	4.2	1.8	8.5	225[a]
3-Br	1.04		1.44	1.04	2.37	1.63		2.0		0.9	4.3	2.0	8.6	292

B. 1,6-NAPHTHYRIDINES

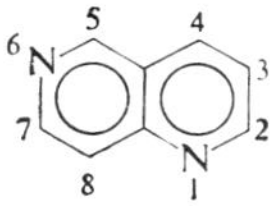

Substituents	$\tau 2$	$\tau 3$	$\tau 4$	$\tau 5$	$\tau 7$	$\tau 8$	J_{23}	J_{24}	J_{34}	J_{48}	J_{58}	J_{78}	Ref.
None	0.90	2.54	1.77	0.78	1.22	2.09	4.2	1.8	8.4	0.9		5.8	225[a]
	0.90	2.48	1.72	0.72	1.24	2.07	4.1	1.9	8.2	0.45	0.45	6.0	289
2-Me	7.24	2.63	1.86	0.81	1.26	2.17			8.2	0.9	0.9	6.0	291

TABLE 4.70 *(continued)*

Substituents	τ2	τ3	τ4	τ5	τ7	τ8	J_{23}	J_{24}	J_{34}	J_{48}	J_{58}	J_{78}	Ref.
3-Me	1.07	7.46	2.01	0.80	1.29	2.12		2.1	0.9	0.9	0.9	6.0	291
4-Me	1.03	2.68	7.24	0.47	1.20	2.06	4.5		0.9		0.8	6.0	291
2-NH_2		2.40	1.42	0.52	1.08	1.74			9.6	0.5	0.5	6.9	292
4-NMe_2	1.40	3.41		0.60	1.41	2.26	5.5					6.0	289
3-NO_2	0.09		0.60	0.27	0.80	1.71		2.6		0.9	0.8	5.9	225[a]
8-NO_2	0.47	2.06	1.34	0.30	0.60		4.2	1.9	8.5				225[a]
4-OMe	1.17	3.25		0.45	1.25	2.20	5.5				0.45	6.0	289
2-Cl		2.48	1.76	0.74	1.22	2.16			8.2			5.2	295
4-Cl	1.08	2.48		0.38	1.17	2.12	5.0				0.45	6.0	289
5-Cl	0.88	2.37	1.38		1.50	2.12	4.0		8.1			5.1	295
3-Br	0.97		1.70	0.80	1.22	2.12		1.5		0.8	0.8	6.0	292
4-Br	1.25	2.35		0.52	1.28	2.21	5.0				0.45	6.0	289
8-Br	0.83	2.40	1.70	0.83	1.02		4.0	1.5	8.3				292
1-Oxide	1.37	2.55	2.15	0.70	1.22	1.54	5.2		8.8			5.8	295
6-Oxide	0.95	2.37	1.58	1.14	1.84	1.96	4.2		8.2				295

C. 1,7-NAPHTHYRIDINES

Substituents	τ2	τ3	τ4	τ5	τ6	τ8	J_{23}	J_{24}	J_{34}	J_{48}	J_{56}	Ref.
None	1.00	2.39	1.83	2.42	1.41	0.48	4.1	1.8	8.3	0.8	5.8	225[a]
8-NH_2	1.02	2.10	1.75	2.82	2.30		4.0	1.9	8.2		7.0	292
5-Br	0.95	2.31	1.57		1.20	0.57	4.0	1.5	8.5	1.0		292

D. 1,8-NAPHTHYRIDINES

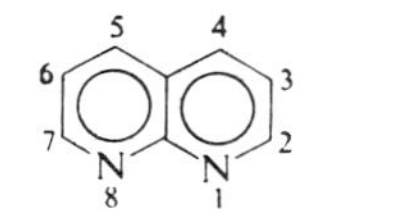

Substituents	$\tau 2$	$\tau 3$	$\tau 4$	$\tau 5$	$\tau 6$	$\tau 7$	J_{23}	J_{24}	J_{34}	J_{56}	J_{57}	J_{67}	Ref.
None	0.89	2.55	1.83	1.83	2.55	0.89	4.1	1.8	8.3	8.3	1.8	4.1	225[a]
	0.83	2.50	1.76	1.76	2.50	0.83	4.3	2.1	8.3	8.3	2.1	4.3	297
2-Me	7.24	2.69	1.92	1.99	2.65	0.97			8.4	8.4	2.0	4.4	291
	7.20	2.65	1.92	1.85	2.69	0.91			8.5	8.2	2.1	4.3	297
3-Me	1.05	7.52	2.12	1.90	2.58	0.95		2.2	1.0	8.0	2.0	4.1	291
4-Me	1.02	2.68	7.35	1.62	2.47	0.88	4.2		0.9	8.2	2.0	4.2	291
2-NH_2		2.67	1.70	1.49	2.31	1.17			9.5	8.0	1.7	5.3	292
2-OMe	5.82	3.01	1.96	1.88	2.65	1.00			8.9	8.1	2.2	4.6	297
3-Br	0.90		1.67	1.89	2.50	0.90		2.0		8.0	2.0	4.0	292

E. 2,6-NAPHTHYRIDINES (294)

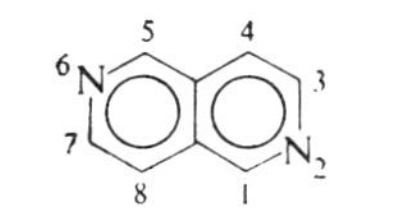

Substituent	$\tau 1, 5$	$\tau 3, 7$	$\tau 4, 8$
None	0.61	1.25	2.22

[a] Data in Ref. 225 have been adjusted to correct a calibration error.

TABLE 4.71 SPECTRAL DATA FOR NAPHTHYRIDINE SALTS (296)

Isomer[a]	Solvent	$\tau 2$	$\tau 3$	$\tau 4$	$\tau 5$	$\tau 6$	$\tau 7$	$\tau 8$	J_{23}	J_{24}	J_{34}	J_{48}	J_{56}	J_{57}	J_{58}	J_{67}	J_{68}	J_{78}
1,6-N	TFA	0.25	1.65	0.68	−0.15		0.82	1.13	4.8	1.8	8.7	0.8		0.8	0.8			6.7
MeI	D_2O	0.52	1.89	0.97	0.00		1.09	1.42	4.3	1.7	8.6	0.5		1.2	0.5			7.0
1,7-N	TFA	0.52	1.73	1.23	1.38	1.13		1.05	4.3	1.7	8.8	0.8	6.5		0.7		0.7	
MeI	D_2O	0.49	1.69	1.09	1.07	1.24		0.30	4.2	1.9	9.0	0.9	7.0		0.8		0.7	
1,8-N	TFA	0.39	1.62	0.71	0.71	1.62	0.39		4.5	2.0	8.0		8.0	2.0		4.5		
MeI	D_2O	0.23	1.58	0.57	0.97	1.77	0.51		5.5	1.6	8.5		8.5	1.8		4.4		

[a] N is naphthyridine.

J. The Triazanaphthalene Systems

1. *1,2,4-Benztriazine*

The spectra of a number of 7-substituted 3-hydroxy-1,2,4-benztriazine-1-oxides (**151**) have been reproduced in a paper dealing with the synthesis of these compounds (299).

2. *The Diazoquinolizines*

The 3-phenyl derivative of 4*H*-pyrido[2,1-*c*]-*as*-triazine (**152**) in $CDCl_3$ gives a singlet at τ 4.81 from the 4-methylene group, and a broad band from

151 **152** **153**

τ 6.15 to 7.90 containing the signals from all of the remaining protons (300). Also, the spectrum of the pyrimido[1,2-*b*]pyridaz-4-one (**153**) in $CDCl_3$ has been reported (301): $\tau 2$ 1.00, $\tau 7$ 1.12, $\tau 8$ 2.27, $\tau 9$ 1.96, J_{78} 4.5, J_{79} 2.5, J_{89} 9.5.

3. *The Pyridopyrimidine Systems*

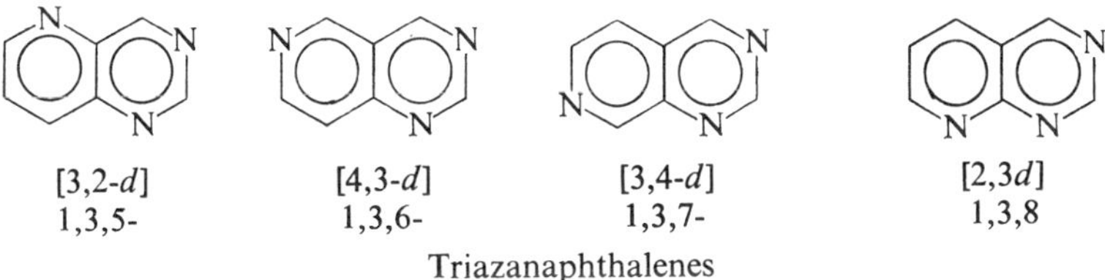

Triazanaphthalenes

Spectra of the parent compounds of these ring systems have been analyzed in a general study of di- and triazanaphthalenes (225). Unequivocal assignments of peaks to H-2 and H-4 were made by examining the spectra of the four 4-deutero derivatives. Data for these heterocycles are collected in Table 4.72.

The spectrum of 4-hydroxy-1,3,6-triazanaphthalene shows it to exist in the 3,4-dihydro-oxo form **154** (302), and that of the pyridouracil (**155**) has been reported (303).

TABLE 4.72 SPECTRAL DATA FOR THE PYRIDOPYRIMIDINES (225)

Isomer	Solvent	$\tau 2$	$\tau 4$	$\tau 5$	$\tau 6$	$\tau 7$	$\tau 8$	J_{48}	J_{56}	J_{57}	J_{58}	J_{67}	J_{68}	J_{78}
[3,2-*d*]	$CDCl_3$	0.45	0.20		0.78	2.05	1.50	0.8				4.1	1.7	8.3
	Me_2CO	0.51	0.27		0.74	1.89	1.48	0.8				4.2	1.8	8.3
	DMSO	0.50	0.25		0.76	1.84	1.44	0.8				4.1	1.8	8.4
[4,3-*d*]	$CDCl_3$	0.36	0.28	0.43		0.90	1.98	0.7			0.8			5.8
	Me_2CO	0.39	0.09	0.30		0.90	2.02	0.7			0.8			5.8
[3,4-*d*]	$CDCl_3$	0.33	0.32	2.11	1.05		0.35		5.8					
[2,3-*d*]	$CDCl_3$	0.33	0.33	1.46	2.18	0.57			8.3	1.9		4.2		

4. *The Pyridopyrazine Systems*

[2,3-*b*]
1,4,5-

[3,4-*b*]
1,4,6-

Triazanaphthalenes

Spectra of the cations of the two parent compounds have been reported in a study of "covalent hydration" of condensed pyrazine heterocycles (304). In anhydrous TFA, the normal cations were formed. However, with DCl/D_2O as solvent, hydration occurred across the 1,2- and 3,4-bonds of the 1,4,6-triazanaphthalene cation to give the 2,3-dihydroxy-1,2,3,4-tetrahydro derivative **156**. The cation of the 1,4,5-triazanaphthalene, previously thought to be anhydrous, was shown to hydrate to a small extent in a similar manner. The very large changes in spectra which occur when these cations are hydrated comprise an extremely good method for detecting this reaction. Data for the anhydrous and hydrated cations are given in Table 4.73. Spectra of substituted 1,4,6-triazanaphthalenes have also been reported (305).

TABLE 4.73 SPECTRAL DATA FOR CATIONS OF 1,4,5- AND 1,4,6-TRIAZANAPHTHALENES (304)

Compound	Species[a]	τ2	τ3	τ5	τ6	τ7	τ8
1,4,5-Triazanaphthalene	AC	0.50	0.50		0.50	1.42	0.50
	DH	4.72,	4.83		2.45	2.90	2.50
1,4,6-Triazanaphthalene	AC	0.29,	0.36	−0.30		0.80	1.00
	DH	4.85,	4.93	2.22		2.16	3.04
3-Me, 1,4,6-triazanaphthalene	AC	0.62	6.95	0.10		0.92	1.50
	DH	4.93	8.23	2.10		2.08	2.98

[a] AC—anhydrous cation in TFA; DH—dihydrated cation in DCl/D_2O.

5. *The Pyridopyridazine Systems*

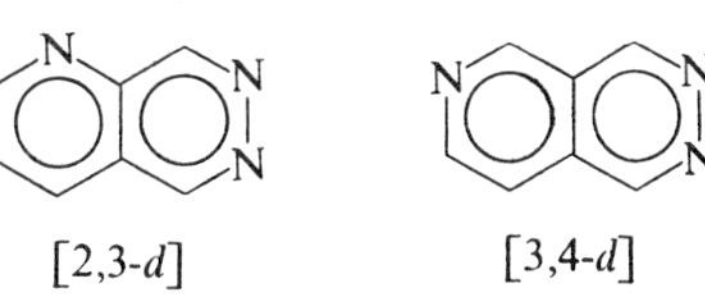

[2,3-*d*]

[3,4-*d*]

A number of reports have appeared recently on the synthesis of these ring systems (306–308) and, in these, spectra were used as the main means of

TABLE 4.74 SPECTRAL DATA FOR THE PYRIDOPYRIDAZINES

A. [2,3-*d*] ISOMER

Substituents	Solvent	τ2	τ3	τ4	τ5	τ8	τNH	J_{23}	J_{24}	J_{34}	J_{48}	J_{58}	Ref.
None	$CDCl_3$	0.69	2.12	1.60	0.37	0.21		4.5	1.7	8.5	0.75	1.5	306
5-Me	$CDCl_3$	0.77	2.18	1.57	6.97	0.32		4.3	1.7	8.5	0.7		306
8-Me	$CDCl_3$	0.70	2.18	1.70	0.52	6.85		4.3	1.8	8.5			306
5,8-Di-Cl	$CDCl_3$	0.58	1.97	1.34									307
6-Oxide	$CDCl_3$	0.98	2.27	1.88	1.37	0.71		4.2	1.7	8.5	0.7	1.2	310
	DMSO	1.20	2.40	1.90	1.30	0.80							308
7-Oxide	$CDCl_3$	0.88	2.42	1.72	0.88	1.25		4.3	1.7	8.3	0.8	1.4	310
	DMSO	0.95	2.40	1.55	0.65	1.26							308
5-Me, 5,6-di-H	$CDCl_3$	1.5	2.7	2.7	5.4, 8.5	2.4	3.0	4.5	1.8				306
8-Me, 7,8-di-H	$CDCl_3$	1.62	2.7	2.7	2.7	5.5	3.9						306
4-Oxo, 5,6-di-H	DMSO	0.80	2.11	1.35		1.53	−2.99	4.6	1.8	8.1			309
8-Oxo, 7,8-di-H	DMSO	0.86	2.02	1.52	1.53			4.4	1.8	8.2			309

B. [3,4-*d*] ISOMER

Substituents	Solvent	τ1	τ4	τ5	τ7	τ8	J_{15}	J_{48}	J_{58}	J_{78}	Ref.
None	$CDCl_3$	0.27	0.27	0.47	0.90	2.11	0.5	0.75	0.5	5.5	306
1,4-Di-Cl	$CDCl_3$			0.25	0.75	1.92					307

determining the position of substitution. In addition, as part of a study of the ring system by Paul and Rodda, spectra of the pyrido[2,3-*d*]pyridazine-5- and 8-ones (**157** and **158**) (309) and the 6- and 7-*N*-oxides (**159** and **160**)

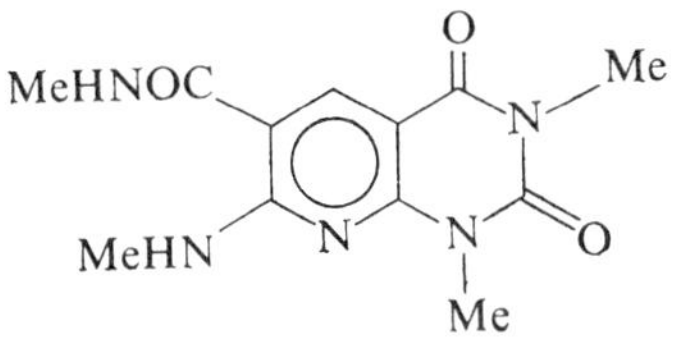

MeHNOC, MeHN, O, Me, N, N, N, O, Me

155

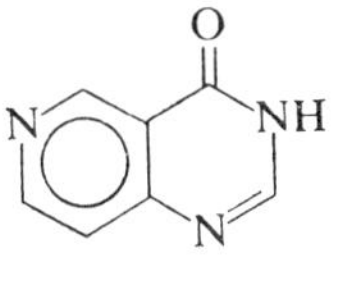

O, N, NH, N

154

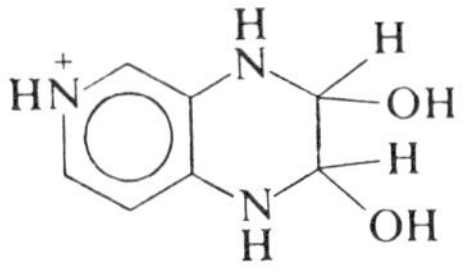

H, H, $H\overset{+}{N}$, N, OH, H, N, H, OH

156

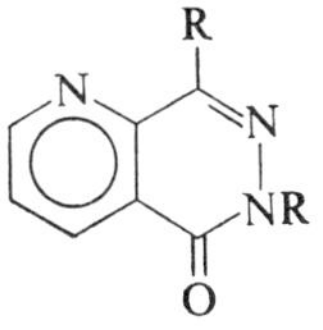

R, N, N, NR, O

157

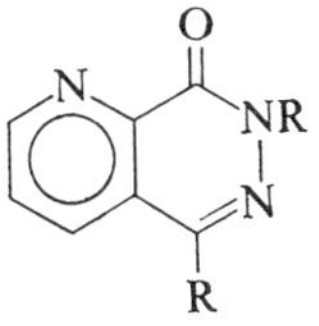

O, N, NR, N, R

158

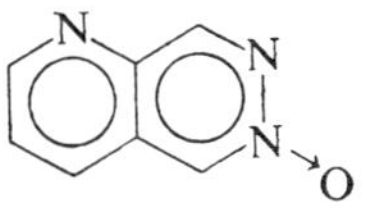

N, N, N, O

159

N, N, O, N

160

(310) have been described. Information is also available for a number of dihydro (308) and tetrahydro (283) derivatives. Data for these compounds are given in Table 4.74.

K. The Pyridazino, Pyrimido, and Pyrazinopyridazine Systems

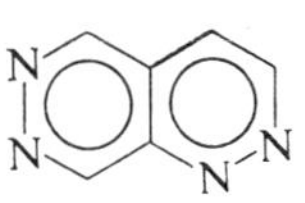

Pyridazino [4,5-*c*]

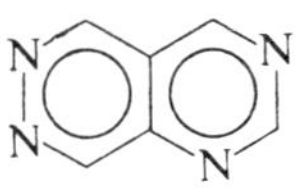

Pyrimido [4,5-*d*]

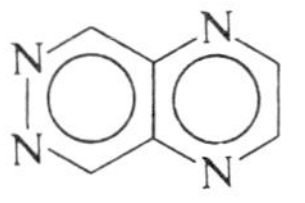

Pyrazino [2,3-*d*]

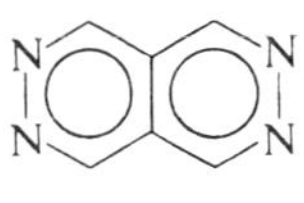

Pyridazino [4,5-*d*]

A few data for substituted members of these series are spread throughout a number of publications dealing with the synthesis of these ring systems (311–315).

L. The Pyrimido[4,5-*b*]pyrazine or Pteridine System

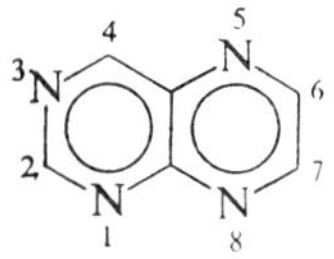

The spectrum of pteridine contains two singlets which can be assigned to the protons of the pyrimidine ring and a pair of doublets from the adjacent H-6 and H-7 (316). Further assignment of peaks to specific protons was made, in the pyrimidine ring by examining specifically deuterated derivatives (145), and in the pyrazine ring by evaluating the effects of methyl substituents in quinoxalines and in the available 7-methylpteridine. Thus, it was shown that in $CDCl_3$ solutions the protons absorbed in the order: H-6, H-7, H-2, H-4, from highest to lowest field. Calculations of this order from π-electron densities have not been successful (148).

TABLE 4.75 SPECTRAL DATA FOR PTERIDINES AND THEIR CATIONS

A. NEUTRAL MOLECULES. SOLVENT—$CDCl_3$

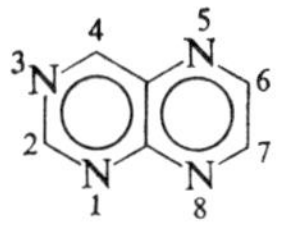

Substituents	$\tau 2$	$\tau 4$	$\tau 6$	$\tau 7$	J_{67}	Ref.
None	0.35	0.20	0.85	0.67	1.7	316
2-Me	6.97	0.33	0.97	0.77	1.7	316
4-Me	0.57	6.87	0.92	0.73	1.7	316
7-Me	0.43	0.32	1.02	7.04		316
2-Ph		0.34	1.15	0.88		323
2-Ph, 4-Me		6.85	1.08	0.83		323
2-Ph, 7-Me		0.32	1.19	7.14		323

B. 3,4-HYDRATED CATIONS. SOLVENT—DCl IN D_2O (317)

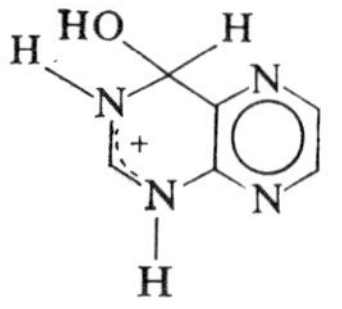

Substituents	$\tau 2$	$\tau 4$	$\tau 6^{a}$	$\tau 7^{a}$
None	1.23	3.40	1.26	1.33
2-Me	7.33	3.48	1.22	1.28
7-Me	1.25	3.43	1.32	7.32
5,6-Di-Me	1.20	3.42	7.30	7.30
2,6,7-Tri-Me	7.30	3.52	7.30	7.30
2-NH_2		3.70	1.41	1.41
6-NH_2	1.10	3.66		1.42
2-OH		3.68	1.40	1.50
6-Cl[b]	1.20	3.42		1.18
7-Cl[c]	1.20	3.38	1.18	

TABLE 4.75 (*continued*)

C. 5,6,7,8-DIHYDRATFD CATIONS. SOLVENT—DCl IN D_2O

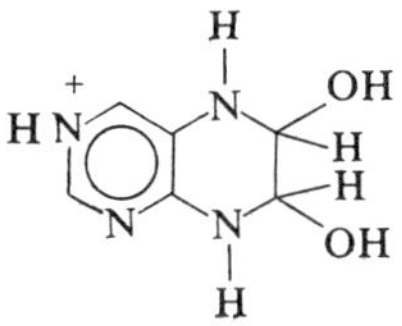

Substituents	τ2	τ4	τ6[a]	τ7[a]	Ref.
None	1.53	2.12	4.65	4.80	317
2-Me	7.40	2.23	4.70	4.84	317
4-Me	1.54	7.48	4.72	4.80	317
7-Me	1.55	2.16	5.00	8.26	317
2-Ph		2.38	4.85	4.68	323
2-OMe	5.10	2.38	4.68	4.82	317
2-Cl		2.23	4.79	4.87	317

D. ANHYDROUS CATIONS. SOLVENT—DCl IN D_2O (317)

Substituents	τ2	τ4	τ6[a]	τ7[a]
4-Me	0.65	6.94	0.65	0.80
2-NH_2, 4-Me		6.85	0.72	0.82
4-NH_2	1.20		0.75	0.86
4-OH	0.78		0.78	0.78
2-Cl		0.11	0.55	0.70

[a] Values for τ6 and τ7 may be interchanged.
[b] Values for τ2 and τ7 may be interchanged.
[c] Values for τ2 and τ6 may be interchanged.

Apart from the parent compound and its monomethyl derivatives, almost all of the simple pteridines whose spectra have been reported come from studies of the addition of water or other nucleophiles across the C=N bonds of this ring system. As mentioned in the discussion of triazanaphthalenes (Section VIII.J.3), NMR gives a direct and unequivocal measure of the amount and position of "covalent hydration" in these compounds. The available data (316–323), collected in Table 4.75, require little discussion.

Most of the remaining information on pteridine spectra involves studies of the structure, tautomerism, protonation, and hydration of the many oxo and imino derivatives of this ring system. The major contribution to this field is an exhaustive report of the spectra of 2-amino-4-oxo-3,4-dihydropteridine cations **161** (324). Protonation always occurs on N-1 or N-3 and a reasonable correlation was obtained between the protonation shift of the 2-NH_2 group and the pK_a values for the monoprotonation step. The preferred tautomeric structures of 9-acetonylxanthopterin (**162**) and erythropterin (**163**) were

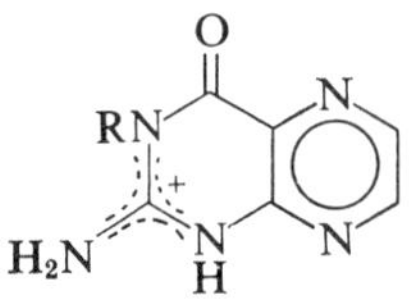

161

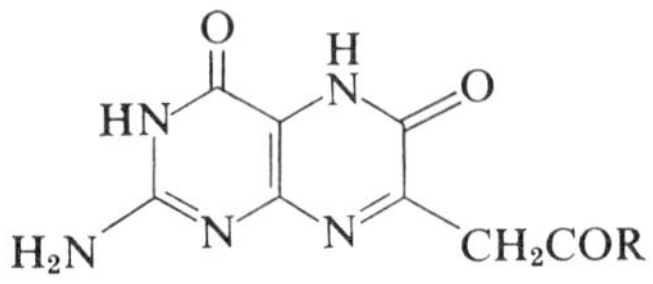

162, R = Me
163, R = CO_2H

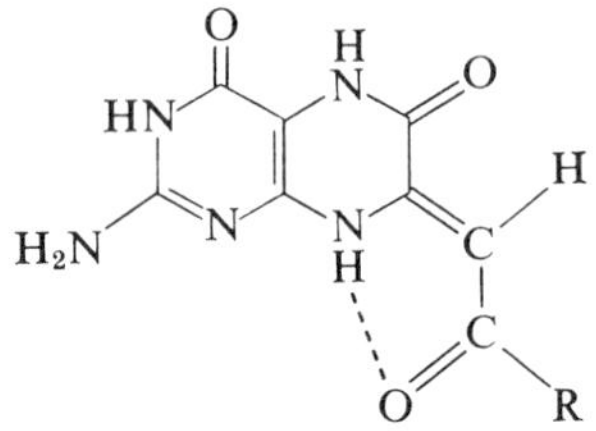

164, R = Me
165, R = CO_2H

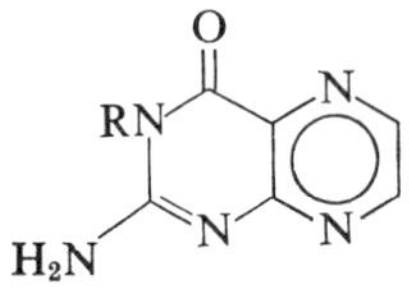

166 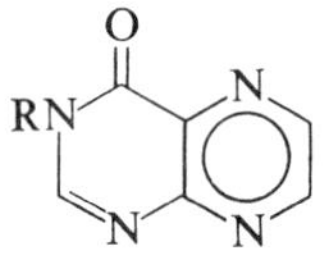 **167** 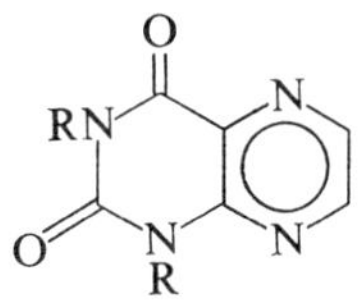 **168**

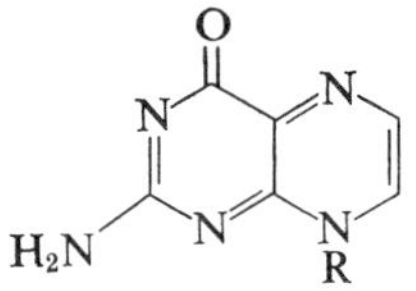

169 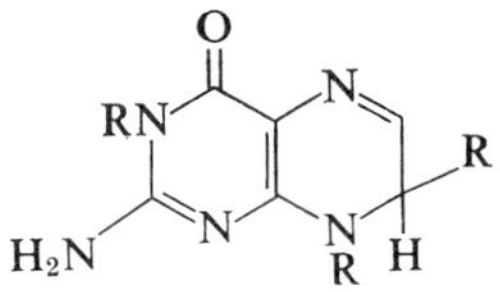 **170** 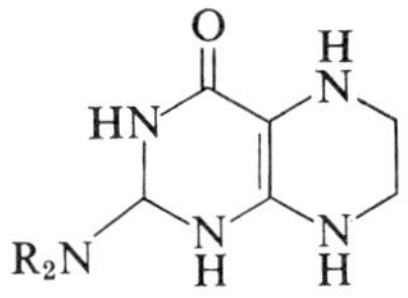 **171**

shown to be **164** and **165**, respectively (325, 326). Data are also available for pterins (**166**) (327), 4-oxo-3,4-dihydropteridines (**167**) (328), lumazines (**168**) (329), and for the anhydrous and hydrated intermediates in the complex protonation–deprotonation equilibria of 8-substituted pterins (**169**) (330). Characteristic data for these compounds are given in Table 4.76.

Apart from the occasional compound in the references above, a number of publications are available which deal specifically with 7,8-dihydropterins (**170**) (331, 332) or 5,6,7,8-tetrahydropteridines (**171**) (283, 333).

TABLE 4.76 SPECTRAL DATA FOR OXO- OR IMINOPTERIDINES[a] (324). SOLVENT—TFA

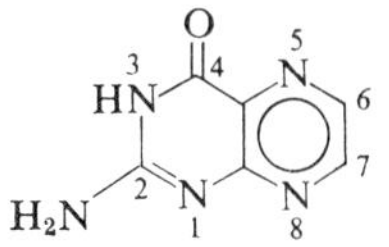

Substituents	τ6, 7	J_{67}	τ2-N*H*R	τMe
None	0.97		1.28	
2-NHMe	0.98		1.18	6.58, 6.53
2-NMe_2	0.96, 1.03	2.2		6.36, 6.32
6-Me	1.08		1.42	7.14
7-Me	1.08		1.39	7.10

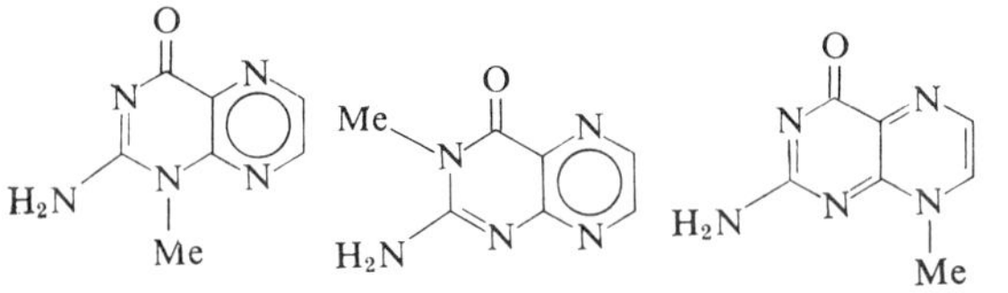

A B C

Compound	τ6, 7	J_{67}	τ2-N*H*R	τMe
A	0.85, 0.98		1.55	5.86
B	0.94		1.42	6.15
C	1.15, 1.22	3.5	2.67	5.61

[a] These compounds are only slightly soluble as neutral molecules. Data for simple oxo- and iminopteridines as cations are included in Table 4.75.

M. The Pyrazo[2,3-*b*]pyrazine System (1,4,5,8-Tetraazanaphthalene)

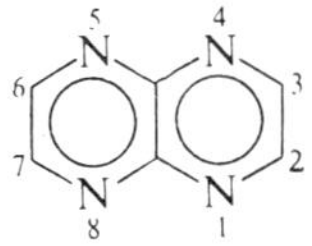

The spectra of the hydrated and anhydrous cations of the parent heterocycle and some of its methyl derivatives were included in an investigation of covalent hydration of condensed pyrazines (304). Data are collected in Table 4.77.

TABLE 4.77 SPECTRAL DATA FOR THE CATIONS OF PYRAZO[2,3-*b*]-PYRAZINES (304)

Substituents	Species[a]	τ2	τ3	τ7	τ8
None	AC	0.51	0.51	0.51	0.51
	DC	4.63	4.63	2.48	2.48
2-Me	AC	7.03	0.62	0.62	0.62
	DC	7.60	2.63	4.65	4.65
2,3-Di-Me	AC			0.80	0.80
	DC			4.60	4.60

[a] AC—anhydrous cation in TFA; DC—dihydrated cation in DCl/D_2O.

N. The Pyrimido[5,4-*e*]-*as*-triazine System

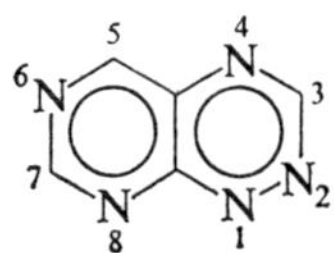

Interest in this system stems from its presence in the antibiotics toxoflavin (**172**) and fervenulin (**173**). Unfortunately, attempts to prepare the parent

172

173

TABLE 4.78 SPECTRAL DATA FOR PYRIMIDO [5,4-*e*]-*as*-TRIAZINES. SOLVENT—DMSO

Substituents	τ3	τ5	τ7	τNH	J_{23}	Ref.
5-NH_2	−0.11	1.12	1.32			336
5-(6*H*)-One	0.08		1.52	2.0		336
5,7-(6*H*,8*H*)-Dione	0.32			2.13		336
1,2-Di-H	3.61	3.08	2.25	1.92, 1.20	3.0	334
1,2-Di-H, 3-Me	8.44	3.10	2.27	1.83, 1.20		334
1,2-Di-H, 7-Me	3.74	3.30	7.99	2.12, 1.41	3.0	344

compound were foiled by the extreme ease with which it adds nucleophiles (334). However, data are available for a number of dihydro derivatives (334) and for amino and dihydro-oxo compounds related to the antibiotics (335, 336). Representative data are collected in Table 4.78.

IX. THE 5,7 SYSTEMS

A. The 1,3-Diazaazulene System

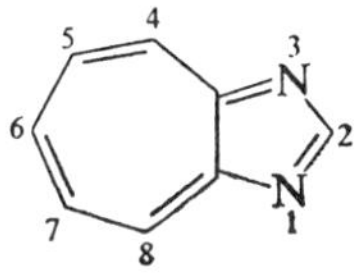

Spectra of a number of 4-substituted 4*H*- (**174**) and 6-substituted 6*H*- (**175**) derivatives have been reproduced in a report on the action of Grignard reagents with the parent heterocycle (337).

X. THE 6,7 SYSTEMS

A. The Benzazepine Systems

Data from these classes of heterocycles are scarce and spread widely through literature dealing mainly with other types of compounds. For this reason it has been extremely difficult to collect information on the various benzazepines, and while the following discussion contains the leading references in this field, it cannot be classed as complete.

The 1-benzazepines are represented by the 2-ethoxy-1-methyl-3,4-bismethoxycarbonyl-1*H* derivative **176** (338). Hindered rotation of the ethoxy group was shown by the nonequivalence of the methylene protons at low temperatures.

In the 2-benzazepine series, the spectrum of the tetrahydro derivative **177** has been reported (339). Also, the spectra of a number of 1*H*- and 3*H*-isomers of 3-benzazepines, **178** and **179**, respectively, have been reproduced, but very few numerical data were given (340).

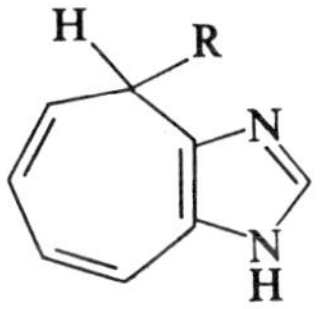

174 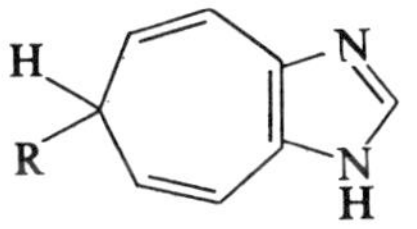 **175** 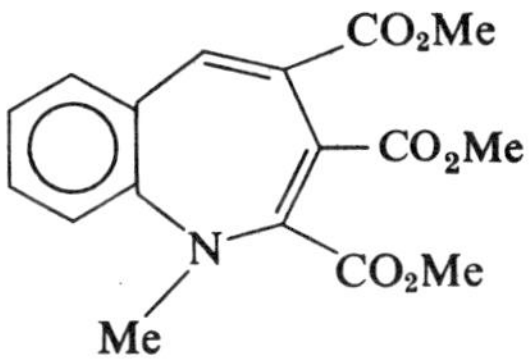 **176**

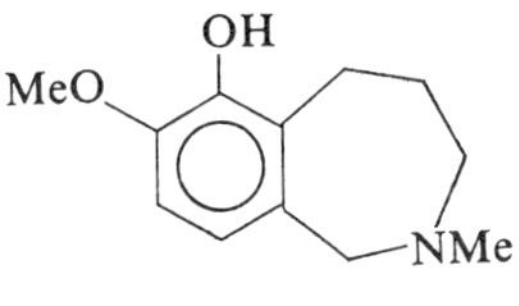

177 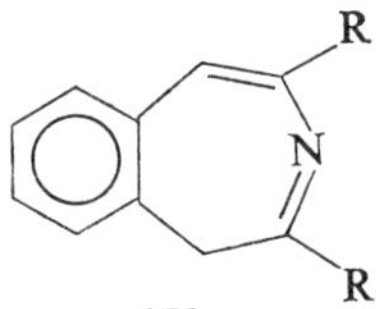 178 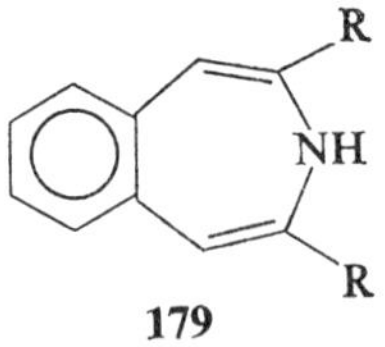 179

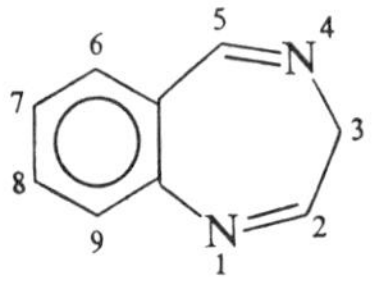

181 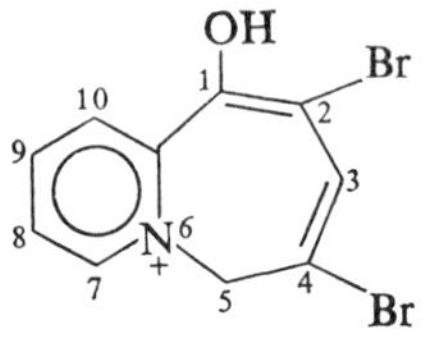 180 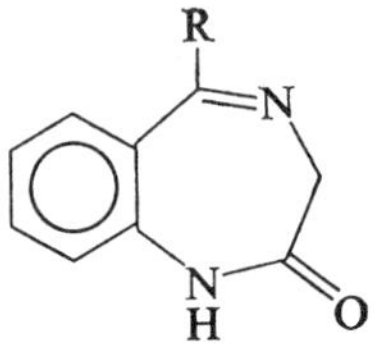 182, R = H
184, R = Ph

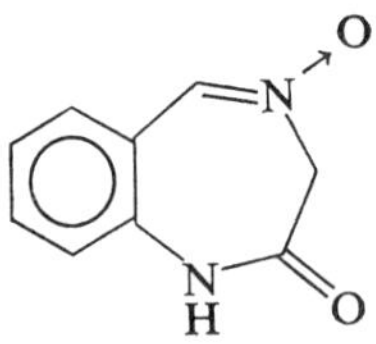

183 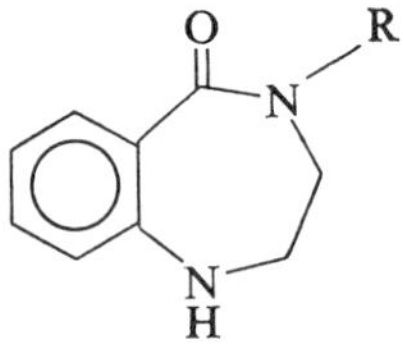 185 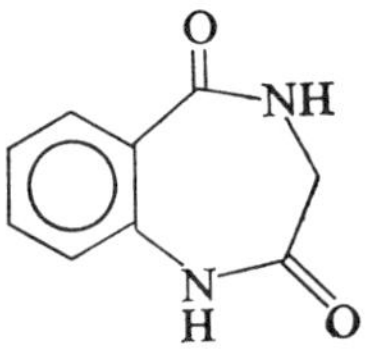 186

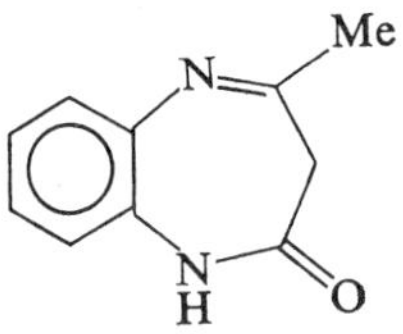

187 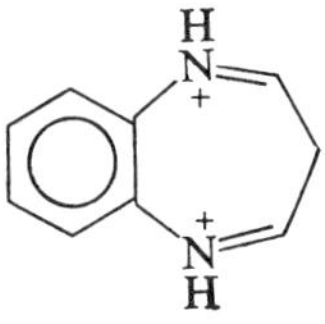 188 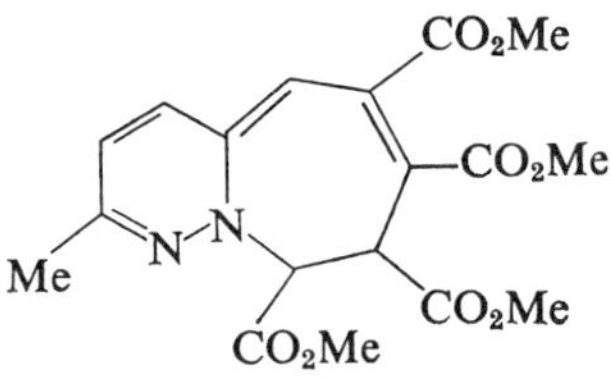 189

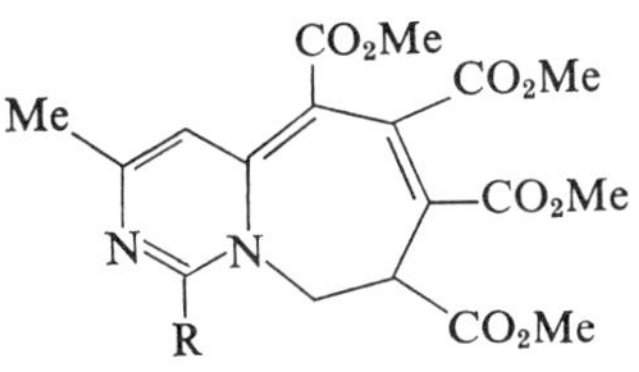

190 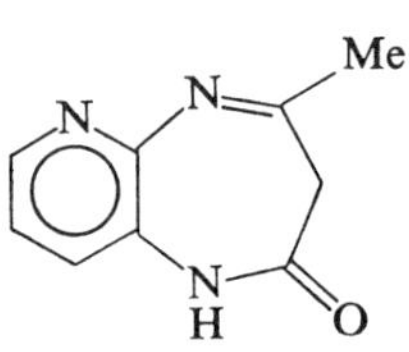 191

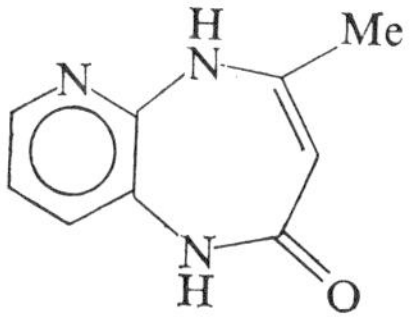

192

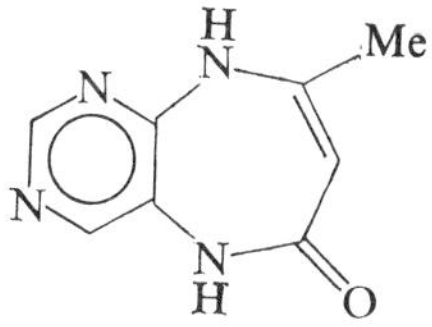

193

B. The Pyrido[1,2-*a*]azepine System

The spectrum of the 2,4-dibromo-1-hydroxy-5*H*-pyrido[1,2-*a*]azepinium bromide (**180**) in D_2O has been reported (341): τ3 2.85, τ5 4.6, τ7 1.0, τ8–10 1.5–2.0.

C. The Benzodiazepine Systems

Within this class of compounds, the NMR method has been applied mainly to the study of conformational equilibria involving the seven-membered rings of various amides, particularly those of the 1,4-benzodiazepine system **181**.

Data have been reported for the simple 1,2-dihydro-2-oxo-(3*H*)-1,4-benzodiazepine (**182**), its 4-oxide (**183**), and a number of reduced derivatives (342). Two variable-temperature studies have been made of the spectra of the biologically active 5-phenyl derivatives of this type of amide (**184**) (343, 344). Thermodynamic constants for the inversion were calculated and it was suggested that, in the preferred conformation, the seven-membered ring assumed a boat shape.

Some information is available for the 4-aralkylamino-1,2,3,4-5-oxo-tetrahydro-(5*H*)-1,4-benzodiazepines (**185**) (345) and for 3-substituted 2,5-dioxo-1,2,3,4-tetrahydro-(5*H*)-1,4-benzodiazepines (**186**) (346).

The spectrum of 1,2-dihydro-4-methyl-2-oxo-3*H*-1,5-benzodiazepine (**187**) has been reported (347). Also, the dication of 2,4-dimethyl-1,5-benzodiazepine, in concentrated H_2SO_4, was shown to exist predominantly as the 3*H*-tautomer **188** (348). Data for benzodiazepines are summarized in Table 4.79.

D. 6,7 Systems with Nitrogen Atoms in Both Rings

Acheson and his co-workers (349, 350), in their study of the addition of dimethylacetylenedicarboxylate to heterocyclic systems, have prepared the substituted dihydropyridazino[2,3-*a*]azepine (**189**) and dihydropyrimido-[3,4-*a*]azepine (190).

TABLE 4.79 SPECTRAL DATA FOR BENZODIAZEPINES AND THEIR ANALOGS

A. 1,4-ISOMER (342)

Substituents	Solvent	$\tau 1$	$\tau 3$	$\tau 5$	$\tau 4$
None	DMSO	−0.75	5.93	1.25	
4-Oxide	DMSO	−0.2	5.50	1.88	
3-OAc	DMSO		4.28, 7.78	1.37	
4,5-Di-H	$CDCl_3$	0.97	6.17, 5.98	6.17, 5.98	8.0
4-OH, 5-H	DMSO	0.02	6.07, 6.60	6.07, 6.60	1.60

B. 1,5-ISOMER (347). SOLVENT—$CDCl_3$

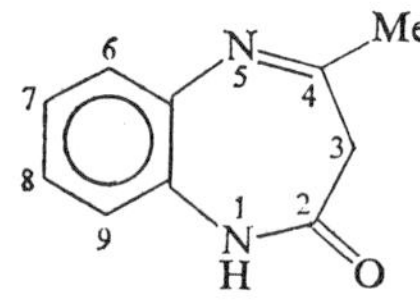

$\tau 1$	$\tau 3$	$\tau 4$	$\tau 6$–8
0.35	6.9	7.64	2.85

C. (352)SOLVENT— DMSO

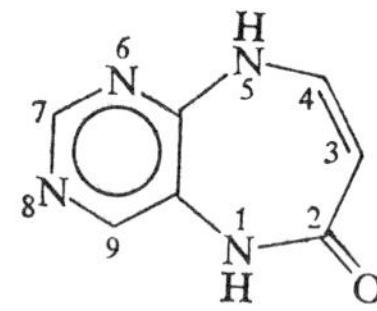

$\tau 3$	$\tau 4$	$\tau 7, 9$
5.42	8.13	1.70, 1.97

The 3*H*- and 5*H*-isomers of 1,2-dihydro-4-methyl-2-oxo-pyrido[2,3-*b*]-diazepine (**191** and **192**) have been isolated and are easily distinguished by the NMR signal from the 3-protons (351). Also, the spectrum of the 8-aza derivative of this amide (**193**) has been recorded (352). The available data for these compounds are included in Table 4.79.

REFERENCES

1. Funke, *Angew, Chem., Int. Ed. Engl.*, **8,** 70 (1969); *Chem. Ber.*, **102,** 3148 (1969).
2. Hortmann and Robertson, *J. Am. Chem. Soc.*, **89,** 5975 (1967).
3. Woerner, Reimlinger, and Arnold, *Angew. Chem., Int. Ed. Engl.*, **7,** 130 (1968).
4. Logothetis, *J. Am. Chem. Soc.*, **87,** 749 (1965).
5. Heine, Weese, and Cooper, *J. Org. Chem.*, **32,** 2708 (1967).

5a. Oida and Ohki, *Chem. Pharm. Bull.*, **17,** 980 (1969).

6. Paquette and Kuhla, *J. Org. Chem.*, **34,** 2885 (1969); Childs and Johnson, *J. Chem. Soc., C*, **1966,** 1950.
7. Paquette, *J. Am. Chem. Soc.*, **86,** 500 (1964).
8. Chapman and Hoganson, *J. Am. Chem. Soc.*, **86,** 498 (1964).
9. Moore, Marascia, Medieros, and Wineholt, *J. Org. Chem.*, **31,** 34 (1966).
10. Moore, Medieros, and Williams, *J. Org. Chem.*, **31,** 52 (1966).
11. Flitsch, Heidhues, and Paulson, *Tetrahedron Lett.*, **1968,** 1181.
12. Schweizer and Light, *J. Org. Chem.*, **31,** 870 (1966).
13. Meinwald and Meinwald, *J. Am. Chem. Soc.*, **88,** 1305 (1966).
14. Culvenor, Heffernan, and Woods, *Aust. J. Chem.*, **18,** 1605 (1965).
15. Culvenor and Woods, *Aust. J. Chem.*, **18,** 1625 (1965).
16. Skvortsov and Elvidge, *J. Chem. Soc., B*, **1968,** 1589.

16a. Meyer and Sapianchiay, *J. Am. Chem. Soc.*, **86,** 3343 (1964).

17. (a) Trofimenko, *J. Am. Chem. Soc.*, **87,** 4394 (1965); (b) Solomons and Voigt, *ibid.*, **87,** 5256 (1965); (c) Solomons, Fowler, and Calderazzo, *ibid.*, **87,** 528 (1965).
18. Felix and Eschenmoser, *Angew. Chem., Int. Ed. Engl.*, **7,** 224 (1968).
19. (a) Maier, *Chem. Ber.*, **98,** 2438, 2446 (1965); (b) Amiet, Johns, and Markham, *Chem. Commun.*, **1965,** 128; (c) Sauer and Heinricks, *Tetrahedron Lett.*, **1966,** 4979; (d) Chapman, Kane, Lassila, Loeschen, and Wright, *J. Am. Chem. Soc.*, **91,** 6858 (1969).
20. Cohen, Daly, Kny, and Witkop, *J. Am. Chem. Soc.*, **82,** 2184 (1960).
21. Jardine and Brown, *Can. J. Chem.*, **41,** 2067 (1963).
22. Hutton and Waters, *J. Chem. Soc.*, **1965,** 4253.
23. Elvidge and Foster, *J. Chem. Soc.*, **1964,** 981.
24. Black and Heffernan, *Aust. J. Chem.*, **18,** 353 (1965).
25. Black, Brown, and Heffernan, *Aust. J. Chem.*, **20,** 1325 (1967).
26. Reinecke, Johnson, and Sebastian, *Chem. Ind.*, **1964,** 151.
27. Noland and Sundberg, *J. Org. Chem.*, **28,** 884 (1963).
28. Powers, *J. Org. Chem.*, **31,** 2627 (1966).
29. Daly and Witkop, *J. Am. Chem. Soc.*, **89,** 1032 (1967).
30. Heacock, Hutzinger, Scott, Daly, and Witkop, *J. Am. Chem. Soc.*, **85,** 1825 (1963); Allen, Pidacks, and Weiss, *ibid.*, **88,** 2536 (1966).
31. Noland and Rush, *J. Org. Chem.*, **31,** 70 (1966).
32. Brooke and Rutherford, *J. Chem. Soc., C*, **1967,** 1189.
33. McDonald and Phillips, *J. Am. Chem. Soc.*, **89,** 6332 (1967).

34. Marki, Robertson, and Witkop, *J. Am. Chem. Soc.*, **83,** 3341 (1961).
35. Mousseron-Canet and Boca, *Bull. Soc. Chim. Fr.*, **1965,** 2438.
36. Mousseron-Canet and Boca, *Bull. Soc. Chim. Fr.*, **1967,** 1296.
37. Kessler and Zeeh, *Tetrahedron*, **24,** 6825 (1968).
38. Hinman and Whipple, *J. Am. Chem. Soc.*, **84,** 2534 (1962).
39. Jackson and Smith, *J. Chem. Soc.*, **1964,** 5510.
40. Reinecke, Johnson, and Sebastian, *Tetrahedron Lett.*, **1963,** 1183.
41. Sebastian, Reinecke, and Johnson, *J. Phys. Chem.*, **73,** 455 (1969).
42. Kasparek and Heacock, *Can. J. Chem.*, **44,** 2805 (1966).
43. Allen, Pidacks, and Weiss, *J. Am. Chem. Soc.*, **88,** 2536 (1966).
44. Daisley and Walker, *J. Chem. Soc.*, *B*, **1969,** 146.
45. Bourdais, *Bull. Soc. Chim. Fr.*, **1968,** 1506.
46. Hino, Nakagawa, and Akaboshi, *Chem. Commun.*, **1967,** 656.
47. Hinman and Bauman, *J. Org. Chem.*, **29,** 2431 (1964).
48. Baumgarten, Creger, and Zey, *J. Am. Chem. Soc.*, **82,** 3977 (1960).
49. Rees and Sabet, *J. Chem. Soc.*, **1965,** 870.
50. Hudson and Robertson, *Aust. J. Chem.*, **20,** 1935 (1967).
51. Nagarajan, Nair, and Pillai, *Tetrahedron*, **23,** 1683 (1967).
52. Nagarajan and Nair, *Tetrahedron*, **23,** 4493 (1967).
53. McLean, *Can. J. Chem.*, **38,** 2278 (1960).
54. Hassner and Haddadin, *J. Org. Chem.*, **28,** 224 (1963).
55. Mishra and Swan, *J. Chem. Soc.*, *C*, **1967,** 1424.
56. Remers, Gibs, Pidacks, and Weiss, *J. Am. Chem. Soc.*, **89,** 5513 (1967).
57. Acheson, *J. Chem. Soc.*, **1965,** 2630.
58. Veber and Lwowsky, *J. Am. Chem. Soc.*, **85,** 646 (1963).
59. Fletcher, *Tetrahedron*, **22,** 2481 (1966).
60. Bergel and Peutherer, *J. Chem. Soc.*, **1964,** 3973.
61. Bonnett and White, *J. Chem. Soc.*, **1963,** 1648.
62. Bender and Bonnett, *J. Chem. Soc.*, *C*, **1968,** 2186.
63. Gerig, *Tetrahedron Lett.*, **1967,** 4625.
64. Wynberg and Klunder, *Rec. Trav. Chim.*, **88,** 328 (1969).
65. Black, Heffernan, Jackman, Porter, and Underwood, *Aust. J. Chem.*, **17, 1128 (1964).**
66. Higham and Richards, *J. Chem. Soc.*, **1960,** 1696.
67. Bragg and Wibberly, *J. Chem. Soc.*, *C*, **1966,** 2120.
68. Acheson and Robinson, *J. Chem. Soc.*, *C*, **1968,** 1633.
69. Acheson, Snaith, and Vernon, *J. Chem. Soc.*, **1964,** 3229.
70. Acheson and Tulley, *J. Chem. Soc.*, *C*, **1968,** 1623.

71a. Fraser, Melera, Molloy, and Reid, *J. Chem. Soc.*, **1962,** 3288.

71b. Fraser, McKenzie, and Reid, *J. Chem. Soc.*, *B*, **1966,** 44.

72. Armarego, *J. Chem. Soc.*, *B*, **1966,** 191.
73. Rader, Young, and Aaron, *J. Org. Chem.*, **30,** 1536 (1965).
74. Anderson, Harrison, and Anderson, *J. Am. Chem. Soc.*, **85,** 3448 (1963).
75. Anderson and Harrison, *J. Am. Chem. Soc.*, **86,** 708 (1964).
76. Black and Heffernan, *Aust. J. Chem.*, **16,** 1051 (1963).
77. Buu-Hoi, Hoeffinger, and Jacquignon, *Bull. Soc. Chim. Fr.*, **1964,** 2019.
78. Elguero, Fruchier, and Jacquier, *Bull. Soc. Chim. Fr.*, **1966,** 3041.
79. Elguero, Fruchier, and Jacquier, *Bull. Soc. Chim. Fr.*, **1967,** 2619.
80. Elguero, Fruchier, and Jacquier, *Bull. Soc. Chim. Fr.*, **1969,** 2064.
81. Dennler, Portal, and Frasca, *Spectrochim. Acta*, **23A,** 2243 (1967).
82. Elguero, Riviere-Baudet, and Satge, *Compt. Rend.*, *Series C*, **266,** 44 (1968).

83. Elguero and Wolf, *Compt. Rend.*, *Series C*, **265**, 1507 (1967).
84. Evans, Whelan, and Johns, *Tetrahedron*, **21**, 3351 (1965).
85. Albright and Goldman, *J. Org. Chem.*, **31**, 273 (1966).
86. Jacquier and Maury, *Bull. Soc. Chim. Fr.*, **1967**, 306.
87. Reddy, Mandell, and Goldstein, *J. Chem. Soc.*, **1963**, 1414.
88. Acheson, Foxton, Abbott, and Mills, *J. Chem. Soc.*, *C*, **1967**, 882.
89. Raines and Kovacs, *J. Heterocycl. Chem.*, **4**, 305 (1967).
90. Garner and Suschitzky, *J. Chem. Soc.*, *C*, **1967**, 2536.
91. Casey and Wright, *J. Chem. Soc.*, *C*, **1966**, 1511.
92. Turner and Wood, *J. Chem. Soc.*, **1965**, 5270.
93. Takahashi and Kano, *Tetrahedron Lett.*, **1965**, 3789.
94. Stacy, Wollner, and Oakes, *J. Heterocycl. Chem.*, **3**, 51 (1966).
95. Paudler and Blewitt, *Tetrahedron*, **21**, 353 (1965).
96. Paudler and Blewitt, *J. Org. Chem.*, **31**, 1295 (1966).
97. Paudler and Blewitt, *J. Org. Chem.*, **30**, 4081 (1965).
98. Paudler and Kuder, *J. Heterocycl. Chem.*, **3**, 33 (1966).
99. Paudler and Shin, *J. Org. Chem.*, **33**, 1638 (1968).
100. Paolini and Robins, *J. Org. Chem.*, **30**, 4085 (1965).
101. Paolini and Robins, *J. Heterocycl. Chem.*, **2**, 53 (1965).
102. Lombardino, *J. Org. Chem.*, **30**, 2403 (1965).
103. Crabb and Newton, *J. Heterocycl. Chem.*, **6**, 301 (1969).
104. Paudler and Dunham, *J. Heterocycl. Chem.*, **2**, 410 (1965); Potts, Singh, and Bhattacharyya, *J. Org. Chem.*, **33**, 3766 (1968).
105. Flitsch and Kramer, *Tetrahedron Lett.*, **1968**, 1479.
106. Wong, Brown, and Rapoport, *J. Org. Chem.*, **30**, 2398 (1965).
107. Le Berre and Renault, *Bull. Soc. Chim. Fr.*, **1969**, 3133.
108. Le Berre and Renault, *Bull. Soc. Chim. Fr.*, **1969**, 3139.
109. Le Berre and Renault, *Bull. Soc. Chim. Fr.*, **1969**, 3146.
110. Roberts, *J. Chem. Soc.*, **1963**, 5556.
111. Black and Heffernan, *Aust. J. Chem.*, **15**, 862 (1962).
112. Rondeau, Rosenberg, and Dunbar, *J. Mol. Spectrosc.*, **26**, 139 (1968).
113. Rondeau, Rosenberg, and Dunbar, *J. Mol. Spectrosc.*, **29**, 305 (1969).
114. Gunther, *Tetrahedron Lett.*, **1967**, 2967.
115. Mallory, Wood, and Hurwitz, *J. Org. Chem.*, **29**, 2605 (1964).
116. Potts, Burton, Crawford, and Thomas, *J. Org. Chem.*, **31**, 3522 (1966).
117. Potts, Burton, and Roy, *J. Org. Chem.*, **31**, 265 (1966).
118. Kauffmann, Vogt, Barck, and Schulz, *Chem. Ber.*, **99**, 2593 (1966).
119. Paudler and Brumbaugh, *J. Heterocycl. Chem.*, **5**, 29 (1968).
120. Kobe, Stanovnik, and Tisler, *Tetrahedron*, **24**, 239 (1968); Pollak, Stanovnik, and Tisler, *ibid.*, **24**, 2623 (1968).
121. Lombardino, *J. Heterocycl. Chem.*, **5**, 35 (1968).
122. Japelj, Stanovnik, and Tisler, *J. Heterocycl. Chem.*, **6**, 559 (1969).
123. Noell and Robins, *J. Heterocycl. Chem.*, **1**, 34 (1964).
124. Ueda and Fox, *J. Org. Chem.*, **29**, 1762 (1964).
125. Paudler and Kuder, *J. Org. Chem.*, **31**, 809 (1966).
126. Hart, Johns, and Lamberton, *Chem. Commun*, **1969**, 1484.
127. de Roos and Salemink, *Rec. Trav. Chim.*, **88**, 1263 (1969).
128. Stocker, Fordice, Larson, and Thorstenson, *J. Org. Chem.*, **31**, 2380 (1966).
129. Southwick, Madhaw, and Fitzgerald, *J. Heterocycl. Chem.*, **6**, 507 (1969).
130. Gerston, Hinshaw, Robins, and Townsend, *J. Heterocycl. Chem.*, **6**, 207 (1969).

131. Klutchko, Hansen, and Meltzer, *J. Org. Chem.*, **30,** 3454 (1965).
132. Imai, *Chem. Pharm. Bull.*, **12,** 1030 (1964).
133. Montgomery and Hewson, *J. Org. Chem.*, **30,** 1528 (1965).
134. Pollak, Stanovnik, and Tisler, *J. Heterocycl. Chem.*, **5,** 513 (1968); Pollak and Tisler, *Tetrahedron*, **22,** 2073 (1966).
135. Paudler and Helmick, *J. Heterocycl. Chem.*, **3,** 269 (1966).
136. Makisumi, Watanabe, and Tori, *Chem. Pharm. Bull.*, **12,** 204 (1964).
137. Reddy, Hobgood, and Goldstein, *J. Am. Chem. Soc.*, **84,** 336 (1962).
138. Papesch and Dodson, *J. Org. Chem.*, **30,** 199 (1965).
139. Bauer, Dhawan, and Mahajanshetti, *J. Org. Chem.*, **31,** 2491 (1966).
140. Bauer and Mahajanshetti, *J. Heterocycl. Chem.*, **4,** 325 (1967).
141. Biffin, Brown, and Porter, *Tetrahedron Lett.*, **1967,** 2029.
142. Jardetzky and Jardetzky, *J. Am. Chem. Soc.*, **82,** 222 (1960)
143. Reddy, Mandell, and Goldstein, *J. Chem. Soc.*, **1963,** 1414.
144. Matsuura and Goto, *Tetrahedron Lett.*, **1963,** 1499.
145. Matsuura and Goto, *J. Chem. Soc.*, **1965,** 623.
146. Schweizer, Chan, Helmkamp, and Ts'o, *J. Am. Chem. Soc.*, **86,** 696 (1964); Chan, Schweizer, Ts'o, and Helmkamp, *ibid.*, **86,** 4182 (1964).
147. Bullock and Jardetzky, *J. Org. Chem.*, **29,** 1988 (1964).
148. Veillard, *J. Chim. Phys.*, **59,** 1056 (1962).
149. Miller and Lykos, *Tetrahedron Lett.*, **1962,** 493.
150. Lykos and Miller, *Tetrahedron Lett.*, **1963,** 1743.
151. Pullman, *Tetrahedron Lett.*, **1963,** 231.
152. Jardetsky, *Biopolym. Symp.*, No. **1,** 501 (1964).
153. Helmkamp and Kondo, *Biochim. Biophys. Acta*, **145,** 27 (1967); **157,** 242 (1968).
154. Danyluk and Hruska, *Biochem.*, **7,** 1038 (1968).
155. Schmid and Krenmayr, *Z. Phys. Chem.*, **51,** 297 (1966).
156. Coburn, Thorpe, Montgomery, and Hewson, *J. Org. Chem.*, **30,** 1110 (1965).
157. Coburn, Thorpe, Montgomery, and Hewson, *J. Org. Chem.*, **30,** 1114 (1965).
158. Temple, Thorpe, Coburn, and Montgomery, *J. Org. Chem.*, **31,** 935 (1966); Temple, Kussner, and Montgomery, *ibid.*, **31,** 2210 (1966).
159. Townsend and Robins, *J. Heterocycl. Chem.*, **3,** 241 (1966).
160. Montgomery, Hewson, Clayton, and Thomas, *J. Org. Chem.*, **31,** 2202 (1966).
161. Brown and Ford, *J. Chem. Soc., C*, **1969,** 2620.
162. Leonard and Deyrup, *J. Am. Chem. Soc.*, **84,** 2148 (1962).
163. Fraenkel and Asahi, *Takeda Kenkyusho Nempo*, **24,** 209 (1965); through *Chem. Abstr.*, **64,** 4909*d* (1966).
164. Lira and Huffman, *J. Org. Chem.*, **31,** 2188 (1966).
165. Alexander and Maienthal, *J. Pharm. Sci.*, **53,** 962 (1964).
166. Ottinger, Boulvin, Reisse, and Chiurdoglu, *Tetrahedron*, **21,** 3435 (1965).
167. Taylor and Garcia, *J. Am. Chem. Soc.*, **86,** 4721 (1964).
168. Stamm, *Arch. Pharm.*, **302,** 174 (1969).
169. Montgomery and Thomas, *J. Org. Chem.*, **31,** 1411 (1966).
170. Read and Goldstein, *J. Am. Chem. Soc.*, **87,** 3440 (1965).
171. Pugmire, Grant, Robins, and Rhodes, *J. Am. Chem. Soc.*, **87,** 2225 (1965).
172. Butula, *Ann.*, **729,** 73 (1969).
173. Patel, Rich, and Castle, *J. Heterocycl. Chem.*, **5,** 13 (1968).
174. Martin and Castle, *J. Heterocycl. Chem.*, **6,** 93 (1969).
175. Rutner and Spoerri, *J. Heterocycl. Chem.*, **3,** 435 (1966).
176. Bunting and Perrin, *J. Chem. Soc., B*, **1966,** 433.

177. Albert, *J. Chem. Soc., C*, **1968**, 2076; Albert, Pfleiderer, and Thacker, *ibid.*, **1969**, 1084.
178. Daunis, Jacquier, and Viallefont, *Bull. Soc. Chim. Fr.*, **1969**, 2492, 3670.
179. Schaefer and Schneider, *J. Chem. Phys.*, **32**, 1224 (1960).
180. Schaefer, *Can. J. Chem.*, **39**, 1864 (1961).
181. Ronayne and Williams, *J. Chem. Soc., B*, **1967**, 805.
182. Reeves and Strømme, *Can. J. Chem.*, **39**, 2318 (1961).
183. Chakrabarty and Hanrahan, *J. Mol. Spectrosc.*, **30**, 348 (1969).
184. Seiffert, *Angew. Chem.*, **74**, 250 (1962).
185. Black and Heffernan, *Aust. J. Chem.*, **17**, 558 (1964).
186. Harris, Katritzky, Øksne, Bailey, and Paterson, *J. Chem. Soc.*, **1963**, 197.
187. Wells, *J. Chem. Soc.*, **1963**, 1967.
188. Albert and Catterall, *J. Chem. Soc., C*, **1967**, 1533.
189. Haigh, Palmer, and Semple, *J. Chem. Soc.*, **1965**, 6004.
190. Diehl, *Helv. Chim. Acta*, **45**, 568 (1962); *J. Chim. Phys.*, **1964**, 199.
191. Renault and Cartron, *Compt. Rend., Series C*, **262**, 1161 (1966).
192. Craig and Pearson, *J. Heterocycl. Chem.*, **5**, 631 (1968).
193. Paterson and Bigam, *Can. J. Chem.*, **41**, 1841 (1963).
194. Tori, Ohtsuru, Aono, Kawazoe, and Ohnishi, *J. Am. Chem. Soc.*, **89**, 2765 (1967).
195. Crepaux, Lehn, and Dean, *Mol. Phys.*, **16**, 225 (1969).
196. Lynden-Bell and Sheppard, *Proc. R. Soc., A*, **269**, 385 (1962).
197. Pople and Santry, *Mol. Phys.*, **8**, 1 (1964).
198. Anet, *J. Chem. Phys.*, **32**, 1274 (1960).
199. Beimer and Fernando, *Anal. Chem.*, **41**, 1003 (1969).
200. Barlin and Brown, *J. Chem. Soc., B*, **1967**, 736.
201. Palmer and Semple, *Chem. Ind.*, **1965**, 1766.
202. Chambers, Hole, Musgrave, Storey, and Iddon, *J. Chem. Soc., C*, **1966**, 2331.
203. Franz, Hall, and Kaslow, *Tetrahedron Lett.*, **1967**, 1947.
204. Tori and Nakagawa, *J. Phys. Chem.*, **68**, 3163 (1964).
205. Malinowski, *J. Am. Chem. Soc.*, **83**, 4479 (1961); Malinowski, Pollara, and Larmann, *ibid.*, **84**, 2649 (1962).
206. Hampson and Mathias, *Chem. Commun.*, **1967**, 371.
207. Taylor and Heindel, *J. Org. Chem.*, **32**, 3339 (1967).
208. Rosowsky and Modest, *J. Org. Chem.*, **30**, 1832 (1965), and references cited therein.
209. Elliott and Yates, *J. Org. Chem.*, **26**, 1287 (1961).
210. Elliott and Dunathan, *Tetrahedron*, **19**, 833 (1963).
211. Easton and Cassady, *J. Org. Chem.*, **27**, 4713 (1962).
212. Brown and Jackman, *J. Chem. Soc.*, **1964**, 3132.
213. Metzger, Larive, Vincent, and Dennilauler, *Bull. Soc. Chim. Fr.*, **1967**, 46.
214. Bramley and Johnson, *J. Chem. Soc.*, **1965**, 1372.
215. Huckings and Johnson, *J. Chem. Soc., B*, **1966**, 63.
216. Tokuyama, Senoh, Sakan, Brown, and Witkop, *J. Am. Chem. Soc.*, **89**, 1017 (1967).
217. Brown, *J. Am. Chem. Soc.*, **87**, 4202 (1965).
218. Katritzky and Ternai, *J. Heterocycl. Chem.*, **5**, 745 (1968).
219. Booth, *J. Chem. Soc.*, **1964**, 1841.
220. Nagarajan, Nair, and Pillai, *Tetrahedron*, **23**, 1683 (1967).
221. Munro and Sewell, *Tetrahedron Lett.*, **1969**, 595.
222. Meyers and Garcia-Munoz, *J. Org. Chem.*, **29**, 1435 (1964).
223. Booth and Bostock, *Chem. Commun.*, **1967**, 177.
224. Black and Heffernan, *Aust. J. Chem.*, **19**, 1287 (1966).

225. Armarego and Batterham, *J. Chem. Soc.*, *B*, **1966,** 750.
226. Cox, Elvidge, and Jones, *J. Chem. Soc.*, **1964,** 1423; Win and Tieckelmann, *J. Org. Chem.*, **32,** 59 (1967).
227. Boyce and Levine, *J. Org. Chem.*, **31,** 3807 (1966); Moriconi, Creegan, Donovan, and Spano, *ibid.*, **28,** 2215 (1963).
228. Johnson, *J. Chem. Soc.*, **1964,** 200.
229. Brown, Dyke, and Sainsbury, *Tetrahedron*, **25,** 101 (1969).
230. Rachlin, Worning, and Enemark, *Tetrahedron Lett.*, **1968,** 4163.
231. Garside and Ritchie, *J. Chem. Soc.*, *C*, **1966,** 2140.
232. Dalton, Ramey, Gisler, Lendray, and Abraham, *J. Am. Chem. Soc.*, **91,** 6367 (1969).
233. Fraenkel, Cave, and Dalton, *J. Am. Chem. Soc.*, **89,** 329 (1967).
234. Abdullaev, Bystrov, Rumsh, Antonov, *Tetrahedron Lett.*, **1969,** 5287.
235. Okamoto, *Chem. Pharm. Bull.*, **15,** 168 (1967).
236. Kimoto and Okamoto, *Chem. Pharm. Bull.*, **15,** 1045 (1967).
237. Durand-Henchoz and Moreau, *Bull. Soc. Chim. Fr.*, **1966,** 3428.
238. Wohl, *Helv. Chim. Acta*, **49,** 2162 (1966).
239. Grob and Wohl, *Helv. Chim. Acta*, **49,** 2434 (1966).
240. Abramovitch and Struble, *Tetrahedron*, **24,** 705 (1968).
241. Acheson, Taylor, Higham, and Richards, *J. Chem. Soc.*, **1960,** 1691; Acheson, Hands, and Woolven, *ibid.*, **1963,** 2082.
242. Acheson, Gagan, and Taylor, *J. Chem. Soc.*, **1963,** 1903.
243. Acheson, Goodall, and Robinson, *J. Chem. Soc.*, **1965,** 2633.
244. Acheson, Feinberg, and Gagan, *J. Chem. Soc.*, **1965,** 948.
245. Higham and Richards, *J. Chem. Soc.*, **1960,** 1696.
246. Fozard and Jones, *J. Chem. Soc.*, **1964,** 3030.
247. Jackman, Johnson, and Tebby, *J. Chem. Soc.*, **1960,** 1579.
248. Miyadera and Kishida, *Tetrahedron*, **25,** 397 (1969).
249. Aaron, Wicks, and Rader, *J. Org. Chem.*, **29,** 2248 (1964).
250. Moynehan, Schofield, Jones, and Katritzky, *J. Chem. Soc.*, **1962,** 2637.
251. England and Sam, *J. Heterocycl. Chem.*, **3,** 482 (1966).
252. Matsunaga, Kawasaki, and Kaneko, *Tetrahedron Lett.*, **1967,** 2471.
253. Williamson, Howell, and Spencer, *J. Am. Chem. Soc.*, **88,** 325 (1966).
253a. Fodor, *J. Am. Chem. Soc.*, **88,** 1040 (1966).
253b. Bohlmann and Schumann, *Tetrahedron Lett.*, **1965,** 2435.
254. Black and Heffernan, *Aust. J. Chem.*, **18,** 707 (1965).
255. Elkins and Brown, *J. Heterocycl. Chem.*, **5,** 639(1968)
256. Ogata, Kano, and Tori, *Chem. Pharm. Bull.*, **11,** 1527 (1963).
257. Suzuki, Nakashima, and Nagasawa, *Tetrahedron Lett.*, **1966,** 2899.
258. Ames, Boyd, Ellis, and Lovesey, *Chem. Ind.*, **1966,** 458.
259. Ames and Novitt, *J. Chem. Soc.*, *C*, **1969,** 2355.
260. Katritzky, Lunt, Ternai, and Tiddy, *J. Chem. Soc.*, *B*, **1967,** 1243.
261. Bruce, Knowles, and Besford, *J. Chem. Soc.*, **1964,** 4044.
262. Ellis and Lovesey, *J. Chem. Soc.*, *B*, **1967,** 1285.
263. Ellis and Lovesey, *J. Chem. Soc.*, *B*, **1968,** 1393.
264. Besford, Allen, and Bruce, *J. Chem. Soc.*, **1963,** 2867.
265. Besford and Bruce, *J. Chem. Soc.*, **1964,** 4037.
266. Haddlesey, Mayor, and Szinai, *J. Chem. Soc.*, **1964,** 5269.
267. Bruce and Knowles, *J. Chem. Soc.*, **1964,** 4046.
268. Katrizky, Reavill, and Swinbourne, *J. Chem. Soc.*, *B*, **1966,** 351.
269. Armarego and Willette, *J. Chem. Soc.*, **1965,** 1258.

270. Brown and England, *Aust. J. Chem.*, **21**, 2813 (1968).
271. Armarego, unpublished results.
272. Brown and England, *Israel J. Chem.*, **6**, 569 (1968).
273. Tori, Ogata, and Kano, *Chem. Pharm. Bull.*, **11**, 681 (1963).
274. Smith, *J. Heterocycl. Chem.*, **3**, 535 (1966).
275. Armarego and Kobayashi, *J. Chem. Soc.*, *C*, **1969**, 1635.
276. Brignell, Katritzky, Reavill, Cheeseman, and Sarsfield, *J. Chem. Soc.*, *B*, **1967**, 1241.
277. Horton and Miller, *J. Org. Chem.*, **30**, 2457 (1965).
278. Reisser and Pfleiderer, *Chem. Ber.* **99**, 547 (1966).
279. Wiemann, Vinot, and Pinson, *Compt. Rend.*, *Series C*, **263**, 608 (1966).
280. Burton, Hughes, Newbold, and Elvidge, *J. Chem. Soc.*, *C*, **1968**, 1274.
281. Allison, Chambers, Macbride, and Musgrave, *Chem. Ind.*, **1968**, 1402.
282. Blears and Danyluk, *Tetrahedron*, **23**, 2927 (1967).
283. Archer and Mosher, *J. Org. Chem.*, **32**, 1378 (1967).
284. Aguilera, Duplan, and Nofre, *Bull. Soc. Chim. Fr.*, **1968**, 4491.
285. Junge and Staab, *Tetrahedron Lett.*, **1967**, 709.
286. Wilson and Bottomly, *J. Heterocycl. Chem.*, **4**, 360 (1967).
287. Dale and Coulon, *J. Chem. Soc.*, **1965**, 3487.
288. Glover and Loadman, *J. Chem. Soc.*, *C*, **1967**, 2391.
289. Paudler and Kress, *J. Heterocycl. Chem.*, **2**, 393 (1965).
290. Paudler and Kress, *J. Org. Chem.*, **31**, 3055 (1966).
291. Paudler and Kress, *J. Heterocycl. Chem.*, **4**, 284 (1967).
292. Paudler and Kress, *J. Org. Chem.*, **33**, 1384 (1968).
293. Paudler and Kress, *J. Heterocycl. Chem.*, **5**, 561 (1968).
294. Giacomello, Gualtieri, Riccieri, and Stein, *Tetrahedron Lett.*, **1965**, 1117.
295. Kobayashi, Kumadaki, and Sato, *Chem. Pharm. Bull.*, **17**, 1045 (1969).
296. Paudler and Kress, *J. Heterocycl. Chem.*, **5**, 561 (1968).
297. Hawes and Wibberly, *J. Chem. Soc.*, *C*, **1967**, 1564.
298. Armarego, *J. Chem. Soc.*, *C*, **1967**, 377.
299. Brenneisen, Thumm, and Benz, *Helv. Chim. Acta*, **49**, 651 (1966).
300. Brown, *Tetrahedron Lett.*, **1969**, 85.
301. Stanovnik and Tisler, *Tetrahedron Lett.*, **1968**, 33.
302. Ismail and Wibberley, *J. Chem. Soc.*, *C*, **1967**, 2613.
303. Elderfield and Wharmby, *J. Org. Chem.*, **32**, 1638 (1967).
304. Batterham, *J. Chem. Soc.*, *C*, **1966**, 999.
305. Montgomery and Wood, *J. Org. Chem.*, **29**, 734 (1964).
306. Queguiner and Pastour, *Bull. Soc. Chim. Fr.*, **1969**, 2519.
307. Paul and Rodda, *Aust. J. Chem.*, **21**, 1291 (1968).
308. Queguiner and Pastour, *Compt. Rend.*, *Series C*, **266**, 1459 (1968); **262**, 1335 (1966).
309. Paul and Rodda, *Aust. J. Chem.*, **22**, 1759 (1969).
310. Paul and Rodda, *Aust. J. Chem.*, **22**, 1745 (1969).
311. Patel and Castle, *J. Heterocycl. Chem.*, **3**, 512 (1966).
312. Singerman and Castle, *J. Heterocycl. Chem.*, **4**, 393 (1967).
313. Patel, Rich, and Castle, *J. Heterocycl. Chem.*, **5**, 13 (1968).
314. Dorman, *J. Heterocycl. Chem.*, **4**, 491 (1967).
315. DiStephano and Castle, *J. Heterocycl. Chem.*, **5**, 53 (1968).
316. Matsuura and Goto, *J. Chem. Soc.*, **1963**, 1773.
317. Albert, Batterham, and McCormack, *J. Chem. Soc.*, *B*, **1966**, 1105.
318. Albert and McCormack, *J. Chem. Soc.*, *C*, **1968**, 63.
319. Albert and Yamomoto, *J. Chem. Soc.*, *C*, **1968**, 2292.

320. Clark, *Tetrahedron Lett.*, **1967,** 1099.
321. Clark, *J. Chem. Soc., C*, **1967,** 1543.
322. Clark and Pendergast, *J. Chem. Soc., C*, **1968,** 1124.
323. Clark, Murdoch, and Roberts, *J. Chem. Soc. C*, **1969,** 1408.
324. Dieffenbacher, Mondelli, and von Philipsborn, *Helv. Chim. Acta*, **49,** 1355 (1966).
325. von Philipsborn, Stierlin, and Traber, *Helv. Chim. Acta*, **46,** 2592 (1963).
326. Merlini, von Philipsborn, and Viscontini, *Helv. Chim. Acta*, **46,** 2597 (1963).
327. Viscontini and Bobst, *Helv. Chim. Acta*, **49,** 1815 (1966).
328. Nieman, Bergmann, and Meyer, *J. Chem. Soc., C*, **1969,** 2415
329. Bauer, Nambury, and Hershenson, *J. Heterocycl. Chem.*, **3,** 22 (1966).
330. Pfleiderer, Bunting, Perrin, and Nubel, *Chem. Ber.*, **101,** 1072 (1968).
331. Viscontini and Nasini, *Helv. Chim. Acta*, **48,** 452 (1965).
332. Viscontini and Huwyler, *Helv. Chim. Acta*, **48,** 764 (1965).
333. Bobst and Viscontini, *Helv. Chim. Acta*, **49,** 875 (1966).
334. Biffin and Brown, *Tetrahedron Lett.*, 2503 (1968).
335. Temple and Montgomery, *J. Org. Chem.*, **28,** 3038 (1963).
336. Temple, Kussner, and Montgomery, *J. Org. Chem.*, **34,** 2102 (1969).
337. Mukai, Tsuruta, and Momotari, *Bull. Chem. Soc. Japan*, **1967,** 1967.
338. Mannschreck, Rissmann, Vogtle, and Wild, *Chem. Ber.*, **100,** 335 (1967).
339. Caesar and Mondon, *Chem. Ber.*, **101,** 990 (1968).
340. Johnson and Nasutavicus, *J. Heterocycl. Chem.*, **2,** 26 (1965).
341. Fozard and Jones, *J. Org. Chem.*, **30,** 1523 (1965).
342. Stempel, Douvan, and Sternbach, *J. Org. Chem.*, **33,** 2963 (1968).
343. Linscheid and Lehn, *Bull. Soc. Chim. Fr.*, **1967,** 992.

5 OXYGEN HETEROCYCLES

1. THE OXIRAN OR EPOXIDE SYSTEM

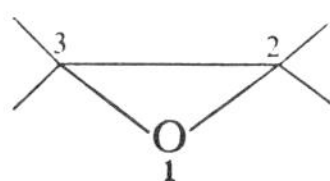

The earliest data for simple oxirans appeared in a general study of geminal couplings (1) but no attempt was made to assign the cis and trans vicinal couplings. In the early 1960s much work was published on the spectra of simple oxirans, and the basic spectral characteristics were well established. During this period the interest of Reilly and Swalen (2–4) in the spectra of various epoxides led to the development of the very popular iterative computer program combination NMREN and NMRIT. The use of these programs and also ABX and ABK tables showed that all couplings in the epoxide system had the same sign, presumed positive. This result was queried by Lauterbur and Kurland (5) but their experiments with ethylene oxide, using double-resonance techniques, proved it to be correct. Substituent effects and general intuition were used by these workers to assign the larger vicinal couplings in monosubstituted expoxides to the cis protons. After nearly ten years of investigation, this assignment, while still unproved, is undoubtedly correct for most epoxides. The spectrum of the parent ethylene oxide (**1**) was fully analyzed by Mortimer (6), using ^{13}C satellites to obtain complete data from the symmetrical system, and full details, including sign determinations, were obtained for styrene oxide (**2**) (7). Lehn and Riehl (8) have studied in some detail the spectra of a series of cis- and trans-disubstituted epoxides (**3**) with stereochemistry known from the method of synthesis. In every case the vicinal coupling for the cis compound was larger than that for the corresponding trans isomer. Also, as expected, the values of both couplings diminished linearly as the electronegativity of the substituents increased.

Similar trends were noted (9) for monosubstituted epoxides. This study was prompted by the observation that the vicinal couplings for oxirans are considerably smaller than those for similar cyclopropanes, a phenomenon

TABLE 5.1 SPECTRAL DATA FOR EPOXIDES

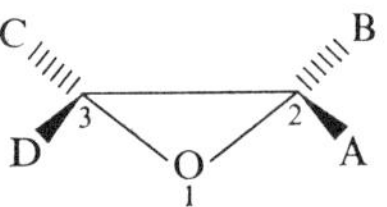

D	Solvent	τA	τB	τC	τD	J_{AB}	J_{AC}	J_{BC}	Ref.
H[a]	Neat		7.46				3.1	4.45	6
Me	Neat					5.5	2.5	4.5	1
CH_2Cl	Neat	7.35	7.16	6.80	6.53, 6.26	5.0	2.4	4.0	4
Ph	C_6H_{12}	7.51	7.14	6.39		6.00	2.38	3.93	11
	CCl_4	7.44	7.03	6.31		5.85	2.40	3.94	11
	C_6H_6	7.65	7.38	6.53		5.81	2.42	4.06	11
	PhMe	7.66	7.36	6.56		5.79	2.39	3.99	11
	$CDCl_3$	7.27	6.93	6.20		5.55	2.55	4.10	11
	C_5H_5N	7.30	7.02	6.17		5.69	2.42	4.11	11
	PhCOMe	7.37	7.06	6.27		5.67	2.41	4.07	11
	Me_2CO	7.27	6.96	6.19		5.56	2.44	4.06	11
	$PhNO_2$	7.27	6.94	6.19		5.53	2.49	4.06	11
	$MeNO_2$	7.24	6.94	6.19		5.42	2.55	4.17	11
	MeCN	7.25	6.95	6.19		5.40	2.56	4.16	11
	DMSO	7.18	6.91	6.09		5.31	2.43	4.13	11
	Neat	7.48	7.19	6.38		5.63	2.48	4.08	11
		7.48	7.18	6.38		5.66	2.52	4.06	7
CN	Neat	6.89	6.98	6.48		5.53	2.51	4.23	2
CHO[b]	Neat	6.90	6.83	6.64	1.08	5.54	2.09	4.87	3
COMe	Neat	7.06	6.98	6.64		5.78	1.95	5.21	2
CO_2H	Me_2CO	7.06	7.01	6.53		6.28	1.86	5.01	2

[a] $J_{^{13}C-H}$ 175.8.
[b] J_{AD} 0.03, J_{BD} −0.31, J_{CD} 6.24.

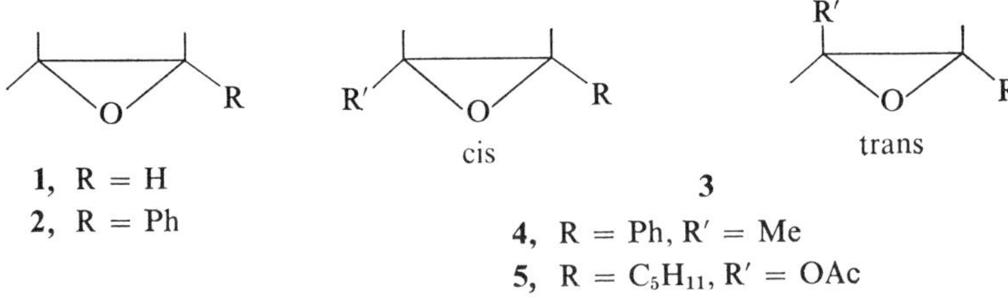

1, R = H
2, R = Ph

3

4, R = Ph, R′ = Me
5, R = C_5H_{11}, R′ = OAc

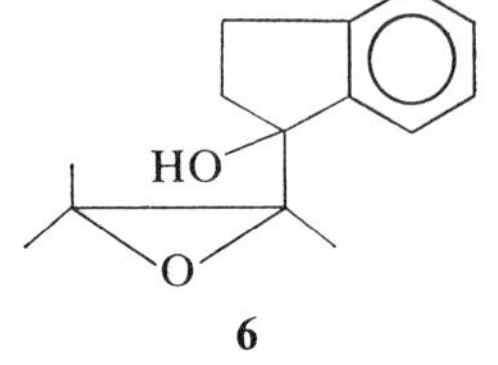

6

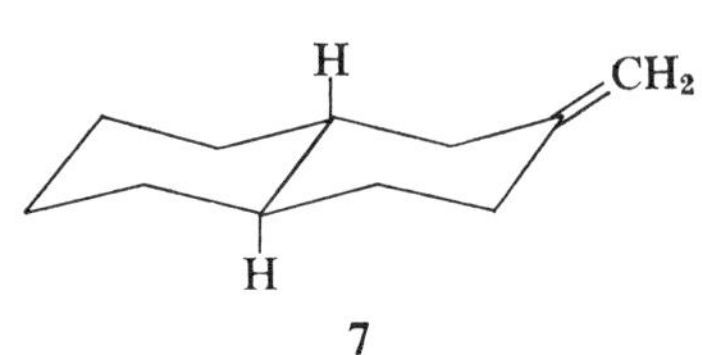

7

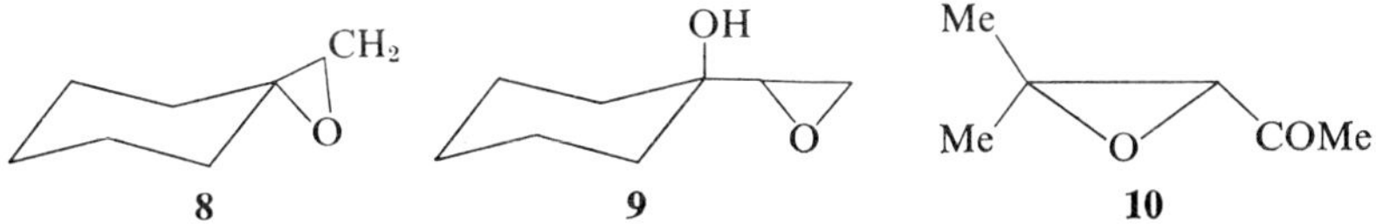

probably due in part to the electronegativity of the ring oxygen atom. Considerable scatter occurred when the coupling constants were plotted against substitutent electronegativities, and it was suggested that delocalization of the oxygen lone pairs took place in a manner which, for most substituents, was not understood. Also, it was noted that a general shift to higher field occurred on going from an open-chain ether to the epoxide system (10).

Smith and Cox (11) have made a thorough study of the spectra of styrene oxide (**2**) in 13 different solvents and observed solvent-induced changes in the gem coupling constant (6.0 Hz in C_6H_{12} to 5.31 Hz in DMSO). These values showed a reasonable linear dependence on the dielectric constant of the solvent. The assignment of positive sign to J_{gem} was based on the usual assumption that J_{vic} and $J_{^{13}C-H}$ are positive. These solvent-induced changes in J_{gem} were thought to depend on the sign of the coupling constant, becoming smaller as the dielectric constant of the solvent increases in this case, but showing the reverse effect in systems where the coupling constants were negative.

Nuclear magnetic resonance techniques have been used to determine configurations and preferred conformations in many more highly substituted oxiran systems. Thus, the fact that J_{cis} was always more than twice J_{trans} was used with the oxirans derived from ephedrine salts (**4**) to support stereochemical assignments (12), and with the epoxides (**5**) to distinguish the cis and trans isomers (13). Spectra of the erythro- and threo-1-methyl-1-(hydroxyindanyl)-oxirans (**6**) distinguished between them and supported conclusions based on infrared data (14). The stereochemical course of epoxidation of rigid methylene cyclohexane systems (**7**) is easily followed by NMR methods (15). Equatorial methylene protons of the epoxide ring give narrow line widths (0.15 $\pm$ 0.07 Hz), whereas axial groups exhibit much broader line widths (1.41 $\pm$ 0.14 Hz). Variable-temperature studies of cyclohexanespirooxiran (**8**) (16) showed that the molecule is fixed on the NMR time scale below $-85°$. The energy difference between the two conformers was calculated as 0.27 $\pm$ 0.04 kcal/mole. The similarity between the conformational free energies calculated for the cyclopropyl and oxiran groups of a series of 1-cyclopropyl- and 1-vinyloxide-cyclohexanols (**9**) suggested that the underlying interactions were mainly governed by steric hindrance (17).

Solvent shifts for epoxides induced by benzene and pyridine (with respect to CCl_4 solutions) have been shown to distinguish the three-membered ring system from four-, five-, or six-membered cyclic ethers (18). In benzene the epoxide protons experience a moderately large upfield shift (0.13–0.38 ppm) and this decreases as the ring size increases. Benzene shifts for a series of simple methyl-substituted epoxy ketones (**10**) have been reported (19). Acetyl methyls, as well as vicinal methyl groups trans to the acetyl group, are shielded in benzene (relative to CCl_4), while cis methyl groups are deshielded.

Magnetic nonequivalence of the side-chain methylene protons in a number of $-CH_2X$-substituted oxirans has been discussed (20). The factors which could affect this phenomenon were considered in some detail and the conclusion was reached that the system was complex and could not be explained solely in terms of conformations.

Inversion at the oxygen atom of the isopropyloxonium salt of oxiran (**11**) in SO_2 was slowed down enough at $-70°$ for the ring protons to produce an AA′BB′ pattern (21), thus proving the tetrahedral geometry of the oxygen atom in this type of compound. At 40°, the rather broad spectrum suggested time-averaging. Spectra of a number of epoxycyclobutanes (**12**) have been reported in a paper dealing with the preparation and stability of this ring system (22).

The rapid reversible equilibria between oxepins (**13**) and benzene oxides (**14**) has been extensively studied by Vogel and his co-workers (23–25). Proton

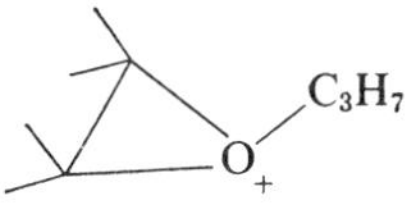

C_3H_7 O+

11

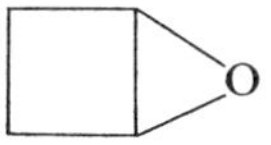

O

12

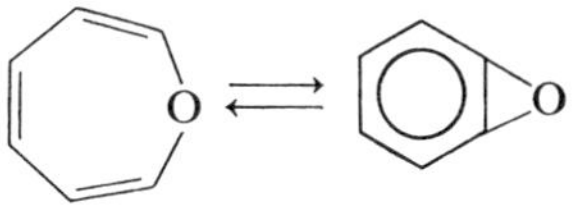

O ⇌ O

13 **14**

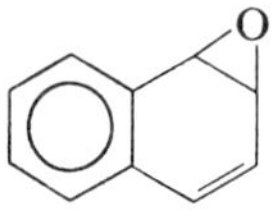

O

15

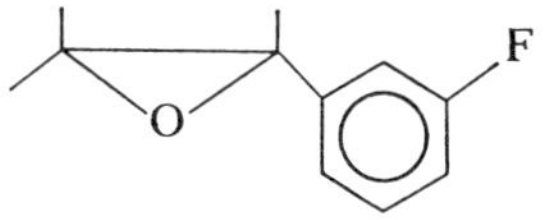

O F

16

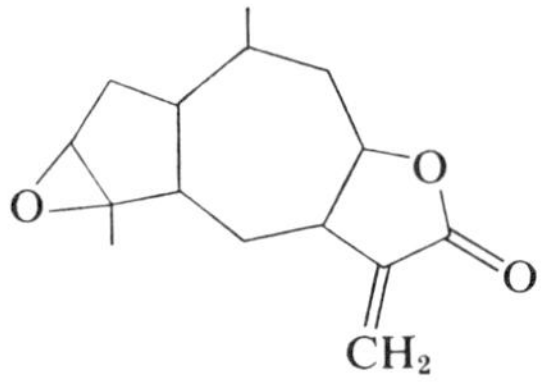

O O CH_2

17

CO_2Me O AcO H

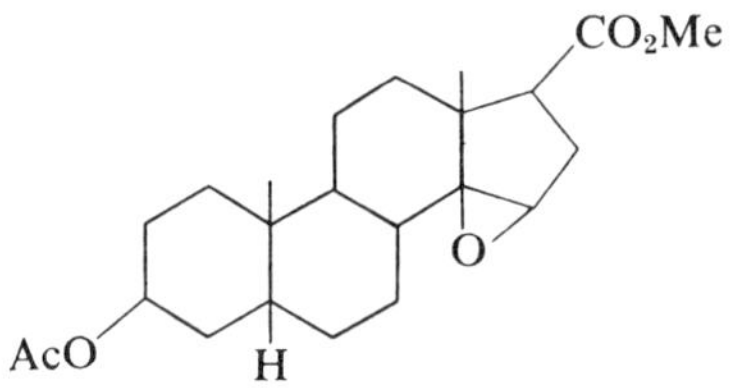

18

magnetic resonance data show the reaction to be solvent dependent and variable-temperature studies were used to obtain approximate kinetic data for the equilibria. 1,2-Naphthalene oxide (**15**) was shown to exist entirely in the epoxide form and to rearrange readily to α-naphthol (25).

Proton and fluorine spectra of a number of *cis*- and *trans*-fluoroepoxides have been reported (26). Also, the ^{19}F spectra of *m*-fluorophenyloxiran (**16**) have been used to determine the inductive and resonance substitution constants for the three-membered cyclic substituent on an aromatic system (27). The oxiran substituent acted in a manner similar to a methoxy group.

Epoxides occur quite often on the five-membered rings of many alicyclic natural products and their derivatives, for example, **17** and **18**. It is important to note that, in these systems, coupling between an epoxide proton and a vicinal proton on the five-membered ring is usually very small and is often not resolved (28, 29). This observation is an extremely useful diagnostic tool in structural studies of these systems.

II. THE OXETAN SYSTEM

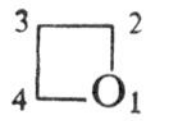

3 2
4 O 1

This system has been little studied. Chemical shifts for the parent compound are given in the Varian Catalog—Spectrum No. 16 (30): τ2,4 5.27; τ3 7.28.

The A_2B_4 spin system encountered here is of great complexity and one approach to its analysis (31) involved complete deuteration of the methylene group at position 2. The AA′BB′ system obtained with this labeled compound was easily analyzed to give all parameters except the now suppressed 2–4 couplings. Values obtained were J_{33} 11.15, J_{44} 6.02, J_{cis} 6.87, and J_{trans} 8.65 Hz. The small values of J_{24} H–D couplings suggest that the corresponding proton–proton couplings will be less than 0.3 Hz.

Data for a number of highly substituted oxetans are available in a report of their preparation by photochemical addition of keto compounds to unsaturated systems (32), of benzophenone to furans (33), and of 1,4-naphthoquinone to benzocyclic olefins (34). β-Propiolactones (**19**) were included in an early study of the methylene group (1).

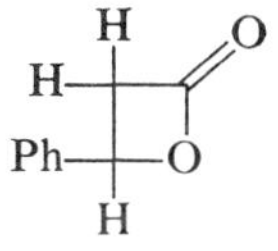

H O
H
Ph O
H

19

A complete analysis for the spectrum of oxetan is given in a recent paper (31a) describing methods for the analysis of the A_2B_4 systems in this compound and in thietan. Composite values (neat) are as follows: $\tau\alpha$ 5.38; $\tau\beta$ 7.38; $^2J_\alpha$ −5.8; $^2J_\beta$ −11.0; $^3J_{\text{cis}}$ 8.70; $^3J_{\text{trans}}$ 6.60; $^4J_{\text{cis}}$ 0.20; $^4J_{\text{trans}}$ 0.14.

III. FIVE-MEMBERED OXYGEN HETEROCYCLES

A. The Furan System

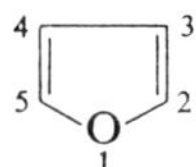

Spectra of substituted furans were reported as early as 1955 (35) and were later used as examples in some of the classical papers by Abraham and Bernstein dealing with ABX and similar spin systems (36–38). The data obtained from substituted furans led to a provisional analysis of the system, but it was not until ^{13}C satellite patterns were used (39) that a full analysis of the AA′BB′ pattern of the parent compound could be produced. More recently, the spectrum of furan oriented in the nematic phase of anisoleazo-phenyl-*n*-capronate has been analyzed (40). Interproton distances compared well with those obtained from microwave spectroscopy. Also, a broad-line study of crystalline furan for both high- and low-temperature phases has been reported (40a).

The aromaticity of furan in terms of its ability to sustain a ring current has been discussed at length (41). A correlation between proton chemical shifts and calculated electron densities for furan was not very successful (41a).

Nuclear magnetic resonance was used as early as 1958 to determine the site of substitution in furans (42). However, the definitive paper on substituted furans comes from the prolific pen of Hoffman and his co-workers (43) who extended earlier observations (38) on substituent effects. A comparison is drawn between the chemical shifts of furan derivatives and those of the corresponding thiophenes, and the differences are discussed at length in terms of the contributions made by different structures to the overall mesomeric system. While the arguments are interesting and a number of valid points are made, there are inconsistencies, and, as in much of this type of work, one is left with the feeling that any results could have been explained by variations of the same themes. The results obtained for adjacent and cross-ring couplings have been used by others (e.g., see Ref. 44) for determining structures of more highly substituted isomers. Another extensive study of the spectra of 2-substituted furans has been made (45) which, although it purports to discuss mesomeric and inductive effects, does not refer to any earlier work, and, in fact, is mainly a mathematical study of the anisotropy of carbonyl substituents. Several authors had previously reported a coupling of 0.8 Hz between the formyl proton and H-5 of furfural (**20**) in the pure liquid (35, 36, 46) and this had been shown to vary with solvent and concentration (43). Such a variation suggested a dependence of the various rotational conformations of the aldehyde group on the nature of the solvent. At temperatures below −80° furfuraldehyde gives two sets of signals which were assigned on the basis of chemical shifts to the two planar rotomers **21** and **22** (47–49). H-3 from the predominant species resonates

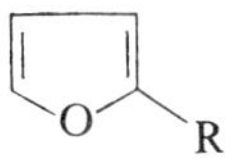

20, R = CHO
23, R = COMe

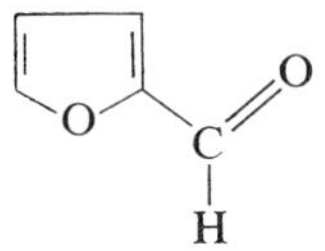

21

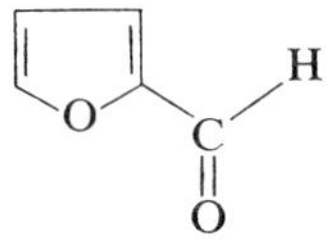

22

downfield from its counterpart in less favored isomer and this was originally explained in terms of deshielding by the carbonyl group in the trans O–O rotamer **21**. The observed deshielding, 0.15 ppm, was in good agreement with a theoretical estimate of 0.17 ppm (45). The long-range couplings for these rotamers, as assigned, did not agree with the extended *W* rule. Roques and his co-workers (45a) very elegantly resolved the differences between the carbonyl deshielding evidence and the extended *W* rule and showed conclusively that the original assignments of conformation to the rotamers were the reverse of those actually present, that is, **22** is the predominant species. Once this was established, the observed long-range couplings had sensible values: $J_{CHO,5} = 1.1$, $J_{CHO,4} = 0.2$ Hz for **22** and $J_{CHO,4} = 0.85$, $J_{CHO,5} = 0.2$ Hz for **21**. For furan-2-aldehyde, the signs of couplings between the ring protons were shown to be the same (50). In a paper mainly devoted to the determination of sign using a triple-resonance technique, the 5-bond couplings, $J_{CHO,4}$ and $J_{CHO,5}$ in furan-2-aldehyde and $J_{CHO,5}$ in the 3-aldehyde, were shown to have the same sign as J_{45}, while the 4-bond coupling, $J_{CHO,4}$ in the 3-aldehyde, had the opposite sign. These results agree with the Dirac vector model for long-range couplings which predicts alternating signs as the coupling paths progressively increase one bond at a time. Calculations of the expected chemical shift for H-3 in the 2-acetylfuran (**23**), taking into account the anisotropies of both the C=O and C–Me bonds, led to the suggestion that the acetyl group spends most of its time distributed evenly between the two planar conjugated rotamers (45). Also in this work, the effect of an electron-donating (or -withdrawing) group at C-2 on the chemical shift of H-4 is ascribed to an augmenting (or diminishing) of the ring current with subsequent deshielding (or shielding) of H-4 and H-5.

Spectra of a number of highly substituted 2-arylfurans have been reported (51). Data for a number of simple furans from the references above and from Ref. 52 are collected in Table 5.2.

Observation of the furan cation in concentrated H_2SO_4 is impossible because of its rapid decomposition, but substitution of the ring with *t*-butyl

TABLE 5.2 SPECTRAL DATA FOR FURANS

(Structure of furan with ring positions numbered: O = 1, then 2, 3, 4, 5)

Substituents	Solvent[a]	τ2	τ3	τ4	τ5	J_{23}	J_{24}	J_{25}	J_{34}	J_{35}	J_{45}	Ref.
None	C_6H_{12}	2.73	3.77	3.77	2.73							43
	Me_2CO	2.70	3.77	3.77	2.70	1.75	0.85	1.4	3.3	0.85	1.75	39
	CCl_4	2.62	3.70	3.70	2.62							44
2-Me[b]	C_6H_{12}		4.20	3.92	2.91				3.05	0.95	1.90	43
	CCl_4		4.12	3.82	2.79							45
2-CH_2NH_2	CCl_4		3.94	3.76	2.72							45
2-CH_2OH	CCl_4		3.81	3.75	2.70							45
2-CH_2SH	CCl_4		3.87	3.77	2.72							45
2-CN	C_6H_{12}		3.14	3.64	2.62				3.55	0.70	1.75	43
	CCl_4		2.85	3.38	2.34							45
2-CHO	C_6H_{12}		2.97	3.58	2.51				3.55	0.80	1.70	43
	CCl_4		2.77	3.39	2.28							45
2-COMe	C_6H_{12}		3.04	3.67	2.67				3.45	0.75	1.70	43
	CCl_4		2.89	3.47	2.43							45
2-CO_2Me	C_6H_{12}		3.02	3.70	2.66				3.40	0.85	1.70	43
	CCl_4		2.85	3.48	2.37							45
2-COCl	CCl_4		2.50	3.31	2.14							45
2-OMe	C_6H_{12}		5.04	3.93	3.30				3.15	1.15	2.15	43
2-SMe	C_6H_{12}		3.82	3.76	2.71				3.20	0.85	1.95	43
2-SCN	C_6H_{12}		3.30	3.64	2.52				3.40	0.85	1.95	43
2-Br	C_6H_{12}		3.85	3.79	2.75				3.25	0.80	2.10	43
	CCl_4		3.72	3.67	2.63							44
2-I	C_6H_{12}		3.58	3.83	2.63				3.25	0.95	1.95	43
3-Me[c]	C_6H_{12}	2.97		3.94	2.86		0.85	1.55			1.75	43
	CCl_4	2.89		3.87	2.77							44
3-CN	C_6H_{12}	2.17		3.48	2.64		0.75	1.55			1.95	43
3-CHO	C_6H_{12}	2.14		3.33	2.69		0.80	1.45			1.90	43
3-COMe	C_6H_{12}	2.16		3.34	2.74		0.75	1.40			1.80	43
3-CO_2Me	C_6H_{12}	2.17		3.37	2.76		0.70	1.55			1.85	43
3-OMe	C_6H_{12}	3.08		3.98	2.99		1.00	1.65			1.90	43
3-SMe	C_6H_{12}	2.80		3.75	2.77		0.80	1.50			1.85	43
3-SCN	C_6H_{12}	2.43		3.51	2.59		0.85					43
3-I	C_6H_{12}	2.75		3.66	2.84		0.70	1.50			1.90	43
3,4-Di-CHO	$CDCl_3$	1.88			1.88							52
2-Br, 3-Me	CCl_4			3.76	2.68						2.0	44
2,5-Di-Br	CCl_4		3.75	3.75								44

[a] The large differences between some values for C_6H_{12} and CCl_4 solutions may be in part due to the use of the solvent peak as internal standard in the former case.
[b] $J_{Me,3}$ 0.95, $J_{Me,4}$ 0.40, $J_{Me,5}$ 0.45.
[c] $J_{Me,2}$ 0.95, $J_{Me,4}$ 0.45, $J_{Me,5}$ 0.40.

groups stabilizes the cationic species. Examination of the spectra of a number of these cations showed that protonation occurred exclusively on the α-carbon atom (53). Representative data are given in Table 5.3.

The ^{13}C spectrum of furan (54, 55) exhibits a number of interesting features. The spectrum of the β-carbon is strictly first order and any permutation of long-range coupling constants leads to the same spectral pattern.

TABLE 5.3 SPECTRAL DATA FOR PROTONATED *t*-BUTYLFURANS (53). SOLVENT—H_2SO_4

Substituents	Protonation site	$\tau 2$(H)	$\tau 3$	$\tau 4$	$\tau 5$(H)	J_{23}	J_{34}	τt-Bu
2,5-*t*-Bu	2	3.27	1.92	0.38		1.2	6.0	7.87, 8.26
2,4-*t*-Bu	5		2.21		3.38			7.93, 8.04
2,3,4-*t*-Bu	2	3.52		2.23				8.00, 8.04, 8.28

Correct assignments were made by observing the patterns from the two β-carbon atoms in 2-methylfuran. Here, C-3 showed two coupling constants, 5.5 and 4.0 Hz, while C-4 showed couplings of 13.1 and 4.2 Hz. Thus, the two approximately 4.0-Hz couplings were assigned to geminal C–H couplings to the "other" β-proton, and the others must involve coupling to H-5. A coupling of 6.8 Hz, found for C-2 but not for C-5, must arise from cross-oxygen, vicinal, C–H interaction. These observations were then extended to furan, and the overall picture for this heterocycle is summarized as follows: chemical shifts relative to CS_2, α-C- +49.8, β-C +82.9 ppm (54); coupling constants, $J_{^{13}C-H}(\alpha)$ 201.4, (β) 175.3 Hz (39, 56), $J_{^{13}C-H}(2,3)$ 14.0, (3,2) 7.0, (4,2) 10.8, (4,3) 4.0, (5,2) 7.0, (5,3) 5.8 Hz (55).

Available NMR data show clearly that all the simple "3-hydroxyfurans" studied exist in the 3-keto form **24** (57–59). Protons on the sp_3-hybridized 2-carbon atom are readily distinguished by their chemical shifts (τ 5.55–5.75).

Spectra of benzofuran (**25**) were first reported in a study of cross-ring couplings in such systems (60). Full analyses of the spectra of this compound and the dibenzofuran (**26**) were later reported by Black and Heffernan as part of their general examination of heterocyclic compounds (61). This information, together with data for substituted benzofurans (62–64), is collected in Table 5.4. Incomplete data are available for some more highly

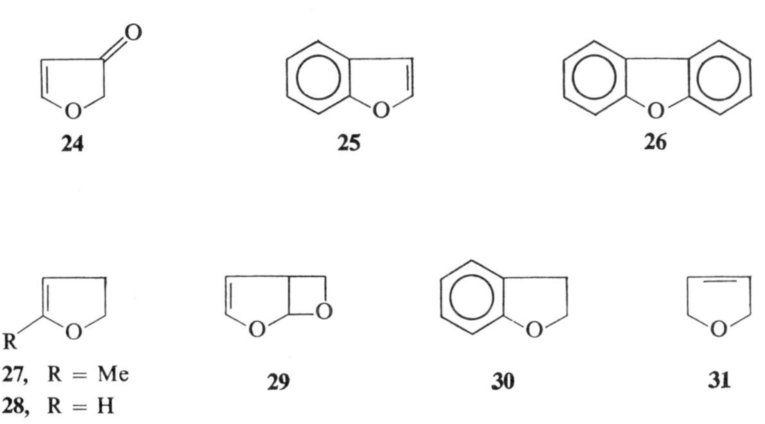

TABLE 5.4 SPECTRAL DATA FOR BENZOFURANS

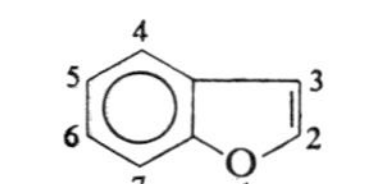

Substituents	Solvent	τ2	τ3	τ4	τ5	τ6	τ7	J_{23}	J_{37}	J_{45}	J_{46}	J_{47}	J_{56}	J_{57}	J_{67}	Ref.
None	Neat	2.73	3.59	2.59	2.90	2.89	2.55	2.3	0.9	8.0	1.0	0.7	7.3	1.2	8.0	61
	Me_2CO	2.21	3.23	2.36	2.77	2.70	2.48	2.1	0.9	7.8	1.3	0.7	7.3	0.9	8.3	61
	CCl_4	2.48	3.34	2.51	2.86	2.80	2.58	2.1	0.8	7.8	1.2	0.8	7.2	0.9	8.4	61
2-SO_2Et	$CDCl_3$		2.06		2.7											62
2-SO_2Et, 5-Cl	$CDCl_3$		1.61	1.95		2.35					1.8	0.8			8.2	62
2-SO_2Et, 5-Br	$CDCl_3$		2.01	2.12		2.68									9.0	62
2-SO_2Et, 5-NO_2	$CDCl_3$		1.51	1.05		1.46	2.13				2.3	0.8			9.9	62
2-SO_2Et, 7-NO_2	$CDCl_3$		1.46	1.60	2.30	1.60				8.5			8.5			62
2-Me, 5-OMe	CCl_4	7.72	3.90	3.2		3.2	3.2	1.0								63
3-Me, 4,6-Di-OMe	CCl_4	2.92	7.73	6.19	3.53	6.26	3.84	1.5								64
3-Me, 4-OH, 6-OMe	$CDCl_3$	2.81	7.67		3.40	6.22	3.80	1.5								64
3-Me, 4-OAc, 6-OMe	CCl_4	2.87	7.84		3.24	6.25	3.49	1.5								64

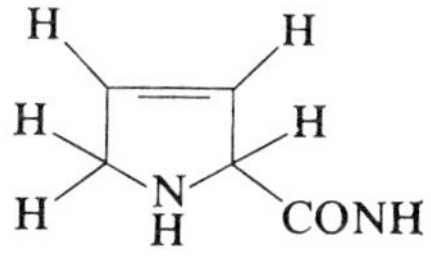

H H
H H
H N CONH
H

32

annelated furans in the accumulation by Martin and his co-workers of spectra of condensed sulfur and oxygen heterocycles (65).

Proton and ^{19}F data have been tabulated for a number of partially fluorinated benzofurans (66). Assignments in the ^{19}F spectra were aided by consideration of methoxy substituent effects where the signal from an ortho fluorine was deshielded by 3 to 7 ppm while those of meta and para fluorines were shielded by only 2 ppm. Data for octafluorodibenzofuran and some of its derivatives have been reported (67). Results for the fluorobenzofurans are summarized in Table 5.5.

5-Methyl-2,3-dihydrofurans (**27**) have been used as models in a study of long-range coupling over five bonds, in this case homoallylic coupling (68). Allylic coupling, $^4J_{Me,4}$, was found to be 1.1 Hz, while $^5J_{Me,3}$ was 2.0 Hz. A full analysis of the spectrum of the parent 2,3-dihydrofuran (**28**) has been reported (69), and information about a number of derivatives is included in papers dealing mainly with preparation and structural determination (70, 71). Also, data for the rather interesting bicyclic system **29** has been presented (68, 72).

In the case of the 2,3-dihydrobenzofurans (**30**), it has been suggested on a number of occasions that, of the two couplings between H-1 and H-2, J_{cis} will be larger than J_{trans} (73, 75). However, with the substituents, *i*-propyl at position 2 and an –OR group at position 3, J_{trans} appears to be larger than J_{cis} (76), an anomaly accredited to an orientation-dependent electronegativity effect of the substituents. However, from the earlier work of Huisgen and his collaborators (77) on the 4,5,6,7-tetrachloro derivatives, J_{trans} varies considerably and not unexpectedly with the bulk and possibly the nature of the substituents at C-2 and C-3. In this case also, where both couplings are available the larger is assigned to J_{cis}, but I would hesitate to assign the stereochemistry in a single compound on the basis of the size of this vicinal coupling. Data for 2,3-dihydrofurans and benzofurans are collected in Table 5.6.

The spectrum of 2,5-dihydrofuran (**31**) is extremely simple, just two broad singlets (vinyl protons τ 4.22, O–CH_2 τ 5.57, CCl_4 solution) (69). Such simplicity brings to mind similar effects in the spectra of dehydroprolinamide (**32**) (78) and is probably due to the same phenomenon. In these ring systems the couplings occur in pairs, vicinal and allylic, with similar values but opposite signs, and hence cancel to give narrow signals. This is borne out

TABLE 5.5 ^{19}F SPECTRAL DATA FOR FLUOROBENZOFURANS (66)

R4	R5	R6	R7	Chemical shifts[a] 4F	5F	6F	7F	Coupling constants J_{45}	J_{46}	J_{56}	J_{57}	J_{67}
F	F	F	F	148.0	162.3, 162.3, 164.8			19.1[b]				
F	F	F	OMe	150.8	166.4	158.3		20.3	2.4	18.5		
OMe	F	F	F		160.4	163.6	165.0			18.0	2.8	19.2
F	F	OMe	F	149.0	159.4[b]		156.8[b]	19.8[b]			2.6[b]	

[a] Parts per million upfield from $CFCl_3$ in CCl_4.
[b] Tentative assignment.

TABLE 5.6 SPECTRAL DATA FOR 2,3-DIHYDROFURANS AND 2,3-DIHYDROBENZFURANS

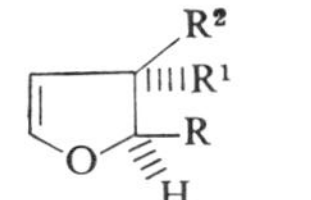

I. 2,3-DIHYDROFURANS

R	R^1	R^2	Others	Solvent	τ2	τ3	τ4	τ5	J_{23}	J_{34}	J_{35}	J_{45}	Ref.
H	H	H		CCl_4	5.80	7.47	5.18	3.78	9.3	2.5	2.6	2.6	69
Ph	H	H	4-COMe, 5-Me	$CDCl_3$	5.59	6.58 (R^1)	7.80	7.69	10.4 (cis)		1.5		71
						7.08 (R^2)			8.6 (trans)		1.5		

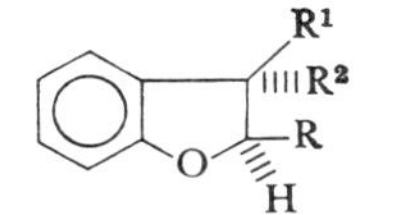

II. 2,3-DIHYDROBENZOFURANS

R	R^1	R^2	Others	Solvent	τ2	τ3	τ Arom.	J_{cis}	J_{trans}	J_{33}	Ref.
Ph	OH	H		$CDCl_3$	4.55, 4.90		2.67	6.0			73
Me	H	Me	4-Me, 5,7-di-Cl		5.53	6.95	2.88 (6)		4.1		75
Ph	H	H	4,5,6,7-Tetra-Cl	CCl_4	4.12	6.28 (R^1)		9.45	8.18	−16.6	77
						6.70 (R^2)					
CO_2Me	H	CO_2Me	4,5,6,7-Tetra-Cl	CCl_4	4.56	5.48			4.63		77
Ph	H	CO_2Me	4,5,6,7-Tetra-Cl	CCl_4	3.99	5.69			6.10		77
Ph	H	Ph	4,5,6,7-Tetra-Cl	CCl_4	4.33	5.43			5.70		77

TABLE 5.7 SPECTRAL DATA FOR 2,5-DIHYDROFURANS

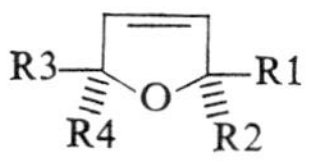

R1	R2	R3	R4	Solvent	τ2	τ3	τ4	τ5	J_{23}	J_{24}	J_{25}	J_{34}	Ref.
H	H	H	H	CCl_4	5.57	4.22	4.22	5.57					69
OMe	H	OMe	H						1.2	−1.2	0.4	6.0	80
				CCl_4	4.45	4.01	4.01	4.55					93
				C_6H_6	4.46	4.28	4.28	4.46					93
OAc	H	OAc	H						1.0	−1.0	0.4	5.6	80
OMe	H	H	OMe						1.2	−1.2	4.0	6.0	80
				CCl_4	4.22	4.01	4.01	4.22					93
				C_6H_6	4.26	4.20	4.20	4.26					93
H	OMe	OAc	OMe		4.92	3.81, 4.00			1.1	−1.1		5.7	79

by later work where it was shown that $J_{23} = -J_{24} = \sim 1$ Hz for both cis and trans 2,5-disubstituted derivatives (79, 80). In the latter of these two references, ^{13}C-satellite spectra were used to analyze the symmetrical AA′BB′ systems found in 2,5-dimethoxy or -diacetoxy derivatives. Also, in agreement with the dehydroprolinamide work mentioned above, a trans 2,5-coupling was shown to be much larger than its cis counterpart. The theoretical basis for this difference was discussed in terms of π and σ contributions to the coupling mechanism but, as with most work of this kind, the results obtained were inconclusive.

A number of simple spirooxetones or 2,5-dehydrofurans, for example, **33**, have been isolated from hop oil (81), and the exocyclic methylene-type structure **34** is sometimes found as an end group in polyacetylenes (82). Also, fusion of a benzene ring to the double bond of this ring system gives the phthalans (**35**) for which some data is available (83). Parameters obtained for 2,5-dihydrofurans are collected in Table 5.7.

In the course of research into donor–acceptor complexes, considerable attention has been devoted to complex formation between maleic anhydride (**36**) and aromatic solvents (84–87). Chemical shifts of protons on maleic anhydride were reported to vary with aromatic solvents (88) and an extremely detailed study of the phenomenon was made by Roberts and his co-workers (89). The chemical shift of the vinyl protons was studied as a function of structure in a wide range of aromatic solvents, of solvent composition in mixed solvents, of concentration, and of temperature. The large changes observed were attributed to stereospecific associations of a charge-transfer type between the maleic anhydride and the solvent molecules.

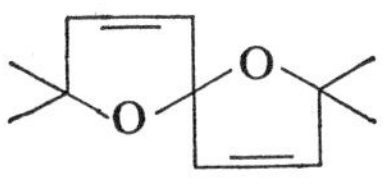

O O
33

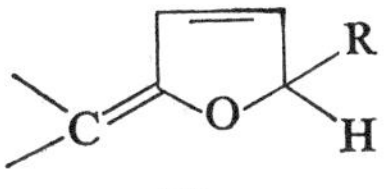

C O R H
34

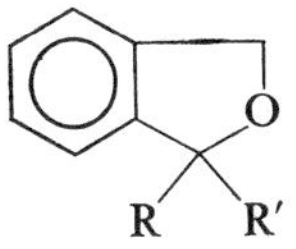

O R R′
35

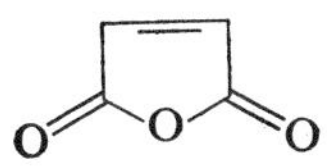

O O O
36

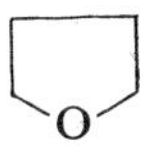

O
37

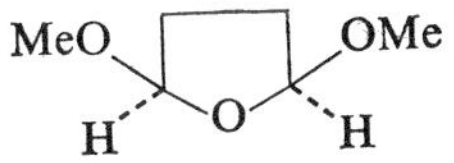

MeO OMe H O H
38

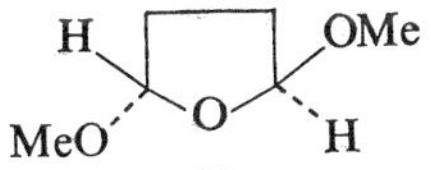

H OMe MeO O H
39

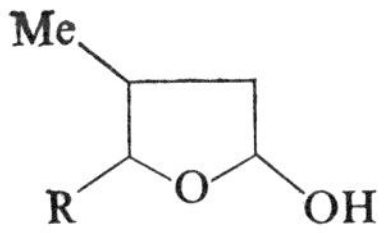

Me R O OH
40

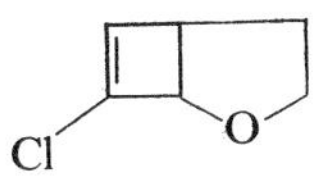

Cl O
41

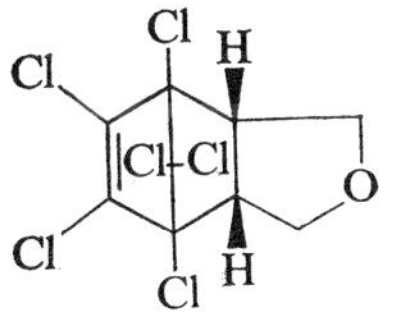

Cl Cl H Cl Cl O Cl H Cl
45

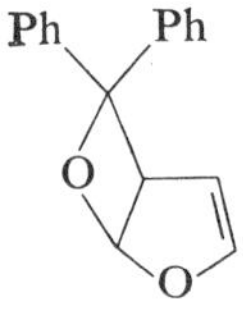

42 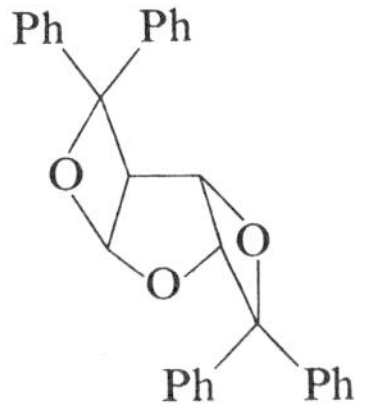 **43** 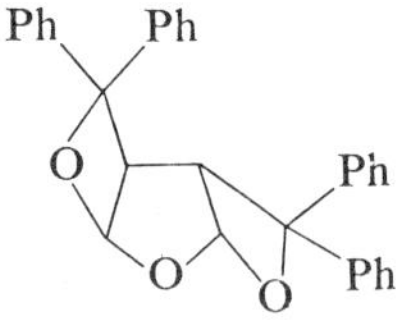 **44**

Information on the spectra of tetrahydrofurans (**37**) can be found in a large number of papers but, unfortunately, most of it is of little use in a discussion of the basic ring system. The most important contribution to this field comes from Gagnaire and his co-workers (80, 90, 91). In the earliest of these papers (90), the Karplus relationship was used successfully to distinguish between the very different coupling patterns expected for the isomers formed on bromination or glycolation of the double bond in *cis*- and *trans*-2,5-dimethoxyl-2,5-dihydrofurans (**38** and **39**). In compounds containing α-CH_2OH groups, these side-chain methylene protons were nonequivalent, and this was attributed to a predominant conformation in which internal hydrogen bonding occurs between the hydroxyl group and the ring oxygen atom (91).

Solvent shifts induced by benzene have been studied for a number of complex tetrahydrofurans, as well as for the parent compound (92). With tetrahydrofuran itself, all protons were shifted upfield when the solvent was changed from CCl_4 to benzene. However, in the case of the cis- and trans-2,5-dimethoxytetrahydrofurans, isomer specific shifts were observed (93). For the cis compound the β-methylene protons absorbed at lower field in benzene while the α-protons were not affected. The methylene protons of the trans isomer experienced no shift, but peaks from H-2 and H-5 moved upfield in the aromatic solvent. Hydrogen-bond formation between tetrahydrofuran and hydroquinone in cyclohexane has also been studied by NMR methods (94).

The stereochemistry of a series of isomeric hemiacetals of the type **40** has been elucidated with the aid of NMR (95) and data have been presented for a number of 3-CH_2X compounds, where X = OH, Cl, Br, or I (96). More highly substituted tetrahydrofurans are represented in a number of preparative papers, for example, Refs. 97–99.

The field of photochemistry provides us with a whole range of tetrahydrofuran derivatives, most with other rings fused onto one or more sides of this heterocyclic system. In the bicyclo compound **41**, the chemical shift of the vinyl proton was successfully predicted (with respect to the nonchlorinated compound) by calculating the anisotropy of the C–Cl bond (100). This class of compounds has also been studied more recently (100a). The photochemical addition of benzophenone to furan has been shown to produce mainly a mixture of one 1:1 adduct **42** and two isomeric 2:1 adducts, **43** and **44**. After some controversy, the structure of the monoadduct and the anti configuration of the diadducts (101–103) were established. Also, benzene-induced solvent effects have been used to analyze the spectra of various adducts of the type **45** (104).

Nuclear magnetic resonance spectroscopy has been used in a study of the copolymerization of tetrahydrofuran and 4-methyl-1,3-dioxene-4 (105) and in structural studies on a wide range of complex natural products containing this ring system, for example, lignans (**46**) (106), the aflatoxins, for example, **47** (107), and polyacetylenes (**48**) (108). Furanose sugars, as a group, have been omitted from this treatment.

Butyrolactones (**49**) have received some attention. Benzene-induced solvent shifts have been shown to be of use in determining the structure of compounds containing this structural feature (109, 92). High-field shifts of the β-methylene protons are usually larger than those of the α-methylene group (110). Riggs and his group (111, 112) have thoroughly investigated the spectra and conformations of butyrolactones and have concluded that these compounds, in solution, undergo rapid inversion between two envelope conformations of the type found in crystalline γ-lactones (113). A large collection of data on γ-lactones has appeared (114). In compounds containing a –CHX–CH_2– group two types of trans coupling were observed, 9–12 Hz for pseudoaxial orientations and less than 1.5 Hz for pseudoequatorial. Also, correlations were obtained between the chemical shifts of methyl groups and the nature of cis substituents on the other ring carbon atoms. A thorough study has been made of the spectra of the isomers of 3,4-disubstituted

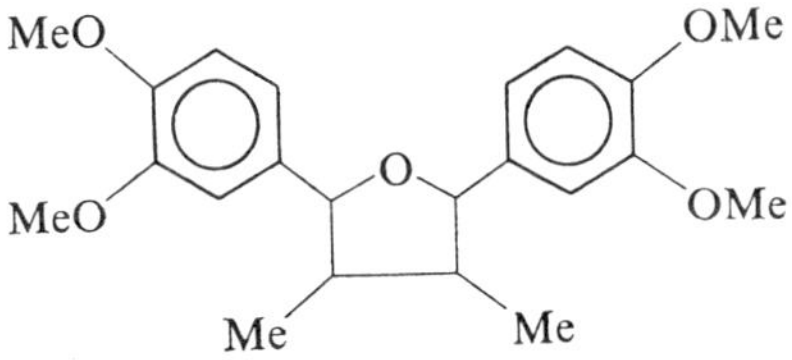

MeO OMe MeO OMe Me Me

46

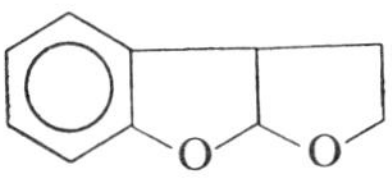

47

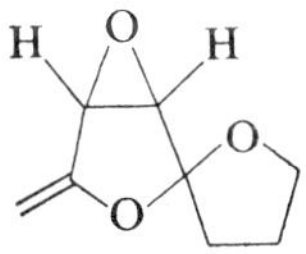

H O H

48

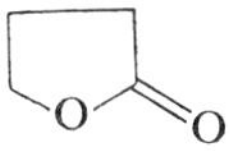

49

R′ R

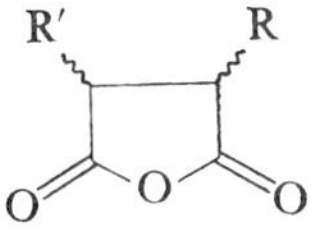

50

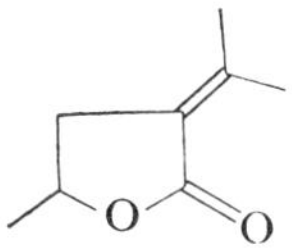

51

TABLE 5.8 SPECTRAL DATA FOR TETRAHYDROFURANS[a]

Substituents	Solvent	τ2, 5	τ3, 4	τ substit.	Ref.
None	$CDCl_3$	6.25	8.15		[b]
2,5-Di-Me (cis)	$CDCl_3$	6.14	8.32	8.80	91
(trans)	$CDCl_3$	5.98	8.32	8.88	91
2,5-Di-CO_2Me (cis)	$CDCl_3$	5.41	7.80	6.25	91
(trans)	$CDCl_3$	5.29	7.80	6.25	91
2,5-Di-CH_2OH (cis)	$CDCl_3$	5.85	8.13	6.35	91
(trans)	$CDCl_3$	5.90	8.15	6.35	91
2,5-Di-OMe (cis)	CCl_4	5.05	8.09	6.69	93
	C_6H_6	5.06	8.12, 8.30	6.69	93
(trans)	CCl_4	5.02	7.9–8.3	6.72	93
	C_6H_6	4.99	8.0–8.3	6.74	93
2,2,5,5-Tetra-CN	MeCN		6.68		97
3,4-Di-Me,					
2,2,5,5-tetra-CN (cis)	Me_2CO		6.22		97
(trans)			6.78		97

[a] For coupling constants and chemical shifts for 2,3,4,5-tetrasubstituted derivatives, see Refs. 80 and 90.

[b] Varian Catalog, No. 77.

succinic anhydrides (**50**) (115) and some information is available for compounds of the type **51** (116). Data for some simple tetrahydrofurans are collected in Table 5.8. ^{13}C chemical shifts for tetrahydrofuran and related heterocyclic systems have been reported (116a).

B. The 1,3-Dioxolan System

Dioxolans can be regarded as cyclic acetals or ketals and can be prepared with great ease. For this reason, a wide range of substituted derivatives is available, particularly with substituents in position 2, and this has led to a considerable amount of work on the conformations of this system.

The parent dioxolan has received little attention. Its spectrum was first reported in 1959 (117) when it was shown that the sum of the time-averaged vicinal couplings for a single proton on C-4 or C-5 was 13.3 Hz; the difference was 1.3 Hz. These results were later confirmed (118) in a study concerned with the application of the Karplus equation to these systems. Up till this time it was generally accepted that the 1,3-dioxolan system was planar but this work, and also some on the spectra of isopropylidene derivatives of sugars (119), led to the conclusion that the ring was puckered and actually existed as a number of rapidly interconverting conformers. Further evidence from 2-substituted dioxolans was interpreted in terms of a progressive rotation between the various half-chair and envelope conformations alternating around the ring (120). Abraham (121) has also been interested in the application of Karplus-type relationships to this ring system and showed that the normal equations are of little use.

Spectra of 2,2-dimethyldioxolans (**52**) have a special place in the history of NMR. It was with 4-substituted derivatives of this compound that Lemieux and his co-workers (122) first showed that geminal and vicinal couplings in aliphatic systems have opposite signs, a result which contradicted the theoretical predictions of the time. Also with these compounds, Anet, in 1962 (123), observed coupling (0.6 Hz) between the protons of the geminal methyl groups on position 2, a coupling which has been "rediscovered" many times since then.

Many papers involving 1,3-dioxolans unsubstituted in positions 4 and 5 have been published, for example, Refs. 118, 121, 124, and 125, and it is unfortunate that a number of these concentrate on the AA′BB′ analysis involved rather than on any real discussion of the results obtained. However, there are exceptions to this and one of the most outstanding of these (126) involves a combination of NMR and dipole moment data to determine the preferred rotamers of OR substituents in the 2-position. These preferences were shown to have a small but quite observable effect on the average conformation of the dioxolan ring. Finegold (127) found variations with change of solvent in vicinal couplings of the ethanic fragments of this and other five-membered rings which are held by substituents in a rotationally invariant conformation. He suggested that field effects due to solute–solvent associations may be important, especially with aromatic solvents.

Most of the remaining publications about 1,3-dioxolans use NMR for the determination of structure of various substituted derivatives (128–132) and hence the stereochemistry of the associated diols. However, a number of interesting phenomena have been observed in the course of this work. With 4-methyl compounds, the anisotropy of the methyl group caused an upfield shift of about 0.15 ppm for the vicinal cis protons (128). Long-range coupling occurred between this methyl group and the adjacent cis proton but not the trans proton, an observation which conflicts with the "extended *W*" coupling scheme. In a study of 4,5-disubstituted acetone ketals (**53**), an additive rule for calculating the chemical shifts of the two 2-methyl groups from the nature and orientation of the 4,5-substituents was developed (125). While it appears to work quite nicely for trans 4,5-substituents it does not for their cis counterparts, and an empirical correction factor was necessary before satisfactory results could be obtained. While this correction factor can be explained in terms of steric interaction between large cis substituents, to my mind it removes much of the value of the approach. Shifts observed for cis diaryl or divinyl substituents were interpreted in terms of structures in which the aromatic rings or the vinyl double bonds are aligned parallel to each other (**54**). A preliminary report on 2,4-cis-5-trisubstituted dioxolans (132) suggests that the syn isomers (H-4, H-5 cis) are normally more con-

52, R = H
53

54

55

56

formationally stable than their anti counterparts. In all cases studied, H-2 for the syn isomer resonates at lower field than that of the corresponding anti isomer.

In strong acids, 2-alkyl-1,3-dioxolans form dioxolenium cations of the type **55**. The various reports of these ions (133–136) use NMR mainly for characterization and to determine the effects of the charge on exocyclic substituents.

1,3-Dioxolan-2-ones (**56**) (cyclic carbonates) have been examined by a range of techniques including NMR (137). The spectral parameters appear to be independent of temperature and this suggests a rigid symmetrical ring system. Data for 1,3-dioxolans are collected in Table 5.9.

C. The Trioxolan Systems

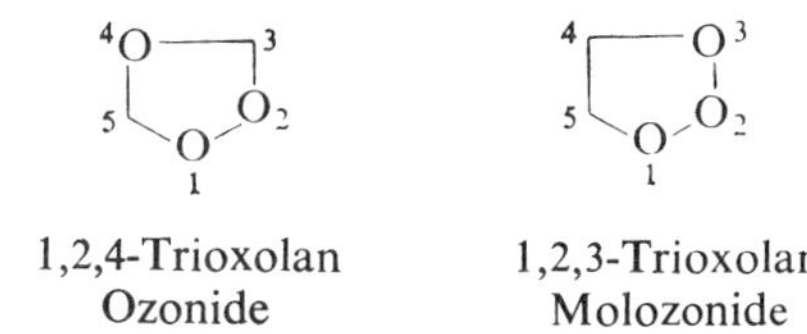

1,2,4-Trioxolan Ozonide — 1,2,3-Trioxolan Molozonide

Nuclear magnetic resonance techniques have been used on a number of occasions to check the structures formed on the ozonolysis of simple alkenes. At normal temperatures these are invariably the ozonides (138, 139) but at temperatures below -100°, extra peaks which were assigned to the

TABLE 5.9 SPECTRAL DATA FOR 1,3-DIOXOLANS

Substituents	Solvent	τ2	τ4, 5	J_{cis}	J_{trans}	J_{gem}	Ref.
None	Neat			7.3	6.0		126
2-Me	CCl_4			7.20	6.06	−7.68	120
2-Ph	Neat			7.16	6.19	−7.26	120
2-CH_2Br	C_6H_6			6.61	6.08	−7.68	126
	$CDCl_3$			6.58	6.02	−7.68	126
2-CCl_3	CCl_4			6.86	6.42	−7.75	126
	$CDCl_3$			6.89	6.30	−7.70	126
	C_6H_6			6.85	6.27	−7.69	126
2-OMe	Neat			6.80	6.24	−7.50	126
	CCl_4			6.70	6.11	−7.31	126
	C_6H_6			6.80	6.28	−7.60	126
2-OAc	C_6H_6			7.15	6.43	−7.64	126
2-Ph	$CDCl_3$	4.35 (2.5, 2.9)	6.18, 6.32	7.00	6.20	−7.75	129
2,2-Di-Me	CCl_4	8.74	6.17				125
2,2,4,5-Tetra-Me (cis)	CCl_4	8.67 (syn)	5.85	5.85			123
		8.78 (anti)					
(trans)	CCl_4	8.73			8.35		123

initially formed molozonide were also present (140, 141). Data for a number of these compounds are given in Table 5.10.

IV. SIX-MEMBERED OXYGEN HETEROCYCLES

A. The Pyran System

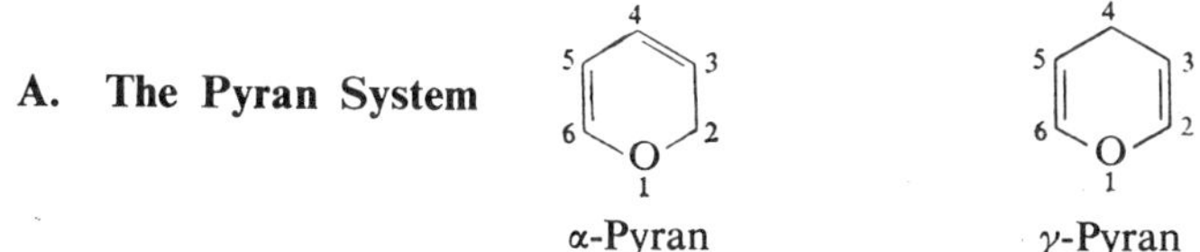

α-Pyran γ-Pyran

The simple pyran ring systems are not as readily available as many of the other oxygen heterocyclic systems discussed in this chapter, and, in fact, the α-pyran system does not exist as such but needs geminal substituents in the α-position to fix the bond structure. However, as will be seen later in this section, the various benzene-fused compounds form the basis of a very large number of natural products whose spectra have been studied exhaustively.

Spectral parameters for γ-pyran were first reported in 1962 (142) and then later in a report concerned with the analysis of the AA′BB′XX′ system

TABLE 5.10 SPECTRAL DATA FOR OZONIDES

A. OZONIDES

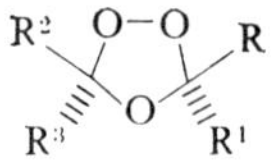

R	R^1	R^2	R^3	Temperature (°C)	Solvent	τR	τR^1	τR^2	τR^3	Ref.
Me	H	Me	H		CCl_4	8.65	4.72	8.65	4.72	138
Me	H	H	Me		CCl_4	8.65	4.78	8.65	4.78	138
Me	H	Me	Me		Me_2CO^a	8.60	4.62	8.51	8.51	139
Me	H	Me	H	−130	CCl_2F_2		5.0		5.1	141
Me	H	H	Me	−130	CCl_2F_2		4.95	5.03		141
Me	H	Et	H	−130	CCl_2F_2		5.05		5.05	141
Me	H	H	Et	−130	CCl_2F_2		5.07	5.07		141

B. MOLOZONIDES

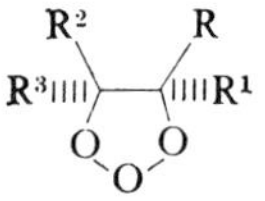

R	R^1	R^2	R^3	Temperature (°C)	Solvent	τR^1	τR^2	τR^3	Ref.
Me	H	Me	H	−130	CCl_2F_2	5.88		5.88	141
Me	H	H	Me	−130	CCl_2F_2	5.48	5.48		141
Me	H	Et	H	−130	CCl_2F_2	5.92		5.92	141
Me	H	H	Et	−130	CCl_2F_2	5.65	5.65		141

[a] External TMS reference.

(143). By current standards these values must be considered somewhat imprecise. Also, data have been given for a number of 3,5-diformyl-γ-pyrans (**57**) (144) and for the dimeric compound **58** (145). Nuclear magnetic resonance data effectively distinguished between 2,2,4,6-tetramethyl-α-pyran (**59**) and its exocyclic methylene isomer **60** (146). Data for pyrans are summarized in Table 5.11.

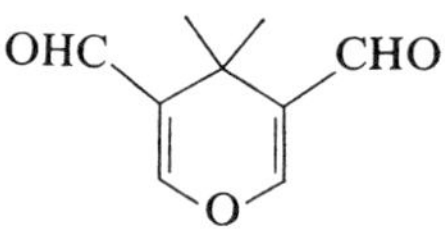

57

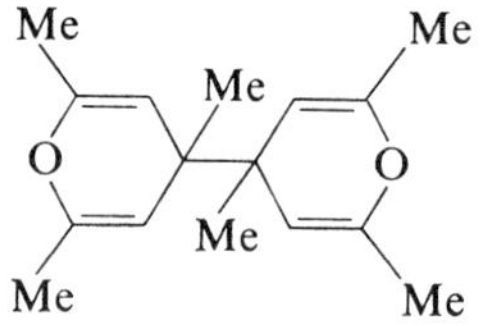

58

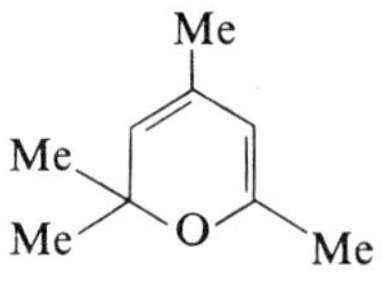

59

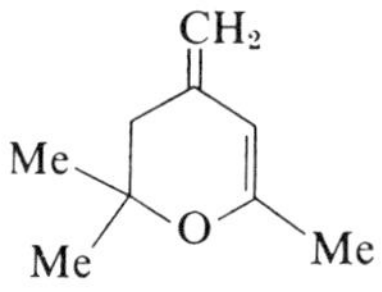

60

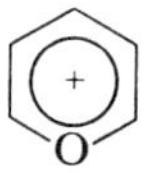

61

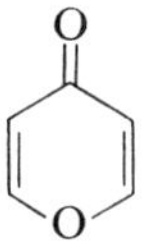

62

TABLE 5.11 SPECTRAL DATA FOR PYRANS AND DIHYDROPYRANS

A. PYRANS

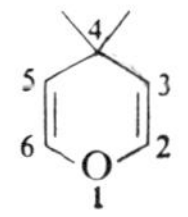

Substituents	Solvent	τ2	τ3	τ4	J_{23}	J_{24}	J_{26}	J_{34}	Ref.
None	CCl_4	3.84	5.37	7.34		1.5		3	142
	Neat	3.86	5.38	7.35	7.0	1.7	1.5	3.4	143
4-Et, 3,5-di-CHO	$CDCl_3$	2.6	(0.5)						144

B. 2,3-DIHYDRO-4*H*-PYRANS

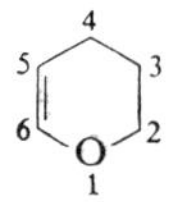

Substituents	Solvent	τ2	τ3	τ4	τ5	τ6	Ref.
2-CO_2Me, 6-OMe (trans)	$CDCl_3$	5.09	4.27	4.02	7.70	5.51	173
2-CO_2Bu, 6-OMe (trans)	$CDCl_3$	5.03	4.27	3.99	7.70	5.52	173
(cis)	$CDCl_3$	5.98	4.33	3.98	7.63	5.63	173

C. 2,3-DIHYDRO-6*H*-PYRANS. NEAT (174)

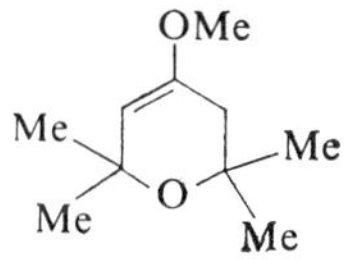

τ2,6(Me) 8.72; τ3 7.98; τ5 5.46; τOMe 6.46

Chemical shifts for a series of substituted pyrylium salts (**61**) were given in the first paper of Katritzky's large series on the NMR of heterocycles (147) and the results were later applied to structural determinations of a number of these salts (148, 149). A full analysis of the spectrum of the perchlorate of the parent ring system has appeared (143) and chemical shifts for the fluoborate are available (150). Information on simple pyrylium salts is given in Table 5.12.

Work on the spectrum of γ-pyrone (**62**) is centered around a continuing controversy as to the aromatic nature of this system [see the references cited in (151) for historical details] and it was only natural that the modern definition of aromaticity in terms of ring current should be applied. In the original paper on the spectrum of this compound (152), it was likened to a vinylogous lactone. After much discusssion of ^{13}C and ^{1}H NMR data, coupled with dipole moment measurements, the great complexity of the

TABLE 5.12 SPECTRAL DATA FOR PYRYLIUM SALTS

Substituents	X^-	Solvent	τ2	τ3	τ4	τ5	τ6	J_{23}	J_{24}	J_{26}	J_{34}	Ref.
None[a]	ClO_4^-	MeCN	0.41	1.60	0.80	1.60	0.41	3.5	2.4	1.5	8.0	143
	BF_4^-		0.5	1.6	0.8	1.6	0.5					150
2,4,6-Tri-Me	ClO_4^-	SO_2	7.10	2.23	7.26	2.23	7.10					147
	$SbCl_6^-$	SO_2	7.03	2.13	7.19	2.13	7.03					147
2,6-Di-Me	$SbCl_6^-$	SO_2	6.96	2.02		2.02	6.96					147

[a] J_{25} and J_{35} assumed zero in this treatment.

situation was emphasized and no real conclusions were reached. The next two pieces of work in this series (153, 154), while involving mainly the thiapyrones (**63**), used the effect of replacing the ring oxygen of pyrone with a sulfur atom to assign the meta couplings J_{26} and J_{35}. Unfortunately, this assignment of these couplings was the reverse of the original. In any AA′BB′ system the values of $J_{AA'}$ and $J_{BB'}$ can be interchanged without altering the pattern. Hence, with such molecules as 4-pyrone, the problem of assigning these two meta couplings from just the proton spectrum is a difficult one. However, as Goldstein and his co-workers pointed out, the problem can be uniquely resolved by consideration of the ^{13}C satellite spectra, and a very thorough analysis of the system (155) showed that the original assignment, $J_{26} < J_{35}$, was indeed correct. Garbisch had also arrived at the correct answer by comparing the values of J_{35} with those for compounds such as cyclohexadienone (**64**). That this disagreement should arise is somewhat disconcerting since it reflects on some of the procedures used in spectral analysis and underlines the difficulties encountered when one tries to use one heterocyclic system as a model for another, no matter how similar these may seem. More recently, the situation with respect to aromaticity has been summarized (156). 4-Pyrone behaves chemically as though it had considerable "aromaticity." Its spectra reveal the presence of a ring current, but this is not substantiated by magnetic susceptibility measurements. Again we are left with uncertainty, and this serves to emphasize how little we really understand about anisotropy and how difficult it is to apply the various experimental definitions of "aromaticity" in borderline cases.

Information is available on the spectra of a number of substituted 4-pyrones in papers dealing mainly with the synthetic aspects of pyrone chemistry (157–161). Also, chemical shifts of all protons in the cis and

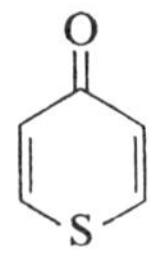

63 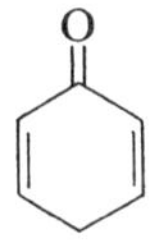 **64**

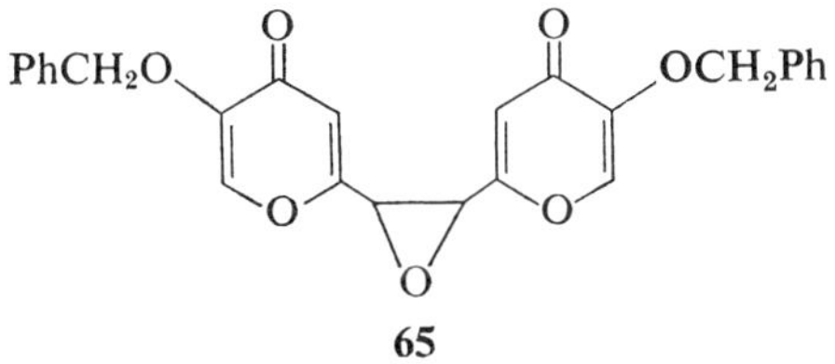

PhCH$_2$O ... OCH$_2$Ph

65

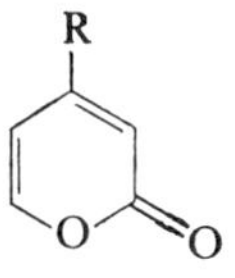

R

66, R = H

67, R = OMe

OH

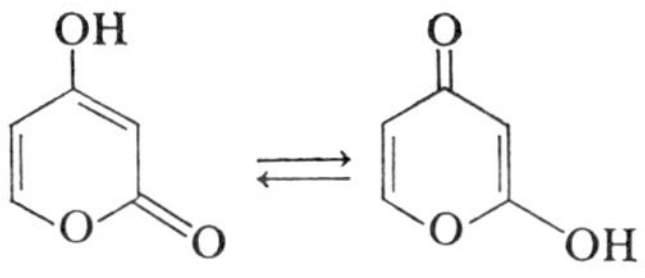

68 ⇌ **69** 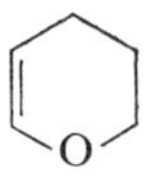 **70**

TABLE 5.13 SPECTRAL DATA FOR 4-PYRONES

Substituents	Solvent	τ2	τ3	τ5	τ6	J_{23}	J_{25}	J_{26}	J_{35}	J_{36}	J_{56}	Ref.
None	$CDCl_3$	2.29	3.65	3.65	2.29	5.88	0.42	1.12	2.68	0.42	5.88	152[a]
		2.12	3.62	3.62	2.12	5.96	0.33	1.05	2.68	0.33	5.96	153
	CCl_4	2.07	3.71	3.71	2.07	5.93	0.26	1.01	2.74	0.26	5.93	153
	C_6D_6	2.81	3.93	3.93	2.81	6.23	0.32	1.14	2.81	0.32	6.23	153
	TFA	1.51	2.94	2.94	1.51	~6.0	~0.4			~0.4	~6.0	153
	D_2O[b]	1.36	3.04	3.04	1.36							153
3-Me	$CDCl_3$	2.50	8.10	3.85	2.47							158
2,6-Di-Me	$CDCl_3$	7.76	3.96	3.96	7.76							159
	D_2O[b]	7.27	3.31	3.31	7.27							152
3,5-Di-Me	D_2O[b]	1.50	7.70	7.70	1.50							158
2-Me, 6-Ph	$CDCl_3$	7.64	3.83	3.31	2.44							158
2-Me, 3-OH	$CDCl_3$	7.62		3.57	2.27							152
2,6-Di-Ph	CCl_4	2.05–2.59	3.31	3.31	2.05–2.59							152
2-Me, 6-OMe	$CDCl_3$	7.75	3.99	4.51					2.0			160
												164

[a] $J_{^{13}C-2,H-2}$ 200.02, $J_{^{13}C-3,H-3}$ 168.54, $J_{^{13}C-2,H-3}$ 6.20, $J_{^{13}C-3,H-2}$ 9.0 Hz (concentrated aqueous solution).

[b] External TMS, infinite dilution.

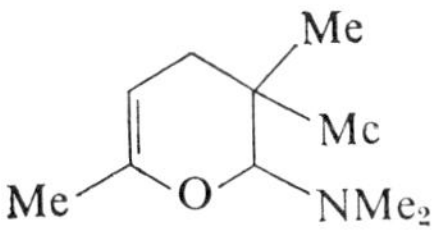

Me Mc NMe₂ O Me **71** RO_2C O OR **72**   OMe Me Me O Me Me **73**

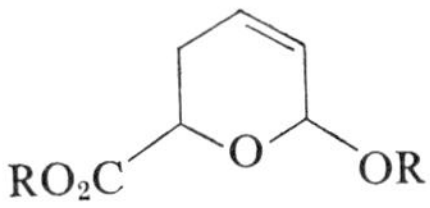

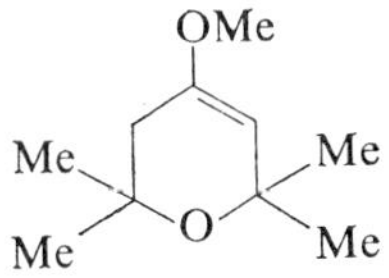

trans isomers of the epoxide **65** have been reported. Data for simple 4-pyrones are collected in Table 5.13.

Until recently, information on the spectra of simple α- or 2-pyrones (**66**) has been difficult to find. However, in 1969 the situation was remedied by Pirkle and Dines (162, 163) who have established the basic parameters for this ring system. A number of formyl derivatives have been reported (164), but the remaining examples are the 4-methoxy derivatives **67** used mainly as "fixed structures" in studies on the tautomeric equilibria between 4-hydroxy-2-pyrones (**68**) and 2-hydroxy-4-pyrones (**69**) (165–169). Table 5.14 contains the available information for these compounds.

The spectrum of 2,3-dihydro-4*H*-pyran (**70**) is included in the Varian Catalog (No. 111) and those of some substituted derivatives have been reported (170, 171). Also, the ^{13}C spectrum of 2-dimethylamino-3,3,6-trimethyl-2,3-dihydro-4*H*-pyran (**71**) has been measured (172). On the basis of their spectra, the preferred conformations of a series of *cis*- and *trans*-2-alkoxy-5,6-dihydro-2-pyran-6-carboxylic acids (**72**) have been determined (173). Data are also available for compound **73** (174). Gélin and Gélin (175, 176) have systematically studied the spectra of substituted 2,3-dihydro-4-pyrones in an effort to differentiate between isomers, and also to follow the tautomerism between these compounds and the corresponding open-chain ethylenic-β-diketones.

As with other reduced heterocyclic systems, most of the work on the fully reduced tetrahydropyrans has been concerned with questions of ring inversion and the effect of the oxygen lone pairs on such processes. The many investigations of pyranose sugars are not included here.

In carbohydrates, the "anomeric effect" has been invoked to explain why groups on C-1 tend to prefer the axial rather than the equatorial conformation, a situation not found in the cyclohexane series. Edward (177) first attributed the anomeric effect to dipolar interaction of the electronegative substituent with the unshared pairs of the adjacent ring oxygen and a number of variations of this definition have appeared. The overall picture is well summarized in Ref. 178 where the general definition of anomeric effect is given: "In any moiety C(X)–C(Y) (X = O, N, or S; Y = O, N, S, Cl, or Br) there are interactions favoring a *gauche* conformation." This statement sums up results published in a number of papers dealing with the spectra

TABLE 5.14 SPECTRAL DATA FOR 2-PYRONES

Substituents	Solvent	$\tau 3$	$\tau 4$	$\tau 5$	$\tau 6$	J_{34}	J_{35}	J_{36}	J_{45}	J_{46}	J_{56}	Ref.
None	$CDCl_3$	3.62	2.44	3.57	2.23	9.4	1.5	1.25	6.3	2.4	5.0	163
4-Me	$CDCl_3$	3.90		3.97	2.71		1.6	1.15			5.20	163
5-CHO	Me_2CO	3.55	2.15	0.21	1.39	9.8		1.0		2.6		164
6-CHO	Me_2CO	3.35	2.20	2.76	0.40	9.5	1.2		6.7			164
5-COCl	$CDCl_3$	3.49	2.13		1.32	10.0		1.0		2.6		163
5-NO_2	$CDCl_3$	3.55	1.90		1.20	10.5		1.0		3.0		163
6-Cl	$CDCl_3$	3.92	2.68	3.79		9.0	0.8		6.8			163
3-Br	$CDCl_3$		2.44	3.77	2.23				6.8	2.0	5.0	163
5-Br	$CDCl_3$	3.8	2.7		2.5	9.7		1.16		2.7		163
3,5-Di-Cl	CCl_4		2.45		2.35					2.25		163
	$CDCl_3$		2.37		2.20					2.5		163

and inversion kinetics of these systems (179–189). Anderson and Sepp (187) found a most interesting effect for some alkyl-substituted 2-carbomethoxy-tetrahydropyrans (**74**) where the carbomethoxy group shows a greater

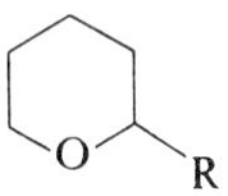

74, R = CO_2Me
75, R = OCH_2Ph

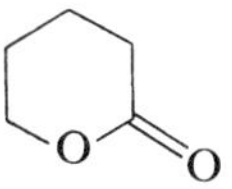

76

preference for the *equatorial* conformation than does this group on a cyclohexane ring, a "reverse anomeric effect." Also, an acetoxy group in 3-acetoxytetrahydropyrans tends to favor the axial position more than one would predict on steric grounds alone (190). This was again attributed to dipole–dipole interactions between the acetyl group and the ring oxygen.

Booth (191) has used tetrahydropyran derivatives to show that the effect of electronegative substituents on the couplings in ethanic fragments depends on the stereochemistry of the system. The ring oxygen atom would be expected to influence couplings across the 2,3-bond but has its maximum effect on the 3*e*-proton. Thus $J_{2a,3e}$ is smaller than $J_{2e,3a}$. Nonequivalence in the benzylic methylene group has been reported for a number of 2-benzyloxy-tetrahydropyrans (**75**) (192, 193). The influence of 4-substituents on the 1-position of a series of fluorinated and nonfluorinated tetrahydropyrans and similar carbocycles has been investigated (194).

Riggs and Johnson (195, 196) have made an extensive study of the conformations of a number of rapidly inverting, substituted valerolactones (**76**).

Fusion of a benzene ring to a pyran, pyrone, or one of their reduced derivatives affords entry into one of the most prolific areas of natural product chemistry. Here can be found the chromens, chromones, xanthones, flavonoids, coumarins, rotenoids, tocopherols, and many other classes of compounds present in bewildering variety in the plant kingdom. It is obviously impossible to deal successfully in a book such as this with the huge compilation of data available in this field and my general remarks will be confined to the more "review"-type publications. However, a large proportion of the available literature is summarized in Tables 5.15 and 5.16, which should allow the reader to find the information pertinent to a given ring system or substitution pattern. In many natural products, a variety of oxygen heterocyclic ring systems occur, fused in various positions onto the basic ring system. Fortunately, from an NMR point of view, most of these can be treated as though they alone were fused to the appropriate benzene ring because very little effect is transferred from more remote ring systems. For instance, the 2,2-dimethylchromene system **77** occurs in many very complex natural products but can always be recognized by its characteristic spectrum.

Double quantum transitions in the spectrum of chromene (**78**) were used to determine the relative signs of couplings between the four protons of the heterocyclic ring (197). Also, data for 2,2-dimethyl-3-chromene are included in a paper dealing mainly with hydroxy derivatives of this system (**79**) (198).

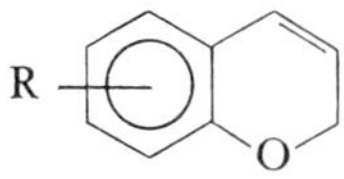

77
79, R = OH

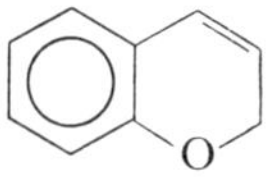

78

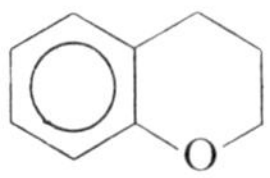

80

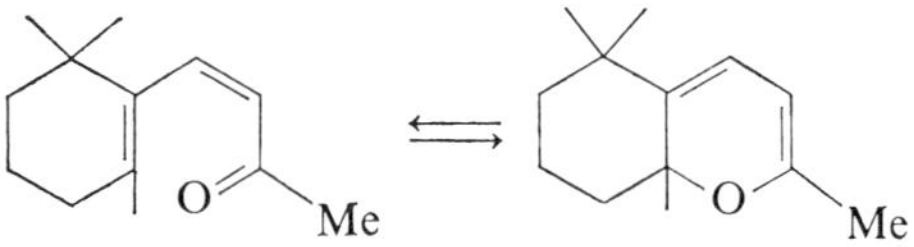

81

82

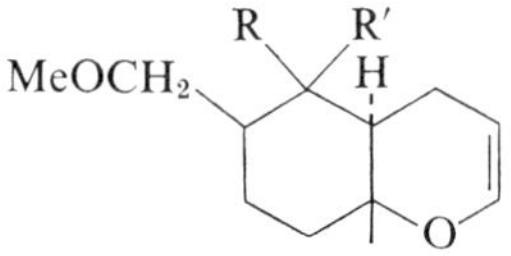

83

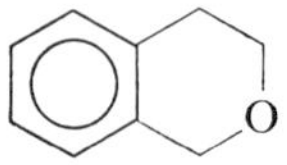

84

Benzene-induced solvent shifts of the various methyl and hydroxyl signals from *o*-hydroxy- or *o*-methoxyacetylchromenes have been studied (199). The characteristic nature of the signals from the 2,2-dimethylchromene ring is quite evident when one peruses the literature involving complex molecules containing this ring system, for example, Refs. 200–204. The gem dimethyl group gives a strong signal at about τ 8.5 and the AB quartet from the ethylenic protons can be easily recognized. As would be expected, the corresponding dimethylchroman system **80** is also found in many natural products, for example, Ref. 202. Information on highly substituted chromene and chroman systems can be found in papers on the tocopherols (205).

Spectra of a number of chromene derivatives reduced in the benzene ring have been reported. Valence isomerism occurs between *cis*-β-ionone (**81**) and the ring-closed α-pyran (**82**), and the equilibrium between these species in tetrachloroethylene has been investigated (206). Data for the rather similar compound **83** are also available (207).

Very little information is available on the isochroman series **84**, but a number of interesting observations were made of couplings in spectra of the cis and trans isomers of the 3,4-diphenyl derivatives (208). In the trans case, J_{34} was 3 Hz, showing conclusively that the phenyl groups strongly prefer the less hindered diaxial conformation. For the cis compound, the value for this coupling is 9 Hz, suggesting considerable distortion from the normally accepted half-chair forms.

Simple 2-chromones or coumarins (**85**) are extremely common natural products and, for this reason, quite a lot is known of their spectra. Dharmatti *et al.* (209) have analyzed the spectrum of the parent compound in tetrahydrofuran, and data are available for $CDCl_3$ solutions (210). Again, the AB system of the heterocyclic ring is clearly visible and the patterns from the benzene ring are typical of the substitution pattern. Also of diagnostic value are the small cross-ring couplings which can be observed, particularly between H-4 and H-8, when these positions are free (200). The relative

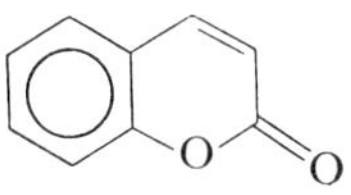

85 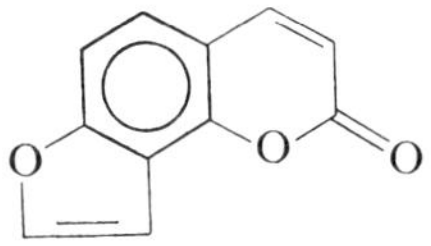 **86** 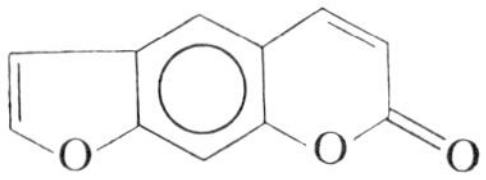 **87**

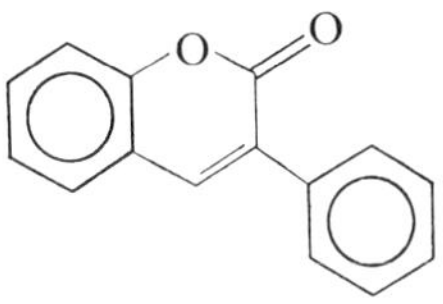

88 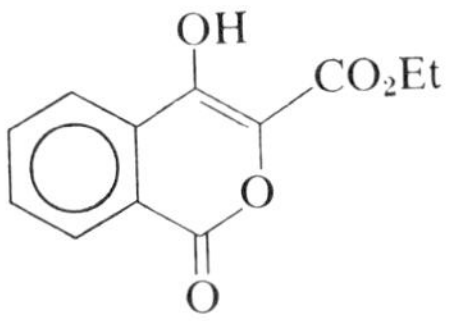 **89**

orientation of methyl, methoxyl, and aromatic protons in coumarins can be detected on the basis of their solvent shifts (211). Spectra of the common furocoumarins, particularly the angelicin types **86** (210) and the linear isomers **87** (212), have been reported (see also Ref. 213). Various members of the 3-arylcoumarin group or coumestans (**88**) are potent phytoestrogens and their spectra have been used to assist in structural elucidation (204, 214, 215). Data for simple coumarins are collected in Table 5.15. Others may be found in Refs. 216–218. Spectra for 3-carbethoxy-4-hydroxyisocoumarin (**89**), its acetoxy and dihydro derivatives have been described (219).

4-Chromones and their aryl derivatives constitute an extremely important and thoroughly investigated group of compounds. The spectrum of the basic ring system was fully analyzed by Mathis and Goldstein (152) as part of their general investigation of pyrone systems. Treatment of 4-chromone with sodium ethoxide produces a dimeric compound which, from its spectrum,

TABLE 5.15 SPECTRAL DATA FOR COUMARINS

Substituents	Solvent	$\tau 3$	$\tau 4$	$\tau 5$	$\tau 6$	$\tau 7$	$\tau 8$	J_{34}	J_{35}	J_{45}	J_{48}	Others	Ref.
None	$CDCl_3$	3.55	2.20	2.37	2.78	2.55	2.80						210
6,8-Di-Br	$CDCl_3$	3.52	2.40	2.48		2.11		9.60	0.35	0.40		J_{37} 0.35 J_{47} 0.15 J_{57} 2.20	200
5,7-Di-OAc	$CDCl_3$	3.62	2.29		3.03		2.98	9.75			0.65	J_{68} 2.05	200
6,7-Di-OAc	$CDCl_3$	3.62	2.37	2.68			2.81	9.60	0.35	0.35	0.68	J_{58} 0.40	200
6,8-Di-OAc	$CDCl_3$	3.64	2.34	2.66	2.89			9.55	0.30	0.35		J_{56} 8.60	200

has structure **90** (220). Spectra of a number of simple chromones with no substituent (221) or with alkyl substituents (222, 223) in the heterocyclic ring have been reported.

The spectrum of 2,3-dihydrochromone or chromanone (**91**) was reproduced in a paper by Schönberg *et al.* in 1966 (224) but unfortunately, their assignment of peaks to H-7 and H-8 needs to be reversed. More recently, the spectrum of this compound has again been pictured (225) and, in the only real NMR paper on the system, spectra of 3-bromo-, 3-chloro-, and 3-methylchromanone have been analyzed (226).

The 2- and 3-aryl derivatives of chromans, chromenes, chromones, and chromanones come under the general heading of flavonoids. It would be impossible in the space available here to discuss in any detail the huge body of literature on these compounds. Luckily, a number of general references on the spectra of various groups of flavonoids exist, and by judicious use of these the reader should be able to predict the spectrum of almost any member of these series. Basic information on flavones (**92**), flavanones (**93**), their

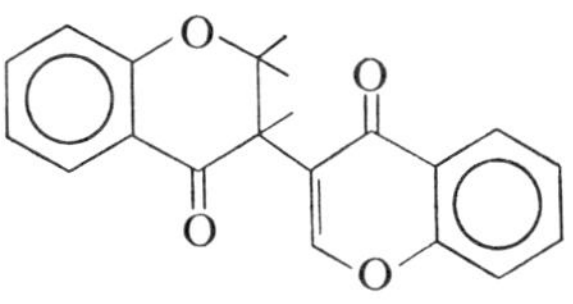

90

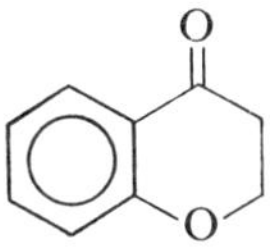

91

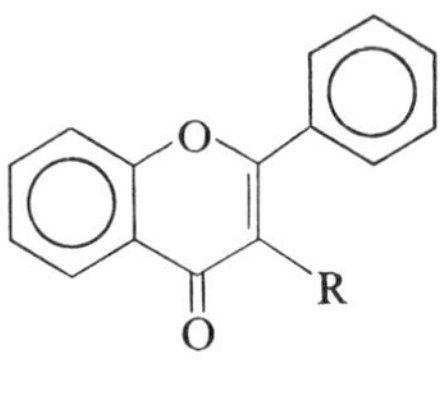

92, R = H
94, R = OH

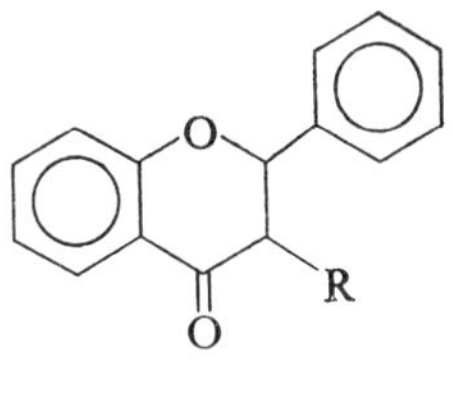

93, R = H
95, R = OH

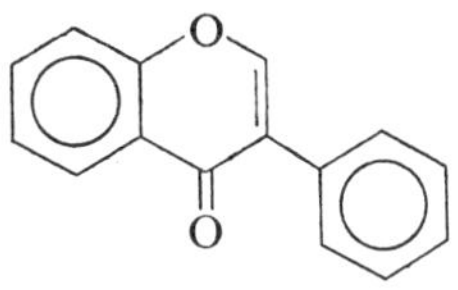

96

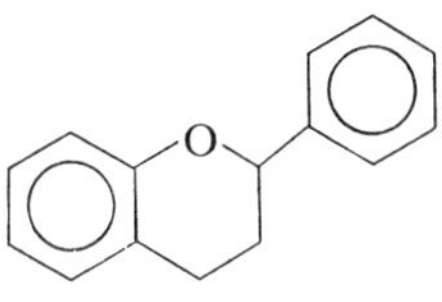

97

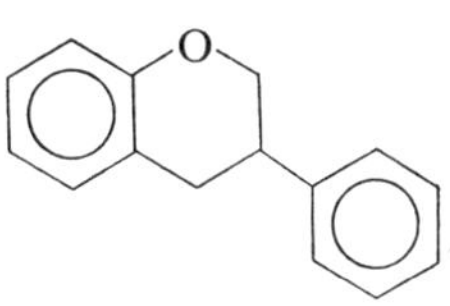

98

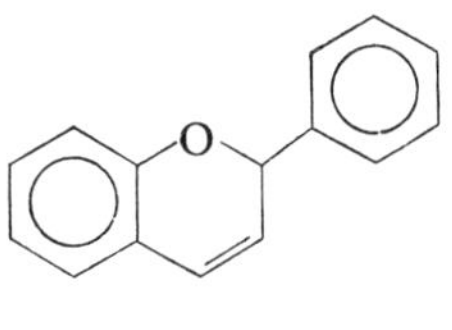

99

3-hydroxy compounds (**94** and **95**), isoflavones (**96**), and a number of reduced systems can be obtained from Refs. 227–229 and from the review 230. For information on the flavans (**97**), isoflavans (**98**), their 2- and/or 3-substituted derivatives, and the flav-3-enes (**99**), the excellent paper by Clark-Lewis (231) stands alone. The use of trimethylsilyl ethers to increase the volatility of flavonoids for gas chromatography also gives enhanced solubility in nonpolar solvents, a property which led to the use of these derivatives in NMR (232–234). In the flavone series, the heterocyclic ring and the adjacent aryl groups are well suited for strong interaction with aromatic and polar solvents, and a discussion of this aspect of the field may be found in papers by Williams and his co-workers (235). Many flavonoids occur naturally as glycosides and information on these can be found in some of the basic papers above as well as in Refs. 236 and 237 for *O*-glycosides and Refs. 238 and 239 for *C*-glycosides. Finally, a lead into the literature on rotenoids, many of which are based on the isoflavone skeleton, can be found in the papers of Crombie and his group, for example, Refs. 240 and 241.

As was mentioned earlier, the patterns produced by each of the carbocyclic rings in flavonoids are dependent on the substitution in that ring and the oxidation state and substitution in the heterocyclic ring. Variations in either of the benzene rings have little effect on the other one. Thus, one can classify these compounds by the substitution patterns in the aromatic rings, and these can be used to piece together the spectrum for almost any flavonoid. This approach has been used in Table 5.16 where references are listed for each substitution pattern of flavones and flavanones, subdivided in terms of the structure of the heterocyclic ring. It must be emphasized that, while Refs. 227–260 cover much of the recent literature, this compilation is not in any way exhaustive, but merely provides an entry into the correct area of the literature. The rings are classified in terms of their *O*-substituents only and the nature of these is indicated in the column headings.

A similar compilation for the more highly reduced flavonoids would be confined almost entirely to the paper by Clark-Lewis (231), and for this reason no table of these is presented. Further information is available on a number of the ring systems: catechins (229, 261), flavan-3,4-diols (262–265), flavans (266), isoflavans (267), dimeric compounds (268–271), and various *O*-bridged compounds (272–274).

As well as work on the natural flavanoids and their derivatives, a number of interesting reports, not included in the above references, have appeared covering aspects of the chemistry and stereochemistry of purely synthetic members of this class of compounds. Thus, configurations and/or preferred conformations of 4-acetoxy-3-bromoflavans (**100**) (275), 3-bromo-6-methylflavanone (**101**) (276), 3-bromo-3-nitroflavanone (**102**) (277), and 2-methyl-4′-methoxyisoflavanones (**103**) (278) have been studied. An investigation of

TABLE 5.16 LITERATURE ON FLAVONOID SUBSTITUTION PATTERNS

A. FLAVONES

RING A SUBSTITUTION PATTERNS

Substituents	X	Type of Substituent[a]					
		OH	OMe	OAc	OG	OTMS	*C*-Substituent
5-OR	H						227
6-OR	H			228			
7-OR	H	229	228	228			227, 229
	OH	229					
	OMe		227, 228 235				
	OAc			228			
5,7-Di-OR	H	229, 239, 242	227, 228, 235	228, 238, 239	229, 236	232, 236	238, 243
	OH	229	229, 235, 237, 244	244	237		
	OMe		228, 235				
	OAC		228	228			
	OG	229, 245			229		
	OTMS					232	
6,7-Di-OR	H		227, 228	228			
6,8-Di-OR	H			228			
7,8-Di-OR	OMe			235			
5,6,7-Tri-OR	H		227, 246, 247		247		227
	OH			248			
	OMe			235, 249			
5,7,8-Tri-OR	H			228			
	OMe		250	250			
5,6,7,8-Tetra-OR	H		251, 252				

TABLE 5.16 *(Continued)*

RING B SUBSTITUTION PATTERNS

Substituents	X	Type of Substituent[a]					
		OH	OMe	OAc	OG	OTMS	Unsubstituted
None	H						227, 228
							229
	OH						229
	OMe						235
4′-OR	H	229, 238, 239, 247	227, 228, 229, 239	228, 239	247	232	
	OH	229	248				
	OMe		228, 235				
	OAc			228			
	OG	229					
2′,4′-Di-OMe	OH	229	229				
	OMe		235				
3′,4′-Di-OMe	H	239, 244	227, 239, 246, 252	239			
	OH	229, 244					
	OMe	249, 250	227, 235				
	OAc			228			
	OTMS					232	
	OG	229, 245					
2′,3′,4′-Tri-OR	H		228				
3′,4′,5′-Tri-OR	H		227, 228		242		
	OH	229, 244					
	OMe		228, 235	228			
	OAc			228			
	OTMS					232	

B. FLAVANONES

RING A SUBSTITUTION PATTERNS

Substituents	X	Type of Substituent[a]				
		OH	OMe	OG	OTMS	*C*-Substituent
5-OR	H	227				
6-OR	H	229	227, 229			
7-OR	H	229		237		
	OTMS				232	
5,7-Di-OR	H	229	229, 253 254	229, 236 237	232	255
	OH	229				
6,7-Di-OR	H	227				
5,7,8-Tri-OR	H	227				

RING B SUBSTITUTION PATTERNS

Substituents	X	Type of substituent[a]				
		OH	OMe	OG	OTMS	Unsubstituted
None	H					229, 254
4′-OR	H	229, 255	237, 253		232	
	OH	229				
3′,4′-Di-OR	H		229, 237	236	232	
	OH	229				
	OTMS				232	

TABLE 5.16 *(continued)*

C. ISOFLAVONES

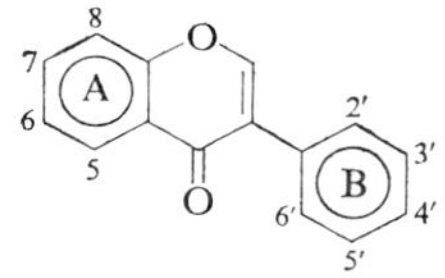

RING A SUBSTITUTION PATTERNS

Substituents	Type of Substituent[a]				
	OH	OMe	OAc	OG	OTMS
7-OR	229, 257			256	256
5,7-Di-OR	229, 258	227, 228, 258	228	259	
6,7-Di-OR		257			

RING B SUBSTITUTION PATTERNS

Substituents	Type of Substituent[a]				
	OH	OMe	OAc	O–CH_2O	Unsubstituted
None					227, 228
4′-OR	229, 258, 259	227, 256, 257	228		
3′,4′-Di-OR				257	
2′,4′,5′-Tri-OR		201		260	
2′,3′,4′,5′-Tetra-OR				203	

[a] Ring substituents: OH, free OH only; OMe, free OH and/or OMe only; OAc, acetate groups present; OG, glycosides; OTMS, trimethylsilyl ethers; *C*-substituent, any *C*-substituent.

the conformations of the syn and anti isomers of the 3-hydroxy- and 3-acetoxyflavanone oximes (**104**) led to the conclusion that acetylation of the 3-hydroxyl group caused a change in the preferred conformation of the oxime (279). Nuclear magnetic resonance has been used to follow the ring-chain

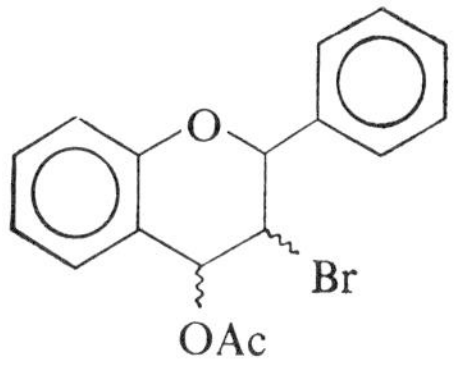

OAc Br
100

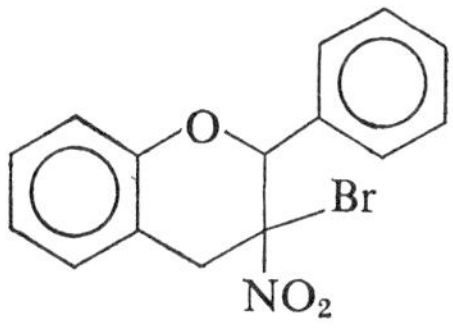

Br
NO_2
102

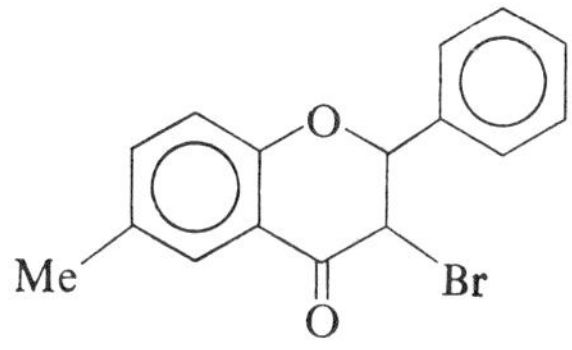

Me Br
O
101

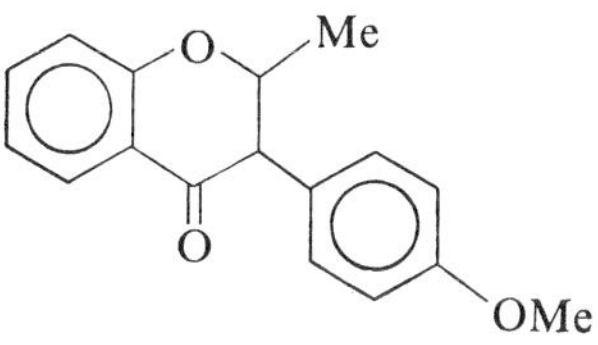

Me
O
OMe
103

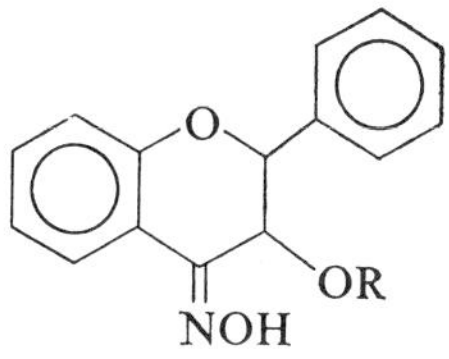

OR
NOH
104

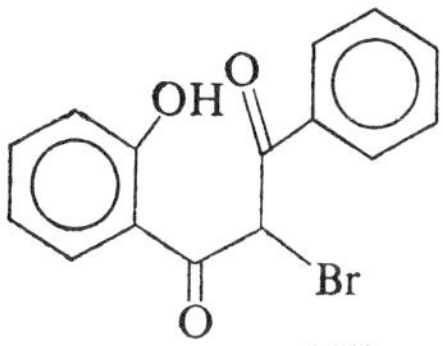

OH O
Br
O
105

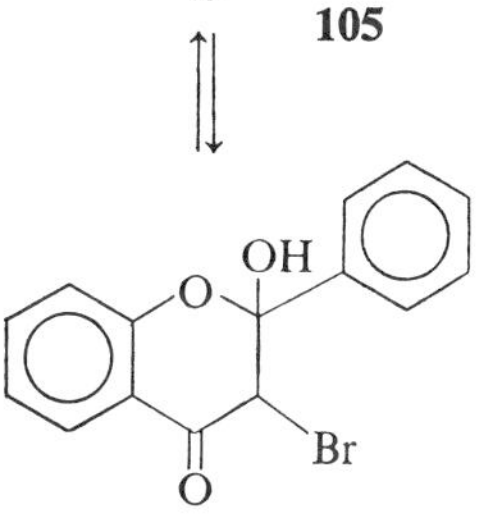

OH
Br
O
106

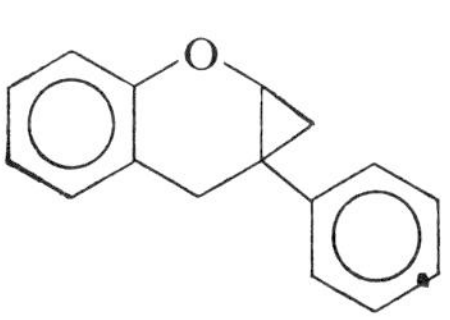

107

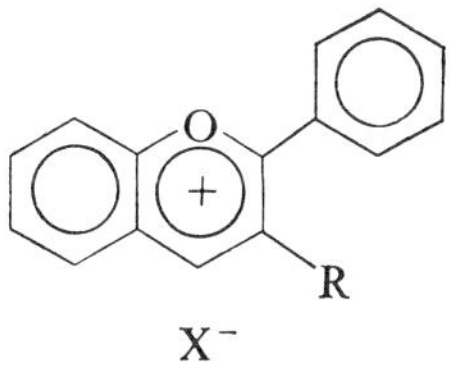

R
X^-
108

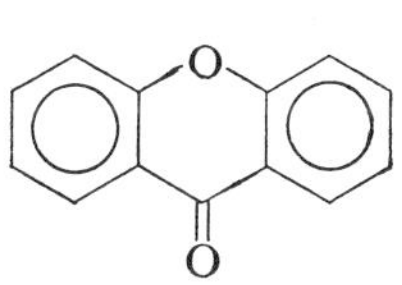

O
109

tautomerism between benzoyl (2-hydroxybenzoyl) methyl bromide (**105**) and its cyclic hemiacetal, 2-hydroxy-3-bromoflavanone (**106**) (280). Data for a number of the most interesting isoflavanones with isopropyl rings fused in the 2,3-position have been reported (**107**) (281).

Flavylium salts (**108**), because of their poor solubilities and relative scarcity in recent years, have been little studied. However, some information is available from Jurd and his co-workers, for example, Ref. 282 (see also Ref. 229).

Dibenzpyrones or xanthones (**109**) can, from an NMR point of view, hardly be classed as heterocycles. The spectral patterns are those which one would expect from appropriately substituted *o*-disubstituted benzenes. The spectrum of the parent xanthone has been analyzed (152) and data for many naturally occurring derivatives are available, for example, Refs. 283–286.

B. The Dioxan Systems

1. 1,2-Dioxan

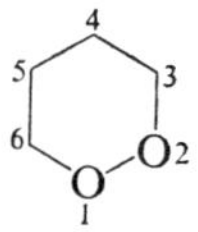

An NMR study of the inversion of 3,3,6,6-tetramethyl-1,2-dioxan (**110**) led to the suggestion that the symmetrical boat form (**111**) of this compound cannot exist and that inversion proceeds via unsymmetrical conformers (**112**) (287).

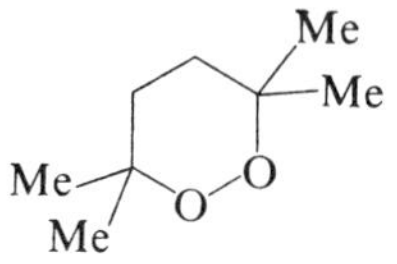

110 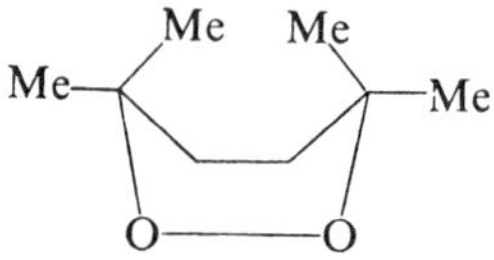 **111** 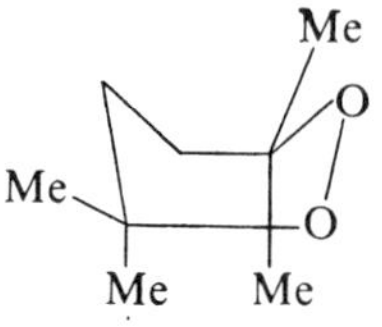 **112**

2. 1,3-Dioxan

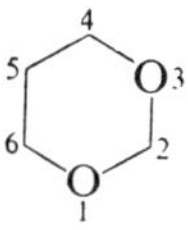

Compounds in this series are cyclic ketals which can be prepared simply and in great variety. Partly for this reason and partly because of their "cyclohexane-like" geometry, a great deal of NMR work has been done on the inversion of this ring system. However, as with the dioxolans, the spectral techniques have been used mainly as tools to determine thermodynamic parameters, and there is really little to be said about the spectra themselves (which are often not even reported in the publications). The difficulties inherent in thermodynamic calculations of this type are great and, almost without exception, very approximate methods have been used.

For this reason (see Ref. 288), many of the values obtained must be regarded as approximate and comparisons between different sets of data must be drawn with great caution.

One of the earliest approaches to the NMR study of the 1,3-dioxan inversion barrier was that of Friebolin *et al.* in 1962 (289). Since then three main groups have studied the ring system in great detail, Anteunis and his co-workers (e.g., Refs. 290–295), Delmau and his co-workers (e.g., Refs. 296–305), and Anderson and his co-workers (306–309), and many papers have appeared from authors with a more short-term approach to the system (e.g., Refs. 310–317). A thorough survey of the thermodynamic and conformational results and arguments in this very large accumulation of literature is beyond the scope of this chapter, but luckily has been admirably summarized in the extensive discussion of the ring system by Eliel and Knoeber (318). However, a number of specifically NMR phenomena will be discussed here.

A great deal of work with substituted 1,3-dioxans has established that the ring prefers various chair conformations. However, it has been suggested (291, 318) that the normal chair conformation is somewhat less buckled than that of cyclohexane. In this work it was necessary to take into account the influence of the oxygen lone-pair orbitals on coupling phenomena. Anteunis (319) proposes that increments of +2.3 and +1.8 Hz should be added to vicinal and geminal couplings, respectively, each time an α-oxygen or nitrogen atom has one of its lone-pair *p*-orbitals parallel to the CH bond of one of the coupled protons. Enhanced 4-bond couplings ($J_{4e,6e}$) of 2.7 Hz were observed in some asymmetrically substituted dioxans, and it was suggested that the orientation of the electronegative oxygens appears to modify greatly the extended *W* rule (310, 320). The preferences of substituents for equatorial rather than axial orientations are substantially different from those observed for cyclohexane derivatives and depend on the nature and location of the substituent on the ring. As with the tetrahydropyrans, the anomeric effect produces the most marked of these differences, for example, an alkoxyl group in position 2 of a 1,3-dioxan prefers the axial position (187). 5-Hydroxyl groups have been shown to prefer the axial conformation, presumably because of hydrogen bonding to the ring oxygen atoms (321). Also, evidence has been produced for an axial *t*-butyl group in some *cis*-2-alkyl-5-*t*-butyl-1,3-dioxans (**113**) (322). A general discussion of the "size" of the oxygen lone pairs and the "rabbit-ear" effect is given in Ref. 318.

As with carbocyclic compounds, equatorial protons usually absorb at lower field than do axial protons (295). The influence of an axial methyl group on the chemical shifts of axial protons causes a downfield shift of about 0.2 ppm (304) [cyclohexanols 0.18 ppm (323)]. However, such a group shields equatorial protons by 0.18 ppm, an effect opposite to that observed for cyclohexanols.

In the extremely complex spin systems observed for most 1,3-dioxans, analysis has been difficult, and it is not surprising that problems were encountered in detecting and assigning the many long-range couplings present. Discussions of these can be found in a number of papers (315, 301, 300, 320), and the current situation is summarized in Table 5.17. Also, the relative signs of the couplings observable in the spectrum of 4-phenyl-6-methyl-1,3-dioxan (**114**) have been determined (294). The expected 4-bond coupling between protons on geminal methyl groups in position 2 has been observed (see 1,3-dioxolans) (316, 324) and was discussed on the basis of

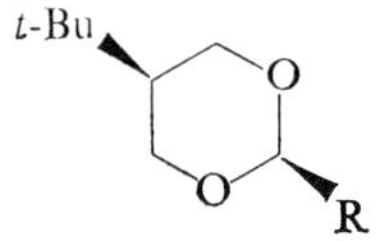

113

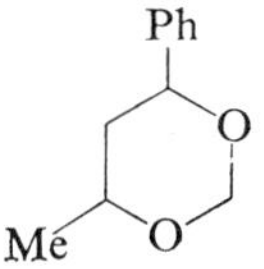

114

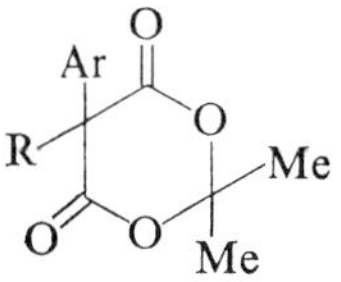

115

competition between "through-space" and "through-bond" contributions. Two short notes have appeared dealing with the effect of hydrogen bonding between hydroxy and hydroxymethylene substituents and the ring oxygen atoms on preferred conformations (312, 325).

Very specific benzene solvent shifts have been found for *m*-dioxan derivatives (307). In aryl-substituted derivatives of the isopropylidene acetals of malonic acid (**115**), the energetically favored conformation appears to have the substituent ring sitting on top of the heterocyclic ring (326). Also, with this compound, coupling between the geminal dimethyl groups was observed and benzene solvent effects gave useful information. Representative data for simple 1,3-dioxans are collected in Table 5.17.

3. *1,4-Dioxan*

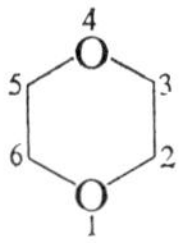

The parent compound, because of its time-averaged symmetry, gives only a singlet, τ 6.29 ($CDCl_3$) (327). Substitution in positions 2 and 3 can, in some instances, produce much more complex spectra, a singlet for H-2,3 and an AA′BB′ pattern for the 5,6-methylene groups. The singlet nature of

TABLE 5.17 SPECTRAL DATA FOR 1,3-DIOXANS

A. CHEMICAL SHIFTS[a]

Substituents	Solvent	$\tau 2e$	$\tau 2a$	$\tau 4e$	$\tau 4a$	$\tau 6e$	$\tau 6a$	$\tau 5e$	$\tau 5a$	Ref.
None	CCl_4	5.3		6.2				8.32		300
2-Me	CCl_4	(8.82)	5.45	6	6.4	6	6.4	8.77	8.02	303
4-Me	CCl_4	5.11	5.43	(8.83)	6.39	6.02	6.42	8.63	8.35	303
5-Me	CCl_4	5.18	5.51	6.1	6.78	6.1	6.78	(9.25)	8.02	303
2-*t*-Bu	CS_2	(9.20)	6.05	6.29				8.80	8.09	291
5-*t*-Bu	CS_2	5.44		6.03	6.60			(9.13)	8.36	291
4-Vinyl	CCl_4	5.05	5.53		6.10	5.98	6.35	8.50	8.10	303
4-Ph	CCl_4	4.9	5.3		5.55	5.98	6.4	8.50	8.10	303
4,5-Di-Me (cis)	CCl_4	5.12	5.41	(8.9)	6.2	6.26	6.26	8.61	(8.93)	303
(trans)	CCl_4	5.19	5.45	(8.8)	6.81	6.15	6.87	(9.30)	8.37	303
4,6-Di-Me (cis)	CCl_4	5.08	5.41	(8.83)	6.40	(8.83)	6.40	8.4–8.7		303
5-Me, 5-COMe	CCl_4	5.21	5.34	5.84	6.51					310
	C_6H_6	5.24	5.51	5.91	6.79					310

TABLE 5.17 *(continued)*

4-Ph, 5-Me (cis)	CCl_4	4.88	5.28		5.35	6.17	6.17	8	(9.18)	303
(trans)	CCl_4	4.9	5.3		6	6	6.7	(9.43)	8.05	303
4-Ph, 6-Me (cis)	CS_2	5.16		(2.85)	5.60	(8.87)	6.37	8.47		294
(trans)	CS_2	5.30		(2.88)	5.27	6.18	(8.73)	8.10		294

B. COUPLING CONSTANTS (Hz)[b]

Geminal		Vicinal		Long Range[c]	
2*e*,2*a*	−6.2	4*a*,5*a*	11.0	2*e*,6*e*	1.5
5*e*,5*a*	−13.2	4*a*,5*e*	2.8	2*e*,5*e*	0.9
6*e*,6*a*	−11.2	6*a*,6*e*	5.0	2*e*,6*a*	0.5
		5*a*,6*a*	11.0	4*a*,6*e*	0.5
		5*e*,6*e*	1.7	2*a*,6*e*	0.4
		5*e*,6*a*	2.8	2*a*,6*a*	0.3
				4*a*,6*a*	0.3

[a] Values in parentheses are for substituent groups.
[b] These can be summarized in the values obtained for 4-phenyl-1,3-dioxan (315).
[c] Signs of long-range couplings not determined.

the signal from H-2,3 is used to suggest that a number of cis-substituted compounds exist predominantly in a boat conformation, while the trans isomers probably have the normal chair structure. This work was extended by Jung (328), who fully analyzed the spectra of the cis and trans isomers of 2,3-dichloro-1,4-dioxan (**116**) and discussed the preferred conformations in terms of chair, boat, and "twist-boat" forms. An even more sophisticated treatment of the conformational problem for the 2,3-dimethyl isomers has been presented (329).

Because 1,4-dioxan is such a good solvent it was inevitable that reports would appear of work where dioxan was used as solvent and where specific solute–solvent interactions were important. The hydroxyl shifts for acetic acid as a function of concentration in rigorously dried acetic anhydride, acetone, and 1,4-dioxan were shown to decrease linearly at low concentrations, allowing accurate infinite dilution shifts to be obtained (330). Shifts of the dioxan peak in aqueous solutions of aluminum perchlorate have been interpreted in terms of solvation of the ClO_4^- ion (331). Proton magnetic resonance evidence has been obtained for the formation of an addition complex between methyl mercurous chloride (MeHgCl) and dioxan (332). The geometry of the ring system has also been studied indirectly using benzene-induced solvent shifts (333).

Data for a number of bicyclic 1,4-dioxans have been reported. Some derivatives of *cis*- and *trans*-tetrahydropyrano[2,3-*b*]-1,4-dioxan (**117**) were

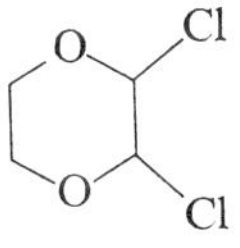

116 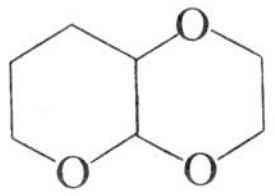 **117** 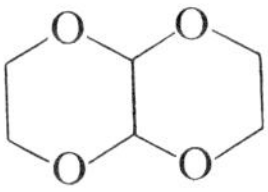 **118**

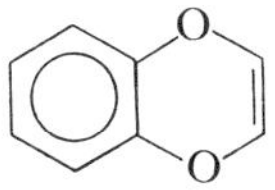

118a

prepared as models and their spectra measured. Coupling constants in the naphthodioxan system **118** led to the conclusion that the rings are cis-fused chairs, each rapidly inverting (334–336). Details of the spectra of simple 1,4-dioxans are summarized in Table 5.18. No evidence could be found, from the magnitudes of vicinal and ^{13}C–H couplings, for aromaticity in the heterocyclic ring of 1,4-benzodioxin (**118a**) (336a).

TABLE 5.18 SPECTRAL DATA FOR 1,4-DIOXANS

R	R^1	R^2	R^3	Solvent	τAA′	τBB′	τCC′[a]	$J_{AA'}$	J_{AB}	$J_{AB'}$	J_{AC}	$J_{BB'}$	$J_{A'C'}$	$J_{CC'}$	Ref.
H	H	H	H											10.2 (*e,e*)	354
														2.8 (*a,e*)	354
Cl	H	H	Cl	CCl_4	5.64	6.30	4.05	12.4	3.3	−12.2	−0.7	0.6	−0.9	0.4	328
H	Cl	H	Cl	CCl_4	5.84	6.24	4.38	3.21	6.35	−12.1		3.2			328
Me	H	H	Me	DMSO				11.51	2.7	−11.7		0.6		8.5	329
H	Me	H	Me	C_6H_6				3.2	6.4	−11.8		3.2		3.2	329
Ph	H	H	Ph	$CDCl_3$	5.58		5.22								327
H	Ph	H	Ph	$CDCl_3$			4.88								327

[a] CC′ represents the pair of protons on the substituted side of the ring.

C. The 1,3,5-Trioxan System

Ring inversion of the parent heterocycle has been studied and rate parameters were obtained through line-shape studies of the AB quartets produced by the O–CH_2–O protons at low temperature (311). The similarity of the results to those obtained with cyclohexane led to the conclusion that the oxygen lone pairs behaved similarly to protons in the inversion process.

Investigation of the proton spectrum obtained from *s*-trioxan in the nematic solvent, *p*,*p*′-di-*n*-hexyloxyazoxybenzene, showed that the molecule exists predominantly in two rapidly inverting chair conformations (337). The indirect ^{13}C–H coupling constant was positive. Each methylene group seems to be slightly tilted by about 3° from the O–C–O plane in the direction which slightly increases the distance between axial protons.

The cyclic trimers formed by various aldehydes are in fact symmetrical 2,4,6-trisubstituted 1,3,5-trioxans (**119**) and some work has been done on their relatively simple spectra. The trimethyl derivative, paraldehyde, was used by Diehl and Freeman (338) as a "disk-like" model in their study of reaction-field effects. Conformations of the cis-cis (β) and cis-trans (α) isomers of paraldehyde have been determined from a consideration of shielding effects in their spectra (339). Other aldehydes were shown to always form the cis-cis isomers which exist in the chair conformation with all substituents equatorial. Some data for this ring system are collected in Table 5.19.

D. The 1,2,4,5-Tetroxan System

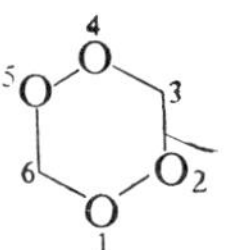

The only member of this ring system whose spectra have been studied is the tetramethyl derivative, acetone diperoxide (**120**) (340). At temperatures below 0°, two clean methyl peaks were observed, separated by 0.44 ppm, a

119 **120**

TABLE 5.19 SPECTRAL DATA FOR 1,35-TRIOXANS (339)

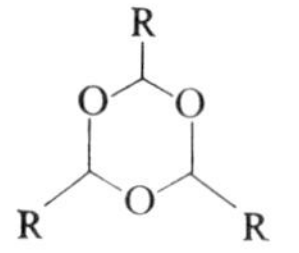

R	Isomer	Solvent	τ Ring H	τR
Me	Cis-cis	CCl_4	5.08 (ax.)	8.74
	Cis-trans	CCl_4	4.79 (ax.)	8.61
			4.72 (eq.)	8.76
$CHCl_2$	Cis-cis	CCl_4	4.36	4.84
CH_2Cl	Cis-cis	CCl_4	4.95	6.48
Et	Cis-cis	CCl_4	5.32	8.39, 9.09
CH_2CH_2Cl	Cis-cis	CCl_4	4.92	7.91, 6.43
CH_2CH_2CN	Cis-cis	$CDCl_3$	4.85	7.95, 7.46
$CH{=}CH_2$	Cis-cis	C_5H_5N	4.41	3.98, 4.68, 4.40

value very similar to that for dimethylcyclohexanes (0.445–0.48 ppm). Models suggest that a true boat form is energetically unfavorable so the suggestion was made that inversion proceeds through a twist-boat form and a narrowly defined barrier form of approximately planar geometry.

V. SEVEN-MEMBERED OXYGEN HETEROCYCLES

A. The Oxepin System

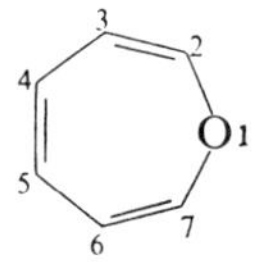

Most work on simple oxepins has been done by various German groups studying the equilibria between benzene oxides and the corresponding oxepins. At −113° in CF_3Br–pentane (2:1), the equilibrium between benzene oxide (**121**) itself and oxepin was slowed enough for the spectra of both isomers to be seen (341). A similar study has been made of the equilibrium between α-methyloxepin (**122**) and toluene-1-oxide (**123**) (342). Spectral parameters of the AA′BB′ system formed by the 3-, 4-, 5-, and 6-protons of oxepin were included in an investigation of this and similar seven-membered rings: δ 27.62, J_{45} 9.68, J_{34} 6.80, J_{35} 0.73, J_{36} −0.16 Hz (343). 4,5-Dicarbomethoxyoxypin (**124**) gives a simple AB quartet: τ 3.73, 4.94 ppm; J 5 Hz (344).

Very little is known of the spectra of reduced oxepins. Data for 3-chloro-2-(2-chloroethoxy)-2,5,6,7-tetrahydro-oxypin (**125**) have been reported (345).

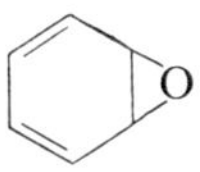

121 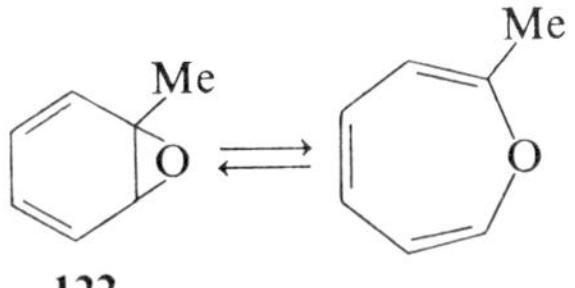 122 123 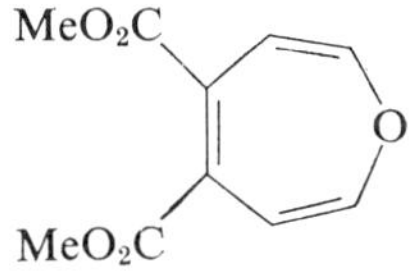 124

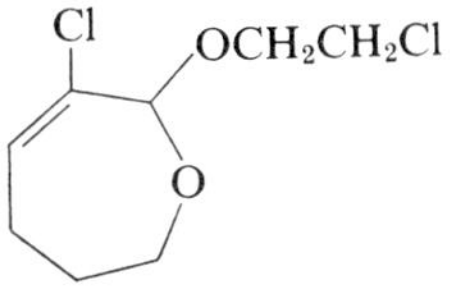

125 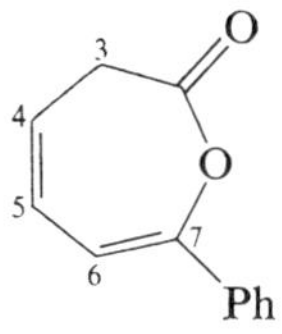 126 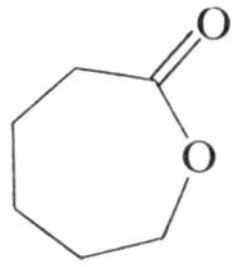 127

A very interesting unsaturated ε-lactone has been reported by Foa *et al.* (346). Parameters for the compound **126** are as follows: $\tau 3$ 3.09, $\tau 4$ 5.80, $\tau 5$ 3.69, $\tau 6$ 3.52 ppm; J_{34} 6.61, J_{35} 0.48, 0.53, J_{36} 0.53, J_{45} 9.29, J_{46} 0.22, J_{56} 5.58 Hz. Benzene-induced solvent shifts in the spectra of ε-caprolactones (**127**) have been shown to be larger for β- than for α-protons (110).

Data for a number of 2,4-disubstituted 3-benzoxepins (**128**) have been published (347). A line-shape analysis has been used to study the effect of hydrogen bonding in the inversion rate of 1-hydroxy-5,7-dihydrodibenz-[*c*,*e*]oxepin (**129**) (348). The sizes of the vicinal coupling, $J_{10,11}$, for a number of cis and trans isomers of 10,11-disubstituted 10,11-dihydrodibenz[*b*,*f*]oxepins (**130**) have been measured. Also, quite a lot of information is available on the spectra of various isomeric tetrahydronaphthoxepins (350, 351).

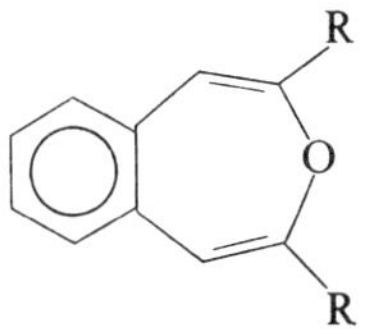

128 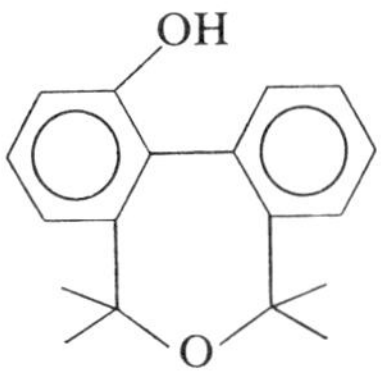 129 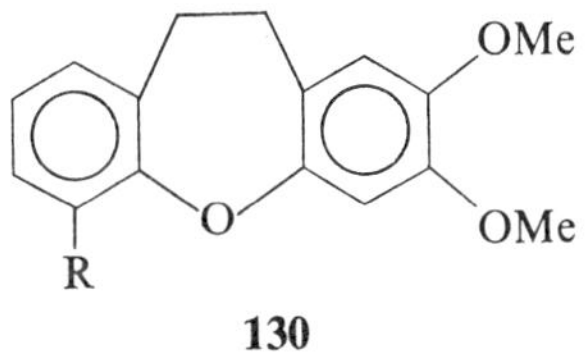 130

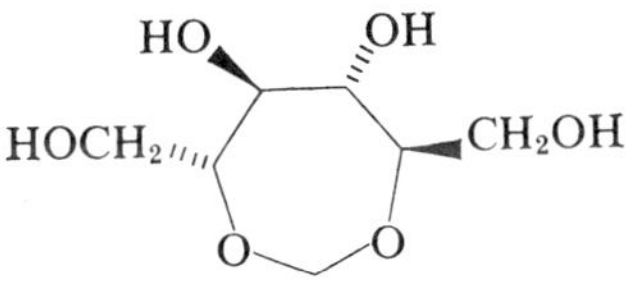

132 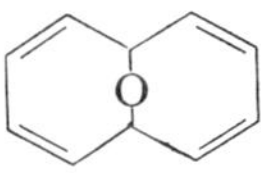 131

It seems reasonable to include here the very interesting "bridged oxepin," 1,6-oxidocyclodecapentaene (**131**). Chemical shifts for the two types of protons in the AA′BB′ system of this symmetrical compound have been reported (τ 2.8, 2.6) (352).

B. The 1,3-Dioxepan System

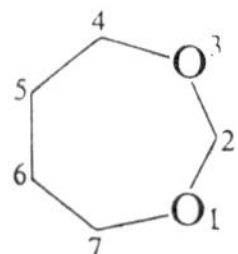

Variable-temperature studies on 1,3:2,5:4,6-tri-*O*-methylene-D-mannitol, which contains the moiety **132**, in thionyl chloride reveal that the 4- and 7-CH_2OH groups are magnetically equivalent. This information was used as evidence for the predominance of a "twist-chair" conformer (353).

REFERENCES

1. Gutowsky, Karplus, and Grant, *J. Chem. Phys.*, **31**, 1278 (1959).
2. Reilly and Swalen, *J. Chem. Phys.*, **32**, 1378 (1960).
3. Reilly and Swalen, *J. Chem. Phys.*, **34**, 980 (1961).
4. Reilly and Swalen, *J. Chem. Phys.*, **35**, 1522 (1961).
5. Lauterbur and Kurland, *J. Am. Chem. Soc.*, **84**, 3405 (1962).
6. Mortimer, *J. Mol. Spectrosc.*, **5**, 199 (1962).
7. Elleman and Manatt, *J. Mol. Spectrosc.*, **9**, 477 (1962).
8. Lehn and Riehl, *Mol. Phys.*, **8**, 33 (1964).
9. Williamson, Lanford, and Nicholson, *J. Am. Chem. Soc.*, **86**, 762 (1964).
10. Dailey, Gawer, and Neikam, *Discuss. Faraday Soc.*, **34**, 18 (1962).
11. Smith and Cox, *J. Mol. Spectrosc.*, **16**, 216 (1965).
12. Lyle and Keefer, *J. Org. Chem.*, **31**, 3921 (1966).
13. Riehl, Lehn, and Hemmert, *Bull. Soc. Chim. Fr.*, **1963**, 224.
14. Cheer and Johnson, *J. Am. Chem. Soc.*, **90**, 178 (1968).
15. Carlson and Behn, *J. Org. Chem.*, **32**, 1363 (1967).
16. Carlson and Behn, *Chem. Commun.*, **1968**, 339.
17. Perraud, Pierre, Butolo, and Arnaud, *Compt. Rend., Series C*, **268**, 974 (1969).
18. Williams, Ronayne, Moore, and Shelden, *J. Org. Chem.*, **33**, 998 (1968).
19. Seyden-Penne, Arnaud, Pierre, and Plat, *Tetrahedron Lett.*, **1967**, 3719.
20. Pierre and Arnaud, *Bull. Soc. Chim. Fr.*, **1969**, 2868.
21. Lambert and Johnson, *J. Am. Chem. Soc.*, **90**, 1349 (1968).
22. Ripoll and Conia, *Bull. Soc. Chim. Fr.*, **1965**, 2755.
23. Vogel and Günther, *Angew. Chem., Int. Ed. Engl.*, **6**, 385 (1967).
24. Günther, Schubart, and Vogel, *Z. Naturforsch.*, **22B**, 25 (1967).
25. Vogel and Klärner, *Angew. Chem., Int. Ed. Engl.*, **7**, 374 (1968).
26. Atlani and Leroy, *Compt. Rend., Series C*, **269**, 349 (1969).
27. Pews, *J. Am. Chem. Soc.*, **89**, 5605 (1967).
28. Batterham, Hart, and Lamberton, *Aust. J. Chem.*, **19**, 143 (1966).
29. Lardon and Reichstein, *Helv. Chim. Acta*, **46**, 392 (1963).

30. Bhacca, Johnson, and Shoolery, *NMR Spectra Catalog*, Varian Associates, Palo, Alto, California, 1962.
31. Lustig, Ragelis, and Duy, *Spectrochim. Acta*, **23A,** 133 (1967).
31a. Lozac'h and Braillon, *J. Chim. Phys.*, **67,** 340 (1970), and references cited therein.
32. Shingemitsu, Odaira, and Tsutsumi, *Tetrahedron Lett.*, **1967,** 55.
33. Rivas and Payo, *J. Org. Chem.*, **32,** 2918 (1967).
34. Krauch and Farid, *Tetrahedron Lett.*, **1966,** 4783.
35. Leane and Richards, *Trans. Faraday Soc.*, **55,** 518 (1955).
36. Abraham and Bernstein, *Can. J. Chem.*, **37,** 1056 (1959).
37. Abraham and Bernstein, *Can. J. Chem.*, **39,** 216 (1961).
38. Abraham and Bernstein, *Can. J. Chem.*, **39,** 905 (1961).
39. Reddy and Goldstein, *J. Am. Chem. Soc.*, **84,** 583 (1962).
40. Diehl, Khetrapal, and Kellerhals, *Helv. Chim. Acta*, **51,** 529 (1968).
40a. Fried, *Compt. Rend.*, *Series C*, **262,** 1497 (1966).
41. Abraham and Thomas, *J. Chem. Soc.*, *B*, **1966,** 127.
41a. Black, Brown, and Heffernan, *Aust. J. Chem.*, **20,** 1325 (1967).
42. Corey, Slomp, Dev, Tobinaga, and Glazier, *J. Am. Chem. Soc.*, **80,** 1204 (1958).
43. Gronowitz, Sörlin, Gestblom, and Hoffman, *Ark. Kemi*, **19,** 483 (1962).
44. Prugh, Huitric, and McCarthy, *J. Org. Chem.*, **29,** 1991 (1964).
45. Pascal, Morizur, and Wiemann, *Bull. Soc. Chim. Fr.*, **1965,** 2211.
45a. Roques, Combrisson, Riche, and Pascard-Billy, *Tetrahedron*, **26,** 3555 (1970); also Martin, Roze, Martin, and Fournari, *Tetrahedron Lett.*, **1970,** 3407.
46. Gronowitz and Hoffman, *Acta Chem. Scand.*, **13,** 1687 (1959).
47. Forsén, Akermark, and Alm, *Acta Chem. Scand.*, **18,** 2313 (1964).
48. Dahlqvist and Forsén, *J. Phys. Chem.*, **69,** 1760 (1965).
49. Dahlqvist and Forsén, *J. Phys. Chem.*, **69,** 4062 (1965).
50. Freeman and Whiffen, *Mol. Phys.*, **4,** 321 (1961).
51. Eugster and Bosshard, *Helv. Chim. Acta*, **46,** 815 (1963).
52. Trofimenko, *J. Am. Chem. Soc.*, **85,** 1357 (1963).
53. Wiersum and Wynberg, *Tetrahedron Lett.*, **1967,** 2951.
54. Page, Alger, and Grant, *J. Am. Chem. Soc.*, **87,** 5333 (1965).
55. Weigert and Roberts, *J. Am. Chem. Soc.*, **90,** 3543 (1968).
56. Tori and Nakagawa, *J. Phys. Chem.*, **68,** 3163 (1964).
57. Rosenkranz, Allner, Good, von Philipsborn, and Eugster, *Helv. Chim. Acta*, **46,** 1259 (1963).
58. Hofmann, von Philipsborn, and Eugster, *Helv. Chim. Acta*, **48,** 1322 (1965).
59. Willhalm, Stoll, and Thomas, *Chem. Ind.*, **1965,** 1629.
60. Elvidge and Foster, *J. Chem. Soc.*, **1963,** 590.
61. Black and Heffernan, *Aust. J. Chem.*, **18,** 353 (1965).
62. Oftedahl, Baker, and Dietrich, *J. Org. Chem.*, **30,** 296 (1965).
63. Darling and Wills, *J. Org. Chem.*, **32,** 2794 (1967).
64. Cavell and MacMillan, *J. Chem. Soc.*, *C*, **1967,** 310.
65. Martin, Defay, Geerts-Evrard, Given, Jones, and Wedel, *Tetrahedron*, **21,** 1833 (1965).
66. Brooke, Furniss, and Musgrave, *J. Chem. Soc.*, *C*, **1968,** 580.
67. Chambers, Cunningham, and Spring, *J. Chem. Soc.*, *C*, **1968,** 1560.
68. Gagnaire and Payo-Subiza, *Bull. Soc. Chim. Fr.*, **1963,** 2623.
69. Korver, Van der Haak, Steinberg, and de Boer, *Rec. Trav. Chim.*, **84,** 129 (1965).
70. Gianturio and Friedel, *Can. J. Chem.*, **44,** 1083 (1966).
71. Ichikawa and Uemura, *J. Org. Chem.*, **32,** 493 (1967).

72. Rivas and Payo, *J. Org. Chem.*, **32,** 2918 (1967).
73. Pappas and Blackwell, *Tetrahedron Lett.*, **1966,** 1171.
74. Nakazaki, Hirose, and Ikematsu, *Tetrahedron Lett.*, **1966,** 4735.
75. Brust, Tarbell, Hecht, Hayward, and Colebrook, *J. Org. Chem.*, **31,** 2192 (1966).
76. Zalkow and Ghosal, *Chem. Commun.*, **1967,** 922.
77. Binsch, Huisgen, and Konig, *Chem. Ber.*, **97,** 2893 (1964).
78. Batterham, Riggs, Robertson, and Simpson, *Aust. J. Chem.*, **22,** 725 (1969).
79. Greene and Lewis, *Tetrahedron Lett.*, **1966,** 4759.
80. Barbier, Gagnaire, and Vottero, *Bull. Soc. Chim. Fr.*, **1968,** 2330.
81. Naya and Kotake, *Tetrahedron Lett.*, **1967,** 1715.
82. Bohlmann and Arndt, *Chem. Ber.*, **99,** 135 (1966).
83. Vaulx, Jones, and Hauser, *J. Org. Chem.*, **29,** 505 (1964).
84. Barb, *Trans. Faraday Soc.*, **49,** 143 (1953).
85. Andrews and Keefer, *J. Am. Chem. Soc.*, **75,** 3776 (1953).
86. Chowdhury, *J. Phys. Chem.*, **66,** 353 (1962).
87. Nepras and Zahradnik, *Collect. Czech. Chem. Commun.*, **29,** 1545 (1964).
88. Arbuzov, Samitov, and Konovalov, *Izv. Akad. Nauk SSSR, Ser. Fiz.*, **27,** 82 (1963), through *Chem. Abstr.*, **59,** 150*c* (1963).
89. Ganter, Newman, and Roberts, *Tetrahedron*, Suppl. 8, Part II, 507 (1966).
90. Gagnaire and Vottero, *Bull. Soc. Chim. Fr.*, **1963,** 2779.
91. Gagnaire and Monzeglio, *Bull. Soc. Chim. Fr.*, **1965,** 474.
92. Narayanan and Bhadane, *Tetrahedron Lett.*, **1968,** 1557.
93. Aito, Matsuo, and Aso, *Bull. Chem. Soc. Japan*, **40,** 130 (1967).
94. Bundschuh, Takahashi, and Li, *Spectrochim. Acta*, **24A,** 1639 (1968).
95. Zysman, Dana, and Wiemann, *Bull. Soc. Chim. Fr.*, **1967,** 1019.
96. Colonge and Infarnet, *Comp. Rend., Series C*, **264,** 894 (1967).
97. Linn and Benson, *J. Am. Chem. Soc.*, **87,** 3657 (1965).
98. Hostettler, *Helv. Chim. Acta*, **49,** 2417 (1966).
99. Turro and Southam, *Tetrahedron Lett.*, **1967,** 545.
100. Paquette, Barrett, Spitz, and Pitcher, *J. Am. Chem. Soc.*, **87,** 3417 (1965).
100a. Kaneko, Yamada, and Ishikawa, *Chem. Pharm. Bull.*, **17,** 1294 (1969).
101. Ogata, Watanabe, and Kano, *Tetrahedron Lett.*, **1967,** 533.
102. Evanega and Whipple, *Tetrahedron Lett.*, **1967,** 2163, and references cited therein.
103. Toki and Sakurai, *Tetrahedron Lett.*, **1967,** 4119.
104. Singer and Ballschmiter, *Chem. Ber.*, **101,** 17 (1968).
105. Kawai, *J. Polym. Sci.*, **6A,** 137 (1968).
106. Crossley and Djerassi, *J. Chem. Soc.*, **1962,** 1459.
107. Roberts, Sheppard, Knight, and Roffey, *J. Chem. Soc., C*, **1968,** 22.
108. Bohlmann, Fanghanel, Kleine, Kramer, Monch, and Schuber, *Chem. Ber.*, **98,** 2596 (1965).
109. Narayanan and Venkatasubramanian, *Tetrahedron Lett.*, **1966,** 5865.
110. Ichikawa and Matsuo, *Bull. Chem. Soc. Japan*, **40,** 2030 (1967).
111. Lowry and Riggs, *Tetrahedron Lett.*, **1964,** 2911.
112. Johnson, Lowry, and Riggs, *Tetrahedron Lett.*, **1967,** 5113.
113. Mathieson and Taylor, *Tetrahedron Lett.*, **1961,** 590.
114. Savostianoff and Pfau, *Bull. Soc. Chim. Fr.*, **1967,** 4162.
115. Erikson, *J. Am. Chem. Soc.*, **87,** 1867 (1965).
116. Rioult and Vialle, *Bull. Soc. Chim. Fr.*, **1968,** 4477.
116a. Maciel and Savitsky, *J. Phys. Chem.*, **69,** 3925 (1965).
117. Sheppard and Turner, *Proc. R. Soc., A*, **252,** 506 (1959).

118. Lemieux, Stevens, and Frazer, *Can. J. Chem.*, **40,** 1955 (1962).
119. Abraham, McLauchlan, Hall, and Hough, *Chem. Ind.*, **1962,** 213.
120. Alderweireldt and Anteunis, *Bull. Soc. Chim. Belges*, **74,** 488 (1965).
121. Abraham, *J. Chem. Soc.*, **1965,** 256.
122. Frazer, Lemieux, and Stevens, *J. Am. Chem. Soc.*, **83,** 3901 (1961).
123. Anet, *J. Am. Chem. Soc.*, **84,** 747 (1962).
124. Mathiasson, *Acta Chem. Scand.*, **17,** 2133 (1963).
125. Chuche, Dana, and Monot, *Bull. Soc. Chim. Fr.*, **1967,** 3300.
126. Altona and van der Veek, *Tetrahedron*, **24,** 4377 (1968).
127. Finegold, *J. Phys. Chem.*, **72,** 3244 (1968).
128. Anteunis and Alderweireldt, *Bull. Soc. Chim. Belges*, **73,** 889, 903 (1964).
129. Gagnaire and Robert, *Bull. Soc. Chim. Fr.*, **1965,** 3646.
130. Chuche, *Comp. Rend., Series C*, **263,** 779 (1966).
131. Thuan and Wiemann, *Bull. Soc. Chim. Fr.*, **1968,** 4550.
132. Eliel and Willy, *Tetrahedron Lett.*, **1969,** 1775.
133. Hart and Tomalia, *Tetrahedron Lett.*, **1966,** 3383.
134. Tomalia and Hart, *Tetrahedron Lett.*, **1966,** 3389.
135. Hart and Tomalia, *Tetrahedron Lett.*, **1967,** 1347.
136. Pittman and McManus, *Tetrahedron Lett.*, **1969,** 339.
137. Pethrick and Wyn-Jones, *J. Chem. Soc., A*, **1969,** 1852.
138. Loan, Murray, and Story, *J. Am. Chem. Soc.*, **87,** 737 (1965).
139. Murray, Story, and Loan, *J. Am. Chem. Soc.*, **87,** 3025 (1965).
140. Durham and Greenwood, *Chem. Commun.*, **1968,** 24.
141. Durham and Greenwood, *J. Org. Chem.*, **33,** 1629 (1968).
142. Masamune and Castellucci, *J. Am. Chem. Soc.*, **84,** 2452 (1962).
143. Degani, Lunazzi, and Taddei, *Estrado Boll. Sci. Fac. Chim. Ind. Bologna*, **23,** 131 (1965).
144. Winterfeldt, *Chem. Ber.*, **97,** 1959 (1964).
145. Canrow and Radlick, *J. Org. Chem.*, **26,** 2260 (1961).
146. Hinnen and Dreux, *Bull. Soc. Chim. Fr.*, **1964,** 1492.
147. Balaban, Bedford, and Katritzky, *J. Chem. Soc.*, **1964,** 1646.
148. Balaban, Romas, and Rentia, *Tetrahedron*, **22,** 1 (1966).
149. Balaban, Katritzky, and Semple, *Tetrahedron*, **23,** 4001 (1967).
150. Dimroth, Kinzebach, and Soyka, *Chem. Ber.*, **99,** 2351 (1966).
151. Smitherman and Ferguson, *Tetrahedron*, **24,** 923 (1968).
152. Mathis and Goldstein, *Spectrochim. Acta*, **20,** 871 (1964).
153. Brown and Bladon, *Spectrochim. Acta*, **21,** 1277 (1965).
154. Jonas, Derbyshire, and Gutowsky, *J. Phys. Chem.*, **69,** 1 (1965).
155. Mayo and Goldstein, *Spectrochim. Acta*, **23A,** 55 (1967).
156. Smitherman and Ferguson, *Tetrahedron*, **24,** 923 (1968).
157. Woods and Dix, *J. Org. Chem.*, **26,** 1028 (1961).
158. Beak and Carls, *J. Org. Chem.*, **29,** 2678 (1964).
159. Mayo, Sapienza, Lord, and Phillips, *J. Org. Chem.*, **29,** 2682 (1964).
160. Rao, Singh, and Bhide, *Tetrahedron Lett.*, **1967,** 719.
161. Mullock and Suschitzky, *J. Chem. Soc., C*, **1967,** 828.
162. Pirkle and Dines, *J. Org. Chem.*, **34,** 2239 (1969).
163. Pirkle and Dines, *J. Heterocycl. Chem.*, **6,** 1 (1969).
164. Kurek and Vogel, *J. Heterocycl. Chem.*, **5,** 275 (1968).
165. Butt and Elvidge, *J. Chem. Soc.*, **1963,** 4483.
166. Beak and Abelson, *J. Org. Chem.*, **27,** 3715 (1962).

167. Edwards, Page, and Pianka, *J. Chem. Soc.*, **1964,** 5200.
168. Bartle, Edwards, Jones, and Mir, *J. Chem. Soc., C*, **1967,** 413.
169. Marcus, Stephen, and Chan, *J. Heterocycl. Chem.*, **6,** 13 (1969).
170. Lutz and Roberts, *J. Am. Chem. Soc.*, **83,** 2198 (1961).
171. Kosower and Sorensen, *J. Org. Chem.*, **28,** 692 (1963).
172. Fleming and Karger, *J. Chem. Soc., C*, **1967,** 226.
173. Achmatowiez, Jurezak, Konowal, and Zamojski, *Org. Magnetic Resonance*, **2,** 55 (1970).
174. Kropp and Gibson, *J. Chem. Soc., C*, **1967,** 143.
175. Gélin and Gélin, *Compt. Rend., Series C*, **263,** 1029 (1966).
176. Gélin and Gélin, *Bull. Soc. Chim. Fr.*, **1968,** 288.
177. Edward, *Chem. Ind.*, **1955,** 1102.
178. de Hoog, Buys, Altona, and Havinga, *Tetrahedron*, **25,** 3365 (1969).
179. Bakassian and Descotes, *Compt. Rend., Series C*, **262,** 1691 (1966).
180. Aguilera and Descotes, *Bull. Soc. Chim. Fr.*, **1966,** 3318.
181. Booth and Ouellette, *J. Org. Chem.*, **31,** 544 (1966).
182. Sweet and Brown, *Can. J. Chem.*, **45,** 1007 (1967).
183. Baldwin and Brown, *Can. J. Chem.*, **45,** 1195 (1967).
184. Anderson and Sepp, *J. Org. Chem.*, **32,** 607 (1967).
185. Lambert, Keske, and Weary, *J. Am. Chem. Soc.*, **89,** 5921 (1967).
186. Pierson and Runquist, *J. Org. Chem.*, **33,** 2572 (1968).
187. Anderson and Sepp, *J. Org. Chem.*, **33,** 3272 (1968).
188. Eliel and Giza, *J. Org. Chem.*, **33,** 3754 (1968).
189. Gatti, Segre, and Morandi, *J. Chem. Soc. B*, **1967,** 1203.
190. Anderson, Sepp, Geis, and Roberts, *Chem. Ind.*, **1968,** 1805.
191. Booth, *Tetrahedron Lett.*, **1965,** 411.
192. Fraser, Hanbury, and Reyes-Zamora, *Can. J. Chem.*, **45,** 2481 (1967).
193. Franck and Auerbach, *Can. J. Chem.*, **45,** 2489 (1967).
194. Jullien and Stahl-Larivière, *Bull. Soc. Chim. Fr.*, **1967,** 99.
195. Riggs, *Tetrahedron Lett.*, **1967,** 5109.
196. Johnson and Riggs, *Tetrahedron Lett.*, **1967,** 5119.
197. Lunazzi and Taddei, *J. Mol. Spectrosc.*, **25,** 113 (1968).
198. Arnone, Cardillo, Merlini, and Mondelli, *Tetrahedron Lett.*, **1967,** 4201.
199. Anthonsen, *Acta Chem. Scand.*, **22,** 352 (1968).
200. Lassak and Pinhey, *J. Chem. Soc., C*, **1967,** 2000; Jarvis and Moritz, *Aust. J. Chem.*, **21,** 2445 (1968).
201. Harper and Underwood, *J. Chem. Soc.*, **1965,** 4203.
202. Jefferson, Moore, and Scheinmann, *J. Chem. Soc., C*, **1967,** 151.
203. Highet and Highet, *J. Org. Chem.*, **32,** 1055 (1967).
204. Johnson and Pelter, *J. Chem. Soc., C*, **1966,** 606.
205. Finegold and Slover, *J. Org. Chem.*, **32,** 2557 (1967).
206. Marvell, Caple, Gosink, and Zimmer, *J. Am. Chem. Soc.*, **88,** 619 (1966).
207. Smit, Semenovskii, Mursakulov, and Kucherov, *Dokl. Akad. Nauk SSSR*, **177,** 1355 (1967); through *Chem. Abstr.*, **68,** 77405*x* (1968).
208. Randall, Vaulx, Hobbs, and Hauser, *J. Org. Chem.*, **32,** 2035 (1965).
209. Dharmatti, Govil, Kanekar, Khetrapal, and Virmani, *Proc. Indian Acad. Sci.*, **56,** 71 (1962).
210. Batterham and Lamberton, *Aust. J. Chem.*, **17,** 1305 (1964).
211. Grigg, Knight, and Roffey, *Tetrahedron*, **22,** 3301 (1966).
212. Abu-Mustafa and Fayez, *Can. J. Chem.*, **45,** 325 (1967).

213. Krauch, Farid, and Schenck, *Chem. Ber.*, **98,** 3102 (1965); Kaufman, Kelly, and Eaton, *J. Org. Chem.*, **32,** 504 (1967); Pozzi, Sánchez, and Comin, *Tetrahedron*, **23,** 1129 (1967).
214. Livingston, Bickoff, Lundin, and Jurd, *Tetrahedron*, **20,** 1963 (1964).
215. Livingston, Witt, Lundin, and Bickoff, *J. Org. Chem.*, **30,** 2353 (1965).
216. Wittmann, Orlinger, and Ziegler, *Monatsh. Chem.*, **96,** 1200 (1965).
217. Crombie, Games, and Knight, *J. Chem. Soc.*, *C*, **1967,** 763.
218. Chow, Duffield, and Jefferies, *Aust. J. Chem.*, **19,** 483 (1966).
219. Molho and Aknin, *Bull. Soc. Chim. Fr.*, **1967,** 2224.
220. Schönberg and Singer, *Chem. Ber.*, **96,** 1529 (1963).
221. McCabe, McCrindle, and Murray, *J. Chem. Soc.*, *C*, **1967,** 145.
222. Pachler and Roux, *J. Chem. Soc.*, *C*, **1967,** 604.
223. Mercier, Mentzer, and Billet, *Compt. Rend.*, *Series C*, **265,** 945 (1967).
224. Schönberg, Praefcke, and Kohtz, *Chem. Ber.*, **99,** 3076 (1966).
225. Grandolini, Ricci, Buu-Hoi, and Périn, *J. Heterocycl. Chem.*, **5,** 133 (1968).
226. Katritzky and Ternai, *J. Heterocycl. Chem.*, **5,** 745 (1968).
227. Massicot and Marthe, *Bull. Soc. Chim. Fr.*, **1962,** 1962.
228. Massicot, Marthe, and Heitz, *Bull. Soc. Chim. Fr.*, **1963,** 2712.
229. Batterham and Highet, *Aust. J. Chem.*, **17,** 428 (1964).
230. Grouiller, *Bull. Soc. Chim. Fr.*, **1966,** 2405.
231. Clark-Lewis, *Aust. J. Chem.*, **21,** 2059 (1968).
232. Waiss, Lundin, and Stern, *Tetrahedron Lett.*, **1964,** 513.
233. Mabry, Kagan, and Rösler, *Phytochem.*, **4,** 177 (1965).
234. Mabry, Kagan, and Rösler, *Phytochem.*, **4,** 487 (1965).
235. Wilson, Bowie, and Williams, *Tetrahedron*, **24,** 1407 (1968); Wilson and Williams, *J. Chem. Soc.*, *C*, **1968,** 2477.
236. Rösler, Mabry, Cranmer, and Kagan, *J. Org. Chem.*, **30,** 4346 (1965).
237. Grouiller and Pacheco, *Bull. Soc. Chim. Fr.*, **1967,** 1938.
238. Hillis and Horn, *Aust. J. Chem.*, **18,** 531 (1965).
239. Gentili and Horowitz, *J. Org. Chem.*, **33,** 1571 (1968).
240. Crombie and Lown, *J. Chem. Soc.*, **1962,** 775.
241. Adam, Crombie, and Whiting, *J. Chem. Soc.*, *C*, **1966,** 542, 544, 550.
242. Lamer, Malcher, and Grimshaw, *Tetrahedron Lett.*, **1968,** 1419.
243. Horn and Lamberton, *Chem. Ind.*, **17,** 691 (1963).
244. Kawano, Miura, and Matsuishi, *Chem. Pharm. Bull.*, **15,** 711 (1967).
245. Courbat, Favre, Guerne, and Uhlmann, *Helv. Chim. Acta*, **49,** 1203 (1966).
246. Kupchan, Sigel, Hemingway, Knox, and Udayamurthy, *Tetrahedron*, **25,** 1603 (1969).
247. Takeda, Mitsui, and Hayashi, *Bot. Mag.* (*Tokyo*), **79,** 578 (1966).
248. Kiang, Sim, and Goh, *J. Chem. Soc.*, **1965,** 6371.
249. Herz, Farkas, Sudarsanam, Wagner, Hörhammer, and Rüger, *Chem. Ber.*, **99,** 3539 (1966).
250. Henrick and Jefferies, *Tetrahedron*, **21,** 3219 (1965).
251. Thomas and Mabry, *J. Org. Chem.*, **32,** 3254 (1967).
252. Norgradi and Farkas, *Kem. Kozl.*, **29,** 25 (1968); through *Chem. Abstr.*, **69,** 96397*h* (1968).
253. Seshadri and Sood, *Tetrahedron Lett.*, **1967,** 853.
254. Mongkolsuk and Dean, *J. Chem. Soc.*, **1964,** 4654.
255. Barton, *Can. J. Chem.*, **45,** 1020 (1967).
256. Markham, Rahman, Jehan, and Mabry, *J. Heterocycl. Chem.*, **4,** 61 (1967).

257. Bevan, Ekong, Obasi, and Powell, *J. Chem. Soc., C*, **1966,** 509.
258. Chopin, Bouillant, and Lebreton, *Bull. Soc. Chim. Fr.*, **1964,** 1038.
259. Rosler, Mabry, and Kagan, *Chem. Ber.*, **98,** 2193 (1965).
260. Schwarz, Cohen, Ollis, Kaczka, and Jackman, *Tetrahedron*, **20,** 1317 (1964).
261. Brown and Whiteoak, *J. Chem. Soc.*, **1964,** 6084.
262. Clark-Lewis, Jackman, and Williams, *J. Chem. Soc.*, **1962,** 3858.
263. Clark-Lewis and Katekar, *J. Chem. Soc.*, **1962,** 4502.
264. Vickars, *Tetrahedron*, **20,** 2873 (1964).
265. Drewes and Roux, *J. Chem. Soc., C*, **1966,** 1644.
266. Bowman, Hewgill, and Kennedy, *J. Chem. Soc., C*, **1966,** 2274.
267. Kurosawa, Ollis, Redman, Sutherland, Gottlieb, and Alves, *Chem. Commun.*, **1968,** 1265.
268. Drewes, Roux, Eggers, and Feeney, *J. Chem. Soc., C*, **1967,** 1217.
269. Drewes, Roux, Saaymann, Eggers, and Feeney, *J. Chem. Soc., C*, **1967,** 1302.
270. Birch, Dahl, and Pelter, *Tetrahedron Lett.*, **1967,** 481.
271. Jackson, Locksley, Scheinmann, and Wolstenholme, *Tetrahedron Lett.*, **1967,** 787.
272. Drewes and Roux, *Chem. Commun.*, **1965,** 500.
273. Hassall and Weatherson, *J. Chem. Soc.*, **1965,** 2844.
274. Nair, Parthasarathy, Radhakrishnan, and Venkataraman, *Tetrahedron Lett.*, **1966,** 5357.
275. Bolger, Marathe, Philbin, Wheeler, and Lillya, *Tetrahedron*, **23,** 341 (1967).
276. Reichel and Weber, *Z. Chem.*, **7,** 62 (1967).
277. Michalska, *Bull. Acad. Pol. Sci. Chim.*, **16,** 575 (1968).
278. Dudley, Corley, Miller, and Wall, *J. Org. Chem.*, **32,** 2312 (1967).
279. Janzsó, Kállay, Koczor, and Radics, *Tetrahedron*, **23,** 3699 (1967).
280. Obara and Onodera, *Bull. Chem. Soc. Japan*, **41,** 2798 (1968).
281. Caplin, Ollis, and Sutherland, *J. Chem. Soc., C*, **1968,** 2302.
282. Jurd and Waiss, *Tetrahedron*, **21,** 1471 (1965); Jurd and Bergot, *ibid.*, **21,** 3697 (1965).
283. Haynes and Taylor, *J. Chem. Soc., C*, **1966,** 1685.
284. Locksley, Moore, and Scheinmann, *J. Chem. Soc., C*, **1966,** 430, 2186, 2265.
285. Jackson, Locksley, and Scheinmann, *J. Chem. Soc., C*, **1967,** 785.
286. Moron and Polonsky, *Bull. Soc. Chim. Fr.*, **1967,** 130.
287. Claeson, Androes, and Calvin, *J. Am. Chem. Soc.*, **83,** 4357 (1961).
288. Eliel and Martin, *J. Am. Chem. Soc.*, **90,** 682 (1968).
289. Friebolin, Kabuss, Maier, and Lüttringhaus, *Tetrahedron Lett.*, **1962,** 683.
290. Anteunis, *Bull. Soc. Chim. Belges*, **73,** 731 (1964).
291. Anteunis, Tavernier, and Borremans, *Bull. Soc. Chim. Belges*, **75,** 396 (1966).
292. Tavernier and Anteunis, *Bull. Soc. Chim. Belges*, **76,** 157 (1967).
293. Anteunis, Vandenbroucke, and Schamp, *Bull. Soc. Chim. Belges*, **76,** 552 (1967).
294. Feeney, Anteunis, and Swaelens, *Bull. Soc. Chim. Belges*, **77,** 121 (1968).
295. Anteunis, Coene, and Tavernier, *Tetrahedron Lett.*, **1966,** 4579.
296. Delmau, Davidson, Parc, and Hellin, *Bull. Soc. Chim. Fr.*, **1964,** 241.
297. Barbier, Davidson, and Delmau, *Bull. Soc. Chim. Fr.*, **1964,** 1046.
298. Delmau and Barbier, *J. Chim. Phys.*, **41,** 1106 (1964).
299. Delmau, *Rev. Inst. Fr. Pet. Ann. Combust. Liq.*, **20,** 94 (1965); through *Chem. Abstr.*, **62,** 14066*c* (1965).
300. Barbier, Delmau, and Ranft, *Tetrahedron Lett.*, **1964,** 3339.
301. Delmau and Duplan, *Tetrahedron Lett.*, **1966,** 559, 2693.
302. Delmau, *Compt. Rend. Congr. Natl. Soc. Savantes, Sect. Sci.*, **90,** 123 (1964); through *Chem. Abstr.*, **64,** 17387*d* (1966).

303. Delmau, Duplan, and Davidson, *Tetrahedron*, **23,** 4371 (1967).
304. Delmau, Duplan, and Davidson, *Tetrahedron*, **24,** 3939 (1968).
305. Duplan, Delmau, and Davidson, *Bull. Soc. Chim. Fr.*, **1968,** 4081.
306. Anderson and Brand, *Trans. Faraday Soc.*, **62,** 39 (1966).
307. Anderson, *Tetrahedron Lett.*, **1965,** 4713.
308. Anderson, Riddell, and Robinson, *Tetrahedron Lett.*, **1965,** 4713.
309. Anderson, Riddell, Fleury, and Morgen, *Chem. Commun.*, **1966,** 128.
310. Cookson, Crabb, and Vary, *Tetrahedron*, **24,** 4559 (1968).
311. Pedersen and Schaug, *Acta Chem. Scand.*, **22,** 1705 (1968).
312. Maffrand and Maroni, *Tetrahedron Lett.*, **1969,** 4201.
313. Riddell and Robinson, *Tetrahedron*, **23,** 3417 (1967).
314. Samilov and Aminova, *J. Struct. Chem. USSR*, **5,** 189 (1964).
315. Ramey and Messick, *Tetrahedron Lett.*, **1965,** 4423.
316. Mijs, *Rec. Trav. Chim.*, **86,** 220 (1967).
317. Friebolin, Schmid, Kabuss, and Faisst, *Org. Magnetic Resonance*, **1,** 67 (1969).
318. Eliel and Knoeber, *J. Am. Chem. Soc.*, **90,** 3444 (1968).
319. Anteunis, *Bull. Soc. Chim. Belges*, **75,** 413 (1966).
320. Anderson, *J. Chem. Soc.*, *B*, **1967,** 712.
321. Baggett, Bukhari, Foster, Lehmann, and Webber, *J. Chem. Soc.*, **1963,** 4157, and references cited therein.
322. Eliel and Knoeber, *J. Am. Chem. Soc.*, **88,** 5347 (1966).
323. Eliel, Gianni, Williams, and Stothers, *Tetrahedron Lett.*, **1962,** 741.
324. Anteunis and Tavernier, *Bull. Soc. Chim. Belges*, **76,** 432 (1967).
325. Duplan and Delmau, *Bull. Soc. Chim. Fr.*, **1968,** 4081.
326. Schuster and Schuster, *Tetrahedron*, **25,** 199 (1969).
327. Caspi, Wittstruck, and Piatak, *J. Org. Chem.*, **27,** 3183 (1962).
328. Jung, *Chem. Ber.*, **99,** 566 (1966).
329. Gatti, Segre, and Morandi, *Tetrahedron*, **23,** 4385 (1967).
330. Muller and Rose, *J. Phys. Chem.*, **69,** 2564 (1965).
331. Hinton, McDowell, and Amis, *Chem. Commun.*, **1966,** 776.
332. Ogimura, Shirasawa, Nagashima, and Fujiwara, *Nippon Kagaku Zasshi*, **88,** 1330 (1967).
333. Alexandrou and Hadjimihalakis, *Org. Magnetic Resonance*, **1,** 401 (1969).
334. Altona and Havinga, *Tetrahedron*, **22,** 2277 (1966).
335. Altona and Havinga, *Tetrahedron*, **23,** 528 (1967).
336. Fraser and Reyes-Zamora, *Can. J. Chem.*, **45,** 1012 (1967).
336a. Katritzky, Kingsland, Rudd, Sewell, and Topsom, *Aust. J. Chem.*, **20,** 1773 (1967).
337. Cocivera, *J. Chem. Phys.*, **47,** 3061 (1967).
338. Diehl and Freeman, *Mol. Phys.*, **4,** 39 (1961).
339. Jungnickel and Reilly, *J. Mol. Spectrosc.*, **16,** 135 (1965).
340. Murray, Story, and Kaplan, *J. Am. Chem. Soc.*, **88,** 526 (1966).
341. Guenther, *Tetrahedron Lett.*, **1965,** 4085.
342. Guenther, Schubart, and Vogel, *Z. Naturforsch.*, *B*, **22,** 25 (1967).
343. Guenther and Hinrichs, *Tetrahedron Lett.*, **1966,** 789.
344. Prinzbach, Arguelles, and Druckrey, *Angew. Chem., Int. Ed. Engl.*, **5,** 1039 (1966).
345. Nerdel, Buddrus, Brodowski, and Weyerstahl, *Tetrahedron Lett.*, **1966,** 5385.
346. Foa, Cassar, and Tacchi-Venturi, *Tetrahedron Lett.*, **1968,** 1357.
347. Jorgenson, *J. Org. Chem.*, **27,** 3224 (1962).
348. Oki, Iwamura, and Nishida, *Bull. Chem. Soc. Japan*, **41,** 656 (1968).

349. Kametani, Shibuya, and Ollis, *J. Chem. Soc., C*, **1968,** 2877.
350. Cagniant and Charaux, *Bull. Soc. Chim. Fr.*, **1966,** 3249.
351. Cagniant, Charaux, and Cagniant, *Bull. Soc. Chim. Fr.*, **1966,** 3644.
352. Gerson, Heilbronner, Böll, and Vogel, *Helv. Chim. Acta*, **48,** 1494 (1965).
353. Grindley, Stoddart, and Szarek, *J. Chem. Soc., B*, **1969,** 172.
354. Smith and Shoulders, *J. Phys. Chem.*, **69,** 579 (1965).

6 SULFUR HETEROCYCLES

I. THREE-MEMBERED SULFUR HETEROCYCLES

A. The Thiiran System

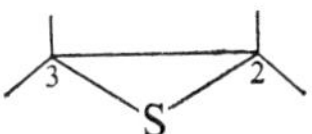

The thiiran or episulfide system, in contrast with its nitrogen and oxygen analogs, has been the subject of very few NMR investigations. Most of the available data come from papers dealing in the main with other heterocyclic systems.

The earliest mention of the spectrum of the parent thiiran comes from a study of reduced nitrogen, oxygen, and sulfur heterocycles by Gutowski and his co-workers in 1954 (1) but it was not till eight years later that the full spectral details of this ring system were reported. Mortimer (2) fully analyzed the thiiran spectrum, using ^{13}C satellites to overcome the problems imposed by the symmetry of the molecule, and the parameters obtained compared favorably with those from the unsymmetrical propylene sulfide (**1**) (3). However, the assignment of peaks in the latter spectrum depended on the assumption that J_{cis} was greater than J_{trans}, as suggested for propylene oxide (4).

A number of papers have appeared (5, 6) which take up Gutowsky's original theme (1) and examine the effects of ring-size and the nature of the heteroatom on the chemical shifts of protons or ^{13}C atoms adjacent to the heteroatom. The general upfield trends as rings became smaller have been correlated with electron-density variations, *s*-character of the involved carbon atoms, and the possible presence of ring currents in the three-membered ring systems. Throughout this work, cyclopropane was used as a reference and the anomalous behavior of the episulfides with respect to that of other threemembered heterocycles was attributed to the large geometrical alteration caused when a methylene group of cyclopropane is replaced by a sulfur atom.

Relative signs of coupling constants in styrene sulfide (**2**) were shown to be "normal," that is, J_{gem} negative and both J_{vic} couplings positive (7). In this sense, styrene sulfide is unlike styrenimine (**3**) in which all couplings have the same sign (presumed positive). The spectrum of stilbene sulfide (**4**) is illustrated in the Varian Catalog (8).

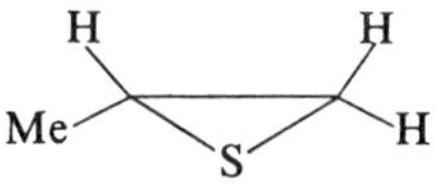

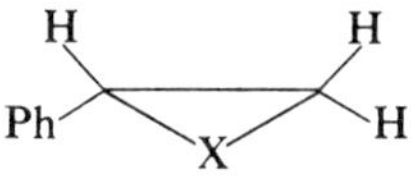

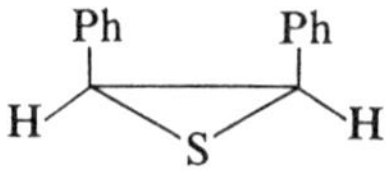

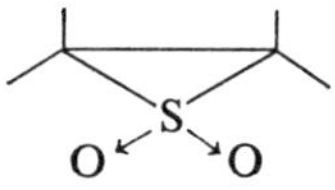

Conversion of the sulfide group of the episulfide ring to a sulfoxide (**5**) has been shown to have a substantial effect on the chemical shifts of the ring protons, an effect which is not observed for reduced sulfur heterocycles with five or more ring atoms (9). Data for the thiiran protons in a series of steroid episulfides have been reported (10). In general, the signals from the episulfide protons of α-isomers lie at higher field than those from the respective β-compounds. Data for thiirans are collected in Table 6.1.

II. FOUR-MEMBERED SULFUR HETEROCYCLES

A. The Thiete System

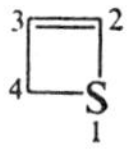

The only representative of this ring system whose spectrum has been reported is the 1,1-dioxide (**6**) (11). Assignment of signals to the ethylenic protons, H-2 and H-3, seems to be the reverse of that expected for such a heterocyclic system. Data for this compound are included in Table 6.2.

Thietan, the 2,3-dihydrothiete (**7**), and its derivatives have been studied much more thoroughly, but the work has often been marred by a seeming reluctance to publish data. The unsubstituted ring system is included in the general studies of proton and ^{13}C spectroscopy discussed in the previous section for episulfides (1, 5, 6), and the conclusions drawn for the effects of ring size and variation of the heteroatom were similar to those expressed for the three-membered ring system. Unfortunately, the spectral pattern obtained from the strongly coupled A_2B_4 system of thietan at 60 MHz contains many unresolved transitions. While the two reported analyses of this system agree

TABLE 6.1 SPECTRAL DATA FOR THIIRANS

		Chemical shifts					Coupling constants				
Substituents	Solvent	τA	τB	τC	τD	^{13}C[a]	J_{cis}	J_{trans}	J_{gem}	$J_{^{13}C-H}$	Ref.
None	Neat	—— 7.73 ——					7.15	5.65		170.5	2
		—— 7.73 ——								170.6	5
						109.8					6
2-Me	Neat	7.22	7.66	8.04	(8.62)		6.3	5.4	0.4		3[b]
		7.17	7.61	7.98			6.0	5.1	1.5		5
2-Ph	Neat	6.44	7.53	7.70			6.60	5.55	−1.37		7
2,3-Di-Ph (cis)	$CDCl_3$	6.04	6.04								8

[a] ^{13}C Chemical shift in ppm from external benzene.
[b] Chemical shifts from internal hexamethyldisiloxane.

TABLE 6.2 SPECTRAL DATA FOR THIETANS

A. THIETAN

$\tau\alpha$	$\tau\beta$	$\tau\beta-\tau\alpha$	$J_{gem}(\alpha)$	$J_{gem}(\beta)$	J_{23}		J_{24}		Ref.
					Cis	Trans	Cis	Trans	
6.79	7.06	0.27	−8.7	−11.7	8.9	6.3	1.2	−0.2	12
		0.27	−8.0	−11.0	8.9	6.3	0.6	−0.5	13

B. 3-CHLOROTHIETAN (NEAT) (14)

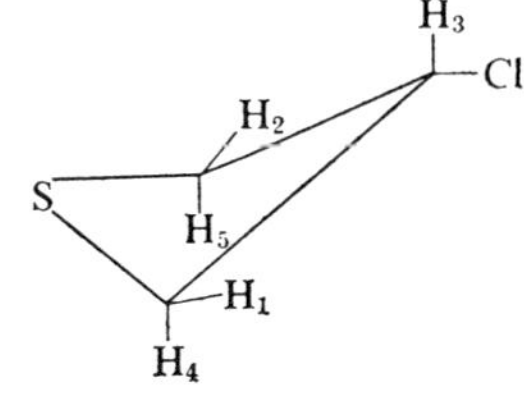

$\tau 1(2)$	$\tau 3$	$\tau 4(5)$	J_{12}	$J_{13(23)}$	$J_{14(25)}$	$J_{15(24)}$	$J_{34(35)}$	J_{45}
8.02	4.87	6.34	3.12	7.67	−8.63	−0.75	9.34	−0.48

C. 2,4-DIMETHYLTHIETANS (16)

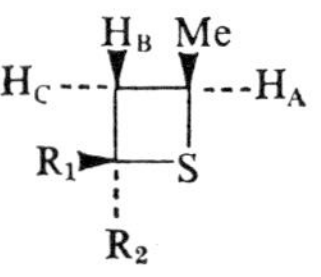

R_1 = H; R_2 = Me (TRANS ISOMER) (NEAT)

τA, τR_1	τB, τC	J_{AB}
6.38	7.39	6.9

R_1 = Me; R_2 = H (CIS ISOMER) (NEAT)

τA, τR_2	τB	τC	J_{AB}	J_{AC}	J_{BC}
6.48	7.87	7.01	8.2	7.3	−11.0

very well for the larger spectral parameters, the values obtained for the small cross-ring couplings between protons on C-2 and C-4 are quite different (12, 13) (see Table 6.2).

3-Chlorothietan (**8**) was chosen as a model compound in an extensive study of computer methods for locating and assigning spectral lines in complex, unresolved multiplets (14, 15). These efforts led to the development of the programs DECOMP and ASSIGN which will, in theory, give every possible line assignment for complex spectrum. In practice, however, the use of these programs is restricted to fairly simple systems because of the large amount of computer time required for anything above a four-spin system. The thietan ring appears to be puckered in a similar manner to other four-membered heterocycles, and this means that one proton on each α-carbon atom will assume a quasiaxial conformation while the other will be quasi-equatorial. In the case of 3-chlorothietan, the chlorine substituent fixes the ring predominantly in the conformation with the chlorine quasiequatorial. The 2- and 4-quasiaxial protons should appear downfield from their quasi-equatorial counterparts, and this assumption makes possible the assignment of all spectral parameters. Throughout the work on this ring system cis vicinal couplings have usually been assumed to be larger than the corresponding trans couplings. In simple monosubstituted thietans this appears to be true, but in more complex derivatives it is often difficult to predict the geometry of the ring and such a generalization must be treated with caution.

The isomeric cis- and trans-2,4-dimethylthietans (**9, 10**) may be differentiated by the symmetry observable in the spectrum of the trans isomer (16). In the corresponding 1-methylthietanonium salts (**11, 12**), inversion of the sulfur atom does not occur up to the decomposition temperature, 80°. Spectra of the isomeric 2,4-diphenylthietans (**13, 14**) and their oxides have also been investigated (17).

The stereochemistry of thietan-1-oxides (**15**) has been elucidated mainly by the use of NMR spectroscopy (18). If a large substituent is present on C-3, the conformation of the ring system is more or less fixed and the cis (**16**) and trans (**17**) sulfoxides can be separated. In the trans isomer, the β-proton is cis to the sulfoxide group and is strongly deshielded. Unfortunately, an earlier report on the cage compound **18** (19) led to the opposite conclusion, but the chemical shift data in this work seem to be inconsistent with the proposed structures. The symmetry of the spectral pattern from trans-2,4-diphenylthietan-1-oxide and -1,1-dioxide has been used to distinguish them from the unsymmetrical cis isomers (20, 21). Also, two papers have appeared which, while they are devoted mainly to the synthesis of some highly substituted thietan dioxides, contain some interesting data for derivatives with various oxygen substituents in position 3 (22, 23). Data for thietans are collected in Table 6.2.

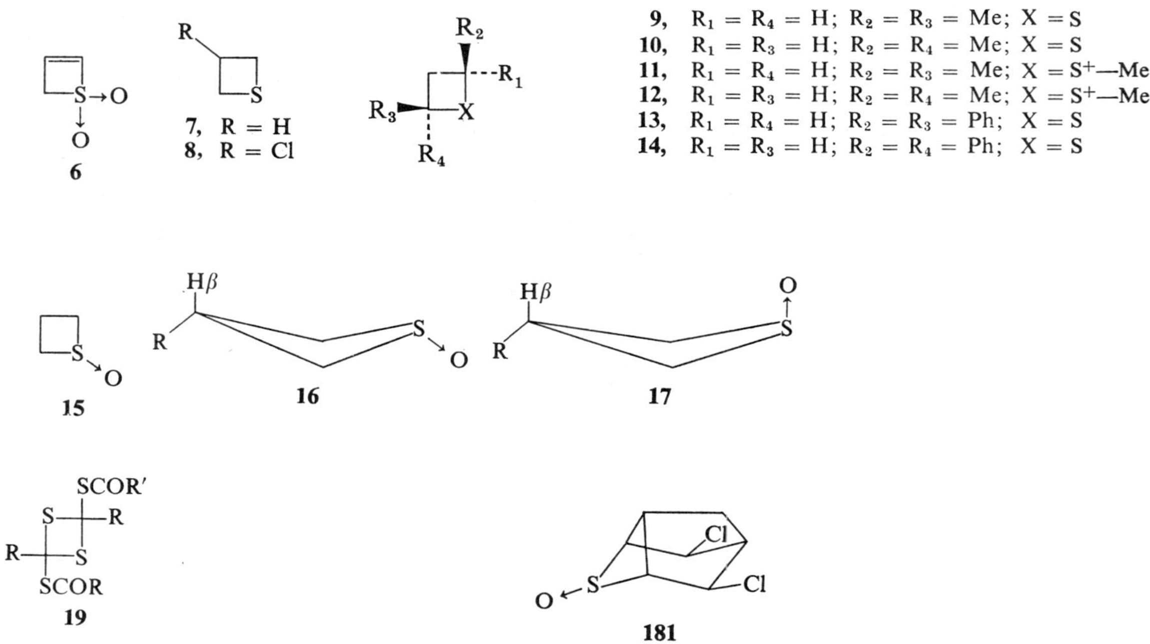

6
7, R = H
8, R = Cl
9, $R_1 = R_4 = H$; $R_2 = R_3 = Me$; X = S
10, $R_1 = R_3 = H$; $R_2 = R_4 = Me$; X = S
11, $R_1 = R_4 = H$; $R_2 = R_3 = Me$; $X = S^+—Me$
12, $R_1 = R_3 = H$; $R_2 = R_4 = Me$; $X = S^+—Me$
13, $R_1 = R_4 = H$; $R_2 = R_3 = Ph$; X = S
14, $R_1 = R_3 = H$; $R_2 = R_4 = Ph$; X = S
15
16
17
19
181

B. The 1,2-Dithietan System

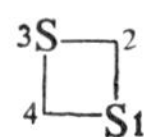

Highly substituted members of this very rare system were isolated during work on polythiaadamantanes (24). In the tetrasubstituted compounds **19**, when R = CH_2Cl, NMR showed the presence of cis and trans isomers.

III. FIVE-MEMBERED SULFUR HETEROCYCLES

A. The Thiophen System

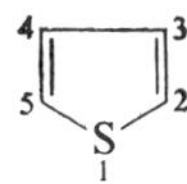

The NMR spectra of thiophens have attracted considerable interest since 1958 when the first systematic study of substituted thiophens appeared (25). This has been due mainly to the interests of Gronowitz and his co-workers in the chemistry and spectral properties of this ring system, but some credit must be given to the relative ease with which most of the spectra can be analyzed and to the presence of very interesting long-range couplings between substituents and hydrogens on the ring.

Protons of the parent ring system for a symmetrical AA′BB′ system which, in the early days of NMR spectroscopy, could only be analyzed by indirect means. Values of parameters obtained for various deuterothiophens were applied to the unsubstituted ring system, and their acceptability was checked from the fit obtained between the calculated and experimental spectra (26). Thiophen was also used by Abraham and Bernstein as an example to check their AA′BB′ tables (27) and by Dischler and his co-workers (28, 29) in their general study of the A_2B_2 system. In a report dealing mainly with selenophen (**20**), ^{13}C satellite spectra were used to obtain a complete analysis of the thiophen system (30). Thiophen figured predominantly in the development of the ring-current definition of aromaticity and in the arguments which took place over this (31–33). More recently (34), thiophen has been used as a model to test the analysis of oriented four-spin systems with c_{2v} symmetry (AA′BB′). From the spectrum of thiophen in the nematic phase of *p,p*′-di-*n*-hexyloxyazoxybenzene the following parameters were obtained. Spin–spin couplings, $J_A = 3.3$, $J_B = 2.8$, $J = 4.7$, $J' = 1.0$ Hz; direct couplings, $D_A = -175.3$, $D_B = -33.0$, $D = -394.4$, $D' = -55.8$ Hz; ratios of inter-proton distances, $r_B/r_A = 1.745 \pm 0.01$, $r/r_A = 0.995 \pm 0.005$, $r'/r_A = 1.653 \pm 0.006$; orientation parameters, $s_1 = 0.0265 \pm 0.0003$, $s_2 = 0.0639 \pm 0.0002$, $s_3 = -0.0904$; and chemical shift, $\delta_{AB} = 10$ Hz, were obtained for the system. Values for s and r can be obtained from the equation

$$D_{ij} = s_{ij}(-h\gamma^2/4\pi^2 r_{ij}^{\ 2})$$

where s_{ij} is the degree of orientation along the axis joining nuclei i and j, separated by r_{ij}. In a molecule with c_{2v} symmetry, any two of the experimentally obtained D values suffice to describe the molecular orientation and the other two can be used to study the geometry of the system.

A large amount of data is available for substituted thiophens (35–77) but, unfortunately, much of it was reported early in the history of NMR and the standards used make it difficult to refer the chemical shifts to the τ scale. However, the early work does make up perhaps the most thorough study of substituent effects in any ring system except benzene.

Coupling constants for monosubstituted (45, 46) and disubstituted thiophens (47) normally fall into four distinct groups: J_{35} 1.25–1.70, J_{25} 3.20–3.65, J_{34} 3.45–4.35, and J_{45} 4.90–5.80 Hz. Extensive calculations led to the conclusion that all of these couplings have the same sign (30) and multiple-resonance studies confirmed this hypothesis (48).

A review of substituent effects on the chemical shifts of ring protons of thiophen has appeared (78), covering the literature to 1961. However, much has been published since then and a number of Gronowitz's original postulates have to be qualified in the light of new and better data.

The chemical shifts, at infinite dilution in cyclohexane, of protons in monosubstituted thiophens relative to those of the α- and β-protons of thiophen are given in Table 6.3. The observed shifts are in accord with the electron-withdrawing or -donating nature of the substituent and with the various resonance paths present in the compounds. The fact that $-$M substituents cause no shift of H-4 in 2-substituted thiophens or H-5 in 3-substituted thiophens, whereas $+$M substituents cause a marked upfield shift, has been ascribed to a relay of the charge from the 3- and 5-positions in canonical forms of the type **21** and **22** to the 4-position in $+$M substituted compounds (46). In the $-$M substituted compounds, an alternative relay of the positive charge to the sulfur atom exists (**23**), and this would appear to predominate. The much larger shift of H-2 compared with H-4 in 3-substituted thiophens is in agreement with the importance attached to resonance forms of the types **24** and **25**. However, it is important to note the high-field shift of H-4 in 3-methoxythiophen (**26**) which suggests the importance of the resonance form **27** in which *pd* overlap is involved. Attempts to correlate substituent shifts with those obtained for substituted benzenes (46) or furans (36) were unsuccessful. Gronowitz and Hoffman (46) suggested that the small substituent effects of halogen, SH, and SMe groups were largely due to magnetic anisotropy phenomena and were additive. In

TABLE 6.3 CHEMICAL SHIFTS (PPM) OF SUBSTITUTED THIOPHENS, RELATIVE TO THE SHIFTS OF THE α- AND β-HYDROGENS IN THIOPHEN (46)

Sub-stituents	Character	2-Substituted thiophens H-3	2-Substituted thiophens H-4	2-Substituted thiophens H-5	3-Substituted thiophens H-2	3-Substituted thiophens H-4	3-Substituted thiophens H-5
NO_2	−I − M	−0.82	+0.03	−0.30	−0.95	−0.60	−0.03
SO_2Cl	−I − M	−0.73	−0.06	−0.45			
CN	−I − M	−0.47[a]	0.00	−0.28[a]	−0.63	−0.20	−0.15
CHO	−I − M	−0.65[a]	−0.10	−0.45[a]	−0.79	−0.45	−0.03
$COCH_3$	−I − M	−0.57	0.00	−0.28	−0.68	−0.47	+0.02
COCl	−I − M	−0.88	−0.06	−0.44	−1.05	−0.50	−0.03
CO_2CH_3	−I − M	−0.70	+0.05	−0.20	−0.78	−0.47	+0.05
SCN	−I(±M)	−0.30	+0.05	−0.28	−0.25	−0.05	−0.05
C≡CH	−I − M	−0.15	+0.16	+0.12			
I	−I + M	−0.13	+0.33	−0.01	−0.06	0.00[a]	+0.19[a]
Br	−I + M	+0.05	+0.27	+0.11	+0.12	+0.08	+0.10
SH	−I + M	0.00	+0.20	+0.07	+0.22[a]	+0.20[a]	+0.10[a]
SCH_3	−I + M	+0.03	+0.18	+0.05	+0.33	+0.10	+0.03
CH_3	+I + M	+0.37	+0.24	+0.28	+0.45	+0.22	+0.14
OCH_3	−I + M	+0.94	+0.43	+0.82	+1.10	+0.38	+0.20
NH_2	−I + M	+0.95	+0.45	+0.85	+1.25	+0.53	+0.25

[a] These resonances were not resolved.

more recent work on dibromothiophen (49) and halogen-substituted formylthiophens (57, 58), these additivity rules were used successfully to predict chemical shifts of protons in various isomeric compounds, but examination of the spectra of all possible chlorosubstituted thiophens (56) led to the conclusion that the additivity rules do not hold for the dichloroderivatives. In this latter report, a discussion is given of chemical shift variations in terms of electronic contributions from charged canonical forms. However, the additivity rules appear to hold for iodo derivatives of methylthiophens (49a) and for silicon-substituted thiophens (68).

As with other aromatic systems, the variations of coupling constants with substitution were largely neglected until instrumentation and computational methods reached a stage where accurate spectral parameters could be obtained without the expenditure of an exorbitant amount of effort. Bulman (69) analyzed the spectra of a series of 2-substituted thiophens with substituents varying from Li to F. The sum of the three couplings correlated linearly with the electronegativity of the substituent. In cases where mesomeric contributions were not large, the values obtained for each of the coupling constants also varied linearly with the substituent electronegativity. However, in the case of 2-nitrothiophen (**28**) the individual couplings deviated from the correlations. Also in this work, data reported by Hoffman and

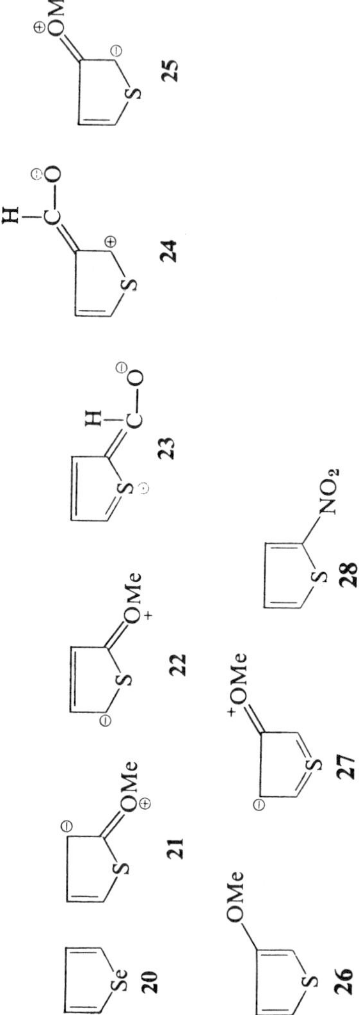
Se
20
S
21
OMe
S
22
OMe
S
23
H
C—O
S
24
H
C—O
S
25
OMe
OMe
S
26
OMe
S
27
S
28
NO_2

Gronowitz (45) for 3-substituted thiophens were examined. While the sum of the couplings varied with electronegativity of the substituent in a manner similar to the 2-substituted compounds, effects on the individual coupling constants were too diverse for any conclusions to be drawn. Data for the lithium derivatives of thiophen, used in this work and reported earlier (59), are most interesting. Somewhat unexpectedly, the lithium atom caused a downfield substituent shift, particularly of the adjacent ring protons. The coupling constants obtained are much smaller than those of "normal" substituted thiophens.

In 1959, observations were reported of couplings between ring protons and those of substituent groups on the thiophen ring. These marked the beginning of a very extensive and fruitful area of research. Spin–spin coupling of the order of 1 Hz between protons of the substituent and certain ring hydrogens have been observed for methylthiophens (**29**) (39, 42, 47), thiophenaldehydes (**30**) (40, 41, 39, 47), thiophenaldoximes (**31**) (39), thiophenthiols (**32**) (47, 43), and 2-thienylacetylene (**33**) (45). The values obtained are summarized: $J_{\mathrm{Me},3} = 1.00$–1.15, $J_{\mathrm{Me},4} = 0.2$–0.5 Hz in 2-methylthiophens (**34**); $J_{\mathrm{Me},2} = 0.9$–1.25, $J_{\mathrm{Me},4}$ and $J_{\mathrm{Me},5} = 0.4$–0.5 Hz in 3-methylthiophen–(**35**); $J_{\mathrm{CHO},5} = 1.05$–$1.40$ Hz in 2-thiophenaldehydes (**36**); $J_{\mathrm{CHO},5} = (<0.4)$–$0.80$ Hz in 3-thiophenaldehydes (**37**); $J_{\mathrm{SH},3} = 1.4$–1.6, $J_{\mathrm{SH},5} = 0.9$–1.0 Hz in 2-thiophenthiols (**38**); $J_{\mathrm{S,H2}} = 0.65$–1.05, $J_{\mathrm{SH},5} = 0.3$ Hz in 3-thiophenthiols (**39**). Accurate analyses of the spectra of the two methylthiophens confirmed the above ranges (79) for these compounds. Coupling between a 2-methyl group and H-5 is also mentioned by Gronowitz but has not been confirmed in two later reports (42, 79). However, a very small coupling of this type ($\geqslant 0.2$ Hz) has been reported for some dimethylthiophens (50). The characteristic coupling pathways observed for different types of substituents led to considerable discussion about the mechanisms of these long-range effects. Also, the specificity of these couplings has been a useful aid for proton assignments in disubstituted thiophens, although at times the expected couplings have been absent [e.g., in 2-bromo- and 2-iodothiophens (**40, 41**) (57)]. Long-range couplings between protons of different side chains have been observed in spectra of 3- and 5-methyl-2-thiophenthiols (**42, 43**) 2-methyl-3-thiophenthiols (**44**), and 2,3-dimethylthiophen (**45**) and its derivatives (75). In the case of the dimethyl compounds the dominant contribution to the mechanism of this coupling was thought to involve π-electron interactions.

The highly resolved three-spin systems of monosubstituted thiophens and the presence of observable long-range couplings have led spectroscopists to choose thiophens as models on which to test sophisticated multiple-resonance techniques (50, 80–85) and methods of resolution enhancement (54), as well as for the observation of double and triple quantum transitions (86–88).

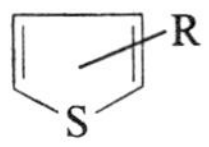

S R

29, R = Me
30, R = CHO
31, R = —CH=NOH
32, R = SH

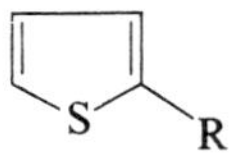

S R

33, R = —C≡CH
34, R = Me
36, R = CHO
38, R = SH
40, R = Br
41, R = I
50, R = COR
51, R = CSR

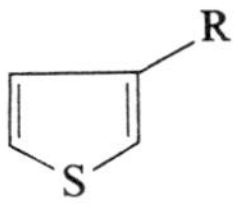

S R

35, R = Me
37, R = CHO
39, R = SH

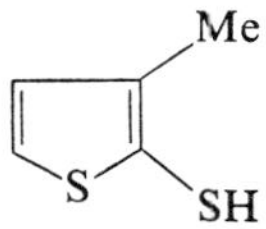

Me S SH

42

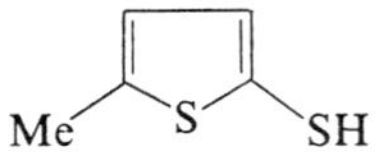

Me S SH

43

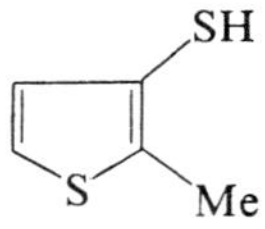

SH S Me

44

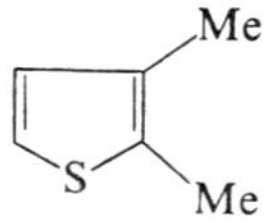

Me S Me

45

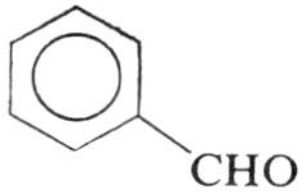

CHO

46

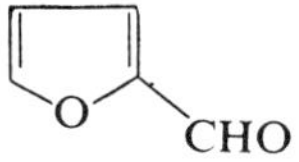

O CHO

47

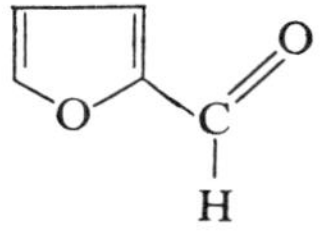

O C O H

48

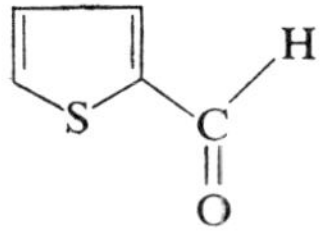

S C H O

49

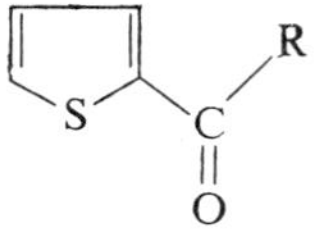

S C R O

52

S C O C Me Me Me

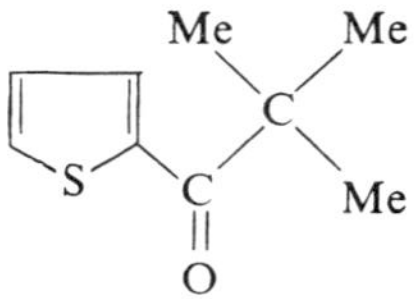

53

These experiments have established the relative signs of couplings between ring protons and also those of a number of the observed long-range couplings. Coupling constants involving any two ring protons are all the same, assumed positive. Couplings between methyl groups and ortho ring protons are opposite in sign to the ring couplings, while couplings between methyl protons and meta protons carry the same sign as the ring coupling constants. The coupling between the two methyl groups of 2,3-dimethylthiophen (**45**) was shown to have the same sign as the couplings between the ring protons (50). In 2-thiophenaldehydes (**36**), the sign of $J_{CHO,5}$ has been shown to be the same as that of J_{45}.

Low-temperature measurements of the spectra of benzaldehydes (**46**) (89) and 2-formylfuran (**47**) (90) showed the presence of two rotamers which were assigned to the two possible conformations in which the formyl group was in the same plane as the aromatic ring. In the case of 2-formylfuran (see Chapter 5), the predominant isomer was thought to be **48** with the two oxygen atoms in the trans configuration. The observation of a long-range coupling between the formyl proton and H-5 (1.2 Hz) contradicted the extended *W* rule which holds for many aromatic systems. Luckily, Roques and his co-workers (91) have been able to demonstrate conclusively that conformations assigned to these two species have to be reversed, and to rationalize the conflict between chemical shift and long-range coupling inherent in the original choice. The spectrum of 2-formylthiophen is unaltered at low temperatures ($-80°$) and this has been interpreted in terms of a very large predominance of one rotamer over the other (62, 77, 90). The conformation of this preferred species was shown to be the same as that of the preferred form of 2-formylfuran, that is, with the sulfur and oxygen atoms cis to each other (91) (**49**). Evidence from benzene-induced solvent shifts also support this arrangement (77). In the process of investigating isomerism in the formylthiophens, spectra were measured of a large number of 2-thienylketones (**50**) and thioketones (**51**) (62, 63, 77, 91). Chemical shift data for 2-benzyl- and 2-thiobenzylthiophens show that the systems are not planar. Nuclear Overhauser effects (NOE) were used to show that a number of the ketones with bulky side chains (*t*-butyl, mesityl, etc.) exist predominantly in the conformation **52** with the sulfur and oxygen atoms cis to each other. Gentle irradiation of the *t*-butyl signal from *t*-butyl-2-thienylketone (**53**) led to a 32% enhancement in the intensity of the signals from H-3 with only minor changes to other signals. This indicates that the methyl groups of the *t*-butyl side chain must be able to interact strongly with H-3, a situation only possible in the conformation suggested above. Similar results were obtained for other thienylketones containing bulky acyl or aroyl moieties. Spectral data for simple thiophens, not involved in tautomeric studies, are collected in Table 6.4.

TABLE 6.4 SPECTRAL DATA FOR SIMPLE THIOPHENS

(Thiophene ring numbering: S 1, then 2, 3, 4, 5)

Substituents	Solvent	τ2	τ3	τ4	τ5	J_{23}	J_{24}	J_{25}	J_{34}	J_{35}	J_{45}	Ref.
one	CCl_4	2.80	3.04	3.04	2.80	4.92	1.32	3.45	2.95	1.32	4.92	56
2-Me	TMS[a]	7.59	3.38	3.25	3.10	1.0	0.24		3.46	1.16	5.20	79
3-Me	TMS[a]	3.25	7.80	3.27	2.95	1.06	1.28	2.94	0.40	0.35	4.88	79
2-CHO									3.7		5.0	37
3-CHO							1.5	2.4				37
2-COMe									3.7		5.1	37
3-COMe							1.4	2.7			5.1	37
2-Cl	CCl_4		3.29	3.24	3.02				3.7	1.79	5.01	56
3-Cl	CCl_4	3.02		3.15	2.83		1.5	3.16			4.81	56
2-Br	Et_2O		2.96	3.17	2.74				3.7	1.7	5.5	59
3-Br	Et_2O	2.68		2.99	2.68							59
2-Li	Et_2O		2.48	2.82	2.34				2.8	0.5	4.3	59
2-C≡C—CHO	CS_2		2.50	2.92	2.50							51
2-C≡C—CH_2OH	CS_2		2.72	3.03	2.72							51
2-CO_2H, 3-Ph	Me_2CO			2.35 ↔	2.90						5.2	52
2,3-Di-CN				3.60	3.20						5	71
2,4-Di-CN			3.20		1.84					2		71
2,5-Di-CN			2.39	2.39								71
3,4-Di-CN		1.95			1.95							71
2-Br, 3-Ph	Me_2CO			2.56 ↔	2.94						5.4	52
2-CHO, 3-Br	$CDCl_3$	0.18		3.00	2.40			1.4			5.15	57
2-CHO, 4-Br	$CDCl_3$	0.17	2.42		2.39			1.2		1.4		57
2-CHO, 5-Br	$CDCl_3$	0.25	2.52	2.85					4.0	0.9		57
3-CHO, 2-Br	$CDCl_3$		0.12	2.71	2.77						5.3	57
3-CHO, 4-Br	$CDCl_3$	1.90	0.04		2.68			3.3				57
4-CHO, 2-Br	$CDCl_3$		2.60	0.28	2.07				0.45	1.45		57
2-CHO, 3-I	$CDCl_3$	0.33		2.75	2.35			1.3			4.7	58
2-CHO, 4-I	$CDCl_3$	0.30	2.34		2.29			1.3		1.3		58
2-CHO, 5-I	$CDCl_3$	0.26	2.64	2.64								58
3-CHO, 2-I	$CDCl_3$	0.96		2.90	2.70					0.9	5.5	58
3-CHO, 4-I	$CDCl_3$	2.00	0.27		2.50			3				58
2-I, 4-CHO	$CDCl_3$		2.35	0.10	1.95					1.3		58
2-Li, 3-Br	Et_2O			2.96	2.41						4.4	59
2-Li, 4-Br	Et_2O		2.73		2.44					0		59
2-Li, 5-Br	Et_2O		2.71	2.94					2.6			59
3-Li, 4-Br	Et_2O	2.68			2.73			2.7				59
2,3-Di-Cl	CCl_4			3.23	3.01						5.8	56
2,4-Di-Cl	CCl_4		3.22		3.17					1.8		56
2,5-Di-Cl	CCl_4		3.40	3.40								56
3,4-Di-Cl	CCl_4	2.87			2.87							56
2,3-Di-Br	Et_2O			3.08	2.62						5.8	59
2,4-Di-Br	Et_2O		2.98		2.66							59
2,5-Di-Br	Et_2O		3.12	3.12								59
3,4-Di-Br	Et_2O	2.51			2.51							59
2-OMe, 3-NO_2	DMSO	5.85		2.71	2.77						6	65
2-NO_2, 3-OMe	DMSO			2.77	2.99						6.1	65

[a] Sample dissolved in tetramethylsilane.

Nuclear magnetic resonance has often been used to provide definitive information about the predominant forms of tautomeric heterocyclic compounds, and this is certainly so for a number of amino-, hydroxy-, and mercaptothiophens. As early as 1960, Gronowitz and Hoffman (43) used this technique to show that 2- or 3-amino- or -mercaptothiophens (**54** to **55**) exist mainly as such while the "2-hydroxy" derivative prefers the 2,5-dihydro-2-oxo form **56**. In this same report, the compound first formed from levulinic acid and P_2S_5 was shown to be the 2,3-dihydro-2-oxo compound **57**, which rearranged on standing to give the 2,5-dihydro isomer **58**. This tautomerism was later studied in detail using NMR as one technique in many (92, 93). Since the original paper on the tautomeric species, the predominant structures have been confirmed a number of times, for example, aminothiophens (94, 95), 2,5-dihydro-2-oxothiophens (96, 97). A most interesting effect has been observed in the spectra of aminothiophens in which the amino group is part of a reduced heterocyclic ring (**59**) (95). The bulk of the substituent causes hindered rotation about the thiophen–nitrogen bond to give predominant rotamers in which the lone pair of the nitrogen is in the plane of the thiophen ring. These preferred orientations of the lone pair are reflected in the chemical shifts of the adjacent ring protons. Data are also available for a number of urea and urethan derivatives of 3-aminothiophen (**60**) (72). The signs of J_{34} and J_{45} in 5-bromo- or 5-chloro-2,5-dihydro-2-oxothiophen (**61**) have been shown to be the same (97).

3-Hydroxythiophen is extremely unstable and attempts to prepare it for NMR studies failed. However, the 2- and 5-acyl- or -ethoxycarbonyl derivatives **62** were available and some very interesting results were obtained on tautomerism in these compounds (53, 98). In these compounds, as well as in the 2-hydroxy-3-acyl isomers, the hydroxy form is stabilized by hydrogen bonding between the –OH and the adjacent carbonyl group and, within the limits imposed by the NMR method, the compounds exist entirely in the aromatic hydroxy form. It is interesting to note that in the solid phase this hydrogen bond does not exist, intermolecular bonds being favored (99). In the case of 2-formyl-3-hydroxythiophen (**63**) (91), the intramolecular hydrogen bond is present when the compound is dissolved in apolar solvents, but with solutions in acetone or DMSO, bonding between the hydroxyl group and the solvent predominates. This can be seen clearly from the long-range couplings between the formyl group and the ring protons. With CCl_4 solutions, $J_{CHO,4}$ is observed and this confirms the trans relationship of the sulfur and formyl-oxygen atoms necessary for internal hydrogen bonding. In acetone solution, this coupling is missing but $J_{CHO,5}$ is present, showing conclusively that the sulfur and formyl-oxygen atoms are cis to each other, the only conformation in which an "all trans" pathway exists between the formyl proton and H-5. Data for potentially tautomeric thiophens are

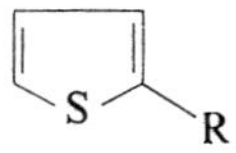

54
R = NH_2 or SH

55
R = NH_2 or SH

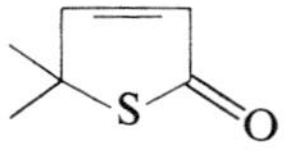

56

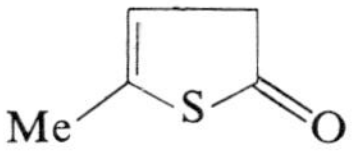

57

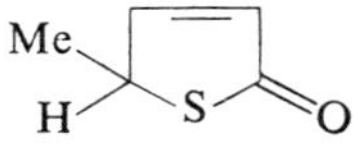

58

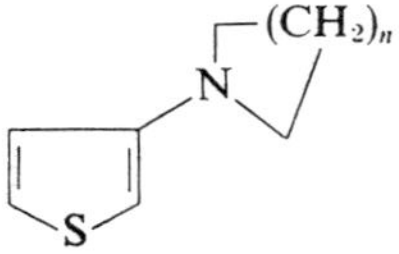

59

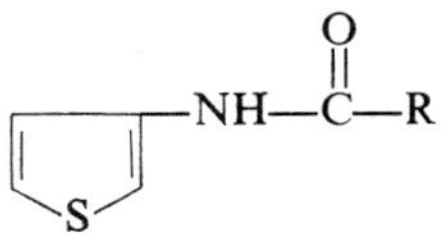

60

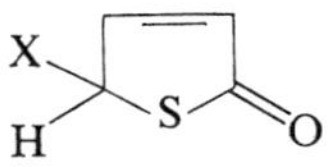

61
X = Cl or Br

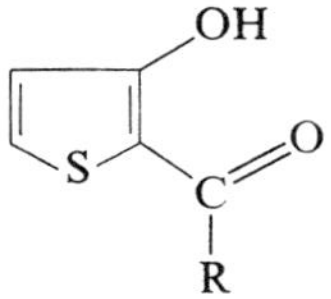

62
R = alkyl or ethoxy
63, R = H

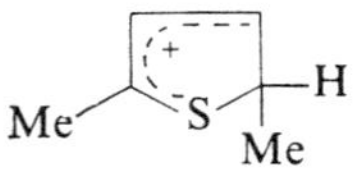

64

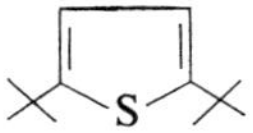

65

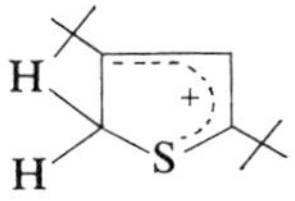

66

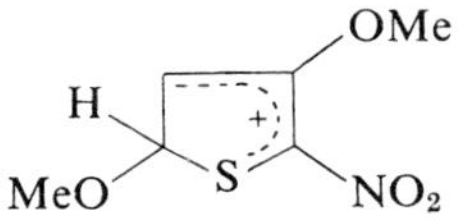

67

TABLE 6.5 SPECTRAL DATA FOR POTENTIALLY TAUTOMERIC THIOPHENS

A. AROMATIC TYPES

Substituents	Solvent	τ2	τ4	τ5	J_{24}	J_{25}	J_{34}	J_{35}	J_{45}	Ref.
2-OMe							3.7	1.8	5.2	37[a]
3-OMe					1.6	3.1			5.0	37[a]
2-NH_2							3.6	1.5	5.9	95
3-NH_2					1.3	3.0			5.5	95
2-Ac, 3-O-*t*-Bu	CCl_4		3.13	2.62					5.5	53
2-Ac, 3-OH	CCl_4		3.30	2.66					5.3	53
3-Ac, 2-O-*t*-Bu	CCl_4		2.91	3.49					6.2	53
3-Ac, 2-OH	CCl_4		3.35	3.75					6.7	53
2-CHO, 3-OH	CCl_4	0.46	3.27	2.49	0.65				5.2	91
	DMSO	0.07	3.21	2.12		1.2			5.2	91

B. 2,5-DIHYDRO-2-OXO TYPES

Substituents	Solvent	τ3	τ4	τ5	J_{34}	J_{35}	J_{45}	Ref.
None	$CDCl_3$	3.76	2.45	5.90				96
					5.9	2.0	2.8	43[a]
3-Me	$CDCl_3$	8.07	2.70	6.03				96
4-Me	$CDCl_3$	3.89	7.77	5.98				96
5-Me					6.0	1.9	2.6	43[a]
5-Br	CCl_4	3.63	2.39	3.66	6.1	1.2	2.9	97
	C_6D_6	4.32	3.34	4.49	5.8	1.1	3.0	97
5-Cl	CCl_4	3.63	2.49	3.63	6.0	1.3	2.8	97

[a] Chemical shifts referred to external cyclohexane.

collected in Table 6.5. Protonation of thiophen and its methyl derivatives appears to occur mainly on the α-carbon atoms (100). At low temperatures, in HF or $HF{\cdot}BF_3$, the appropriate species were observed, for example, **64**. When 2,5-di-*t*-butylthiophen (**65**) was dissolved in concentrated sulfuric acid, the spectrum obtained showed that it was completely converted into the cation of 2,4-di-*t*-butylthiophen (**66**) (101). Also, when 1 equiv of methoxide ion is added to solutions of methoxynitrothiophens in DMSO, Meisenheimer-type adducts are formed (102), for example, **67**, and their structures are easily determined from their spectra.

^1H and ^{19}F spectra of 2- and 3-fluorothiophens have been fully analyzed (103) and the results for the fluorine parameters, together with those for

TABLE 6.6 SPECTRAL DATA FOR FLUOROTHIOPHENS

Substituents	Solvent	Chemical shifts[a]				Coupling constants						Ref.
		2	3	4	5	J_{23}	J_{24}	J_{25}	J_{34}	J_{35}	J_{45}	
2-F	Neat	28.30				1.62	3.07	3.10				103
3-F	Neat		32.05			1.08			−0.81	3.30		103
2,3-Di-F	C_6H_6	−6.31	−13.71	(3.90)	(4.22)	0.77	3.12	4.47	−0.12	4.61	6.58	104
	C_6H_{12}	−6.14	−13.33	(3.48)	(3.57)	0.22	3.09	4.30	−0.13	4.57	6.60	104
2,4-Di-F	C_6H_6	−37.25	(3.98)	−39.83	(4.37)	1.35	9.23	3.45	−0.27	2.32	2.92	104
	C_6H_{12}	−37.27	(3.82)	−39.85	(4.14)	1.29	9.28	3.40	−0.33	2.29	2.81	104
2,5-Di-F	C_6H_6	−26.53	(4.42)	(4.42)	−26.53		3.66	22.91	4.42			104
	C_6H_{12}	−26.50	(4.09)	(4.09)	−26.50							104
3,4-Di-F	C_6H_6		−23.68	−23.68		1.35	3.26	3.92	−13.19	3.26	1.36	104
	C_6H_{12}	(3.49)	−23.80	−23.80	(3.49)	1.23	3.17	3.94	−12.88	3.17	1.23	104
2,3,5-Tri-F	Neat	2.70	−14.62	(3.87)	−26.69	4.78	3.62	27.70	0.88	15.22	3.40	104
	CCl_4	2.69	−14.37	(3.88)	−26.41	4.85	3.60	27.76	0.80	15.26	3.30	104
Tetra-F	Neat	164.9	155.6	155.6	164.9	7	17	31	7	17	7	105

[a] ^1H chemical shifts in parentheses as τ values. ^{19}F chemical shifts in ppm from internal C_6F_6 (or $CFCl_3$ for tetra-F).

other fluorothiophens, are included in Table 6.6. A set of small peaks 0.014 ppm upfield from the main fluorine resonance of 2-fluorothiophen was ascribed to the ^{19}F signal from the ^{34}S isotopomer. ^{34}S has a 4% natural abundance, but this secondary isotope effect seems to be rather large. Data were also obtained for a number of di- and trifluorothiophens (104). J_{34} in 34-difluorothiophen was shown to be of opposite sign to the proton coupling J_{25}. In polyfluorothiophens with fluorines in both the 2- and 3-positions a considerable increase of the fluorine chemical shifts compared with those of 2- and 3-fluorothiophen was observed. The ^{13}C–F couplings were measured for these compounds. In tetrafluorothiophen and the two monomethoxytrifluorothiophens, the α-fluorines resonate at higher field than the β-fluorines, an order opposite to that observed for most other five-membered fluoroheterocycles (105).

The ^{13}C spectrum of thiophen appeared in an extremely fragmented form and the total picture must be drawn from the work of a number of authors. Although ^{13}C statellites were used a number of times in the analysis of the spectra of thiophens and ^{13}C–H couplings measured (30, 76, 106), it was only in 1965 that the actual ^{13}C spectrum of thiophen was reported (107). Since then, Weigert and Roberts (108) have studied the spectrum in detail and have determined the long-range C–C coupling constants. Chemical shifts followed variations in local charge densities and long-range coupling constants correlated well with C–C and H–H couplings in substituted ethylenes. Extended Hückel calculations did not correlate well with spin–spin couplings in this system. ^{13}C coupling constants have been used to assign protons in various chloro derivatives of thiophen (64). ^{13}C data for thiophens are collected in Table 6.7.

Thiophen and α,α-dithienyls (**68**) are present in certain polyacetylene-type natural products and data for some of these, as well as for a few simple dithienyls, are available, for example, in Refs. 109–114. Spectra of a number of 1,2-dithienylethylenes (**69**) have also been reported (115). Nuclear magnetic resonance techniques were used to sort out keto-enol tautomerism in a series of dimers of 3-hydroxydihydrothiophen (115a).

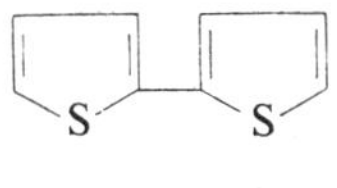

68 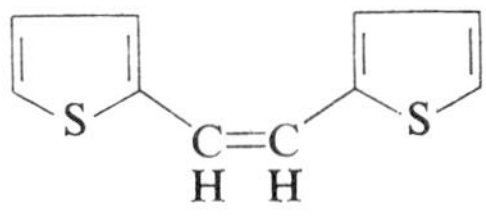 **69** 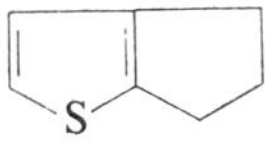 **70**

Thiophens with fused cyclopentane rings are comparatively rare. One paper, dealing with the preparation of cyclopenta[*b*]- and [*c*]thiophens, **70** and **71**, respectively, describes the spectra of a number of these compounds (116). Also, data for the corresponding cyclopentanonethiophens have been

TABLE 6.7 ^{13}C SPECTRAL PARAMETERS FOR THIOPHENS

A. CHEMICAL SHIFTS[a]

Substituents	C-2	C-3	C-4	C-5	CMe
None	67.9	66.4	66.4	67.9	
2-Me	53.4	67.7	66.2	69.6	178.1
2,5-Di-Me	55.4	67.5	67.5	55.4	177.6

B. DIRECT ^{13}C–H COUPLING CONSTANTS (HZ)

Substituents	C-2	C-3	C-4	C-5	CMe	Ref.
None	184.5	167.5	167.5	184.5		106
	185.25	166.92	166.92	185.25		30
	187.0	180.0	180.0	187.0		76
	189	168	168	189		107
2-Me		163.56	165.50	184.92	128.26	79
3-Me	182.36		164.48	184.92	127.10	79
2,5-Di-Me		161.24	161.24		128.20	79
2-Cl		171.54	169.77	188.26		30
2-Br		172.49	169.91	188.11		30
2-I		172.22	169.64	187.50		30
3-Cl	191		174.5	189		64
2,3-Di-Cl			176	191		64
2,4-Di-Cl		178.5		193		64
2,5-Di-Cl		171.61	171.61			30
2,5-Di-Br		171.42	171.42			30
2,3,4,5-Tetrahydro	142.1	126.3	126.3	142.1		5

C. LONG-RANGE ^{13}C–H COUPLINGS (HZ) (108)

Substituents	C-2			C-3			C-4			C-5		
	H-3	H-4	H-5	H-2	H-4	H-5	H-2	H-3	H-5	H-2	H-3	H-4
None	7.35	10.0	5.15	4.7	5.9	9.5						
2-Me					5.8	8.2		5.55	3.8		6.8	9.8
3-Me		8.95	4.3				7.2		5.5	4.3		8.95
2,5-Di-Me	7.3	7.3			4.6							

[a] Parts per million from CS_2. Benzene is +64.8 ppm from CS_2. Neat solutions (107).

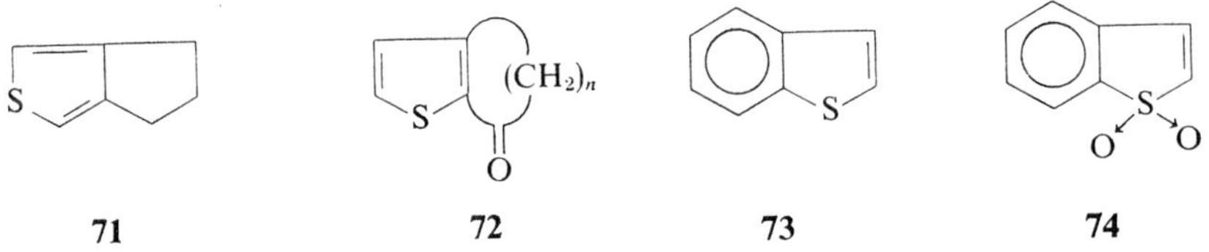

71 **72** **73** **74**

included in a study of the spectral properties of thienocyclenones (**72**) (117).

The benzo[*b*]thiophen or thianaphthene ring system **73** has been the subject of a number of papers over the past few years (118–134). Although most of these reports dealt with synthetic and structural problems, spectral data for the parent compound and many of its simple derivatives were included. However, an accurate analysis of the spectrum of benzo[*b*]thiophen is not available at the time of writing this book. As with many benz-fused compounds, the spin system can be analyzed with relative ease, but two assignments are possible for the four protons on the benzene ring. Two groups of workers (118, 127) have estimated the chemical shifts of the six protons, using partially deuterated compounds as models. Unfortunately, the two assignments for H-5 and H-6 do not agree, although the more recent work (127) would seem to be the most reliable. Chapman *et al.* (130) have also reported data for this compound but they frankly admit that their assignment of chemical shifts to the closely coupled H-5 and H-6 was quite arbitrary. The very small chemical shift between these protons places the ABCD system of the benzene ring in the deceptively simple area and it is doubtful whether more sophisticated methods would give more accurate results. Substituent effects on chemical shifts and coupling constants have been examined for a large number of mono- (mainly 5-) and disubstituted benzothiophens (128, 130, 131, 133). In all cases the observations could be explained by invoking reasonable combinations of inductive, mesomeric, steric, or anisotropic effects. Long-range couplings in this system are particularly interesting. As well as the expected cross-ring coupling ($J_{37} \sim$ 0.8 Hz), one of about 0.5 Hz occurred between H-2 and H-6, and this was attributed to "extended *W*" interactions across the sulfur atom (122, 123, 127, 130, 131). Small couplings ($\sim$ 1 Hz) between methyl groups in positions 2 or 3 and adjacent protons and between the two methyl groups of 2,3-dimethyl derivatives were thought to be transmitted mainly by a π-electron contact mechanism, as in similar thiophens (126). Faller (134) has made a study of orientation phenomena in a series of acetyl- and aroylbenzo[*b*]-thiophens containing other bulky substituents. Electronic and hindered rotation effects could be discerned, but individual rotamers were not detected. Oxidation of the sulfur atom of this ring system to the dioxide **74** caused consistent changes in the NMR parameters (130): H-2 and H-4 were shielded, H-5 and H-6 deshielded, while H-3 and H-7 were relatively insensitive to the change; J_{23} was increased, but the cross-ring coupling J_{26} was decreased to a point where it could no longer be resolved. These changes were discussed in terms of inductive, mesomeric, and anisotropic effects of the sulfonyl group, together with the loss of ring current associated with sulfone formation. Few potentially tautomeric benzo[*b*]thiophens have been investigated by NMR. However, the 2-amino (**75**) (125) and 3-hydroxy (**76**) (124) derivatives were shown to exist predominantly in their aromatic forms. Data for simple benzo[*b*]thiophens are collected in Table 6.8.

TABLE 6.8 SPECTRAL DATA FOR BENZO[*b*]THIOPHENS

		Chemical shifts						
Substituents	Solvent	τ2	τ3	τ4	τ5	τ6	τ7	Ref.
None	CCl_4	2.71	2.74	2.29	2.73	2.70	2.23	118
		2.67	2.78	2.28	2.74	2.76	2.21	127
				2.30	2.74	2.76	2.22	130
	$CDCl_3$	2.60	2.71	2.22	2.67	2.69	2.14	130
3-Me	CCl_4	3.11		2.50	2.83	2.73	2.33	118
		3.07		2.39	2.74	2.76	2.25	130
2-Me	CCl_4	7.5	3.2	2.5	2.9	2.9	2.5	123
4-Me	CCl_4	2.7	2.7	7.4	3.0	2.9	2.4	123
5-Me	CCl_4	2.8	2.9	2.6	7.60	3.0	2.4	123
6-Me	CCl_4	2.8	2.9	2.5	3.0	7.6	2.5	123
7-Me	CCl_4	2.8	2.8	2.5	2.9	3.0	7.5	123
2-NO_2	DMSO		2.13	2.37	2.78	2.88	2.50	132
3-NO_2	DMSO	1.53		1.95	2.82	2.89	2.40	132
4-NO_2	C_6D_6	2.86	1.97		1.99	3.19	2.52	132
5-NO_2	C_6D_6	2.17	2.31	1.69		2.55	2.81	132
6-NO_2	C_6D_6	2.76	3.06	2.66	2.05		1.63	132
7-NO_2	C_6D_6	2.51	3.36	2.67	2.93	2.78		132

	Coupling constants (132)								
	J_{23}	J_{26}	J_{37}	J_{45}	J_{46}	J_{47}	J_{56}	J_{57}	J_{67}
2-NO_2			0.38	8.49	1.13	0.82	7.21	1.05	8.31
3-NO_2		0.36		8.47	1.15	0.77	7.24	1.06	8.33
4-NO_2	5.81	0.50	0.94				8.07	0.86	8.03
5-NO_2	5.55	0.15	0.20		1.71	1.30			6.33
6-NO_2	5.55		0.74	9.07		0.62		2.16	
7-NO_2	5.20	0.45		7.97	1.03		7.83		

The ^{19}F spectra of a number of polyfluorobenzo[*b*]thiophens, fluorinated in the benzene ring, have been measured, and some interesting couplings were observed (135). Both H-2 and H-3 showed long-range coupling to F-7 and, in all compounds studied, the para F–F coupling J_{47} was large, 15 to 19 Hz. Table 6.9 contains ^{19}F and ^{1}H parameters for these compounds.

TABLE 6.9 SPECTRAL DATA FOR POLYFLUORO DERIVATIVES OF BENZO[*b*]THIOPHENS (135)

	Substituents		Solvent	Chemical shifts[a]					
	Benzene ring	Hetero-ring		τ2	τ3	4	5	6	7
I	Tetra-F	None	CCl_4	2.57	2.61	145.18	160.18	160.19	141.68
II		2-COOH	Me_2CO	−0.45	2.04	142.77	159.80	156.83	141.68
III	5,6,7-Tri-F	None	CCl_4	2.58	2.82	2.70	137.15	164.29	136.3
IV		2-COOH	Me_2CO	−0.72	1.96	2.35	136.97	161.23	137.35
V	4,6,7-Tri-F	None	CCl_4	2.68	2.68	120.37	3.12	140.00	145.3
VI		2-COOH	Me_2CO	0.98	2.15	117.38	2.78	136.13	145.75
VII	4,5,7-Tri-F	None	CCl_4	2.58	2.61	148.78	140.36	3.13	117.40
VIII		2-COOH	Me_2CO	1.55	1.89	147.01	139.5	2.61	117.0
IX	4,5,7-Tri-F, 6-OMe	None	CCl_4	2.64	2.67	146.47	155.52	5.93	136.06

Coupling constants

	J_{23}	J_{25}	J_{27}	J_{37}	J_{45}	J_{46}	J_{47}	J_{56}	J_{57}	J_{67}
I			0.95	2.8	20.6	1.2	16.2	19.1	0.3	19.1
II				3.3	18.8	−0.4	16.1	18.4	0.7	18.7
III	5.55	0.75	0.7	3.4	8.8	6.0	1.7	18.9	4.7	19.4
IV				3.28	10.0	6.5	1.55	19.2	4.6	19.2
V			1.9	1.9	8.7	1.5	18.2	10.2	5.7	20.0
VI				3.35	9.2	3.2	18.8	10.8	5.6	19.7
VII			1.45	2.75	19.1	5.4	18.85	10.3	1.4	8.8
VIII				3.47	19.1	6.2	19.0	10.7	2.0	9.4
IX				4.35	19.0		14.6	0.8	3.2	1.8

[a] The numbers 4, 5, 6, and 7 refer to ^{1}H or ^{19}F shifts, ^{1}H in τ units, ^{19}F in ϕ^* units (ppm from internal $CFCl_3$).

Cagniant *et al.* (136), in a paper dealing with the syntheses of a number of ring systems, reported spectral data for some 4-oxo-4,5,6,7-tetrahydrobenzo-[*c*]thiophens (**77**).

Information is also available on the spectra of some more highly annelated thiophens, for example, dibenzothiophen (**78**) (137, 138), two isomers of naphthothiophen (**79** and **80**) (137), and a number of more highly substituted systems (139).

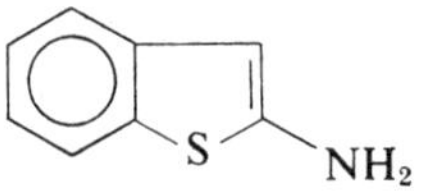

NH$_2$ S **75**   OH S **76**

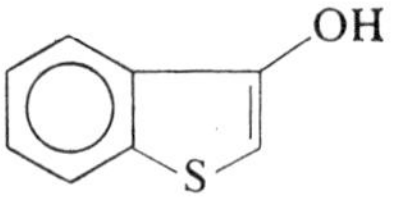

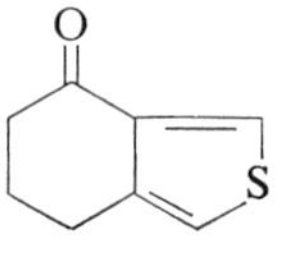

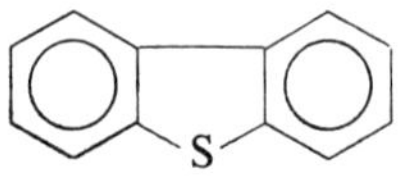

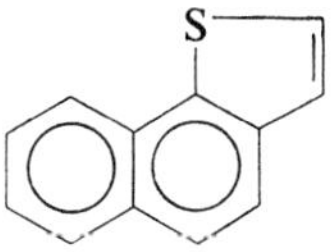

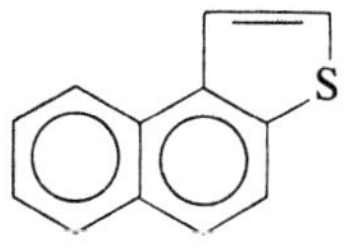

O S **77**   S **78**

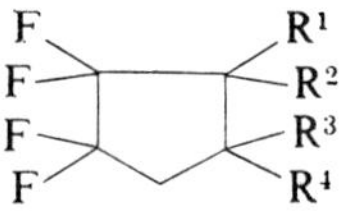

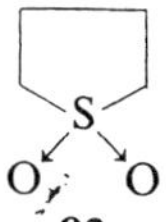

S **79**   S **80**

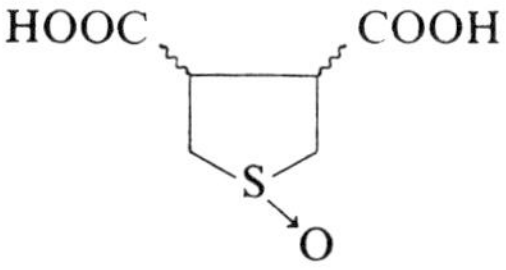

S **81**   F F F F R^1 R^2 R^3 R^4 **82**

S O O **83**   HOOC COOH S O **84**   COOH S **84a**

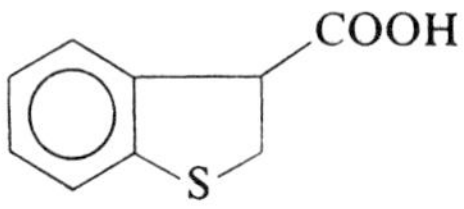

Data for reduced thiophens are very scarce. Tetrahydrothiophen (**81**) is included in the Varian Catalog (Spectrum No. 80) and in the study of carbon hybridization in reduced heterocycles by Lippert and Prigge (5). The parameters listed were $\tau\alpha$ 7.25, $\tau\beta$ 8.12 ppm; $J_{^{13}\text{CH}}(\alpha) = 142.1$, $J_{^{13}\text{CH}}(\beta) = 126.3$ Hz. A large number of highly substituted 2,2,3,3-tetrafluorothiolan (**82**) have been prepared and NMR was used to help determine isomeric structures (140). Spectra of some simple thiolan-1,1-dioxides (**83**) have been reported (141) and the spectrum of the parent can be found in the Varian Catalog (No. 416). In the case of the isomeric 3,4-dicarboxythiolan sulfoxides (**84**), the S–O bond behaved normally, the β-proton being deshielded by the cis sulfoxide group (142). The α-methylene groups of 2,6-dihydrothiophen-1,1-dioxide absorb at τ 6.36 and the β-protons at τ 3.92 (Varian Catalog, No. 406).

TABLE 6.10 SPECTRAL DATA FOR 1,2-DITHIOLIUM SALTS IN TRIFLUOROACETIC ACID (143)

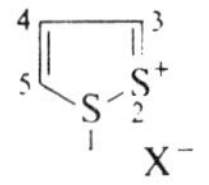

Substituents	X	τ3	τ4	τ5	J_{45}
3-Me	ClO_4	6.71	1.44	−0.26	4.9
3,5-Di-Me	Br	6.84	1.73	6.84	
3-Me, 5-Ph	ClO_4	6.79	1.38		
3-Me, 4-Ph	ClO_4	6.90		−0.03	

Spectra of 2,3-dihydrobenzo[*b*]thiophen-3-carboxylic acid (**84a**) and its cis and trans sulfoxides are reproduced in a paper dealing with the absolute stereochemistry of the sulfoxides (142a). Unfortunately, numerical data are not given.

B. The 1,2-Dithiolium System

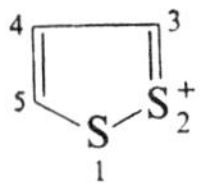

This ring system is relatively rare and few spectra have been reported (143). Representative data are given in Table 6.10. However, NMR has proved to be a powerful tool in the study of structures of uncharged 2,3-dihydro derivatives with an exocyclic double bond on C-3. These include the 3-ones (**85**), the 3-thiones (**86**), and a number of compounds with general structure **87**.

The basic spectral patterns for the dithiolones and thiones were established in 1965 (144) and little has been added since. Spectra of 4- and 5-phenyl-1,2-dithiol-3-thione have been used as models in an attempt to illustrate the

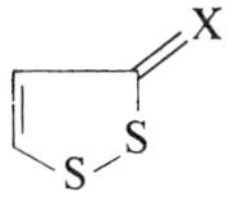

85, X = O
86, X = S

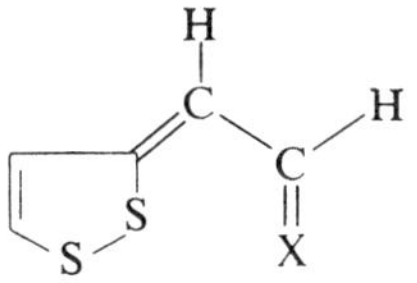

87

88

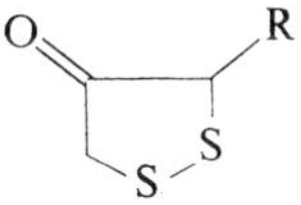

88a

TABLE 6.11 SPECTRAL DATA FOR 1,2-DITHIOL-3-ONE, -3-THIONE, AND SIMILAR COMPOUNDS

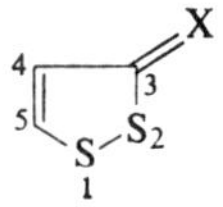

Substituents	Solvent	X = O $\tau 4$	X = O $\tau 5$	X = O J_{45}	X = S $\tau 4$	X = S $\tau 5$	X = S J_{45}	Ref.
None	$CDCl_3$	3.36	1.60	5.3	2.80	1.65	5.4	144
4-Me	$CDCl_3$	7.93	2.03	1.13	7.78	1.90	0.92	144
5-Me	$CDCl_3$	3.60	7.52	1.05	3.00	7.52	0.97	144
5-CO_2Et	$CDCl_3$	2.7			2.35			144
4-Ph	$CDCl_3$		1.64			1.61		144
5-Ph	$CDCl_3$	3.23			2.65			144
4-Iso-Pr	CCl_4					1.87		146
5-Iso-Pr	CCl_4				3.04			146
4-NH_2	$CDCl_3$		2.95			2.57		144
4-NHMe	$CDCl_3$		3.53			3.03		144

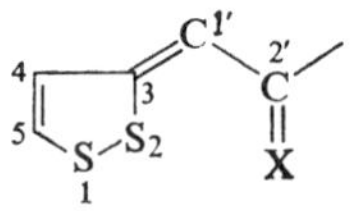

Substituents	X	Solvent	$\tau 4$	$\tau 5$	$\tau 1'$	$\tau 2'$	$J_{1'2'}$	Ref.
5-Ph	O	$CDCl_3$	2.51		3.18	0.59	1.6	143
4-Ph	O	$CDCl_3$		2.12	3.28	0.61	1.6	143
1'-Me, 5-Ph	O	$CDCl_3$	2.51		7.75	0.77		143
2'-Me, 4-Ph	O	$CDCl_3$		2.23	3.33	7.76		143
		CCl_4		2.29	3.41	7.82		147
2',5-Di-Me	O	$CDCl_3$	3.18	7.55	3.37	7.76		143

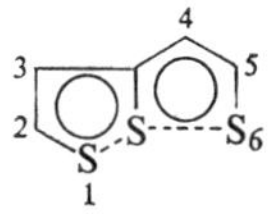

Substituents	Solvent	$\tau 2$	$\tau 3$	$\tau 4$	$\tau 5$	J_{23}	J_{45}	Ref.
None	$CDCl_3$	0.82	2.04	2.04	0.82	6.3		143
2-Me	$CDCl_3$	7.33	2.28	2.23	0.87		6.3	143
2-Ph	$CDCl_3$		1.74	2.06	0.82		6.4	143
3-Ph	$CDCl_3$	1.16		2.21	0.86		6.4	143
2-CO_2Et	$CDCl_3$		1.45	1.93	0.74		6.2	143

aromaticity of the dithiole ring (145), and data are available for an added series of alkyl-substituted 3-thiones (146). With compounds of type **87**, an interesting problem arose (143). If **X** = S or Se, the expected thio- or selenoformyl resonance ($\tau = 0$ to -2) was not observed, but a new signal was observed at about $\tau = 2$ ppm. This was in agreement with the bicyclic structure **88** established by other methods (see Ref. 143). The oxygen analogs, however, were monocyclic (see also Ref. 147). Data for these dithiole derivatives are collected in Table 6.11. Claeson and Thalen (148) have used NMR as an aid to determine the structures of 3-substituted 1,2-dithiolan-4-ones **(88a)**.

C. The 1,3-Dithiolium System

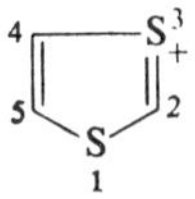

This cationic system has been studied mainly by Campaigne and his co-workers (148–150). In the case of 2-dialkylamino derivatives with a substituent at position 4 (**89**), the chemical shift of H-5 was discussed in terms of contributions from 2-imino-type structures (**90**). Spectra of the 2-phenyl-4-hydroxydithiolium salt **91** in TFA or 70% perchloric acid showed that it

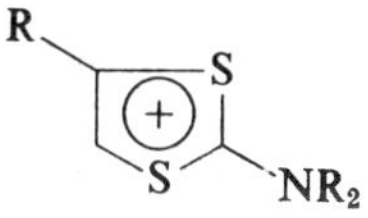

89 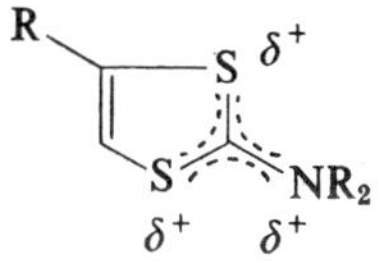 **90** 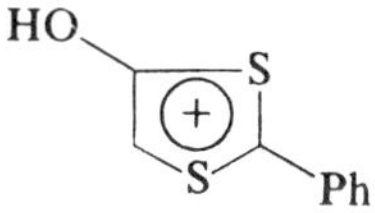 **91** 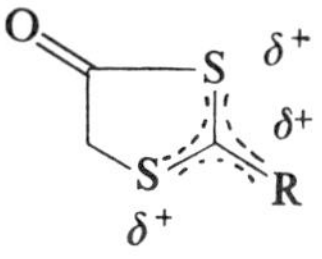 **92**

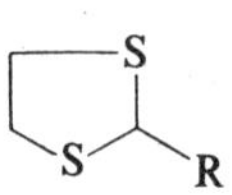

93, R = H
96, R = NMe_2
94 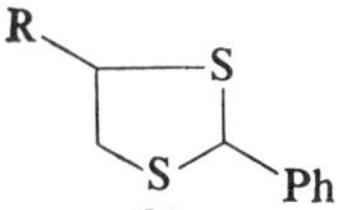 **95** 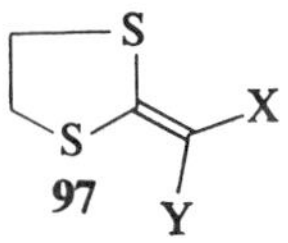 **97** 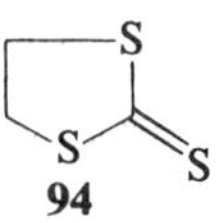

exists in the hydroxy form. However, when the 2-substituent was NMe_2 or SMe, the "oxo" tautomers were predominant and structures **92** were suggested for these. Nuclear magnetic resonance was also used to follow deuteration at position 2 of a number of 4-substituted 1,3-dithiolium salts (151).

The fully reduced members of this ring system, the 1,3-thiolans (**93**), are thioacetals or thioketals and as such are widely used to protect carbonyl groups. It is surprising that very little information is available on the NMR spectra of these compounds. The spectrum of the parent compound was reported by Martin (152) in a note dealing with its preparation by the hydrogenolysis of trithiocarbonates (**94**) .The spectrum (presumably of neat liquid) consisted of two singlets, one at τ 6.17 (H-2) and the other at τ 6.87 (H-4, 5). Atkinson *et al.* (153) studied the ring contraction of 2-phenyl-1,3-

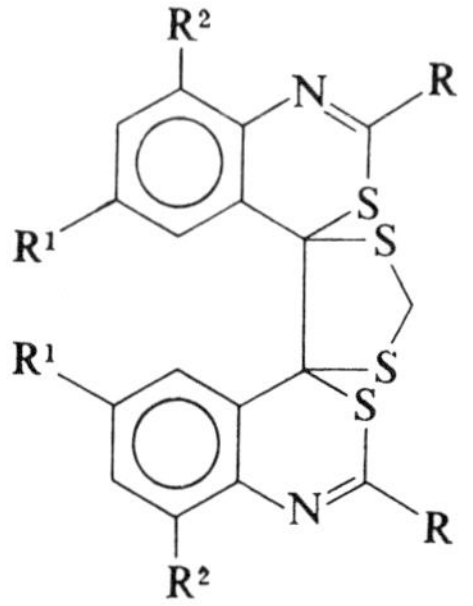

98

dithians (**95**) to 2-phenyl-1,3-dithiolans and used NMR to assign structures to their products. Unfortunately, the only data given are the chemical shifts of H-2 in the cis and trans isomers of the various 2,4-disubstituted compounds formed in these reactions. The general downfield shift of this benzylic proton on going from the six-membered ring system is similar to that observed for the 1,3-dioxan to 1,3-dioxolan contraction (154). Data are also available for a number of 2-dimethylamino derivatives **96** (155) and stick diagrams were presented of spectra of some highly substituted 1,3-dithiolans (156). Chemical shifts of the thioketal protons have been used to help distinguish between C-3, C-17, and C-20 thioketals in the steroid series (157). 1,3-Dithiolans with an exocyclic double bond in position 2 (**97**) were included in an NMR investigation of rotation about such double bonds (158). A detailed consideration of the relevant chemical shifts and coupling constants, as well as dipole moment data, led to an unambiguous assignment of stereochemistry to the cis and trans dispiranic 1,3-dithiolans (**98**) (159).

D. The Thienothiophen Systems

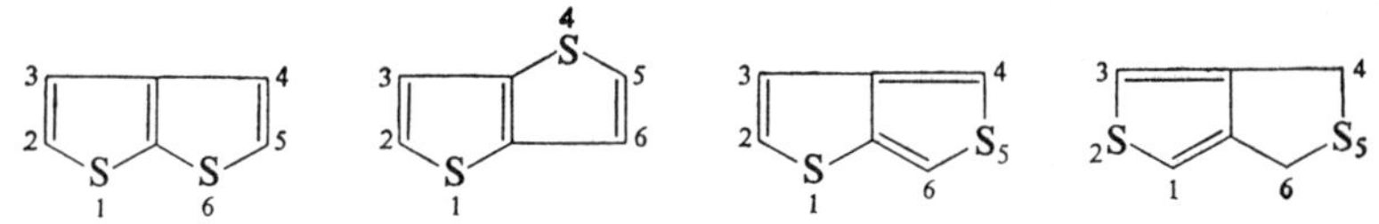

[2,3-*b*] [3,2-*b*] [3,4-*b*] [3,4-*c*] 4,6-Dihydro-

Little information is available on the spectra of thieno[2,3-*b*]thiophens but excellent data for the parent system and certain bromo derivatives have been reported recently (160). Long-range coupling (+1.2 Hz) was observed between H-2 and H-5, across six bonds via an extended *W* pathway. Also, in this paper, data are presented for thieno[3,2-*b*]thiophen and some of its bromo derivatives. The largest cross-ring couplings were again across an extended *W* pathway containing six bonds (J_{25} = 1.5 Hz), but the five-bond coupling J_{36} was also resolved (0.7 Hz). Previously, thieno[3,2-*b*]thiophen was used as a model to test perturbation approaches to the analysis of the A_2B_2 spin system (161). Spectra of the parent system and its 3-methyl derivative have also been reported by Litvinov and Fraenkel (162).

Wynberg and his co-workers (163–165) have been studying the chemistry and spectra of the thieno[3,4-*b*]thiophen system for a number of years, but the data reported in the first two of these papers are of little use. However, the situation was rectified in their most recent report (165) in which long-range couplings were found to be specially useful in structural assignments. Information is also available on the 4,6-dihydro derivatives of thieno[3,4-*b*]- and [3,4-*c*]thiophens (163, 166, 167). Data for the thienothiophens are collected in Table 6.12.

IV. SIX-MEMBERED SULFUR HETEROCYCLES

A. Thiabenzene and the Thiopyrans

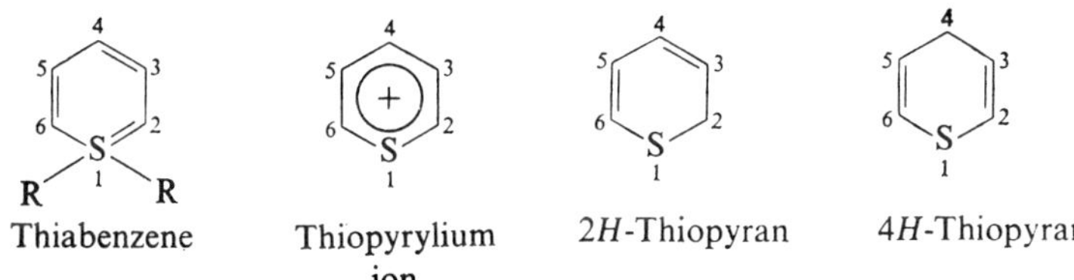

Thiabenzene   Thiopyrylium ion   2*H*-Thiopyran   4*H*-Thiopyran

Fully "aromatic" members of this system can exist only when the sulfur atom is either pentavalent as in the thiabenzenes or positively charged as in the thiopyrylium salts. The thiabenzenes are extremely rare compounds, but spectral data have been reported for the 3,5-diphenyl derivative **99** in $CDCl_3$ (168): τ2,6 4.25, τ4 3.81, τMe 6.50, τPh 2.45–2.85 ppm; J_{24} = 1.1 Hz. The spectrum of the thiopyrylium ion has been analyzed in detail in a theoretical study of the A_2B_2X system (169).

2*H*-Thiopyrans have been more thoroughly investigated. A complete set of spectral parameters are available for 6-ethylthio-2*H*-thiopyran (**100**) (170) and chemical shifts have been reported for the parent system and a number of simple derivatives (170–172). 4*H*-Thiopyran was used as a model compound to study the $A_2B_2X_2$ spin system (169). Data for thiopyrans and the thiopyrylium ion are collected in Table 6.13.

TABLE 6.12 SPECTRAL DATA FOR THIENOTHIOPHENS

A. THIENO[2,3-*b*]THIOPHEN

Substituents	Solvent	τ2	τ3	τ4	τ5	J_{23}	J_{24}	J_{25}	J_{34}	J_{35}	J_{45}	Ref.
None	Me_2CO	2.57	2.76	2.76	2.57	5.23	−2.02	1.20	−0.18	−0.02	5.23	160
2-Br	Me_2CO		2.60	2.73	2.41				−0.13	0	5.26	160
3-Br	Me_2CO	2.44		2.79	2.42		−0.04	1.17			5.21	160
2,4-Di-Br	Me_2CO		2.75		2.41							160
2-Me	CCl_4	7.64	3.35	2.96	3.12	1.3					5.1	165a

B. THIENO[3,2-*b*]THIOPHEN

Substituents	Solvent	τ2	τ3	τ5	τ6	J_{23}	J_{25}	J_{26}	J_{35}	J_{36}	J_{56}	Ref.
None	Me_2CO	2.54	2.68	2.54	2.68	5.25	1.55	−0.20	−0.20	0.75	5.25	160
2-Br	Me_2CO		2.52	2.41	2.68				−0.12	0.69	5.30	160
3-Br	Me_2CO	2.46		2.38	2.58		1.55	−0.20			5.25	160
2,6-Di-Br	Me_2CO		2.39	2.31					−0.15			160
2-Me	CCl_4	7.69	3.38	3.10	2.99	1.3				0.5	5.2	165a

C. THIENO[3,4-*b*]THIOPHEN

Substituents	Solvent	$\tau 2$	$\tau 3$	$\tau 4$	$\tau 6$	J_{23}	J_{36}	J_{46}	Ref.
None	CCl_4	2.84	3.25	2.95	2.85	5.5	0.7	2.5	165
4-CHO	$CDCl_3$	2.34	2.61	0.19	2.32	5.5	0.7	0.6	165
6-CHO	$CDCl_3$	2.53	2.99	2.18	0.30	5.5			165
2-COOH	DMSO		2.20	2.30	2.03			2.5	164

D. 4,6-DIHYDROTHIENO[3,4-*c*]THIOPHEN. SOLVENT—ACETONE (167)

$\tau 1,3$ 3.02; $\tau 4,6$ 6.12

TABLE 6.13 SPECTRAL DATA FOR THE THIOPYRYLIUM CATION AND FOR THIOPYRANS

A. THIOPYRYLIUM PERCHLORATE. SOLVENT—ACETONITRILE (169)

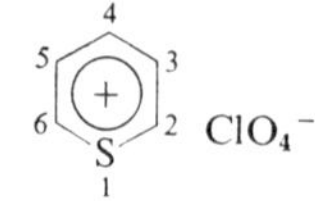

τ2,6	τ3,5	τ4	$J_{23(56)}$	$J_{24(46)}$	$J_{25(36)}$	J_{26}	$J_{34(45)}$	J_{35}
−0.13	1.03	1.00	7.7	1.7	0	1.6	6.1	0

B. 2*H*-THIOPYRANS. SOLVENT—CCl_4

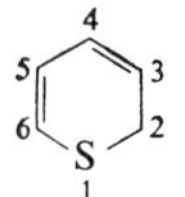

Substituents	τ2	τ3	τ4	τ5	τ6	J_{23}	J_{24}	J_{34}	J_{35}	J_{45}	Ref.
None	6.81		3.82–4.52								171
4-Me	6.84	5.01	8.25	3.92							171
6-SEt	6.82	4.55	4.06	3.77		5.5	1.1	9.4	0.8	6.0	170
4-Me, 6-SMe	6.84	4.80	8.23	3.80							170
5-Me, 6-SMe	6.86	4.47	4.05								170

C. 4*H*-THIOPYRAN. NEAT LIQUID (169)

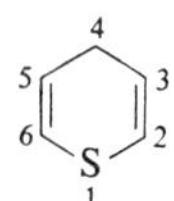

τ2,6	τ3,5	τ4	$J_{23(56)}$	$J_{24(46)}$	$J_{25(36)}$	J_{26}	$J_{34(45)}$	J_{35}
4.03	4.45	7.16	10.0	1.1	0	2.9	3.9	0

The general appearance of the spectrum of 2*H*-thiopyran-1,1-dioxide has been reported by Molenaar and Strating (173): multiplets centered at τ 3.33, 3.70, and 6.00 (each for two protons, $CDCl_3$). The structure of this compound was previously assigned incorrectly as the 4*H*-isomer **101** (174) which is now thought to rearrange spontaneously to the 2*H*-isomer **101a**. 1-Thio-4-pyrone (**102**) in $CDCl_3$ gives a normal AA′BB′ spectrum with the two multiplets centered at τ 2.11 and 2.91 ppm (175). A study has been made of the spectra of a number of 1-thio-4-pyrones and -4-thiopyrones with polymethylene bridges between C-3 and C-5 (**103**) (176). Unfortunately, these compounds contain few resonances which could be classed as typical of the heterocyclic system. Some data are available for 5-NR and 5-SR derivatives of 5,6-dihydro-2*H*-thiopyran-1,1-dioxide (**104**) (177).

The sulfur analogs of the very important natural products based on various benzpyran ring systems have not evoked much interest, and very few attempts have been made to analyze fully the spectra of any of these systems. Chemical shifts are available for the protons of the heterocyclic ring of 2*H*-benzo[*b*]-thiopyran (**104**) and 4*H*-benzo[*b*]thiopyran (**105**) (178), and of 1*H*-benzo[*c*]-

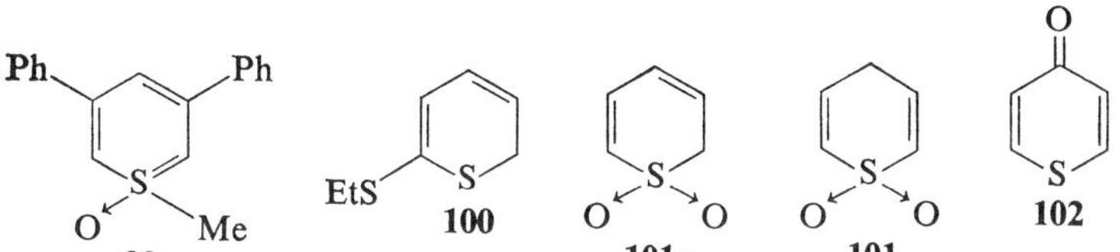

99 **100** **101a** **101** **102**

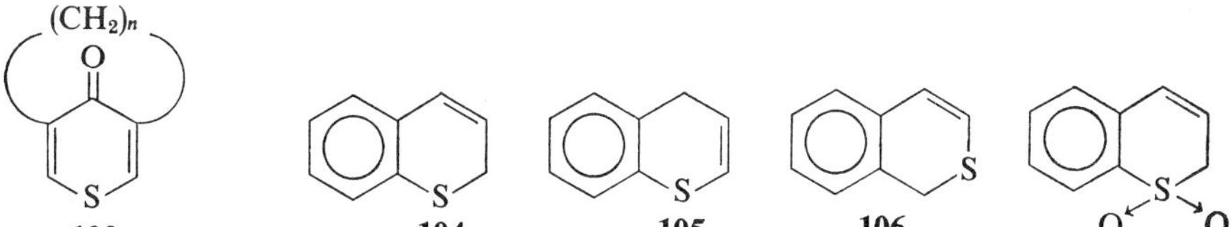

103 **104** **105** **106** **107a**

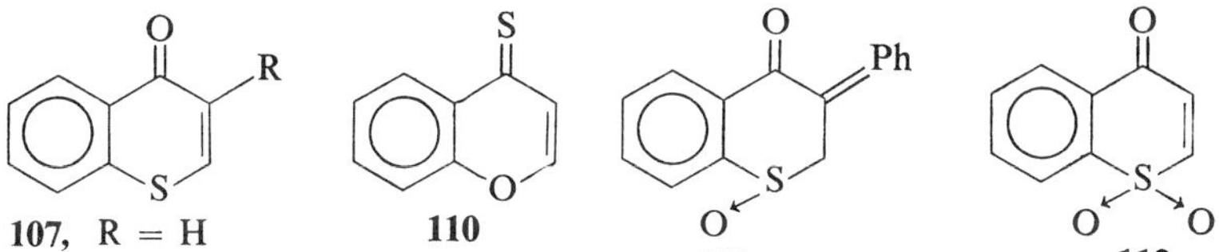

107, R = H
108, R = Br
109, R = I
110 **111** **112**

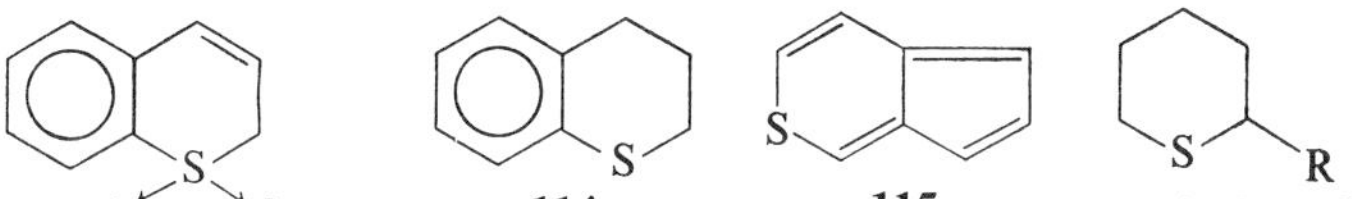

113 **114** **115**
116, R = H
117, R = Cl

thiopyran (**106**) (179), but no attempt was made to analyze the complex patterns from the protons of the benzene ring. J_{24} of **104** was shown to be of opposite sign to the other couplings in the heterocyclic ring (180). The spectrum of 2*H*-benzo[*b*]thiopyran-1,1-dioxide (**107a**) has been reported (183). Data for thiochromanone (benzo[*b*]thiopyranone) (**107**) have appeared in two very short and rather confusing papers. In one (181), the AA′BB′ system of the heterocyclic ring was analyzed but only coupling constants were reported. Complete sets of parameters for the heterocyclic rings of the 3-bromo (**108**) and 3-iodo (**109**) derivatives were used to suggest that the

TABLE 6.14 SPECTRAL DATA FOR BENZOTHIOPYRANS AND SIMILAR COMPOUNDS

A. 2*H*-BENZOTHIOPYRAN

Substituents	Solvent	τ2	τ3	τ4	τ5–8	Ref.
None	CCl_4	6.65	4.18	3.64	3.18	178
2-CH_2Cl	CCl_4	6.21	3.98	3.33	2.90	178

B. 4*H*-BENZOTHIOPYRAN

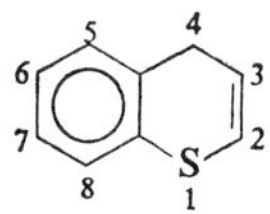

Substituents	Solvent	τ2	τ3	τ4	τ5–8	Ref.
None	CCl_4	3.70	4.11	6.67	2.96	178
2,3-Dihydro	CCl_4	7.18	8.10	7.18	3.18	178

C. ISOTHIOCHROMEN

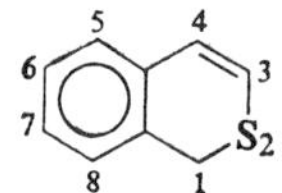

Solvent	τ1	τ3	τ4	τ5–8	Ref.
CCl_4	6.2	3.2–3.7		2.87	179

D. THIOCHROMANONE

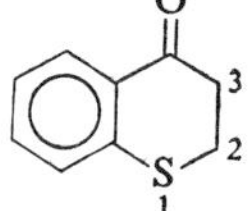

Substituents	Solvent	τ2	τ3	J_{22}	J_{33}	J_{23} Cis	J_{23} Trans	Ref.
None	$CDCl_3$	6.80	7.10					182
	TFA	6.44	6.44					182
	Me_2CO			−15.3	−17.7	3.6	8.2	181
3-Br	CCl_4	5.93,[a] 6.07[b]	5.10	−14.1		3.0	8.6	181
3-I	Me_2CO	6.38,[a] 6.71[b]	4.62	−14.9		2.7	5.9	181
1,1-Dioxide	$CDCl_3$	6.38	6.38					182

[a] Cis to 3-H. [b] Trans to 3-H.

3-halogen group preferred to be in a pseudoaxial conformation. The second study (182) of this and similar ring systems used only the chemical shifts of the methylene groups in the heterocyclic ring. A comparison of these chemical shifts for the neutral molecules and cations of thiochromanone (**107**) and isothiochromanone (**110**) led to the conclusion that protonation occurs on the sulfur atom. Nishio (183a) used 3-benzalthiochromanone-1-oxides (**111**) to study the nonequivalence induced by the sulfoxide configuration on the protons of the 2-methylene group. Thiochromanone-1,1-dioxide (**112**) was shown not to protonate on the 4-oxo function (182). Deuterium exchange in 2*H*-thiochromen-1,1-dioxide (**113**) was followed by NMR (183). Actual spectra were reproduced and spectral parameters were given.

In the thiochroman series **114**, spectra of the 3-bromo derivatives have been reported by Hofmann and Salbeck (184) while NMR methods were used in a very competent study of the stereochemistry of thioflavanols (185, 186). Data for the various benzothiopyran derivatives are collected in Table 6.14.

The spectrum of the sulfur analog of 2-pyrindene (**115**) has been measured and the chemical shift data were used as evidence to show that charge-separated structures contribute more to the ground state in this molecule than in azulene (187).

The fully reduced thiopyran or thian system **116** has been studied mainly by Lambert and Keske (188–190) as part of their work on inversion phenomena in reduced heterocyclic systems. Determination of conformation based on *R* values (*R* value equals J_{trans}/J_{cis} for a $-CH_2-CH_2-$ grouping) for some deuterated thians led to the conclusion that the thian ring is slightly puckered (188). The S–H proton of the conjugate acid of thian was shown to reside exclusively in the axial position (189). These workers (190), and also Martin and Uebel (191), used NMR to show that a 1-oxide group on a thian system exhibits a surprising preference for the axial conformation. Thian was also included in a general study of ^{13}C–H coupling constants in reduced heterocycles (5); $J_{CH}(\alpha) = 135.0$, $J_{CH}(\beta, \gamma) = 123.7$ Hz. The proton resonances of the α-protons occur at about τ 7.48 and the β,γ-protons near τ 8.29 ppm (for the spectrum of thian see Varian Catalog, No. 118). Data are also available for α-chlorothian (**117**) in benzene (192): $\tau\alpha$ 5.0, $\tau\beta,\gamma$ 1.1–3.1 ppm.

B. Dithiens and Dithians

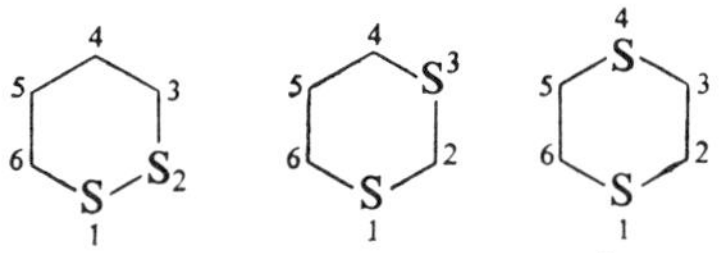

1,2 1,3 1,4

Variable-temperature NMR was used to study the inversion of 1,2-dithians (193). Although 3,3,6,6-tetramethyl-1,2-dioxane was shown to invert abnormally, the corresponding 1,2-dithian appears to invert via a mechanism of the type proposed for cyclohexanes. Chemical shifts have been reported for the ring protons of a number of highly substituted 1,3-dithiens of the type **118** (194, 195). The 1,3-dithians are actually thioketals, used regularly to protect carbonyl groups. Thus, the ring system is readily available with a wide range of substituents, making it ideal for studies of conformational preferences. This system has been shown to exist mainly in a chair conformation and to be very similar to the cyclohexane system, in spite of the size of two sulfur atoms (196). A comparison of the spectra of 1,3-dithians with those of the corresponding 1,3-dioxans (see Chapter 5) showed that the atypical long-range coupling paths found for the oxygen heterocycles were also effective in the dithians (197). Coupling through sulfur was shown to be much greater than through oxygen and the comparative ease of distortion of the dithian ring compared with dioxan was illustrated by the very strong substituent dependence of long-range effects in the dithians. In 2-phenyldithians (**119**), the ortho protons of an axial phenyl group absorb about 0.6 ppm downfield from the meta and para protons while an equatorial phenyl group gives rise to a single peak (196). The effect was absent from the spectra of the corresponding dioxans, and it was suggested that the sulfur lone pair in a p_z orbital could interact strongly with the neighboring ortho protons of the axial phenyl group to give the observed differences. Abraham and Thomas (198) obtained some quite remarkable results for the cis and

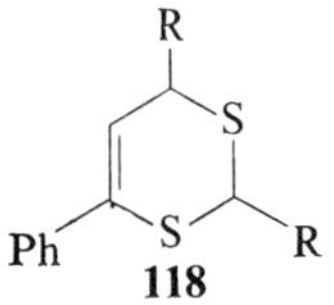

118

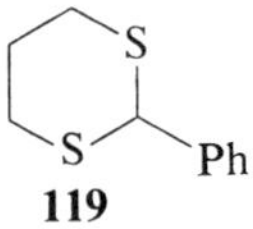

119

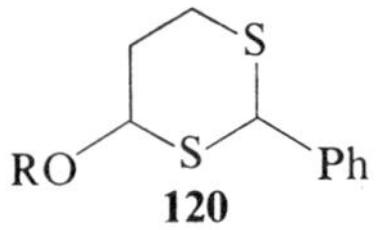

120

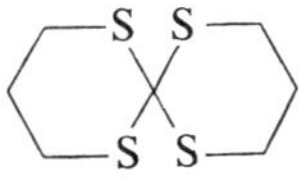

121

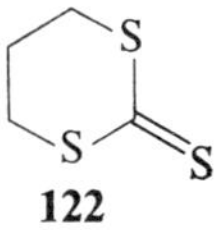

122

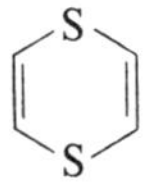

123

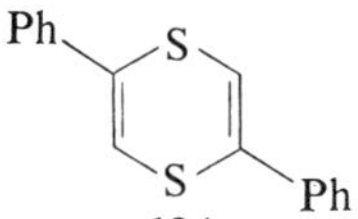

124

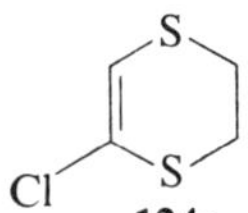

124a

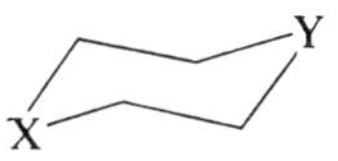

125

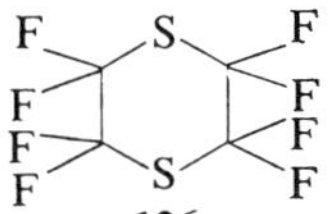

126

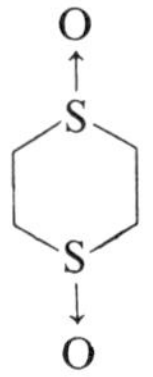

127

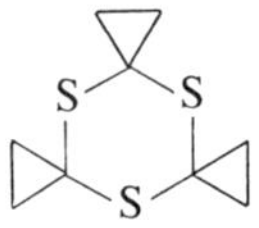

128

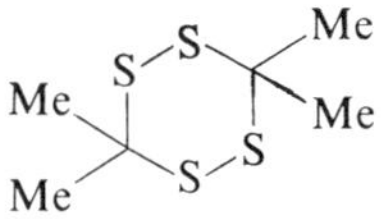

128a

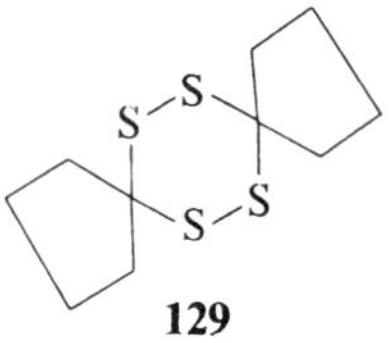

129

trans isomers of 5-hydroxy- or 5-acetoxy-2-phenyl-1,3-dithians (**120**). The cis isomer of the hydroxy compound and the trans isomer of the acetoxy derivative were found to exist in highly preferred conformations, and for these, axial protons absorbed at lower field than the corresponding equatorial protons, an order opposite to that observed for cyclohexanes. These authors, as well as a number of others (199–201), have used NMR to obtain thermodynamic constants for the inversion of this ring system. Spectra of a number of 2-phenyl-1,3-dithians have been reported in an investigation of the ring contraction of these compounds to the corresponding 1,3-dithiolans (153). Nuclear magnetic resonance was used to check the structures of the very unusual spiro dimer of 1,3-dithian (**121**) as well as the cyclic trithiocarbonate (**122**) (202). Data for a few characteristic 1,3-dithians are collected in Table 6.15.

TABLE 6.15 SPECTRAL DATA FOR DITHIANS AND RELATED COMPOUNDS

A. 1,3-DITHIANS

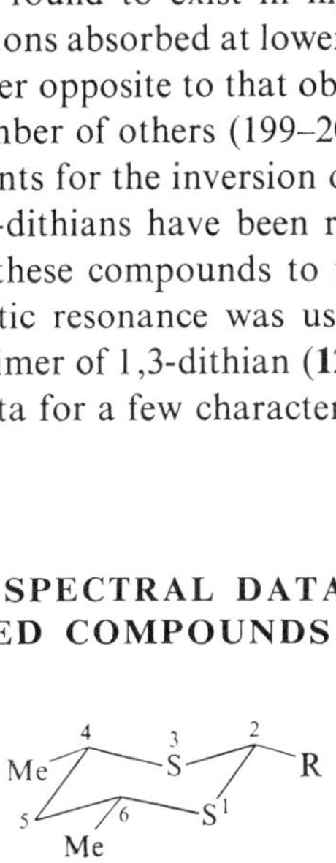

I, 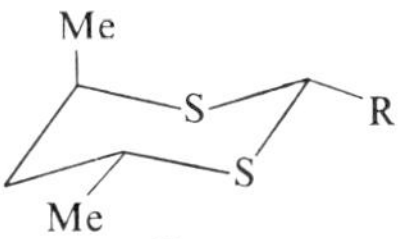 II, 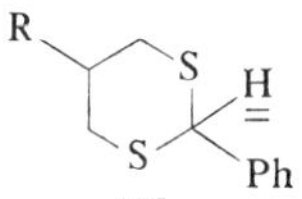 III

Compound	Solvent	τ2-H	τ4-H	τ6-H	τ2-Me	τ4,6-Me	Ref.
I R = H	CS_2	6.31	7.29			8.82	199
II R = H	CS_2	6.32	6.95			8.72	199
I R = Me[a]	CS_2	6.02		7.28	8.60	8.81	199
II R = Me[a]	CS_2	5.78		6.93	8.59	8.66, 8.87	199
III R = H	$CDCl_3$	4.82					153
III R = OH	$CDCl_3$	4.93 (cis)					153
		4.94 (trans)					153
III R = OAc	$CDCl_3$	4.92 (cis)					
		4.96 (trans)					

B. 1,4-DITHIIN (202a)—NEAT

τH 3.87; J_o 6.97; J_m 0.93; J_p 0.65; $^1J_{^{13}C-H}$ 179.76; $^2J_{^{13}C-H}$ 8.12; $^3J_{^{13}C-H}$ 2.15; $^4J_{^{13}C-H}$ −0.48.

[a] Coupling constants (*J*)—nuclei listed:
4*a*,5*a* 11.40; 6*a*,5*a* 9.93; 5*a*,4*e* 3.88; 4*a*,5*e* 2.31; 6*a*,5*e* 3.78; 5*e*,4*e* 3.56 Hz.

Long-range couplings for 1,3-dithians:
2*e*,5*e* 0.8; 2*a*,4*e* 0.6; 2*e*,4*a* 0.2; 2*e*,5*a*, 2*a*,5*e*, 2*a*,4*a* (6*a*), 2*a*,5*a* 0 Hz; 4*a*,6*e*, 4*e*,6*a* (4*a*,6*e*), 4*a*,6*e*, 4*a*,6*a*—small but real.

^{13}C satellite spectra were used to analyze spectra of 1,4-dithiin (**123**) (202a) and data are included in Table 6.15. The spectrum of 2,5-diphenyl-1,4-dithiin (**124**) is included in the Varian Catalog (No. 651), the ring protons giving a singlet at τ 3.47. Chemical shift data are also available for 6-chloro-2,3-dihydro-1,4-dithiin (**124a**) and its 5-methyl derivative (203). Lambert (204) used his *R*-value method to study conformations of systems of the general type **125**. When X and Y are both S, *R* values deviate from those observed for cyclohexanes, and this suggests deviation of ring geometry from the "classical" chair form. Apart from this work, few 1,4-dithians have been studied. Spectra of some highly substituted derivatives were reported (205–207). The width of the broad singlet from the fluorines of perfluoro-1,4-dithian (**126**) was shown to be temperature dependent (208).

Features of the NMR spectrum of the α-isomer of 1,4-dithian-1,4-dioxide (**127**) were in agreement with the suggestion that this molecule exists in a fixed conformation in solution with both sulfoxide bonds axial (209). The β-isomer, which gives only a singlet in the NMR, has been pictured in a chair form with one sulfoxide axial and one equatorial. The NMR spectrum suggests a rapid equilibrium between the two equivalent chair forms.

C. Trithians

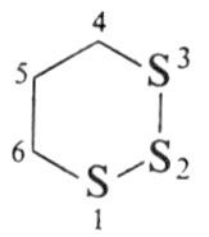

1,2,3 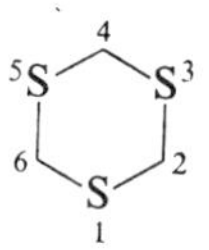 1,3,5

Ring inversion of 1,2,3-trithian and its 5,5-dimethyl derivative has been investigated and thermodynamic parameters for the process were reported (210). The ΔG values obtained were higher than those for the corresponding dithians, but were not high enough to allow separation of the various conformers.

1,3,5- or *sym*-Trithians are trimers of thiocarbonyl compounds and hence are reasonably accessible. Spectra have been reported of stereoisomers of a number of thioaldehyde trimers (211) and of the rather interesting tris-(thiocyclopropanone) (**128**) (212). Campaigne and his co-workers (213–215) have used NMR arguments in their studies of structures and conformations of various thiobenzaldehyde trimers.

D. Tetrathians

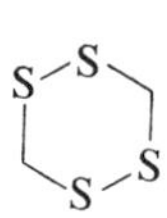

1,2,4,5

Bushweller (216, 217) was able to demonstrate, by NMR techniques, the presence of a stable twist-boat form of duplodithioacetone (**128a**) at $-80°$

in CS_2. At room temperature, a number of peaks were observed in the spectrum, but as the temperature was lowered, these coalesced into one sharp singlet, τ 8.30. These results are consistent with the formation of the twist-boat isomer in which all four methyl groups are equivalent. This work was extended to cover bis(tetramethylene)-1,2,4,5-tetrathian (**129**), which was shown to have a strong preference for the chair conformation (218).

V. SEVEN-MEMBERED SULFUR HETEROCYCLES

The spectrum of thiepan (**130**) and its sulfone (**131**) were reported by Biscarini *et al.* (219) who found that sulfone formation made little difference to the chemical shifts of the α-protons.

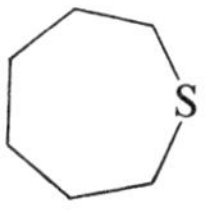

130 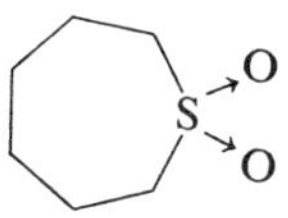 **131** 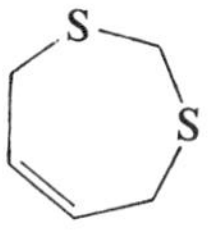 **132** 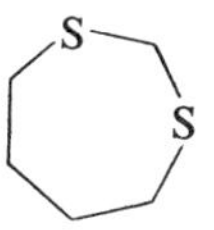 **133** 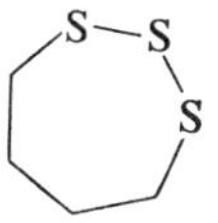 **134**

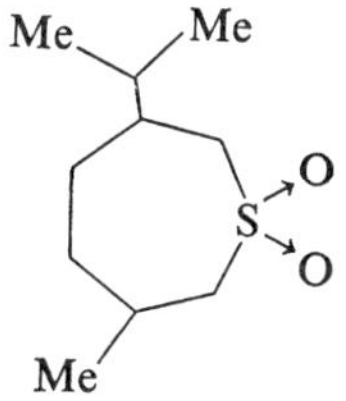

135

Ring inversion in 1,3-dithiep-5-enes (**132**) (220), 1,3-dithiepans (**133**), 1,2,3-trithiepans (**134**), and a number of their benz-fused homologs (210) has been studied by NMR.

Spectra measured at 250 MHz have been used to study the ring inversion of 3-isopropyl-6-methylthiepin-1,1-dioxide (**135**) (221).

REFERENCES

1. Gutowsky, Rutledge, Tamres, and Searles, *J. Am. Chem. Soc.*, **76**, 4242 (1954).
2. Mortimer, *J. Mol. Spectrosc.*, **5**, 199 (1960).
3. Musher and Gordon, *J. Chem. Phys.*, **36**, 3097 (1962).
4. Gutowsky, Karplus, and Grant, *J. Chem. Phys.*, **31**, 1278 (1959)
5. Lippert and Prigge, *Ber. Bunsenges. Phys. Chem.*, **67**, 415 (1963).
6. Maciel and Savitzky, *J. Phys. Chem.*, **69**, 3925 (1965).
7. Manatt, Elleman, and Brois, *J. Am. Chem. Soc.*, **87**, 2220 (1965).

8. *Varian Spectra Catalog*, Vol. 2, Varian Associates, Palo Alto, California, 1963, No. 629.
9. Biscarini, Taddei, and Zauli, *Boll. Sci. Fac. Chim. Ind. Bologna*, **21,** 169 (1963).
10. Tori, Komeno, and Nakagawa, *J. Org. Chem.*, **29,** 1136 (1964).
11. Dittmer and Christy, *J. Org. Chem.*, **26,** 1324 (1961); *Varian Spectra Catalog*, Vol. 1, Varian Associates, Palo Alto, California, 1962, No. 22.
12. Lozac'h and Braillon, *J. Chim. Phys.*, **67,** 340 (1970).
13. Ferretti, Ph.D. Thesis, *Diss. Abstr.*, **26,** 7060 (1966).
14. Keller, Lusebrink, and Sederholm, *J. Chem. Phys.*, **44,** 782 (1966).
15. Lusebrink, U.S. A.E.C., UCRL-16344.
16. Trost, Schinski, and Mantz, *J. Am. Chem. Soc.*, **91,** 4320 (1969).
17. Jancis, Ph.D. Thesis, *Diss. Abstr.*, **29,** 941 (1968).
18. Johnson and Siegel, *Tetrahedron Lett.*, **1969,** 1879.
19. Lautenschlaeger, *J. Org. Chem.*, **31,** 1679 (1966).
20. Dodson and Klose, *Chem. Ind.*, **1963,** 450.
21. Dodson and Hammen, *Chem. Commun.*, **1968,** 1294.
22. Truce and Norell, *J. Am. Chem. Soc.*, **85,** 3231 (1963).
23. Truce, Abraham, and Son, *J. Org. Chem.*, **32,** 990 (1967).
24. Olsson, Baeckström, and Engwall, *Ark. Kemi*, **26,** 219 (1967).
25. Gronowitz, *Ark. Kemi*, **13,** 269 (1958).
26. Hoffman and Gronowitz, *Ark. Kemi*, **15,** 45 (1960).
27. Abraham and Bernstein, *Can. J. Chem.*, **37,** 2095 (1959).
28. Dischler and Maier, *Z. Naturforsch.*, **16a,** 318 (1961).
29. Dischler and Englert, *Z. Naturforsch.*, **16a,** 1180 (1961).
30. Read, Mathis, and Goldstein, *Spectrochim. Acta*, **21,** 85 (1965).
31. Elvidge and Jackman, *J. Chem. Soc.*, **1961,** 859.
32. Abraham, Sheppard, Thomas, and Turner, *Chem. Commun.*, **1965,** 43.
33. Abraham and Thomas, *J. Chem. Soc.*, *B*, **1966,** 127.
33a. Dharmatti, Dhingra, Govil, and Khetrapal, *Proc. Nucl. Phys. Solid State Phys. Symp., Chandigarh, India, 1964* (Pt. B), p. 410.
34. Diehl, Khetrapal, and Lienhard, *Can. J. Chem.*, **46,** 2645 (1968).
35. Anderson, *Phys. Rev.*, **102,** 151 (1956).
36. Gronowitz, Sörlin, Gestblom, and Hoffman, *Ark. Kemi*, **19,** 483 (1962).
37. Gronowitz and Hoffman, *Ark. Kemi*, **17,** 279 (1958).
38. Gronowitz, *Ark. Kemi*, **13,** 295 (1958).
39. Gronowitz and Hoffman, *Acta Chem. Scand.*, **13,** 1687 (1959).
40. Takahashi, Matsuki, Mashiko, and Hazato, *Bull. Chem. Soc. Japan*, **32,** 156 (1959).
41. Leane and Richards, *Trans. Faraday Soc.*, **55,** 518 (1959).
42. Corio and Weinberg, *J. Chem. Phys.*, **31,** 569 (1959).
43. Gronowitz and Hoffman, *Ark. Kemi*, **15,** 499 (1961).
44. Hoffman and Gronowitz, *Ark. Kemi*, **16,** 501 (1961).
45. Hoffman and Gronowitz, *Ark. Kemi*, **16,** 515 (1961).
46. Gronowitz and Hoffman, *Ark. Kemi*, **16,** 539 (1961).
47. Hoffman and Gronowitz, *Ark. Kemi*, **16,** 563 (1961).
48. Cohen and McLaughlan, *Discuss. Faraday Soc.*, **34,** 132 (1962).
49. Takahashi, Sone, Matsuki, and Hazato, *Bull. Chem. Soc. Japan*, **38,** 1041 (1965).
49a. Takahashi, Ito, and Matsuki, *Bull. Chem. Soc. Japan*, **40,** 605 (1967).
50. Rodmar, Rodmar, Khan, Gronowitz, and Pavulans, *Acta. Chem. Scand.*, **20,** 2515 (1966).
51. Atkinson, Curtis, and Taylor, *J. Chem. Soc.*, *C*, **1967,** 578.

52. Gronowitz and Gjos, *J. Org. Chem.*, **32,** 463 (1967).
53. Jakobsen and Lawesson, *Tetrahedron*, **23,** 871 (1967).
54. Ernst, Freeman, Gestblom, and Lusebrink, *Mol. Phys.*, **13,** 283 (1967).
55. Kellogg and Wynberg, *J. Am. Chem. Soc* . **89,** 3495 (1967)
56. Kergomard and Vincent, *Bull. Soc. Chim. Fr.*, **1967,** 2197.
57. Fournari, Guilard, and Person, *Bull. Soc. Chim. Fr.*, **1967,** 4115.
58. Guilard, Fournari, and Person, *Bull. Soc. Chim. Fr.*, **1967,** 4121.
59. Gronowitz and Bugge, *Acta Chem. Scand.*, **22,** 59 (1968).
60. Huckerby, *Tetrahedron Lett.*, **1971,** 353.
61. Morel, Paulmier, and Pastour, *Compt. Rend.*, *Series C*, **266,** 1300 (1968).
62. Martin, Andrieu, and Martin, *Bull. Soc. Chim. Fr.*, **1968,** 698.
63. Andrieu, Martin, and Martin, *Bull. Soc. Chim. Fr.*, **1968,** 703.
64. Kergomard and Vincent, *Bull. Soc. Chim. Fr.*, **1968,** 4429.
65. Doddi, Illuminati, and Stegel, *Chem. Commun.*, **1969,** 953.
66. Ebdon, Huckerby, and Thorpe, *Tetrahedron Lett.*, **1971,** 2921.
67. Ewing and Scrowston, *Org. Magnetic Resonance*, **3,** 405 (1971).
68. Egorochkin, Burov, Vyazankin, Savushkina, Anisimova, and Chernyshev, *Dokl. Akad. Nauk SSSR*, **184,** 351 (1969).
69. Bulman, *Tetrahedron*, **25,** 1433 (1969).
70. Kamienski and Krygowski, *Tetrahedron Lett.*, **1971,** 103.
71. Paulmier, Morel, Pastour, and Semard, *Bull. Soc. Chim. Fr.*, **1969,** 2511.
72. Brunnett and McCarthy, *J. Heterocycl. Chem.*, **5,** 417 (1968).
73. Schaumberg, Orsted, and Jakobsen, *J. Magn. Res.*, **2,** 1 (1970).
74. Campaigne and Johnson, *J. Heterocycl. Chem.*, **5,** 235 (1968).
75. Gronowitz, Gestblom, and Hoffman, *Acta Chem. Scand.*, **15,** 1201 (1961).
76. Goldstein and Reddy, *J. Chem. Phys.*, **36,** 2644 (1962).
77. Kaper and de Boer, *Rec. Trav. Chim.*, **89,** 825 (1970).
78. Gronowitz, in *Advances in Heterocyclic Chemistry*, Vol. 1 (Katritzky, Ed.), Academic Press, New York, 1963, p. 7.
79. Mathis and Goldstein, *J. Phys. Chem.*, **68,** 571 (1964).
80. Hoffman, Gestblom, Gronowitz, and Forsén, *J. Mol. Spectrosc.*, **11,** 454 (1963).
81. Cohen, Freeman, McLauchlan, and Whiffen, *Mol. Phys.* **7,** 45 (1963).
82. Freeman and Gestblom, *J. Chem. Phys.*, **47,** 1472 (1967).
83. Gestblom and Mathiasson, *Acta Chem. Scand.*, **18,** 1905 (1964).
84. Forsén, Gestblom, Gronowitz, and Hoffman, *Acta Chem. Scand.*, **18,** 313 (1964).
85. Freeman and Gestblom, *J. Chem. Phys.*, **47,** 2744 (1967).
86. Jakobsen, Nielsen, Schaumburg, and Begtrup, *Mol. Phys.*, **15,** 423 (1968).
87. Jakobsen and Neilsen, *J. Magn. Res.*, **1,** 393 (1969).
88. Bucci, Ceccarelli, and Veracini, *J. Chem. Phys.*, **50,** 1510 (1969).
89. Anet and Ahmad, *J. Am. Chem. Soc.*, **86,** 119 (1964).
90. Dahlqvist and Forsén, *J. Phys. Chem.*, **69,** 4062 (1965).
91. Roques, Combrisson, Riche, and Pascard-Billy, *Tetrahedron*, **26,** 3555 (1970).
92. Hörnfeldt, *Ark. Kemi*, **29,** 455 (1968).
93. Hörnfeldt, *Ark. Kemi*, **29,** 461 (1968).
94. Eck and Stacy, *J. Heterocycl. Chem.*, **6,** 147 (1969).
95. Radeglia, Hartmann, and Scheithauer, *Z. Naturforsch.*, **24b,** 286 (1969).
96. Brown, Rae, and Sternhell, *Aust. J. Chem.*, **18,** 1211 (1965).
97. Jakobsen, *Tetrahedron*, **23,** 3737 (1967).
98. Jakobsen and Lawesson, *Tetrahedron*, **21,** 3331 (1965).
99. Danielsen, *Acta Chem. Scand.*, **23,** 2031 (1969).

100. Hogeveen, *Rec. Trav. Chim.*, **85,** 1072 (1966).
101. Wiersum and Wynberg, *Tetrahedron Lett.*, **1967,** 2951,
102. Spinelli, Armanino, and Corrao, *J. Heterocycl. Chem.*, **7,** 1441 (1970).
103. Rodmar, Rodmar, Sharma, Gronowitz, Christiansen, and Rosén, *Acta Chem. Scand.*, **22,** 907 (1968).
104. Christiansen, Gronowitz, Rodmar, Rodmar, Rosén, and Sharma, *Ark. Kemi*, **30,** 561 (1969).
105. Burdon, Campbell, Parsons, and Tatlow, *Chem. Commun.*, **1969,** 27.
106. Tori and Nakagawa, *J. Phys. Chem.*, **68,** 3163 (1964).
107. Page, Alger, and Grant, *J. Am. Chem. Soc.*, **87,** 5333 (1965), and references cited therein.
108. Weigert and Roberts, *J. Am. Chem. Soc.*, **90,** 3543 (1968).
109. Bohlmann, Kleine, and Arndt, *Chem. Ber.*, **97,** 2125 (1964).
110. Bohlmann, Arndt, Kleine, and Bornowski, *Chem. Ber.*, **98,** 155 (1965).
111. Curtis and Phillips, *J. Chem. Soc.*, **1965,** 5134.
112. Atkinson, Curtis, and Phillips, *J. Chem. Soc.*, **1965,** 7109.
113. Coogan, Horn, and Lamberton, *Aust. J. Chem.*, **18,** 723 (1965).
114. Atkinson, Curtis, and Phillips, *J. Chem. Soc., C*, **1967,** 2011.
115. Campaigne, Fleming, and Dinner, *J. Heterocycl. Chem.*, **5,** 191 (1968).
115a. Mason, Smith, Stern, and Elvidge, *J. Chem. Soc., C*, **1967,** 2171.
116. MacDowell, Patrick, Frame, and Ellison, *J. Org. Chem.*, **32,** 1226 (1967).
117. Cagniant, Cagniant, and Merle, *Bull. Soc. Chim. Fr.*, **1968,** 3816.
118. Angeloni and Tramontini, *Boll. Sci. Fac. Chim. Ind. Bologna*, **21,** 217 (1963).
119. Elvidge and Foster, *J. Chem. Soc.*, **1964,** 981.
120. Martin-Smith, Reid, and Sternhell, *Tetrahedron Lett.*, **1965,** 2393.
121. Takahashi, Kanda, and Matsuki, *Bull. Chem. Soc. Japan*, **38,** 1799 (1965).
122. Takahashi, Kanda, Shoji, and Matsuki, *Bull. Chem. Soc. Japan*, **38,** 508 (1965).
123. Takahashi, Kanda, and Matsuki, *Bull. Chem. Soc. Japan*, **37,** 768 (1964).
124. Buu-Hoi, Bellavita, Ricci, and Grandolini, *Bull. Soc. Chim. Fr.*, **1965,** 2658.
125. Stacy, Villaescusa, and Wollner, *J. Org. Chem.* **30,** 4074 (1965).
126. Takahashi, Kanda, and Matsuki, *Bull. Chem. Res. Inst. Nonaqueous Solutions, Tohuku University*, **16,** 11 (1966).
127. Takahashi, Ito, and Matsuki, *Bull. Chem. Soc. Japan*, **39,** 2316 (1966).
128. Cagniant, Faller, and Cagniant, *Bull. Soc. Chim. Fr.*, **1966,** 3055.
129. Cagniant and Cagniant, *Bull. Soc. Chim. Fr.*, **1966,** 3674.
130. Chapman, Ewing, Scrowston, and Westwood, *J. Chem. Soc., C*, **1968,** 764.
131. Caddy, Martin-Smith, Norris, Reid, and Sternhell, *Aust. J. Chem.*, **21,** 1853 (1968).
132. Boswell, Brennan, Landis, and Rodewald, *J. Heterocycl. Chem.*, **5,** 69 (1968).
133. Cagniant, Cagniant, and Trierweiler, *Bull. Soc. Chim. Fr.*, **1969,** 601.
134. Faller, *Bull. Soc. Chim. Fr.*, **1969,** 934.
135. Castle, Mooney, and Plevey, *Tetrahedron*, **24,** 5457 (1968).
136. Cagniant, Reisse, and Cagniant, *Bull. Soc. Chim. Fr.*, **1969,** 985.
137. Faller, *Bull. Soc. Chim. Fr.*, **1967,** 387.
138. Campaigne and Ashby, *J. Heterocycl. Chem.*, **6,** 875 (1969).
139. Fujiwara, Acton, and Goodman, *J. Heterocycl. Chem.*, **6,** 389 (1969).
140. Krespan, *J. Org. Chem.*, **27,** 3588 (1962).
141. Argyle, Goadby, Mason, Reed, Smith, and Stern, *J. Chem. Soc., C*, **1967,** 2156.
142. Jonsson and Holmquist, *Ark. Kemi*, **29,** 301 (1968).
142a. Jonsson, *Ark. Kemi*, **26,** 357 (1967).
143. Dingwall, McKenzie, and Reid, *J. Chem. Soc., C*, **1968,** 2543.

144. Brown, Rae, and Sternhell, *Aust. J. Chem.*, **18,** 1211 (1965).
145. Landis, *Chem. Rev.*, **65,** 237 (1965).
146. Mouchel and Thuillier, *Comp. Rend., Series C*, **264,** 1552 (1967).
147. Klingsberg, *J. Org. Chem.*, **31,** 3489 (1966).
148. Campaigne and Jacobsen, *J. Org. Chem.*, **29,** 1703 (1964).
149. Campaigne, Hamilton, and Jacobsen, *J. Org. Chem.*, **29,** 1708 (1964).
150. Campaigne and Hamilton, *J. Org. Chem.*, **29,** 1711 (1964).
151. Prinzbach, Berger, and Lüttringhaus, *Angew. Chem.*, **77,** 453 (1965).
152. Martin, *J. Org. Chem.*, **34,** 473 (1969).
153. Atkinson, Beer, Harris, and Royall, *J. Chem. Soc., C*, **1967,** 638.
154. Baggett, Buck, Foster, Randall, and Webber, *J. Chem. Soc.*, **1965,** 3394.
155. Feugeas and Olschwang, *Bull. Soc. Chim. Fr.*, **1969,** 325; Feugeas and Olschwang, *ibid.*, **1969,** 332.
156. Kirrmann, Cantacuzène, Vio, and Martin, *Bull. Soc. Chim. Fr.*, **1963,** 1067.
157. Shroff and Kormas, *Steroids*, **8,** 739 (1966).
158. Isaksson, Sandström, and Wennerbeck, *Tetrahedron Lett.*, **1967,** 2233.
159. Ebel, Legrand, and Lozac'h, *Bull. Soc. Chim. Fr.*, **1968,** 2081.
160. Bugge, Gestblom, and Hartmann, *Acta Chem. Scand.*, **24,** 105 (1970).
161. Gestblom, Hoffman, and Rodmar, *Acta Chem. Scand.*, **18,** 1222 (1964).
162. Litvinov and Fraenkel, *Izv. Akad. Nauk SSSR, Ser. Khim.*, **1968,** 1828; through *Chem. Abstr.*, **70,** 19958*k* (1969).
163. Zwanenburg, de Haan, and Wynberg, *J. Org. Chem.*, **31,** 3363 (1966).
164. Wynberg and Zwanenburg, *Tetrahedron Lett.*, **1967,** 761.
165. Wynberg and Feijen, *Rec. Trav. Chim.*, **89,** 77 (1970).
165a. Brandsma and Schuijl-Laros, *Rec. Trav. Chim.*, **89,** 110 (1970).
166. MacDowell and Patrick, *J. Org. Chem.*, **31,** 3592 (1966).
167. Wynberg and Zwanenburg, *J. Org. Chem.*, **29,** 1919 (1964).
168. Hortmann, *J. Am. Chem. Soc.*, **87,** 4972 (1965).
169. Degani, Lunazzi, and Taddei, *Boll. Sci. Fac. Chim. Ind. Bologna*, **23,** 131 (1965).
170. Schuijl, Bos, and Brandsma, *Rec. Trav. Chim.*, **88,** 597 (1969).
171. Degani, Fochi, and Vicenzi, *Gazz. Chim. Ital.*, **97,** 397 (1967).
172. Parasaran and Price, *J. Org. Chem.*, **29,** 946 (1964).
173. Molenaar and Strating, *Rec. Trav. Chim.*, **86,** 1047 (1967).
174. Molenaar and Strating, *Tetrahedron Lett.*, **1965,** 2941.
175. Pauson, Proctor, and Rodger, *J. Chem. Soc.*, **1965,** 3037.
176. Portail and Vialle, *Bull. Soc. Chim. Fr.*, **1968,** 3790.
177. Molenaar and Strating, *Rec. Trav. Chim.*, **86,** 436 (1967).
178. Parham and Koncos, *J. Am. Chem. Soc.*, **83,** 4034 (1961).
179. Price, Hori, Parasaran, and Polk, *J. Am. Chem. Soc.*, **85,** 2278 (1963).
180. Lunazzi and Taddei, *J. Mol. Spectrosc.*, **25,** 113 (1968).
181. Katritzky and Ternai, *J. Heterocycl. Chem.*, **5,** 745 (1968).
182. Grandolini, Ricci, Buu-Hoi, and Périn, *J. Heterocycl. Chem.*, **5,** 133 (1968).
183. Rossi and Pagani, *Tetrahedron Lett.*, **1966,** 2129.
183a. Nishio, *Chem. Pharm. Bull.*, **17,** 274 (1969).
184. Hofman and Salbeck, *Tetrahedron Lett.*, **1969,** 2587.
185. Katekar, *Aust. J. Chem.*, **19,** 1251 (1966).
186. Katekar and Moritz, *Aust. J. Chem.*, **20,** 2235 (1967).
187. Anderson, Harrison, and Anderson, *J. Am. Chem. Soc.*, **85,** 3448 (1963).
188. Lambert and Keske, *Tetrahedron Lett.*, **1967,** 4755.
189. Lambert, Keske, and Weary, *J. Am. Chem. Soc.*, **89,** 5921 (1967).

190. Lambert and Keske, *J. Org. Chem.*, **31**, 3429 (1966).
191. Martin and Uebel, *J. Am. Chem. Soc.*, **86**, 2936 (1964).
192. Tuleen and Bennett, *J. Heterocycl. Chem.*, **6**, 115 (1969).
193. Claeson, Androes, and Calvin, *J. Am. Chem. Soc.*, **87**, 4357 (1961).
194. Campaigne and Edwards, *J. Org. Chem.*, **27**, 4488 (1962).
195. Pasto and Serve, *J. Org. Chem.*, **27**, 4665 (1962).
196. Kalff and Havinga, *Rec. Trav. Chim.*, **85**, 467 (1966).
197. Gelan and Anteunis, *Bull. Soc. Chim. Belges*, **77**, 447 (1968).
198. Abraham and Thomas, *J. Chem. Soc.*, **1965**, 335.
199. Gelan and Anteunis, *Bull. Soc. Chim. Belges*, **77**, 423 (1968); Gelan and Anteunis, *ibid.*, **78**, 599 (1969).
200. Friebolin, Schmid, Kabuss, and Faisst, *Org. Magnetic Resonance*, **1**, 67 (1969).
201. Kabuss, Friebolin, and Lüttringhaus, *Angew. Chem.*, **76**, 590 (1964).
202. Johnston, Stringfellow, and Gallagher, *J. Org. Chem.*, **27**, 4068 (1962).
202a. Butler, Read, and Goldstein, *J. Mol. Spectrosc.*, **35**, 83 (1970).
203. Mueller and Dines, *J. Heterocycl. Chem.*, **6**, 627 (1969).
204. Lambert, *J. Am. Chem. Soc.*, **89**, 1836 (1967).
205. Kirrmann, Cantacuzène, Vio, and Martin, *Bull. Soc. Chim. Fr.*, **1963**, 1067.
206. Tsuchihashi, Yamauchi, and Fukayuma, *Tetrahedron Lett.*, **1967**, 1971.
207. Ohno, Ohnishi, and Tsuchihashi, *J. Am. Chem. Soc.*, **91**, 5038 (1969).
208. Tiers, *J. Phys. Chem.*, **66**, 764 (1962).
209. Chen and Le Fèvre, *Aust. J. Chem.*, **16**, 917 (1963).
210. Kabuss, Lüttringhaus, Friebolin, and Mecke, *Z. Naturforsch.*, **21b**, 320 (1966).
211. Matlack, Chien, and Breslow, *J. Org. Chem.*, **26**, 1455 (1961).
212. Price and Vittimberga, *J. Org. Chem.*, **27**, 3736 (1962).
213. Campaigne and Georgiadis, *J. Org. Chem.*, **28**, 1044 (1963).
214. Campaigne, Chamberlain, and Edwards, *J. Org. Chem.*, **27**, 135 (1962).
215. Campaigne and Georgiadis, *J. Heterocycl. Chem.*, **6**, 339 (1969).
216. Bushweller, *J. Am. Chem. Soc.*, **89**, 5978 (1967).
217. Bushweller, *J. Am. Chem. Soc.*, **90**, 2450 (1968).
218. Bushweller, *J. Am. Chem. Soc.*, **91**, 6019 (1969).
219. Biscarini, Taddei, and Zauli, *Boll. Sci. Fac. Chim. Ind. Bologna*, **21**, 169 (1963).
220. Friebolin, Mecke, Kabuss, and Lüttringhaus, *Tetrahedron Lett.*, **1964**, 1929.
221. Anet, Bradley, Brown, Mock, and McCausland, *J. Am. Chem. Soc.*, **91**, 7782 (1969).

7 MISCELLANEOUS HETERO-CYCLIC SYSTEMS CONTAINING ONLY ONE TYPE OF HETERO-ATOM. COMPOUNDS CONTAINING P, As, Se, Te, B, AND Si

I. PHOSPHORUS HETEROCYCLES

The proton and ^{31}P spectra of phosphirans (**1**) have been discussed by Chan *et al.* (1). Although the complex patterns of the proton spectra have not been fully analyzed, a number of important conclusions were reached. The inversion rate at the phosphorus atom for 1-substituted derivatives was found to be slow on the NMR time scale. 2-Ethylphosphirinan (**2**) was shown to exist in two stable forms with different ^{31}P shifts and slightly different P–H couplings (see Table 7.1). The authors were careful to point out that, at the present state of knowledge, it is impossible to assign specific peaks to these two isomers. However, on the basis of the effect of a 2-substituent on the C–P–C bond angle, the cis structure was provisionally assigned to the isomer producing the downfield ^{31}P signal. Data for these compounds are collected in Table 7.1.

Nuclear magnetic resonance methods have been used to assign structures to a number of highly substituted phosphetans (**2a**) and their oxides and cations (2–4). Again peaks from cis and trans isomers were evident in the proton spectra, but it was not possible to make firm assignments of spectral

TABLE 7.1 SPECTRAL DATA FOR PHOSPHIRANS (1)

3 2 P 1 H

	Chemical Shifts			Coupling Constants
Substituents	τ1	τ2,3	$\delta^{31}P^{a}$	J_{PH}
None		8.6–9.7	341	
1-Me		8.2–9.7	251	
1-Ph	2.8–3.5	9.0–9.7	234	
2-Me	11.4–12.4	7.5–9.9		
2-Et[b]	11.3–12.2	7.4–9.8	271, 288	158, 159

[a] Parts per million upfield from 85% H_3PO_4 (external).
[b] Mixture of two isomers.

lines to particular species. More recently (4), stereospecific ^{13}C–^{31}P couplings were found for the cis and trans isomers of certain phosphetan oxides and salts although no marked stereospecific one-bond couplings were found for the five-membered 1,2,5-trimethyl-3-phospholene-1-oxide (**3**).

Phosphole (**4**) is of interest as a phosphorus analog of the aromatic pyrrole and to date the simplest available derivative is the 1-methyl compound **5** (5, 6). The proton and ^{31}P spectra of this compound have been analyzed and, although the assignment of peaks to the α- and β-protons was based on rather unsatisfactory models, the conclusions are probably correct. Differences between H–H coupling constants for this compound and those

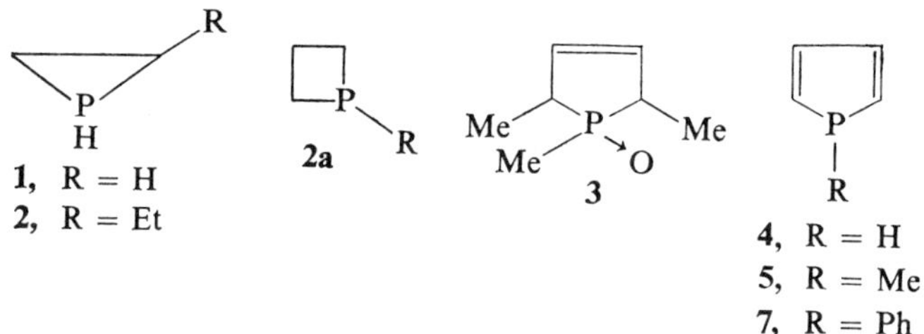

Ph P Me t-Bu **6** P R **8** P R **9** Me P R O **10** Me P R O **11**

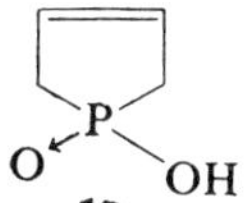

12 **13** **14** **15** **17**

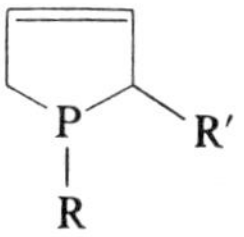

16, R = Ph, R′ = Me

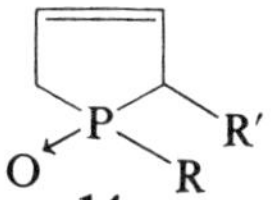

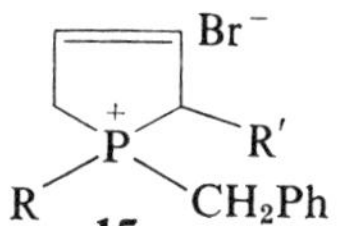

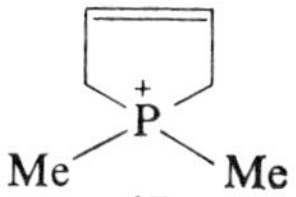

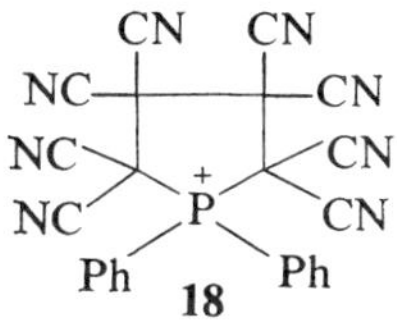

18 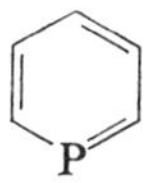 **19**  **20** 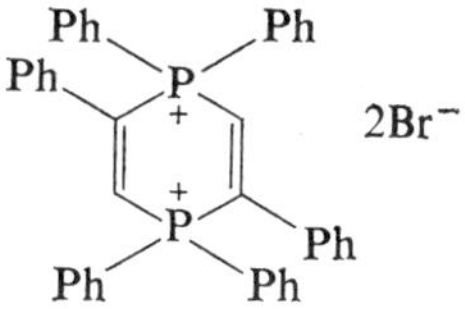 **21**

for other five-membered heterocycles were attributed to differences in bond angles, and chemical shift data were used to suggest that the compound was "aromatic" in the sense that it supported a ring current. The low barrier to inversion at phosphorus, as measured by NMR, of 1-*t*-butyl-2-methyl-5-phenylphosphole (**6**) has been ascribed to the "aromaticity" of the system (7). Märkl and his co-workers (8,9) have also examined the spectra of a number of 1-phenylphospholes (**7**).

Nuclear magnetic resonance spectra have been used to determine the structures of dihydrophospholes or phospholenes in which the double bond can be between C-2 and C-3 (the Δ^2-isomer **8**) or between C-3 and C-4 (the Δ^3-isomer **9**). Coupling constants of the olefinic protons with the ^{31}P nucleus are characteristic of the isomer involved. Weitkamp and Korte (10) have made a thorough study of the spectra of the isomeric phospholene-1-oxides (**10** and **11**) and show that for the Δ^2-isomer **10**, $J_{^{31}P,3}$ (~48 Hz) is considerably larger than $J_{^{31}P,2}$ (16–32 Hz). Proton and ^{31}P resonances have also been investigated as diagnostic tools in the synthesis of 1-hydroxy-Δ^3-phospholene-1-oxides (**12**) (11). The Δ^2- and Δ^3-isomers of 1-halophospholenes have been prepared and again the ^{31}P–H couplings are characteristic of the isomer being studied (12). The configurations about the 1,2-bond of 1,2-disubstituted Δ^3-phospholenes (**13**) and the corresponding 1-oxides (**14**) and benzyl bromide salts (**15**) have been deduced from their NMR properties (13). For the 1-phenyl-2-methyl systems **16**, the cis configurations was assigned to the isomer giving the upfield 2-Me signal (phenyl shielding). This isomer had a much smaller $J_{^{31}P,2}$ (10 Hz) than the trans isomer (18 Hz). Proton magnetic resonance data for 1,1-dimethyl-Δ^3-phospholene (**17**) and related cyclic phosphonium and phosphonyl compounds have been recorded (14). It is interesting to note that $J_{^{31}P,3}$ is typically about 28 Hz, but increases to 36 Hz when the double bond is saturated with bromine. Arbuzov and his co-workers (15, 16) have also used NMR to determine the double-bond position in a number of 1-alkoxyphospholenes.

^{31}P spectra have also been measured for compounds of the type **18** (17). Representative data for five-membered phosphorus heterocycles are collected in Table 7.2.

Most of the work on six-membered phosphorus heterocycles has centered around the quest for phosphabenzene (phosphorin, **19**). Spectra of a number of 1,3,5-trisubstituted derivatives of this heterocycle have been reported (18–20), as well as those of certain of their 1,2-dihydro derivatives (21, 22). Recently the parent heterocycle has been prepared (23) and although the spectra were not analyzed, experimental traces were reproduced. The low-field shift of all ring protons was interpreted as "aromaticity." Nuclear magnetic resonance was also used to study conformational preferences in the 1-alkylphosphorinan-4-ols (**20**) (24). Data for phosphorin and its derivatives are summarized Table 7.3.

The very low field at which the ^{13}P nuclei of 1,1,2,4,4,5-hexaphenyl-1,4-dihydrophospha(*V*)pyrazine dibromide (**21**) resonates has been used as evidence to support the claim that extensive delocalization of the π-electrons occurs in this system (25).

II. ARSENIC HETEROCYCLES

Spectra of this class of compounds have been little studied. In a note dealing with the preparation of some 1,2,5-trisubstituted arsoles (**22**) (8), some

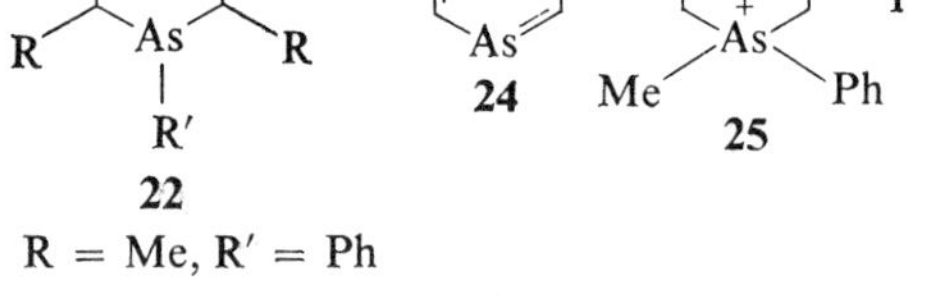

R As R
R′
22
23, R = Me, R′ = Ph

As
24

O
As⁺ I⁻
Me Ph
25

information is given and the spectrum of the 1-phenyl-2,5-dimethyl derivative (**23**) is reproduced. The spectrum of arsabenzene (**24**) is shown in a short communication on the preparation and "aromaticity" of this ring system. The ring protons of this compound absorb at low field (all below τ3) and this was taken as evidence for the presence of a substantial ring current.

Spectra have also been recorded for a number of arsenan derivatives of the type **25** (26).

TABLE 7.2 SPECTRAL DATA FOR FIVE-MEMBERED PHOSPHORUS HETEROCYCLES

A. PHOSPHOLES

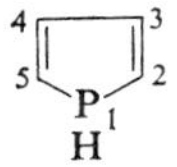

Substituents	Solvent	τ2,5	τ3	τ4	J_{12}	J_{13}	J_{23}	J_{24}	J_{25}	J_{34}	J_{14}	Ref.
1-Me	$CDCl_3$	[a]	[a]	[a]	34.49	13.77	7.24	1.12	3.01	1.96		5
1-Ph, 2,5-Di-Me	$CDCl_3$	8.02	3.61	3.61	10	12.5						8
1-*t*-Bu, 2-Me, 5-Ph	$CFCl_3$		3.46	3.20		11.0				3.0	9.5	7

B. Δ^2-PHOSPHOLENES

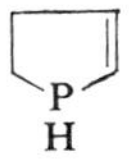

Substituents	Solvent	τ2	τ3	τ4	τ5	P[b]	J_{12}	J_{23}	J_{13}	Ref.
1-Cl	Neat[c]	3.7–2.2		7.8–6.5						12
1-Cl, 3-Me	Neat[c]	3.48	7.54	7.5–6.5		−132.5	46.5			12
1-OEt, 1-oxide		3.80	2.99				23.1	8.48	48.8	10
1-Ph, 1-oxide		3.73	2.83				25.2	8.4	47.7	10
1-Ph, 3-Me, 1-oxide		4.17					25.8			10

TABLE 7.2 (*Continued*)

C. Δ^3-PHOSPHOLENES

Substituents	Solvent	τ2,5	τ3	τ4	J_{12}	J_{13}	P[b]	τPMe	Ref.
1-Br	Neat[c]	6.34	3.40	3.40	20.5	6.5			12
1-Br, 3-Me	Neat[c]	3.02–4.70	7.73	4.08			−127.5		12
1,2-Di-Me (cis)	Neat		4.07–4.60					9.27	13
(trans)	Neat		4.30			6		9.17	13
1-OMe, 1-oxide			4.10			32.25			10

D. Δ^3-PHOSPHOLENIUM SALTS

Substituents	X^-	Solvent	τR ($J_{1,R}$)	τR′ ($J_{1,R'}$)	τ2,5 (J_{12})	τ3,4 (J_{13})	Ref.
R = R′ = Me	Cl	CD_3OD	7.91 (15.4)	7.91 (15.4)	6.87 (10.6)	4.00 (28)	14
R = Me, R′ = CH_2Ph, 2-Me							
Cis Me, Me	Br	$CDCl_3$	7.74 (13.5)	5.29 (CH_2) (17)	8.64 (Me) (17)	4.20 (27)	13
Trans Me, Me	Br	$CDCl_3$	7.87 (13.5)		8.35 (Me) (17)	4.00 (26)	13

[a] $\delta\alpha\beta$ = 7.43 Hz at 100 MHz.
[b] Relative to 85% H_3PO_4.
[c] External TMS.

TABLE 7.3 SPECTRAL DATA FOR PHOSPHABENZENES (PHOSPHORINS) AND THEIR DERIVATIVES

A. PHOSPHABENZENES

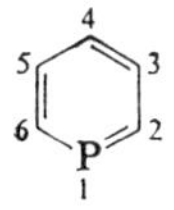

Substituents	Solvent	τ3,5	J_{12}	J_{13}	J_{23}	τOMe	Ref.
None	CCl_4		38		10		23
2,4,6-Tri-*t*-Bu		2.23		6			19
2,4,6-Tri-Ph	$CDCl_3$	1.9		6			18
1,2,4,6-Tetra-Ph, 1-OMe	$CDCl_3$	2.19		30		6.71	20

B. 1,2-DIHYDROPHOSPHABENZENES. SOLVENT—$CDCl_3$

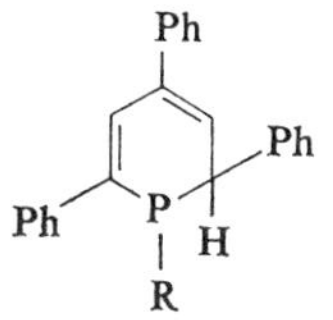

Substituents	τ2	τ3	τP–Me	J_{12}	Ref.
R = Me	5.75	3.75	9.25, 8.87		22
R = Ph	5.79	3.6		13	21

III. SELENIUM HETEROCYCLES

Selenium heterocycles are known to be chemically similar to their sulfur analogs and this similarity carries over into their NMR spectra. The earliest work on the spectrum of selenophen (**26**) (27) was carried out to check the planarity of the ring system which was being criticized on the grounds of infrared and Raman spectra. The general similarity between the spectral parameters obtained for selenophen and those for thiophen and furan led to the suggestion that the system was reasonably planar. Read *et al.* (28) reported a very detailed investigation of the spectra of selenophen and a number of its halo derivatives and again showed the similarity between this ring system and thiophen. Temperature effects on the spectrum of 1,1-diiodotetrahydroselenophen (**27**) were used to show that the molecule was oscillating between two equivalent forms (29).

4*H*-Selenapyran (**28**) and the corresponding selenapyrylium salt (**29**) were used, in conjunction with their oxygen and sulfur analogs, as models for a study of the AA′BB′X and AA′BB′XX′ systems (30). Unfortunately, the work contains many simplifying assumptions which tend to reduce the accuracy of the results.

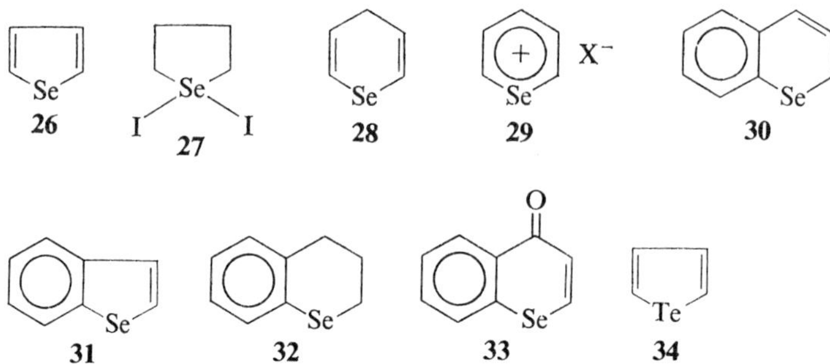

O
Se
31
Se
32
Se
33
Te
34

Nuclear magnetic resonance methods were used to check the structure of 2*H*-selenachromen (**30**) (31) and to characterize benzselenephen (**31**) and selenachroman (**32**) products of the Wolf-Kishner-Minlon reduction of selenachromanone (**33**) (32). Spectral data for selenium heterocycles are collected in Table 7.4.

IV. TELLURIUM HETEROCYCLES

The synthesis and physical and chemical properties of tellurophen (**34**) have been described (33). Chemical shifts of the ring protons [$\tau\alpha$ 1.06, $\tau\beta$ 2.17 ($CDCl_3$)] were taken as an indication of the "aromatic" character of the ring system.

V. BORON HETEROCYCLES

The number of cyclic boron compounds (not containing other heteroatoms) which could be classed as heterocycles is very small. However, an NMR

BH
35
B—Ph
36

study has been made of the aromatic character of 3-benzborepin derivatives **35** (34) and hence this ring system is included as a heterocycle.

Consideration of the chemical shift data (see Table 7.5) for 1-phenyl-4,5-dihydroborepin (**36**), 1-phenyl-3-benzborepin (**37**), and 1-dimethylamino-1-phenyl-3-benzborepin (**38**) leads to the suggestion that **37** does support a

TABLE 7.4 SPECTRAL DATA FOR SELENIUM HETEROCYCLES

A. SELENOPHEN

Substituents	Solvent	$\tau 2$	$\tau 3$	$\tau 4$	$\tau 5$	J_{23}	J_{24}	J_{25}	J_{34}	J_{35}	J_{45}	Ref.
None	Neat	2.30	2.88	2.88	2.30	5.35	1.05	2.47	3.56	1.05	5.35	27[a]
	TMS	2.12	2.77	2.77	2.12	5.40	1.46	2.34	3.74	1.46	5.40	28[b]
2-Cl	TMS		3.07	3.08	2.40				3.87	1.42	6.04	28[b]
2-Br	TMS		3.04	3.10	2.26				3.88	1.34	6.05	28[b]
2,5-Di-Cl	TMS		3.30	3.30					4.26			28[b]
2,5-Di-Br	TMS		3.13	3.13					4.17			28[b]

B. γ-SELENAPYRAN (30)—NEAT

$\tau 2,6$	$\tau 3,5$	$\tau 4$	J_{23}	J_{24}	J_{26}	J_{34}	(J_{25} and J_{35} assumed zero)
3.53	4.08	7.35	9.2	0.9	2.5	4.5	

TABLE 7.4 *(Continued)*

C. SELENAPYRYLIUM PERCHLORATE (30). SOLVENT—MeCN

τ2,6	τ3,5	τ4	J_{23}	J_{24}	J_{26}	J_{34}	(J_{25} and J_{35} assumed zero)
−1.1	1.25	1.06	11.3	2.1	2.5	9.1	

D. 2H-SELENACHROMEN (31). SOLVENT—CCl_4

τ2 6.69,	τ3 4.33,	τ4 3.74,	τ5–8 3.09

[a] $J_{^{77}Se,\alpha H} = 48$, $J_{^{77}Se,\beta H} = 9.5$ Hz.
[b] $J_{^{13}C-H}$

Substituents	C-3	C-4	C-5
None	164.62	164.62	187.23
2-Cl	169.15	167.17	189.96
2-Br	169.69	167.32	189.58
2,5-Di-Cl	171.61	171.61	
2,5-Di-Br	171.42	171.42	

TABLE 7.5 SPECTRAL PARAMETERS FOR PROTONS ON THE SEVEN-MEMBERED RING OF 3-BENZBOREPIN AND RELATED COMPOUNDS (34)

Compound	τH*a*	τH*b*	$J_{a,b}$
Borepin, B—Ph	2.82	3.25	13
37 (3-benzborepin, B—Ph)	1.78	2.28	14
38 (B(Ph)←NHMe$_2$)	2.95	3.64	13.5
39 (SnMe$_2$)	2.50	3.68	14

ring current, but that this can only be sustained if the boron p_z-orbital is free to participate in the π-electron system. No indication of aromaticity was found for the benzstannatropylium ion **39**.

VI. SILICON HETEROCYCLES

The decision to include silicon-containing cyclic systems in a book on heterocyclic chemistry is probably open to criticism, particularly when one considers that the monomeric nature of many of these compounds has not been firmly established. However, a large amount of information has accumulated on the spectra of these compounds, so a short description of their spectral properties appears to be in order. References given are intended to be illustrative rather than complete.

In the four-membered ring systems, the spectrum of 1,1,2-triphenyl-1-silacyclobut-2-ene (**40**) has been reported (35) and a number of workers have presented data for other derivatives of this ring system (35–37). Spectra of the parent silacyclopentadiene (**41**) and its potassium salt (**42**) have been used to postulate the presence of an "aromatic" ring current (38). In the case of the silacyclopentadienide ion **42** such "aromaticity" can be explained in terms of overlap of the free silicon p_z-orbital with the π-electrons of the unsaturated system. Silacyclopentadiene contains no such free p_z-orbital and involvement of the silicon $3d$-orbitals was invoked to produce the "aromaticity" required to cause the ring protons to resonate at low field. A number of references list data on silacyclopent-2-enes (**43**) (35, 39) and silacyclopentanes (**44**) (35, 37, 39, 40).

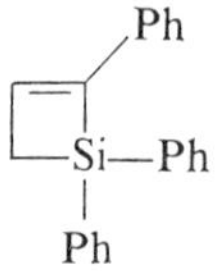

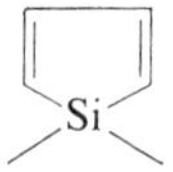

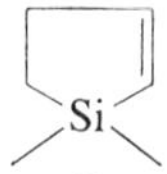

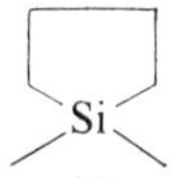

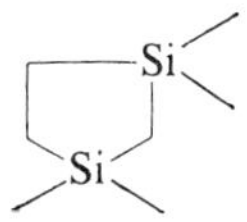

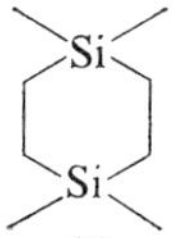

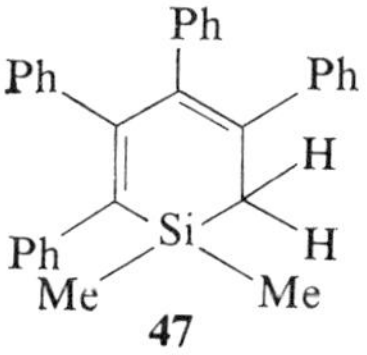

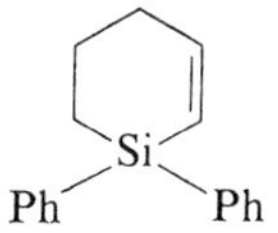

Murray and Kaplan (41) included the 1,3-disilacyclopentane (**45**) and the 1,4-disilacyclohexane (**46**) in their study of inversion of reduced 1,4-diheterosystems. In work on the preparation of the elusive silabenzene derivatives, the spectrum of the 1,2-dihydro compound **47** has been reported (42). Brook and Pierce (43) have recorded data for 1,1-diphenylsilacyclohex-2-ene (**48**).

REFERENCES

1. Chan, Goldwhite, Keyser, Rowsell, and Tang, *Tetrahedron*, **25,** 1097 (1969), and references cited therein.
2. Cremer and Chorvat, *J. Org. Chem.*, **32,** 4066 (1967).
3. McBride, Jungermann, Killheffer, and Clutter, *J. Org. Chem.*, **27,** 1833 (1962).
4. Gray and Cremer, *Tetrahedron Lett.*, **1971,** 3061.
5. Quin, Bryson, and Moreland, *J. Am. Chem. Soc.*, **91,** 3308 (1969).
6. Quin and Bryson, *J. Am. Chem. Soc.*, **89,** 5984 (1967).
7. Egan, Tang, Zon, and Mislow, *J. Am. Chem. Soc.*, **92,** 1442 (1970).
8. Märkl and Hauptmann, *Tetrahedron Lett.*, **1968,** 3257.
9. Märkl and Potthast, *Angew. Chem.*, **79,** 58 (1967).
10. Weitkamp and Korte, *Z. Anal. Chem.*, **204,** 245 (1964).

11. Mathey and Mavel, *Compt. Rend., Series C*, **263,** 855 (1966).
12. Myers and Quin, *J. Org. Chem.*, **36,** 1285 (1971).
13. Quin and Barket, *J. Am. Chem. Soc.*, **92,** 4303 (1970).
14. Bond, Green, and Pearson, *J. Chem. Soc., B*, **1968,** 929.
15. Arbuzov and Vizel, *Dokl. Akad. Nauk SSSR*, **158,** 1105 (1964).
16. Arbuzov, Samitov, Vizel, and Zykova, *Dokl. Akad. Nauk SSSR*, **159,** 1062 (1964).
17. Reddy and Weis, *J. Org. Chem.*, **28,** 1822 (1963).
18. Märkl, *Angew. Chem.*, **78,** 907 (1966).
19. Dimroth and Mach, *Angew. Chem., Int. Ed. Engl.*, **7,** 460 (1968).
20. Märkl, Merz, and Rausch, *Tetrahedron Lett.*, **1971,** 2989.
21. Märkl, Lieb, and Merz, *Angew. Chem.*, **79,** 59 (1967).
22. Märkl and Merz, *Tetrahedron Lett.*, **1971,** 1215.
23. Ashe, *J. Am. Chem. Soc.*, **93,** 3293 (1971).
24. Shook and Quin, *J. Am. Chem. Soc.*, **89,** 1841 (1967).
25. Aguiar, Hansen, and Reddy, *J. Am. Chem. Soc.*, **89,** 3067 (1967).
26. Gallagher and Mann, *J. Chem. Soc.*, **1962,** 5110.
27. Heffernan and Humffray, *Mol. Phys.*, **7,** 527 (1963).
28. Read, Mathis, and Goldstein, *Spectrochim. Acta*, **21,** 85 (1965).
29. Pedersen and Hope, *Acta Crystallogr.*, **19,** 473 (1965).
30. Degani, Lunazzi, and Taddei, *Boll. Sci. Fac. Chim. Ind. Bologna*, **23,** 131 (1965).
31. Degani, Fochi, and Vincenzi, *Gazz. Chim. Ital.*, **94,** 451 (1964).
32. Bellinger, Cagniant, and Cagniant, *Tetrahedron Lett.*, **1971,** 49.
33. Mack, *Angew. Chem.*, **78,** 940 (1966).
34. Leusink, Drenth, Noltes, and van der Kerk, *Tetrahedron Lett.*, **1967,** 1263, and references cited therein.
35. Gilman and Atwell, *J. Am. Chem. Soc.*, **87,** 2678 (1965); Gilman and Atwell, *ibid.*, **86,** 2687 (1964).
36. Laane, *J. Am. Chem. Soc.*, **89,** 1144 (1967).
37. Seyferth, Damrauer, and Washburne, *J. Am. Chem. Soc.*, **89,** 1538 (1967).
38. Benkeser, Grossman, and Stanton, *J. Am. Chem. Soc.*, **84,** 4727 (1962).
39. Benkeser, Noe, and Nagai, *J. Org. Chem.*, **30,** 378 (1965).
40. Benkeser, Nagai, Noe, Cunico, and Gund, *J. Am. Chem. Soc.*, **86,** 2446 (1964).
41. Murray and Kaplan, *Tetrahedron*, **25,** 1651 (1969).
42. Märkl and Merz, *Tetrahedron Lett.*, **1971,** 1303.
43. Brook and Pierce, *J. Org. Chem.*, **30,** 2566 (1965).

8 HETEROCYCLIC SYSTEMS CONTAINING TWO OR MORE DIFFERENT HETEROATOMS

Since the inception of this book in 1967, the number of heterocyclic systems containing two or more different heteroatoms has grown considerably, to a point where a discussion of their NMR spectra would not fit into a single chapter of a book of this size. Fortunately, many of these systems are bicyclic with two different heterocyclic rings fused together. In most cases such compounds can be treated as though they were substituted derivatives of the two parent heterocycles, and their spectral behavior can be predicted from material presented in previous chapters.

Information on these heterocyclic systems should be included in a book such as this, so as a compromise it was decided to tabulate the main references to papers dealing with each ring system (Table 8.1). Also, where reliable data are available for a parent compound or a simple derivative, these are included in Table 8.2. As in previous chapters, these tables include data for monocyclic and bicyclic compounds, but not for tricyclic or larger systems.

The task of arranging 145 heterocyclic systems into some logical order for Table 8.1 proved formidable. Eventually three groups emerged in this order: monocyclic systems, bicyclic systems with all hetero-atoms in only one ring, and bicyclic systems with hetero-atoms in each ring. Each group was then arranged in increasing size of ring(s), and within each such sub-group systems with two hetero-atoms preceded those with three or more. The given references usually cover one or more of the following aspects: configuration, conformation, protonation, or tautomerism. It is emphasised that the list of systems and the references are indicative rather than exhaustive; ten years hence this chapter might need a complete volume.

TABLE 8.1 REFERENCES TO NMR SPECTRA OF HETEROCYCLIC SYSTEMS CONTAINING TWO OR MORE DIFFERENT HETEROATOMS

Ring System[a]	Structure	Contents[b]	ON[c]	Ref.[d]
			^{13}C	1*
Oxaziridine	O–N	C		2, 3
			^{15}N	358
1,2-Azaboretidine	$\bar{B}H_2$, $\overset{+}{N}H_2$			4
1,2-Oxaphosphetan	P, O			105
Isothiazole	S, N			6,* 7, 10, 11*
		A		9
		B		5,* 8
Thiazole	1 S, 2, N 3, 4, 5			7, 16,* 20,* 25,* 29, 36, 42, 44,* 51,* 52,* 53,* 137, 384
		B		8, 21, 54,* 41, 43*
		B	^{13}C	55,* 56*
			^{13}C	17,* 18,* 19,* 48*
		A		27, 28, 45, 46
		A, B		26
			^{15}N	359, 360
2,3-Dihydro				33, 39, 57
4,5-Dihydro				13, 14,* 15,* 22, 23,* 24, 30–32, 49, 90, 389
			^{13}C	47*

TABLE 8.1 (*continued*)

Ring System[a]	Structure	Contents[b]	ON[c]	Ref.[d]
1,5-Dihydro		A		34
Tetrahydro				38, 49, 50, 390
		A		12, 35, 37, 40
Isoxazole	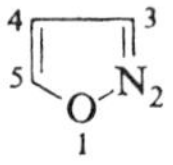 (isoxazole ring, positions 1–5: O 1, N 2)			58*–61, 385
		B		62
Dihydro		A		63–65, 68–70
4,5-Dihydro				66,* 67,* 71, 72,* 385
Tetrahydro				73,* 74–77
Oxazole	 (oxazole ring, O, N)			78,* 96,* 82
		B		78, 79, 80*
			^{15}N	359, 360
4,5-Dihydro				79, 87, 88, 89, 90, 23,* 389
		A		84, 92
2,5-Dihydro				85, 86, 88, 89, 91
Tetrahydro				81, 83, 95, 97, 390
		A		92, 98, 99
			^{19}F	94
1,2-Oxathiolan (sultones)	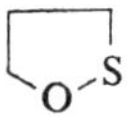 (ring, O, S)			100
1,3-Oxathiolan	(ring, O, S)	A		101
	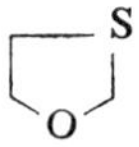	C		102,* 103*
1,2-Oxaphospholan	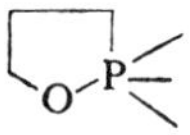 (ring, O, P)		^{31}P	104, 105

TABLE 8.1 *(continued)*

Ring System[a]	Structure	Contents[b]	ON[c]	Ref.[d]
1,2-Boroxolan				136
1,2,3-Oxadiazole		B		115, 116
1,2,4-Oxadiazole				106
			^{19}F	109
		C		107*
4,5-Dihydro		A		107*
1,2,5-Oxadiazole (furazan)				110, 114
1,3,4-Oxadiazole				113
4,5-Dihydro				108, 112
Tetrahydro		C, D		111
1,2,3-Thiadiazole				7, 10
1,2,5-Thiadiazole				110
Tetrahydro		C		117
1,3,4-Thiadiazole				7, 118
		B		8
1,3,4-Oxathiazoline				119, 120
1,3,2-Dioxathiolan				121, 346

TABLE 8.1 (*continued*)

Ring System[a]	Structure	Contents[b]	ON[c]	Ref.[d]
1,3,2-Diazaphospholan	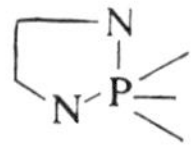			135*
			^{31}P	134,* 338
1,3,2-Oxazaphospholan	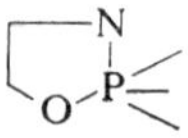			337
1,3,2-Dioxaphospholen	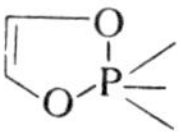		^{31}P	123, 124, 130, 347
1,3,2-Dioxaphospholan	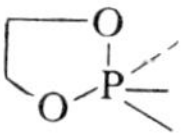			105, 129,* 338
		C, D		126–128, 346, 350–355
			^{31}P	125, 131, 132
		C, D	^{31}P	122, 133, 348, 349, 354, 355
1,4,2-Dioxaphospholan	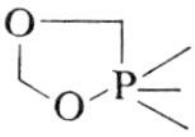			105
1,3,2-Dithiaphospholan	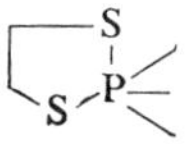			338
1,3,2-Dioxaborolan	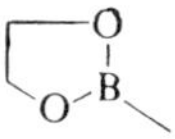			136
1,2,5-Selenadiazole	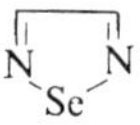			138, 139*
1,2-Oxazine (6H)	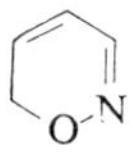			141*
2,3-Dihydro				140,* 142*
Tetrahydro				142*

TABLE 8.1 (*continued*)

Ring System[a]	Structure	Contents[b]	ON[c]	Ref.[d]
1,3-Oxazine (tetrahydro)		C		144, 145, 147, 356
		D		143
			^{15}N	146
1,4-Oxazine (2H)				
5,6-Dihydro			^{19}F	149
Tetrahydro (morpholine)				150, 333, 391
		C		151–155, 156,* 157,* 336, 391
			^{19}F	148,* 158
1,2-Thiazine (6H)				
2,3-Dihydro				159
1,3-Thiazine (6H)				164
2,3-Dihydro				161
4,5-Dihydro				49
(4H) 2,3-Dihydro				162, 163
Tetrahydro		A		49, 160
1,4-Thiazine (2H)				165–167
(4H) 2,3-Dihydro				168, 330
1,3-Oxathian		C		169–171, 340, 341
1,4-Oxathiin				172
5,6-Dihydro				175, 176
Tetrahydro		C, D		173, 174, 176, 177, 333

TABLE 8.1 (*continued*)

Ring System[a]	Structure	Contents[b]	ON[c]	Ref.[d]
Phosphonia-4-pyran	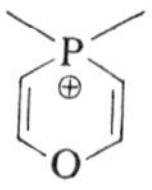		^{31}P	178
1,2,4-Oxadiazine (6H)	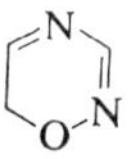			179
4,5-Dihydro				180
Tetrahydro		A		181
1,3,4-Oxadiazine	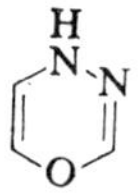			
(4H) 5,6-Dihydro		C		187, 188
Tetrahydro		A		186*
1,2,4-Dioxazine	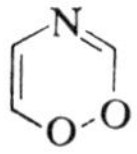			179
1,2,4-Thiadiazine	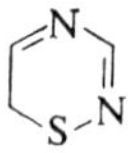	A		182
1,2,6-Thiadiazine				183, 184
1,3,4-Thiadiazine (4H)				
5,6-Dihydro				185
1,3,5-Thiadiazine	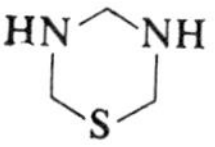			
(tetrahydro)		C		189, 379
3,5-Dithiapyridine	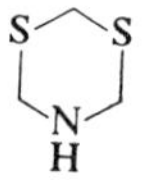			
(tetrahydro)				380
1,3,2-Dioxathian	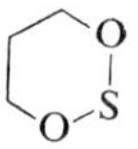			190, 372
		C		191, 192

TABLE 8.1 (*continued*)

Ring System[a]	Structure	Contents[b]	ON[c]	Ref.[d]
1,3,2-Dioxaphosphirinan	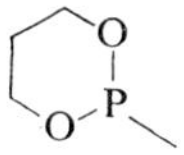			198–200, 375, 392, 93
		C, D		193, 195, 201–203
			^{31}P	194, 196, 197
			^{13}C, ^{31}P	381
1,3,2-Diazaphosphirinan	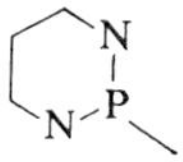			135*
1,3,2-Oxazaphosphirinan	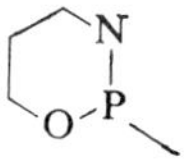	A, D		193
1,2,5-Oxazaphosphirinan	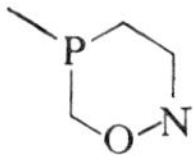			386
1,3,2-Dioxaboronium	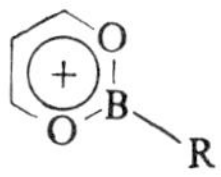			332
1,3,2-Dioxaborinate	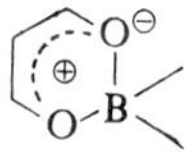			332
1,3,2-Dioxaborinan	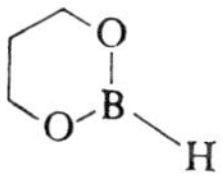			204
1,3,5-Aza-, oxa-, or thiadiphosphirinan	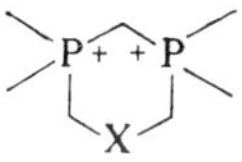			213

TABLE 8.1 (*continued*)

Ring System[a]	Structure	Contents[b]	ON[c]	Ref.[d]
2,5-Diborapiperazine	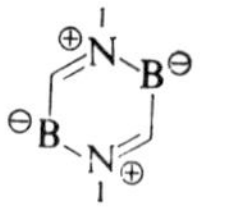			209–212
2,6-Diphospha-1,3,5-triazine			^{31}P	205–208
1,4-Oxazepin (tetrahydro)	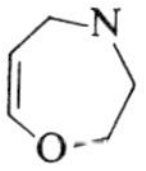			215, 216
1,4,5-Thiadiazepin (dihydro)	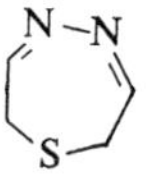			214
4,5-Boroxathiepan	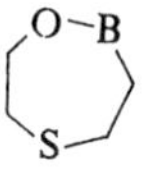			259
Benzoxazole	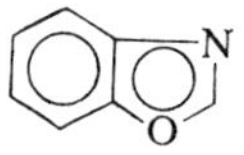			224,* 226,* 233, 234, 342–344
			^{13}C	235
Reduced				236
Benzisothiazole				217, 219
2,3-Dihydro				220
Benzthiazole	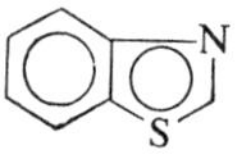			223, 224,* 226,* 227,* 228,* 230–232, 343, 344
		B		221, 222

TABLE 8.1 (*continued*)

Ring System[a]	Structure	Contents[b]	ON[c]	Ref.[d]
			^{14}N	225
			^{13}C	235
2,3-Dihydro				218, 229
Benzsenelazole	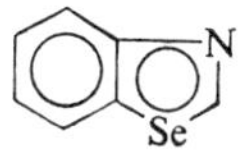 Se, N			224,* 226,* 235, 343, 344
Benz-2,1,3-oxadiazole (benzfurazan)	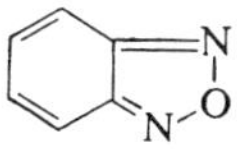 N, O, N	B		155, 237–245, 247, 248,* 249,* 335*
			^{17}O	246
Benz-1,2,3-thiadiazole	 N, N, S			250,* 61
Benz-2,1,3-thiadiazole	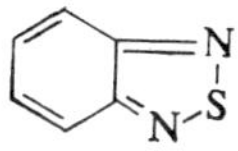 N, S, N			248,* 249,* 335*
Benz-2,1,3-selenadiazole	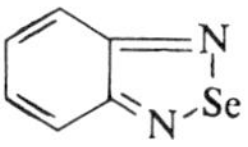 N, Se, N			248,* 249,* 251*
Benz-1,3,2-dithiazolium	 S⊕, N, S			285*
2-Silabenzimidazole (reduced)	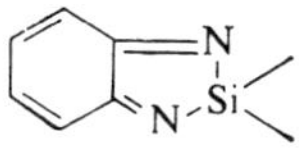 N, Si, N			286
1,3-Benzoxazine	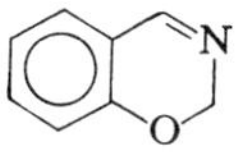 N, O			253
3,4-Dihydro				252

TABLE 8.1 (*continued*)

Ring System[a]	Structure	Contents[b]	ON[c]	Ref.[d]
1,2-Benzthiazine (dihydro)	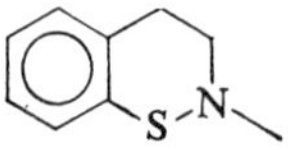			256
1,3-Benzthiazine (dihydro)	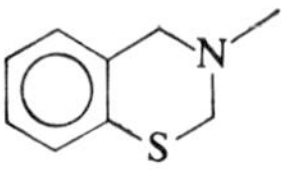			254*
1,4-Benzthiazine 3,4-Dihydro	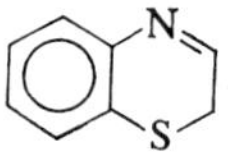			255 165, 257, 258
1,2-Benzoxathian	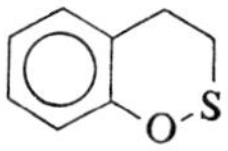			370
1,3-Benzoxathian	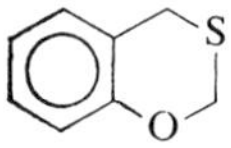			260
2,1-Borazanaphthalene	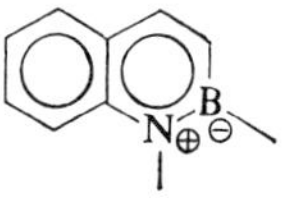		^{11}B	369
2,1,3-Benzthiadiazine	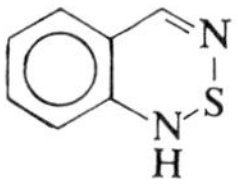			270
1,2,4-Benzthiadiazine (4H) 2,3-Dihydro	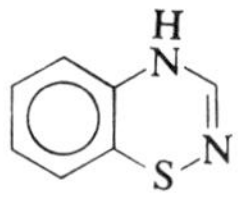			262
4,3-Borazaisoquinoline	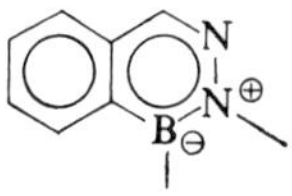		 ^{11}B	263* 369

TABLE 8.1 (*continued*)

Ring System[a]	Structure	Contents[b]	ON.[c]	Ref.[d]
4,3-Boroxaisoquinoline	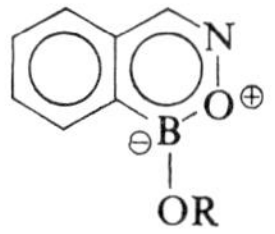		^{11}B	369
1,3,2-Benzdioxaphospholan Reduced	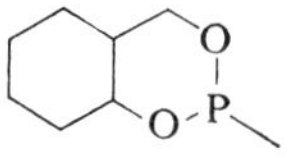	B		382
1,3,2-Benzdioxathiolan Reduced	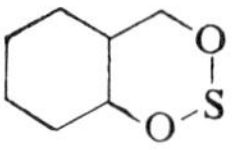	B		383
1,4-Benzoxazepin (4H) 2,3-Dihydro	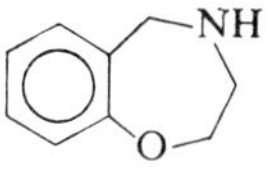			264
3,1-Benzoxazepin	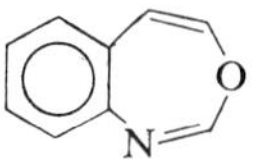			269
1,4-Benzoxathiepin (5H)	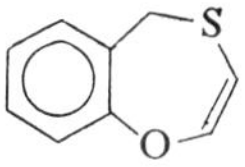			268
1,4-Benzthiazepin Tetrahydro	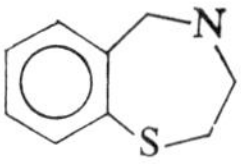			266
2,1,5-Benzthiadiazeipn Tetrahydro	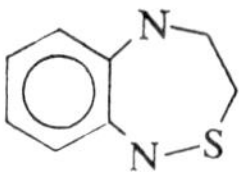			267, 357
5,1,2-Benzoxadiazepin 3,4-Dihydro	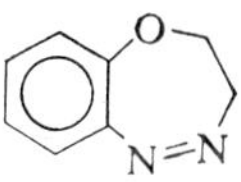			265
Oxaziridino[2,3-*a*]- pyrrolidine	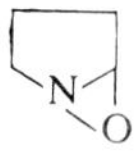			271*

TABLE 8.1 (*continued*)

Ring System[a]	Structure	Contents[b]	ON[c]	Ref.[d]
Aziridino[2,3-*c*]thiophen Reduced	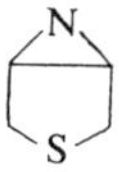			 376
Azetidino[1,2-*b*]- thiazolidine	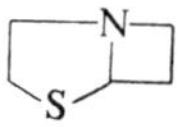			272
Pyrrolo[1,2-*b*]isoxazole Reduced	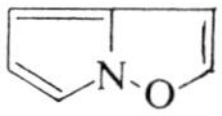			 273, 274
Thieno[2,3-*c*]pyrrole 5,6-Dihydro				289* 289
Thieno[2,3-*b*]pyrrole	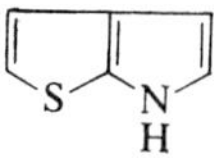			277
Thieno[3,2-*b*]pyrrole	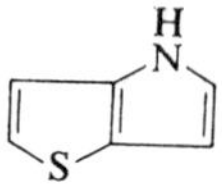			275–284, 287,* 288
Pyrrolo[2,1-*b*]thiazole Hexahydro		 B		291 290* 292
Thieno[3,4-*d*]-1,3-dioxolan Reduced	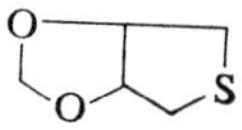			 373, 374
Thieno[3,4-*d*]imidazole Reduced	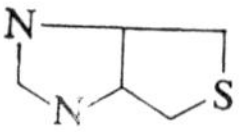			 373, 374
s-Triazolo[3,4-*b*]-1,3,4- thiadiazole	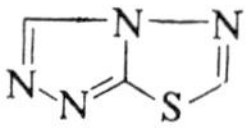			326, 327
Furo[2,3-*b*]pyridine	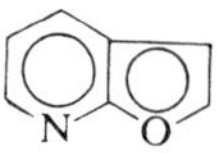			328

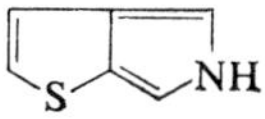

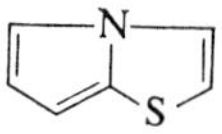

TABLE 8.1 (*continued*)

Ring System[a]	Structure	Contents[b]	ON[c]	Ref.[d]
Furo[2,3-*c*]pyridine (reduced)				294*
Furo[3,2-*b*]pyridine				293
Oxazolo[3,2-*a*]pyridinium				297,* 298
Oxazolo[3,4-*a*]pyridine (3H) Hexahydro				296, 371
Thieno[2,3-*b*]pyridine				300,* 303,* 362
Thieno[3,2-*b*]pyridine 2,3-Dihydro				300* 365
Thieno[2,3-*c*]pyridine		B		299,* 302*
Thieno[3,2-*c*]pyridine		B		299,* 302*
Thieno[3,4-*b*]pyridine				301*
Thieno[3,4-*c*]pyridine				301*
Thiazolo[3,2-*a*]pyridinium				304, 364

TABLE 8.1 (*continued*)

Ring System[a]	Structure	Contents[b]	ON[c]	Ref.[d]
Thiazolo[3,4-*a*]pyridine Reduced	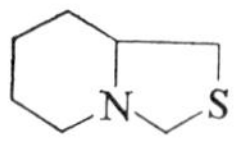			305
Thieno[3,4-*b*]pyrans	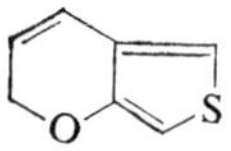			293
Furo[2,3-*d*]pyridizine	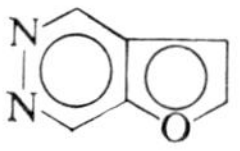			367, 368
Furo[3,4-*d*]pyridazine	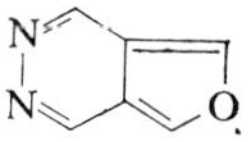			306
Pyrazolo[3,4-*b*]pyran (4H) 2,3-Dihydro	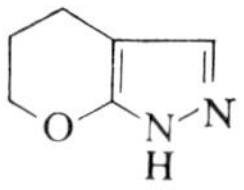			378
Furo[2,3-*d*]pyrimidine Dihydro	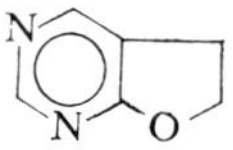			307
Thieno[2,3-*d*]pyridazine	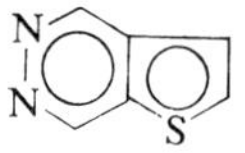			363
Thieno[3,4-*d*]pyridazine	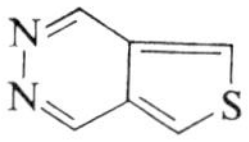			308, 363
Thieno[3,2-*d*]pyrimidine	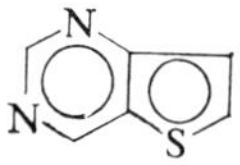			309,* 377*
Thiazolo[3,2-*a*]pyrimidine	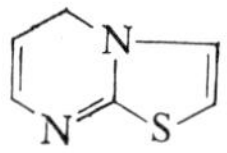			
(5H)				311
(7H)				311
Tetrahydro				310

TABLE 8.1 (*continued*)

Ring System[a]	Structure	Contents[b]	ON[c]	Ref.[d]
Thieno[3,2-*d*]-1,2-oxazine 1,2-Dihydro	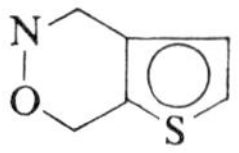			289
4*H*-Pyrano[3,2-*d*]isoxazole 2,3-Dihydro	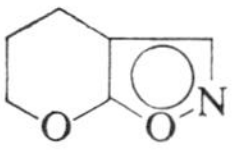	C		312, 378
4*H*-Pyrano[2,3-*c*]pyrazole 2,3-Dihydro	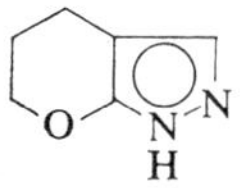	C		312
Thiazolo[5,4-*d*]pyrimidine				314,* 334
1,2,5-Thiadiazolo[3,4-*b*]-pyridine	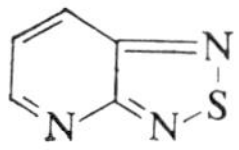			313
1,2,5-Thiadiazolo[3,4-*c*]-pyridine	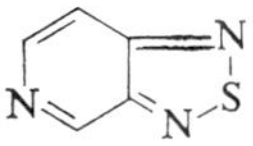			313
1,2,5-Selenadiazolo[3,4-*b*]-pyridine	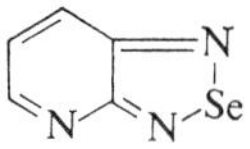			313
1,2,5-Selenadiazolo[3,4-*c*]-pyridine	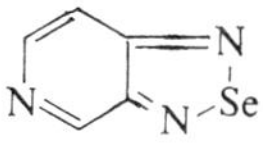			313
Thiazolo[3,2-*b*]-1,2,4-triazine (5H) Tetrahydro	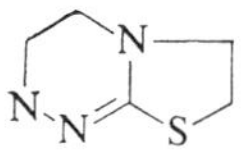			315*
Thiazolo[2,3-*c*]-1,2,4-triazine (2H) Tetrahydro	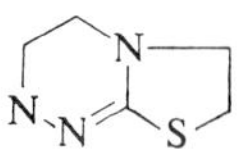			315*

TABLE 8.1 (*continued*)

Ring System[a]	Structure	Contents[b]	ON[c]	Ref.[d]
1,7,8,9-Triazaborahydrindane	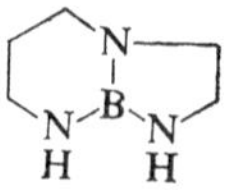			316
4,5-Borazarothieno[2,3-*c*]-pyridine	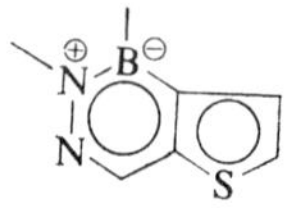			317, 331
7,6-Borazarothieno[3,2-*c*]-pyridine				317
7,6-Borazarothieno[3,4-*c*]-pyridine	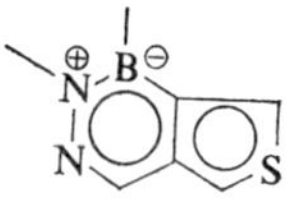			331
4,5-Boroxarothieno[2,3-*c*]-pyridine	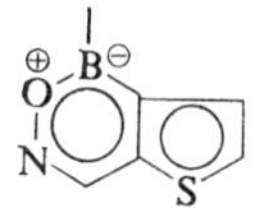			331
Furazano[3,4-*b*]-1,4-diazepin	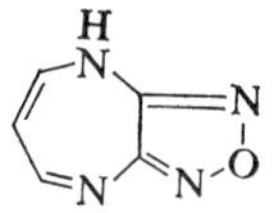			318*
Pyrano[2,3-*b*]pyridine (4H)	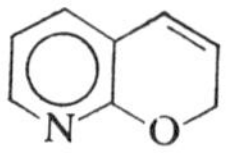			295* 366
Pyrido[1,2-*c*]-1,3-oxazine Reduced	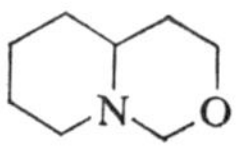			320, 324, 325
Pyrido[1,2-*c*]-1,3-thiazine Reduced	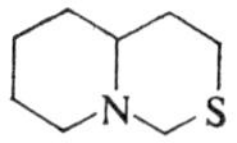			339

TABLE 8.1 (*continued*)

Ring System[a]	Structure	Contents[b]	ON[c]	Ref.[d]
Thiapyrano[2,3-*b*]pyridine	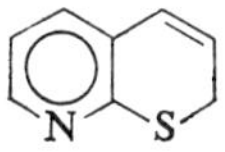			329*
10,9-Borazanaphthalene	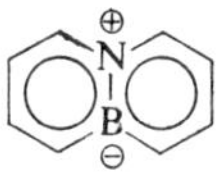			261,* 345
			^{11}B	369
Decahydro				261,* 345
Pyrimido[2,1-*b*]oxazine Reduced	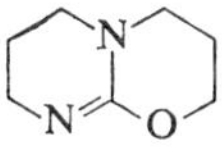			321
Pyrano[3,4-*e*]-1,3-oxazine 4,5-Dihydro	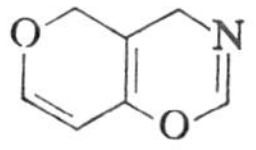			319
Pyrido[3,4-*d*]-1,3-dioxan	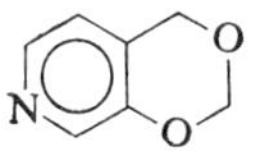			322, 323

[a] Ring index names used in most cases.
[b] Papers include information on the following: A, tautomerism; B, protonation; C, conformation; D, configuration.
[c] ON—other nuclei.
[d] Key papers marked with asterisk.

TABLE 8.2 SPECTRAL DATA FOR SOME HETEROCYCLES CONTAINING TWO OR MORE DIFFERENT HETEROATOMS. FOR STRUCTURES SEE TABLE 8.1

1. 2-*t*-Butyloxaziridine (1) neat
   τ3 6.31, 6.45; J_{gem} 9.7; $J_{^{13}C-H}$ 178.2

2. 1,2-Azaboretidine (4) $CDCl_3$
   τ3,4 7.6, 8.1

3. Thiazole
   Acetone (44)
   τ2 1.00; τ4 2.07; τ5 2.35; J_{25} 2.1; J_{45} 3.2
   C_6H_{12} (43)
   τ2 1.32; τ4 2.17; τ5 2.82; J_{25} 1.9; J_{45} 3.2; J_{24} 0.5

4. Thiazolium ion (44) TFA
   τ2 0.02; τ4 1.55; τ5 1.77; J_{25} 2.2; J_{45} 3.6; J_{24} 0.9

5. Isothiazole (5) CCl_4
   τ3 1.46; τ4 2.74; τ5 1.28; J_{34} 1.7; J_{35} < 0.4; J_{45} 4.7

6. Oxazole (78) CCl_4
   τ2 2.05; τ4 2.91; τ5 2.31; J_{25} 0.8; J_{45} 0.5

7. 1,2,5-Oxadiazole (furazan) (110) CCl_4
   τ3,4 1.81; $J_{^{13}C-H}$ 199

8. 1,2,5-Thiadiazole (110) CCl_4
   τ3,4 1.42; $J_{^{13}C-H}$ 192

9. 4-Methyl-6*H*-1,2-oxazine (141) CCl_4
   τ3 2.60; τ5 4.13; τ6 5.70; τMe 8.23; J_{35} 2

10. Morpholine (333) CCl_4
    $\tau OC\underline{H}_2$ 6.45; $\tau NC\underline{H}_2$ 7.27

11. 1,4-Thioxan (333) CCl_4
    $\tau OC\underline{H}_2$ 6.12; $\tau S\underline{C}H_2$ 7.43

12. 4-Benzylthieno[3,2-*b*]pyrrole (287)
    τ2 3.10; τ3 3.27; τ5 3.15; τ6 3.58; J_{23} 5.3; J_{25} 1.0; J_{56} 3.0

13. Thieno[2,3-*c*]pyridine (302) $CDCl_3$
    τ2 2.35; τ3 2.69; τ4 2.32; τ5 1.54; τ7 0.86; J_{23} 5.2; J_{45} 5.5

14. Thieno[3,2-*c*]pyridine (302) $CDCl_3$
    τ2,3 2.53, 2.57; τ4 0.89; τ6 1.57; τ7 2.72; J_{67} 5.6

15. Thieno[3,4-*c*]pyridine (301) $CDCl_3$
    τ1,3 1.98, 2.36; τ4 0.88; τ6 1.96; τ7 2.62; J_{13} 3; J_{67} 6.5

TABLE 8.2 *(continued)*

16. 2-Methylbenzoxazole (226) acetone
    τ4 2.40; τ5 2.69; τ6 2.68; τ7 2.36; τMe 7.41; J_{45} 8.04; J_{46} 1.19; J_{47} 0.70; J_{56} 7.54; J_{57} 1.10; J_{67} 8.13

17. 2-Methylbenzthiazole (226) acetone
    τ4 2.11; τ5 2.53; τ6 2.62; τ7 2.04; τMe 7.20; J_{45} 8.26; J_{46} 1.10; J_{47} 0.61; J_{56} 7.32; J_{57} 1.16; J_{67} 8.11

18. 2-Methylbenzselenazole (226) acetone
    τ4 2.08; τ5 2.55; τ6 2.70; τ7 1.98; τMe 7.18; J_{45} 8.12; J_{46} 1.32; J_{47} 0.53; J_{56} 7.28; J_{57} 1.26; J_{67} 7.98

19. Furo[2,3-*b*]pyridine (328) CCl_4
    τ2 2.33; τ3 3.28; τ4 2.15; τ5 2.86; τ6 1.77; J_{23} 2.5; J_{45} 7.7; J_{46} 1.8; J_{56} 4.8

20. 2,1,3-Benzoxadiazole (249) acetone
    τ4,7 2.06; τ5,6 2.41; J_{45} 9.15; J_{46} 0.88; J_{47} 1.15; J_{56} 6.40

21. 2,1,3-Benzthiadiazole (249) acetone
    τ4,7 1.95; τ5,6 2.28; J_{45} 8.86; J_{46} 1.10; J_{47} 0.93; J_{56} 6.65

22. 2,1,3-Benzselenadiazole (249) acetone
    τ4,7 2.16; τ5,6 2.40; J_{45} 9.10; J_{46} 1.18; J_{47} 0.92; J_{56} 6.40

23. Thiazolo[5,4-*d*]pyrimidine (314) $CDCl_3$
    τ2 0.81; τ5 0.79; τ7 0.53

24. 1,2,5-Thiadiazolo[3,4-*b*]pyridine (313) CCl_4
    τ5 0.94; τ6 2.45; τ7 1.68; J_{56} 3.9; J_{57} 1.9; J_{67} 8.9

25. 1,2,5-Selenadiazolo[3,4-*b*]pyridine (313) CCl_4
    τ5 0.82; τ6 2.38; τ7 1.61; J_{56} 3.7; J_{57} 1.9; J_{67} 9.0

26. 1,2,5-Selenadiazolo[3,4-*c*]pyridine (313) $CDCl_3$
    τ4 0.61; τ6 1.57; τ7 2.30; J_{46} 0.1; J_{47} 1.2; J_{67} 6.3

27. 2*H*-Pyrano[2,3-*b*]pyridine (295) CCl_4
    τ2 5.0; τ3 4.3; τ4 3.7; τ5 2.9; τ6 3.32; τ7 2.13; J_{23} 3.4; J_{24} 1.8; J_{34} 9.8; J_{56} 7.0; J_{57} 2.1; J_{67} 5.0

28. 2*H*-Thiopyrano[2,3-*b*]pyridine (329) CCl_4
    τ2 6.37; τ3 4.19; τ4 3.70; τ5 2.90; τ6 3.18; τ7 1.92; J_{23} 4.8; J_{24} 1.4; J_{34} 10.0; J_{56} 7.6; J_{57} 2.0; J_{67} 4.8

29. 10,9-Borazanaphthalene (345)
    τ1,8 2.28; τ2,7 3.36; τ4,5 2.67; τ3,6 2.53; J_{12} 6.5; J_{23} 6.0; J_{34} 10.0

REFERENCES

1. Radeglia, *Spectrochim. Acta*, **23A,** 1677 (1967).
2. Mannschreck, Linss, and Seitz, *Ann. Chem.*, **727,** 224 (1969).
3. Boyd, Spratt, and Jerina, *J. Chem. Soc., C*, **1969,** 2650.
4. Akerfeldt and Hellström, *Acta Chem. Scand.*, **20,** 1418 (1966).
5. Staab and Mannschreck, *Chem. Ber.*, **98,** 1111 (1965).
6. Finley and Volpp, *J. Heterocycl. Chem.*, **6,** 841 (1969).
7. Olofson, Landesberg, Houk, and Michelman, *J. Am. Chem. Soc.*, **88,** 4265 (1966).
8. Olofson and Landesberg, *J. Am. Chem. Soc.*, **88,** 4263 (1966).
9. le Coustumer and Mollier, *Bull. Soc. Chim. Fr.*, **1970,** 3076.
10. Lee and Volpp, *J. Heterocycl. Chem.*, **7,** 415 (1970).
11. Anderson, *J. Heterocycl. Chem.*, **1,** 279 (1964).
12. Reeve and Nees, *J. Am. Chem. Soc.*, **89,** 647 (1967).
13. Gronowitz, Mathiasson, Dahlbom, Holmberg, and Jensen, *Acta Chem. Scand.*, **19,** 1215 (1965).
14. Pouzard, Pujol, Roggero, and Vincent, *J. Chim. Phys.*, **61,** 612 (1964).
15. Pouzard, Pujol, Roggero, and Vincent, *J. Chim. Phys.*, **61,** 613 (1964).
16. Vincent, Phan-Tan-Luu, Metzger, and Surzur, *Bull. Soc. Chim. Fr.*, **1966,** 3524.
17. Vincent, Phan-Tan-Luu, and Metzger, *Bull. Soc. Chim. Fr.*, **1966,** 3537.
18. Vincent, Phan-Tan-Luu, and Metzger, *Bull. Soc. Chim. Fr.*, **1966,** 3530.
19. Cottet, Gallo, and Metzger, *Bull. Soc. Chim. Fr.*, **1967,** 4499.
20. Vernin, Aune, Dou, and Metzger, *Bull. Soc. Chim. Fr.*, **1967,** 4523.
21. Babadjamian and Metzger, *Bull. Soc. Chim. Fr.*, **1968,** 4878.
22. Roggero and Pierre-Pierre, *Compt. Rend., Series C*, **266,** 1316 (1968).
23. Weinberger and Greenhalgh, *Can. J. Chem.*, **41,** 1038 (1963).
24. Reynaud, Moreau, and Fodor, *Compt. Rend., Series C*, **266,** 632 (1968).
25. Dou, Friedmann, Vernin, and Metzger, *Compt. Rend., Series C*, **266,** 714 (1968).
26. Vivaldi, Dou, and Metzger, *Compt. Rend., Series C*, **264,** 1652 (1967).
27. Bastianelli, Chanon, and Metzger, *Bull. Soc. Chim. Fr.*, **1967,** 1948.
28. Bastianelli, Chanon, and Metzger, *Compt. Rend., Series C*, **264,** 1704 (1967).
29. Barton, Kenner, and Sheppard, *J. Chem. Soc., C*, **1966,** 2115.
30. Roggerő and Metzger, *Bull. Soc. Chim. Fr.*, **1964,** 1715.
31. Cherbuliez, Baehler, Espejo, Jaccard, Jindra, and Rabinowitz, *Helv. Chim. Acta*, **49,** 2408 (1966).
32. Cherbuliez, Baehler, Jaccard, Jindra, Weber, Wyss, and Rabinowitz, *Helv. Chim. Acta*, **49,** 807 (1966).
33. Dorn, Bedford, Hilgetag, and Katritzky, *J. Chem. Soc.*, **1965,** 1219.
34. Gawron, Fernando, Keil, and Weismann, *J. Org. Chem.*, **27,** 3117 (1962).
35. Stacy and Strong, *J. Org. Chem.*, **32,** 1487 (1967).
36. Friedmann, Bouin, and Metzger, *Bull. Soc. Chim. Fr.*, **1970,** 3155.
37. Gobrovsky and Schmir, *Tetrahedron*, **27,** 1185 (1971).
38. Clarke and Sykes, *Chem. Commun.*, **1965,** 370.
39. Chanon and Metzger, *Bull. Soc. Chim. Fr.*, **1968,** 2868.
40. Chanon and Metzger, *Bull. Soc. Chim. Fr.*, **1968,** 2855.
41. Dou and Metzger, *Bull. Soc. Chim. Fr.*, **1966,** 2395.
42. Anderson, Barnes, and Khan, *Can. J. Chem.*, **42,** 2375 (1964).
43. Bak, Nielsen, Rastrup-Andersen, and Schottländer, *Spectrochim. Acta*, **18,** 741 (1962).

44. Borgen, Gronowitz, Dahlbom, and Holmberg, *Acta Chem. Scand.*, **20,** 2593 (1966).
45. Pereslini, Scheichenko, and Sheinker, *Zh. Fiz. Khim.*, **40,** 38 (1966).
46. Sheinker, Pereslini, Kol'tsov, Bozhenov, and Vol'kenshtein, *Dokl. Akad. Nauk SSSR*, **148,** 878 (1963).
47. Vincent, Phan-Tan-Luu, Roggero, and Metzger, *Compt. Rend.*, *Series C*, **270,** 1688 (1970).
48. Vincent, Phan-Tan-Luu, and Metzger, *Compt. Rend.*, *Series C*, **270,** 666 (1970).
49. Rabinowitz, *Helv. Chim. Acta*, **52,** 255 (1969).
50. Clarke and Sykes, *J. Chem. Soc.*, *C*, **1967,** 1411.
51. Selim, Selim, and Martin, *Bull. Soc. Chim. Fr.*, **1968,** 3272.
52. Selim, Martin, and Selim, *Bull. Soc. Chim. Fr.*, **1968,** 3268.
53. Martin, Selim, and Selim, *Bull. Soc. Chim. Fr.*, **1968,** 3270.
54. Clarke and Williams, *J. Chem. Soc.*, **1965,** 4597.
55. Haake and Miller, *J. Am. Chem. Soc.*, **85,** 4044 (1963).
56. Haake, Bausher, and Miller, *J. Am. Chem. Soc.*, **91,** 1113 (1969).
57. Werbel, *Chem. Ind.*, **1966,** 1634.
58. Bertini, De Munno, Pelosi, and Pino, *J. Heterocycl. Chem.*, **5,** 621 (1968).
59. Stork and McMurry, *J. Am. Chem. Soc.*, **89,** 5461 (1967).
60. Battaglia, Dondoni, and Taddei, *J. Heterocycl. Chem.*, **7,** 721 (1970).
61. Sumimoto, *Kogyo Kagaku Zasshi*, **66,** 1831 (1963).
62. Woodward and Woodman, *J. Org. Chem.*, **31,** 2039 (1966).
63. Boulton, Katritzky, Hamid, and Oksne, *Tetrahedron*, **20,** 2835 (1964).
64. Katritzky, Oksne, and Boulton, *Tetrahedron*, **18,** 777 (1962).
65. Jacquier, Petrus, Petrus, and Verducci, *Bull. Soc. Chim. Fr.*, **1970,** 1978.
66. Aversa, Cum, and Crisafulli, *Gazz. Chim. Ital.*, **96,** 1046 (1966).
67. Aversa, Cum, and Crisafulli, *Gazz. Chim. Ital.*, **98,** 42 (1968).
68. Jacquier, Petrus, Petrus, and Verducci, *Bull. Soc. Chim. Fr.*, **1970,** 2685.
69. Jacquier, Malet, and Petrus, *Bull. Soc. Chim. Fr.*, **1965,** 2702.
70. Jacquier, Petrus, Petrus, and Verducci, *Bull. Soc. Chim. Fr.*, **1967,** 3003.
71. Perotti, Bianchi, and Grunanger, *Chim. Ind. (Milan)*, **48,** 492 (1966).
72. Sustmann, Huisgen, and Huber, *Chem. Ber.*, **100,** 1802 (1967).
73. Cum, Aversa, and Uccella, *Gazz. Chim. Ital.*, **98,** 782 (1968).
74. Riddell, Lehn, and Wagner, *Chem. Commun.*, **1968,** 1403.
75. Griffith and Olson, *Chem. Commun.*, **1968,** 1682.
76. Tartakovskii, Chlenov, Lagodzinskaya, and Novikov, *Dokl. Akad. Nauk SSSR*, **161,** 136 (1965).
77. Tartakovskii, Onishchenko, Lagodzinskaya, and Novikov, *J. Org. Chem. USSR*, **3,** 730 (1967).
78. Brown and Ghosh, *J. Chem. Soc.*, *B*, **1969,** 270.
79. Kille and Fleury, *Bull. Soc. Chim. Fr.*, **1967,** 4619.
80. Staab, Irngartinger, Mannschreck, and Wu, *Ann. Chem.*, **695,** 55 (1966).
81. Kjaer and Thomsen, *Acta. Chem. Scand.*, **16,** 591 (1962).
82. Bowie, Donaghue, and Rodda, *J. Chem. Soc.*, *B*, **1969,** 1122.
83. Herweh, Foglia, and Swern, *J. Org. Chem.*, **33,** 4029 (1968).
84. Howell, Fulmor, Quinones, and Hardy, *J. Org. Chem.*, **29,** 370 (1964).
85. Weygand, Steglich, Mayer, and v. Philipsborn, *Chem. Ber.*, **97,** 2023 (1964).
86. Steglich and Hurnaus, *Tetrahedron Lett.*, **1966,** 383.
87. Lambert and Kristofferson, *J. Org. Chem.*, **30,** 3938 (1965).
88. Iwakura, Toda, and Torii, *Bull. Chem. Soc. Japan*, **40,** 149 (1967).
89. Bassiri, Levy, and Litt, *J. Polym. Sci.*, **5B,** 871 (1967).

90. Potts, Kanaoka, Crawford, and Thomas, *J. Heterocycl. Chem.*, **1,** 297 (1964).
91. Hansen and Boyd, *J. Heterocycl. Chem.*, **7,** 911 (1970).
92. Carson, Poos, and Almond, *J. Org. Chem.*, **30,** 2225 (1965).
93. White, *J. Mol. Struct.*, **6,** 75 (1970).
94. Middleton and Krespan, *J. Org. Chem.*, **32,** 951 (1967).
95. Herweh, *J. Heterocycl. Chem.*, **5,** 687 (1968).
96. Staab, Wu, Mannschreck, and Schwalbach, *Tetrahedron Lett.*, **1964,** 845.
97. King, Babiec, and Karabinos, *J. Heterocycl. Chem.*, **5,** 587 (1968).
98. Paukstelis and Hammaker, *Tetrahedron Lett.*, **1968,** 3557.
99. Farrissey and Nashu, *J. Heterocycl. Chem.*, **7,** 331 (1970).
100. Ohline, Allred, and Bordwell, *J. Am. Chem. Soc.*, **86,** 4641 (1964).
101. Ottmann, Vickers, and Hooks, *J. Heterocycl. Chem.*, **4,** 527 (1967).
102. Wilson, Huang, and Bovey, *J. Am. Chem. Soc.*, **92,** 5907 (1970).
103. Pasto, Klein, and Doyle, *J. Am. Chem. Soc.*, **89,** 4368 (1967).
104. Grayson and Farley, *Chem. Commun.*, **1967,** 830.
105. Ramirez, *Bull. Soc. Chim. Fr.*, **1970,** 3491.
106. Warburton, *J. Chem. Soc., C*, **1966,** 1522.
107. Selim and Selim, *Bull. Soc. Chim. Fr.*, **1969,** 823.
108. Kametani, Sota, and Shio, *J. Heterocycl. Chem.*, **7,** 821 (1970).
109. Stevens, *J. Org. Chem.*, **33,** 2660 (1968).
110. Olofson and Michelman, *J. Org. Chem.*, **30,** 1854 (1965).
111. Zwanenburg, Weening, and Strating, *Rec. Trav. Chim.*, **83,** 877 (1964).
112. Breslow, Yaroslavsky, and Yaroslavsky, *Chem. Ind.*, **1961,** 1961.
113. Lwowski, Hartenstein, Devita, and Smick, *Tetrahedron Lett.*, **1964,** 2497.
114. Coburn, *J. Heterocycl. Chem.*, **5,** 83 (1968).
115. Götz and Grozinger, *J. Heterocycl. Chem.*, **7,** 123 (1970).
116. Lawson, Brey, and Kier, *J. Am. Chem. Soc.*, **86,** 463 (1964).
117. Abel, Bush, and Hopton, *Trans. Faraday. Soc.*, **62,** 3277 (1966).
118. Remers, Gibs, and Weiss, *J. Heterocycl. Chem.*, **6,** 835 (1969).
119. Jakobsen and Senning, *Chem. Commun.*, **1968,** 1245.
120. Deyrup and Moyer, *J. Org. Chem.*, **34,** 175 (1969).
121. Thompson, Crutchfield, and Dietrich, *J. Org. Chem.*, **30,** 2696 (1965).
122. Ramirez, *Acc. Chem. Res.*, **1,** 168 (1968).
123. Ramirez, Tasaka, Desai, and Smith, *J. Am. Chem. Soc.*, **90,** 751 (1968).
124. Ramirez, Patwardhan, Kugler, and Smith, *Tetrahedron Lett.*, **1966,** 3053.
125. Ramirez, Smith, Pilot, and Gulati, *J. Org. Chem.*, **33,** 3787 (1968).
126. Revel and Navech, *Compt. Rend., Series C*, **271,** 650 (1970).
127. Fontal and Goldwhite, *Tetrahedron*, **22,** 3275 (1966).
128. Devillers, Tran, and Navech, *Bull. Soc. Chim. Fr.*, **1970,** 182.
129. Gagnaire, Robert, Verrier, and Wolf, *Bull. Soc. Chim. Fr.*, **1966,** 3719.
130. Pinkus, Waldrep, and Ma, *J. Heterocycl. Chem.*, **2,** 357 (1965).
131. Ramirez, Patwardhan, and Smith, *J. Org. Chem.*, **31,** 3159 (1966).
132. Ramirez, Bhatia, Patwardhan, and Smith, *J. Org. Chem.*, **32,** 2194 (1967).
133. Devillers, Mathis, and Navech, *Compt. Rend., Series C*, **267,** 849 (1968).
134. Scherer and Wokulat, *Z. Naturforsch.*, **22b,** 474 (1967).
135. Ulrich, Tucker, and Sayigh, *J. Org. Chem.*, **32,** 1360 (1967).
136. Laurent, Bonnet, and Commenges, *Bull. Soc. Chim. Fr.*, **1967,** 2702.
137. Taurins and Schneider, *Can. J. Chem.*, **38,** 1237 (1960).
138. Shealy and Clayton, *J. Heterocycl. Chem.*, **4,** 96 (1967).
139. Bucci, Bertini, Ceccarelli, and de Munno, *Chem. Phys. Lett.*, **1967,** 473.

140. Knesze and Firl, *Tetrahedron*, **24,** 1043 (1968).
141. Klamann, Fligge, Weyerstahl, and Kratzer, *Chem. Ber.*, **99,** 556 (1966).
142. Banks, Barlow, and Haszeldine, *J. Chem. Soc.*, **1965,** 4714.
143. Lehn, Linscheid, and Riddell, *Bull. Soc. Chim. Fr.*, **1968,** 1172.
144. Gurne, Stefaniak, Urbanski, and Witanowski, *Nitro Compounds, Proc. Int. Symp., Warsaw, Poland, 1963*, 211.
145. Crabb and Judd, *Org. Magnetic Resonance*, **2,** 317 (1970).
146. Riddell and Lehn, *J. Chem. Soc.*, *B*, **1968,** 1224.
147. Allingham, Cookson, Crabb, and Vary, *Tetrahedron*, **24,** 4625 (1968).
148. Banks, Haszeldine, and Hatton, *J. Chem. Soc.*, *C*, **1967,** 427.
149. Banks and Burling, *J. Chem. Soc.*, **1965,** 6077.
150. Dillard and Easton, *J. Org. Chem.*, **31,** 122 (1966).
151. Cahill and Crabb, *Tetrahedron*, **25,** 1513 (1969).
152. Spragg, *J. Chem. Soc.*, *B*, **1968,** 1128.
153. Brügel, *Org. Magnetic Resonance*, **1,** 425 (1969).
154. Harris and Spragg, *J. Chem. Soc.*, *B*, **1968,** 684.
155. Dischler and Englert, *Z. Naturforsch.*, **16a,** 1180 (1961).
156. Harris and Spragg, *Chem. Commun.*, **1966,** 314.
157. Booth and Gidley, *Tetrahedron*, **21,** 3429 (1965).
158. Lee and Orrell, *Trans. Faraday Soc.*, **63,** 16 (1967).
159. Kataev and Plemenkov, *Zh. Obshch. Khim.*, **32,** 3817 (1962).
160. Campaigne and Nargund, *J. Med. Chem.*, **7,** 132 (1964).
161. Takamizawa and Hirai, *J. Org. Chem.*, **30,** 2290 (1965).
162. Lown and Ma, *Can. J. Chem.*, **45,** 939 (1967).
163. Lown and Ma, *Can. J. Chem.*, **45,** 953 (1967).
164. Cherbuliez, Baehler, Espejo, Jindra, Willhalm, and Rabinowitz, *Helv. Chim. Acta*, **50,** 331 (1967).
165. Sica, Santacroce, and Prota, *J. Heterocycl. Chem.*, **7,** 1143 (1970).
166. Sica, Paolillo, and Ferretti, *Ric. Sci.*, **37,** 629 (1967); through *Chem. Abstr.*, **68,** 110037*d* (1967).
167. Sataty, *J. Org. Chem.*, **34,** 250 (1969).
168. Stoodley, *Tetrahedron Lett.*, **1967,** 941.
169. Gelan and Anteunis, *Bull. Soc. Chim. Belges*, **77,** 447 (1968).
170. Gelan and Anteunis, *Bull. Soc. Chim. Belges*, **77,** 423 (1968).
171. Gelan and Anteunis, *Bull. Soc. Chim. Belges*, **79,** 313 (1970).
172. Bottini and Böttner, *J. Org. Chem.*, **31,** 385 (1966).
173. Foster, Inch, Qadir, and Webber, *Chem. Commun.*, **1968,** 1086.
174. Buck, Fahin, Foster, Perry, Qadir, and Webber, *Carbohydrate Res.*, **2,** 14 (1966).
175. ten Haken, *J. Heterocycl. Chem.*, **7,** 1211 (1970).
176. de Wolf, Henniger, and Havinga, *Rec. Trav. Chim.*, **86,** 1227 (1967).
177. Buck, Foster, Pardoe, Qadir, and Webber, *Chem. Commun.*, **1966,** 759.
178. Filleux-Blanchard, Simalty, Berry, Chahine, and Mebazaa, *Bull. Soc. Chim. Fr.*, **1970,** 3549.
179. Hermes and Braun, *J. Org. Chem.*, **31,** 2568 (1966).
180. Rajagopalan and Talaty, *J. Am. Chem. Soc.*, **88,** 5048 (1966).
181. Kornowski, Trichot, and Delage, *Bull. Soc. Chim. Fr.*, **1966,** 683.
182. Hinman and Hoogenboom, *J. Org. Chem.*, **26,** 3461 (1961).
183. Roe and Harbridge, *Chem. Ind.*, **1965,** 182.
184. Ouchi and Moeller, *J. Org. Chem.*, **29,** 1865 (1964).
185. Trepanier, Reifschneider, Shumaker, and Tharpe, *J. Org. Chem.*, **30,** 2228 (1965).

186. Dorman, *J. Org. Chem.*, **32,** 255 (1967).
187. Trepanier and Sprancmanis, *J. Org. Chem.*, **29,** 2151 (1964).
188. Trepanier, Sprancmanis, and Wiggs, *J. Org. Chem.*, **29,** 668 (1964).
189. Angiolini, Jones, and Katritzky, *Tetrahedron Lett.*, **1971,** 2209.
190. Maroni, Tisnes, and Gorrichon, *Comp. Rend., Series C*, **270,** 1817 (1970).
191. Samitov, *Dokl. Akad. Nauk SSSR*, **164,** 347 (1963).
192. Overberger, Kurtz, and Yaroslavsky, *J. Org. Chem.*, **30,** 4363 (1965).
193. Sanchez, Wolf, Burgada, and Mathis, *Bull. Soc. Chim. Fr.*, **1968,** 773.
194. Katritzky, Nesbit, Michalski, Tulimowski, and Zwierzak, *J. Chem. Soc., B*, **1970,** 140.
195. Bentrude and Hargis, *J. Am. Chem. Soc.*, **92,** 7136 (1970).
196. Bartle, Edmundson, and Jones, *Tetrahedron*, **23,** 1701 (1967).
197. Bartle, Edmundson, and Jones, *Tetrahedron*, **23,** 3226 (1967).
198. Albrand, Gagnaire, Robert, and Haemers, *Bull. Soc. Chim. Fr.*, **1969,** 3496.
199. Gagnaire, Robert, and Verrier, *Bull. Soc. Chim. Fr.*, **1968,** 2392.
200. Gagnaire and Robert, *Bull. Soc. Chim. Fr.*, **1967,** 2240.
201. Edmundson and Mitchell, *J. Chem. Soc., C*, **1970,** 752.
202. Edmundson and Mitchell, *J. Chem. Soc., C*, **1968,** 2091.
203. Edmundson and Mitchell, *J. Chem. Soc., C*, **1968,** 3033.
204. Woods and Strong, *J. Am. Chem. Soc.*, **88,** 4667 (1966).
205. Schmidpeter and Eberling, *Angew. Chem.*, **79,** 534 (1967).
206. Schmidpeter and Eberling, *Angew. Chem.*, **79,** 100 (1967).
207. Schmidpeter and Schindler, *Chem. Ber.*, **102,** 856 (1969).
208. Schmidpeter and Eberling, *Chem. Ber.*, **101,** 3883 (1968).
209. Hesse, Witte, and Haussleiter, *Angew. Chem.*, **78,** 748 (1966).
210. Bresadola, Carraro, Pecile, and Turco, *Tetrahedron Lett.*, **1964,** 3185.
211. Hesse, Witte, and Bittner, *Ann. Chem.*, **687,** 9 (1965).
212. Hesse and Witte, *Ann. Chem.*, **687,** 1 (1965).
213. Aguiar, Hansen, and Mague, *J. Org. Chem.*, **32,** 2383 (1967).
214. Sataty, *J. Heterocycl. Chem.*, **7,** 431 (1970).
215. Easton and Dillard, *Tetrahedron Lett.*, **1963,** 1807.
216. Dillard and Easton, *J. Org. Chem.*, **31,** 122 (1966).
217. Shanta, Scrowston, and Twigg, *J. Chem. Soc., C*, **1967,** 2364.
218. Manning and Strow, *J. Org. Chem.*, **32,** 2731 (1967).
219. Krüger and Hettler, *Ber. Bunsenges. Phys. Chem.*, **73,** 15 (1969).
220. Menger and Mandell, *J. Am. Chem. Soc.*, **89,** 4424 (1967).
221. Vorsanger, *Bull. Soc. Chim. Fr.*, **1968,** 971.
222. Vorsanger, *Bull. Soc. Chim. Fr.*, **1968,** 964.
223. Guglielmetti, Vincent, Metzger, Berger, and Garnier, *Bull. Soc. Chim. Fr.*, **1967,** 4195.
224. Di Modica, Barni, and Gasco, *J. Heterocycl. Chem.*, **2,** 457 (1965).
225. Mathias, *Mol. Phys.*, **12,** 381 (1967).
226. Tobiason and Goldstein, *Spectrochim. Acta*, **23A,** 1385 (1967).
227. Lunazzi, Taddei, and Todesco, *Boll. Sci. Fac. Chim. Ind. Bologna*, **23,** 99 (1965).
228. Todesco, *Boll. Sci. Fac. Chim. Ind. Bologna*, **23,** 107 (1965).
229. Metzger, Larivé, Vincent, and Dennilauler, *J. Chim. Phys.*, **60,** 944 (1963).
230. Di Modica, Barni, and Monache, *Gazz. Chim. Ital.*, **95,** 432 (1965).
231. Todesco and Vivarelli, *Boll. Sci. Fac. Chim. Ind. Bologna*, **20,** 125 (1962).
232. Grandolini, Ricci, Martani, and Mezzetti, *J. Heterocycl. Chem.*, **3,** 299 (1966).
233. Garner, Mullock, and Suschitzky, *J. Chem. Soc., C*, **1966,** 1980.

234. Okuda and Nagai, *Bull. Chem. Soc. Japan*, **40,** 1999 (1967).
235. Barni, Di Modica, and Gasco, *J. Heterocycl. Chem.*, **4,** 139 (1967).
236. Crabb and Williams, *J. Heterocycl. Chem.*, **4,** 169 (1967).
237. Mallory, Manatt, and Wood, *J. Am. Chem. Soc.*, **87,** 5433 (1965).
238. Mallory and Varimbi, *J. Org. Chem.*, **28,** 1656 (1963).
239. Harris, Katritzky, Oksne, Bailey, and Paterson, *J. Chem. Soc.*, **1963,** 197.
240. Boulton, Ghosh, and Katritzky, *J. Chem. Soc.*, *B*, **1966,** 1004.
241. Norris and Osmundsen, *J. Org. Chem.*, **30,** 2407 (1965).
242. Brown and Keyes, *J. Org. Chem.*, **30,** 2452 (1965).
243. Boulton, Gray, and Katritzky, *J. Chem. Soc.*, *B*, **1967,** 909.
244. Boulton, Gray, and Katritzky, *J. Chem. Soc.*, *B*, **1967,** 911.
245. Boulton, Katritzky, Sewell, and Wallis, *J. Chem. Soc.*, *B*, **1967,** 914.
246. Diehl, Christ, and Mallory, *Helv. Chim. Acta*, **45,** 504 (1962).
247. Boulton, Halls, and Katritzky, *J. Chem. Soc.*, *B*, **1970,** 636.
248. Brown and Bladon, *Spectrochim. Acta*, **24A,** 1869 (1968).
249. Tobiason and Goldstein, *Spectrochim. Acta*, **25A,** 1027 (1969).
250. Poesche, *J. Chem. Soc.*, *B*, **1966,** 568.
251. Elvidge, Newbold, Percival, and Senciall, *J. Chem. Soc.*, **1965,** 5119.
252. Bobowski and Shavel, *J. Org. Chem.*, **32,** 953 (1967).
253. Brownstein, Horswill, and Ingold, *Can. J. Chem.*, **47,** 3243 (1969).
254. Legrand and Lozac'h, *Bull. Soc. Chim. Fr.*, **1967,** 2067.
255. Wilhelm and Schmidt, *J. Heterocycl. Chem.*, **6,** 635 (1969).
256. Zinnes, Comes, Zuleski, Caro, and Shavel, *J. Org. Chem.*, **30,** 2241 (1965).
257. Bourdais, *Compt. Rend.*, *Series C*, **262,** 495 (1966).
258. Prota, Petrillo, Santacroce, and Sica, *J. Heterocycl. Chem.*, **7,** 555 (1970).
259. Kropf and Bernert, *Ann. Chem.*, **743,** 151 (1971).
260. Pfitzner, Marino, and Olafson, *J. Am. Chem. Soc.*, **87,** 4658 (1965).
261. Dewar and Jones, *J. Am. Chem. Soc.*, **90,** 2137 (1968).
262. Yale, *J. Org. Chem.*, **33,** 2382 (1968).
263. Dewar and Rosenberg, *J. Am. Chem. Soc.*, **88,** 358 (1966).
264. Perotti, Dall'sta, and Pedrazzoli, *Bull. Soc. Chim. Fr.*, **1968,** 401.
265. Ried and Junker, *Ann. Chem.*, **696,** 101 (1966).
266. Ponci, Baruffini, and Gialdi, *Farm. Ed. Sci.*, **19,** 515 (1964).
267. Sidhu, Thyagarajan, and Bhalerao, *J. Chem. Soc.*, *C*, **1966,** 969.
268. Rynbrandt, *J. Heterocycl. Chem.*, **7,** 191 (1970).
269. Buchardt, *Tetrahedron Lett.*, **1966,** 6221.
270. Wright, *J. Org. Chem.*, **30,** 3960 (1965).
271. Kaminsky and Lamchen, *J. Chem. Soc.*, *C*, **1966,** 2295.
272. Corey and Felix, *J. Am. Chem. Soc.*, **87,** 2518 (1965).
273. Bonnett, Ho, and Raleigh, *Can. J. Chem.*, **43,** 2717 (1965).
274. Murray and Turner, *J. Chem. Soc.*, *C*, **1966,** 1338.
275. Keener, Skelton, and Snyder, *J. Org. Chem.*, **33,** 1355 (1968).
276. Machiele, Witt, and Snyder, *J. Org. Chem.*, **30,** 1012 (1965).
277. Olsen and Snyder, *J. Org. Chem.*, **30,** 184 (1965).
278. Josey, Tuite, and Snyder, *J. Am. Chem. Soc.*, **82,** 1597 (1960).
279. Carpenter and Snyder, *J. Am. Chem. Soc.*, **82,** 2592 (1960).
280. Gale, Scott, and Snyder, *J. Org. Chem.*, **29,** 2160 (1964).
281. Michel and Snyder, *J. Org. Chem.*, **27,** 2689 (1962).
282. Van Dyke and Snyder, *J. Org. Chem.*, **27,** 3888 (1962).
283. Michel and Snyder, *J. Org. Chem.*, **27,** 2034 (1962).

284. Holmes and Snyder, *J. Org. Chem.*, **29,** 2725 (1964).
285. Huestis, Walsh, and Hahn, *J. Org. Chem.*, **30,** 2763 (1965).
286. Stewart, Koepsell, and West, *J. Am. Chem. Soc.*, **92,** 846 (1970).
287. Holmes and Snyder, *J. Org. Chem.*, **29,** 2155 (1964).
288. Gutowsky and Porte, *J. Chem. Phys.*, **35,** 839 (1961).
289. Feijen and Wynberg, *Rec. Trav. Chim.*, **89,** 639 (1970).
290. Molloy, Reid, and McKenzie, *J. Chem. Soc.*, **1965,** 4368.
291. McKenzie, Molloy, and Reid, *J. Chem. Soc.*, *C*, **1966,** 1908.
292. Hiskey and Dominianni, *J. Org. Chem.*, **30,** 1506 (1965).
293. Mladenovic and Castro, *J. Heterocycl. Chem.*, **5,** 227 (1968).
294. Mertes, Borne, and Briden, *Spectrochim. Acta*, **25A,** 1953 (1969).
295. Sliwa, *Bull. Soc. Chim. Fr.*, **1970,** 631.
296. Crabb and Newton, *J. Heterocycl. Chem.*, **3,** 418 (1966).
297. Bradsher and Zinn, *J. Heterocycl. Chem.*, **1,** 219 (1964).
298. Bradsher and Zinn, *J. Heterocycl. Chem.*, **4,** 66 (1967).
299. Dressler and Joullie, *J. Heterocycl. Chem.*, **7,** 1257 (1970).
300. Klemm, Zell, Barnish, Klemm, Klopfenstein, and McCoy, *J. Heterocycl. Chem.*, **7,** 373 (1970).
301. Klemm, Johnson, and White, *J. Heterocycl. Chem.*, **7,** 463 (1970).
302. Eloy and Deryckere, *Bull. Soc. Chim. Belges*, **79,** 301 (1970).
303. Klemm and Zell, *J. Heterocycl. Chem.*, **5,** 773 (1968).
304. Jones and Jones, *J. Chem. Soc.*, *C*, **1967,** 515.
305. Crabb and Newton, *Tetrahedron*, **24,** 2485 (1968).
306. Robba and Zaluski, *Compt. Rend.*, *Series C*, **263,** 429 (1966).
307. Campaigne and Ellis, *J. Heterocycl. Chem.*, **7,** 43 (1970).
308. Robba, Moreau, and Roques, *Compt. Rend.*, **259,** 3783 (1964).
309. Robba, Lecomte, and Cugnon de Sevricourt, *Tetrahedron*, **27,** 487 (1971).
310. Joly, Divorne, and Roggero, *Compt. Rend.*, *Series C*, **271,** 875 (1970).
311. Andrew and Bradsher, *J. Heterocycl. Chem.*, **4,** 577 (1967).
312. Cook and Desumoni, *Tetrahedron*, **27,** 257 (1971).
313. Harts, de Roos, and Salemink, *Rec. Trav. Chim.*, **89,** 5 (1970).
314. Benedek-Vamos and Promel, *Tetrahedron Lett.*, **1969,** 1011.
315. Trepanier and Krieger, *J. Heterocycl. Chem.*, **7,** 1231 (1970).
316. Niedenzu and Weber, *Z. Naturforsch.*, **21b,** 811 (1966).
317. Gronowitz and Namtvedt, *Acta Chem. Scand.*, **21,** 2151 (1967).
318. Gasco, Rua, Menziani, Nano, and Tappi, *J. Heterocycl. Chem.*, **7,** 131 (1970).
319. Davis and Elvidge, *J. Chem. Soc.*, **1962,** 3553.
320. Beyerman, Maat, van Veen, and Zweistra, *Rec. Trav. Chim.*, **84,** 1367 (1965).
321. Gmünder and Lindenmann, *Helv. Chim. Acta*, **47,** 66 (1964).
322. Korytnyk and Paul, *J. Heterocycl. Chem.*, **2,** 481 (1965).
323. Korytnyk and Ahrens, *J. Heterocycl. Chem.*, **7,** 1013 (1970).
324. Crabb, Chivers, Jones, and Newton, *J. Heterocycl. Chem.*, **7,** 635 (1970).
325. Crabb and Newton, *Tetrahedron*, **24,** 4423 (1968).
326. Potts and Huseby, *J. Org. Chem.*, **31,** 3528 (1966).
327. Todeso and Vivarelli, *Bull. Sci. Fac. Chem. Ind. Bologna*, **20,** 125 (1962).
328. Sliwa, *Bull. Soc. Chim. Fr.*, **1970,** 646.
329. Sliwa, *Bull. Soc. Chim. Fr.*, **1970,** 642.
330. Dunn, McMillan, and Stoodley, *Tetrahedron*, **24,** 2985 (1968).
331. Gronowitz and Bugge, *Acta Chem. Scand.*, **19,** 1271 (1965).
332. Trestianu, Niculescu-Majewska, Bally, Barabás, and Balaban, *Tetrahedron*, **24,** 2499 (1968).

333. Smith and Shoulders, *J. Phys. Chem.*, **69,** 579 (1965).
334. Suzuki, Sugiura, Naito, and Inoue, *Chem. Pharm. Bull.*, **16,** 750 (1968).
335. Boulton, Halls, and Katritzky, *Org. Magnetic Resonance*, **1,** 311 (1969).
336. Brügel, *Org. Magnetic Resonance*, **1,** 425 (1969).
337. Devillers, Navech, and Albrand, *Org. Magnetic Resonance*, **3,** 177 (1971).
338. Albrand, Cogne, Gagnaire, Martin, Robert, and Verrier, *Org. Magnetic Resonance*, **3,** 75 (1971).
339. Crabb and Newton, *Tetrahedron*, **26,** 3941 (1970).
340. Friebolin, Schmid, Kabuss, and Faisst, *Org. Magnetic Resonance*, **1,** 67 (1969).
341. Allingham, Crabb, and Newton, *Org. Magnetic Resonance*, **3,** 37 (1971).
342. Kamiya, *Chem. Pharm. Bull.*, **17,** 1815 (1969).
343. Kamiya, Katayama, and Akahori, *Chem. Pharm. Bull.*, **17,** 1821 (1969).
344. Katritzky and Takeuchi, *Org. Magnetic Resonance*, **2,** 569 (1970).
345. Dewar, Gleicher, and Robinson, *J. Am. Chem. Soc.*, **86,** 5698 (1964).
346. Haake, McNeal, and Goldsmith, *J. Am. Chem. Soc.*, **90,** 715 (1968).
347. Ramirez, Patwardhan, and Smith, *J. Org. Chem.*, **30,** 2575 (1965).
348. Ramirez, Pilot, Madan, and Smith, *J. Am. Chem. Soc.*, **90,** 1275 (1968).
349. Ramirez, Kugler, Patwardhan, and Smith, *J. Org. Chem.*, **33,** 1185 (1968).
350. Ramirez, Patwardhan, Ramanathan, Desai, Greco, and Heller, *J. Am. Chem. Soc.*, **87,** 543 (1965).
351. Ramirez, Patwardhan, Desai, and Heller, *J. Am. Chem. Soc.*, **87,** 549 (1965).
352. Ramirez, Madan, and Smith, *J. Am. Chem. Soc.*, **87,** 670 (1965).
353. Ramirez, Madan, and Heller, *J. Am. Chem. Soc.*, **87,** 731 (1965).
354. Ramirez, Gulati, and Smith, *J. Org. Chem.*, **33,** 13 (1968).
355. Ramirez, Gulati, and Smith, *J. Am. Chem. Soc.*, **89,** 6283 (1967).
356. Sterk and Ziegler, *Monatsh. Chem.*, **100,** 739 (1969).
357. Bellasio, Pagani, and Testa, *Gazz. Chim. Ital.*, **94,** 639 (1964).
358. Jerina, Boyd, Paolillo, and Becker, *Tetrahedron Lett.*, **1970,** 1483.
359. Kintzinger and Lehn, *Chem. Commun.*, **1967,** 660.
360. Kintzinger and Lehn, *Mol. Phys.*, **14,** 133 (1968).
361. Poesche, *J. Chem. Soc., B*, **1971,** 368.
362. Klemm, Klopfenstein, Zell, McCoy, and Klemm, *J. Org. Chem.*, **34,** 347 (1969).
363. Robba, Roques, and le Guen, *Bull. Soc. Chim. Fr.*, **1967,** 4220.
364. Bradsher and Bolick, *J. Org. Chem.*, **32,** 2409 (1967).
365. Inouye and Otsuka, *J. Org. Chem.* **26,** 2613 (1961).
366. Bonsall and Hill, *J. Chem. Soc., C*, **1967,** 1836.
367. Robba, Zaluski, and Roques, *Compt. Rend., Series C*, **263,** 814 (1966).
368. Robba and Zaluski, *Bull. Soc. Chim. Fr.*, **1968,** 4959.
369. Davis, Dewar, and Jones, *J. Am. Chem. Soc.*, **90,** 706 (1968).
370. Givens and Hamilton, *J. Org. Chem.*, **32,** 2857 (1967).
371. Crabb and Newton, *Tetrahedron*, **24,** 1997 (1968).
372. Wood and Srivastava, *Tetrahedron Lett.*, **1971,** 2937.
373. Lett and Marquet, *Tetrahedron Lett.*, **1971,** 2851.
374. Lett and Marquet, *Tetrahedron Lett.*, **1971,** 2855.
375. Yee and Bentrude, *Tetrahedron Lett.*, **1971,** 2775.
376. Meyers and Takaya, *Tetrahedron Lett.*, **1971,** 2609.
377. Robba, Lecomte, and de Sevricourt, *Tetrahedron*, **27,** 487 (1971).
378. Cook and Desimoni, *Tetrahedron*, **27,** 257 (1971), and references cited therein.
379. Angiolini, Jones, and Katritzky, *Tetrahedron Lett.*, **1971,** 2209.
380. Ohno, Kito, and Koizumi, *Tetrahedron Lett.*, **1971,** 2421.
381. Bentrude, Yee, Bertrand, and Grant, *J. Am. Chem. Soc.*, **93,** 797 (1971).

382. Haemers, Ottinger, Reisse, and Zimmermann, *Tetrahedron Lett.*, **1971,** 461.
383. Woerden and de Vries-Miedema, *Tetrahedron Lett.*, **1971,** 1687.
384. Kjellin and Sandström, *Spectrochim. Acta*, **25A,** 1865 (1969).
385. Huisgen and Christl, *Angew. Chem., Int. Ed. Engl.*, **6,** 456 (1967).
386. Wulff and Huisgen, *Angew. Chem., Int. Ed. Engl.*, **6,** 457 (1967).
387. Harris and Pyper, *Mol. Phys.*, **20,** 467 (1971).
388. Kumar, Krishna, and Rao, *Mol. Phys.*, **18,** 11 (1970).
389. Abraham, Parry, and Thomas, *J. Chem. Soc., B*, **1971,** 446.
390. Richter and Ulrich, *Ann. Chem.*, **743,** 10 (1971).
391. Cook, Katritzky, and Manas, *J. Chem. Soc., B*, **1971,** 1330.
392. Bodkin and Simpson, *J. Chem. Soc., B*, **1971,** 1136.

9 "SHIFT" REAGENTS AND HETEROCYCLES

Spectra of metal complexes containing heterocyclic ligands are considered to be outside the scope of this book. However, the recent use of paramagnetic metal complexes to produce highly specific variations in chemical shifts warrants inclusion. For basic references on transition-metal complexes and paramagnetic shifts the reader should consult the recent reviews by Schwarzhaus (1) and De Boer and Willigen (2). Brief descriptions of the "shift" phenomena are given below to allow the nonspecialist reader to follow arguments presented in current papers in this field.

The observation of an NMR spectrum for a paramagnetic species depends on the electronic relaxation time T_l, the electronic exchange time T_e, or a combination of both of these. An isotropic hyperfine contact interaction between unpaired electrons and the magnetic nucleus will lead to a large shift in the resonance frequency of this nucleus. If the paramagnetic system is characterized by times such that T_l^{-1} or $T_e^{-1} \gg A_n$, where A_n is the electron spin–nuclear spin coupling constant, then the isotropic nuclear resonance shifts can be observed. For $CDCl_3$ solutions, only isotropic nuclear resonance shifts are important, and these are usually divided into two types, contact shifts and pseudocontact shifts.

The contact interaction results from coupling between the electron spin density at the resonating nucleus and the effective electronic spin magnetization of the unpaired electron in the magnetic field of the NMR experiment to produce a shift of the nuclear resonance frequency from its normal diamagnetic value. This effect is given quantitatively for a proton by the equation

$$\Delta\nu = -A_i \frac{\gamma e}{\gamma H} \frac{S(S+1)}{3kT} g\beta\nu \tag{9.1}$$

where A_i is the electron–proton hyperfine interaction constant of the ith proton, γe and γH are the electronic and nuclear gyromagnetic ratios, $\Delta\nu$ is the isotropic contact shift for the paramagnetic species, and ν is the frequency of resonance (3).

For axially symmetric systems in which the electronic spin relaxation time is short compared with the molecular correlation time for tumbling in solution, a g-tensor anisotropy gives rise to an isotropic shift called the pseudocontact shift and described by the equation

$$\Delta\nu_i = -\nu \frac{\beta^2 S(S-1)}{45kT}\left[\frac{3\cos^2\chi - 1}{r_i^3}\right](3g\|^2 + g\,\|\,g\perp - 4g\perp^2) \quad (9.2)$$

where r_i is the length of a vector joining the ith proton and the metal atom and χ is the angle between this vector and the principal molecular axis (4). The pseudocontact shift is therefore dependent on the geometry of the complex as well as on its anisotropy.

It is often difficult in practice to decide which of these two shielding mechanisms is operating or is predominant in a given complex. The isotropic shifts found for many nickel complexes have been successfully explained when the assumption was made that they arise solely from contact interactions. In cobalt complexes, however, both mechanisms appear to be operating.

The effects of complexing to paramagnetic transition-metal ions have been little used in the study of spectra of heterocycles. However, in an elegant piece of work, the assignment of peaks to H-3 and H-5 of 1-methylpyrazole (1) was made with the aid of contact shifts and relaxation phenomena when the heterocycle was complexed to nickel (see Chapter 3, Section II.A).

The use of lanthanide β-diketone complexes ("shift reagents") to induce shifts in the spectra of many classes of compound is a recent development of great potential. In 1969, Hinckley (5) found that useful paramagnetic shifts could be produced in the spectrum of cholesterol by the dipyridyl adduct of trisdipivoalylmethanato-europium(III), Eu(DPM)$_3$, without significant peak broadening. Then, in 1970, Sanders and Williams (6) reported that larger shifts were obtained if the pyridine was omitted from the shift reagent. Since this original work, many lanthanide complexes have been tested as shift reagents but Eu(DPM)$_3$, the fluorine-containing complex Eu(fod)$_3$ (where fod is 1,1,1,2,2,3,3-heptafluoro-7,7-dimethyl-4,6-octandione), and the corresponding praeseodymium analogs seem to be the only ones suitable for routine use.

The paramagnetic ion of choice appears to be Eu^{3+} because of its anomalously inefficient nuclear spin-lattice relaxation properties. This inefficiency leads to much less broadening of the shifted signals than is observed with other lanthanide ions. Also, Eu^{3+} and the other lanthanide ions induce shifts largely, if not exclusively, by a pseudocontact mechanism. Thus, Eq. (9.2) may be applied to these interactions and the relative shifts will be dependent only on the geometry of the "shift complexes" as described by Hinckley's approximation

$$\Delta\nu_i = K\frac{(3\cos^2\chi - 1)}{r_i^3} \tag{9.3}$$

In a number of cases the observed shifts could be explained satisfactorily if the angle χ in Eq. (9.3) was assumed to be "invariant" (e.g., see Refs. 7 and 8). If this rather surprising assumption is correct, Eq. (9.3) becomes

$$\Delta\nu_i = K'(1/r_i^3) \tag{9.4}$$

More recently (9), *upfield* as well as *downfield* shifts were observed for signals from different protons in the same molecule in the presence of $Eu(DPM)_3$, and the conclusion was reached that the angular dependence term in Eq. (9.3) is important, at least in some cases.

Work with $Eu(DPM)_3$ (7) and $Yb(DPM)_3$ (8) showed that π-deficient nitrogen heterocycles such as pyridine and quinoline form useful complexes with these shift reagents, but that the π-excessive pyrrole and indoles complexed very weakly, if at all. The reduced systems pyrrolidine, piperidine, and quinuclidine behaved in a similar manner to aliphatic amines and formed strong complexes as indicated by large observed shifts. Furan was little affected but the ring protons of tetrahydrofuran were strongly deshielded.

At the time of writing, the use of these shift reagents with heterocyclic systems has been reported almost exclusively in preliminary communications or short notes. However, one systematic study of the shifts induced by $Eu(DPM)_3$ on the signals from the ring protons of π-deficient heterocycles has been carried out (10). Shifts observed for the protons of pyridine were such that the angular dependence term from Eq. (9.3), $3\cos^2\chi$, had to be taken into account before a sensible Eu–N distance could be calculated (2.9 Å). It was found necessary to include this term in calculations of the geometry of all of the symmetrical complexes encountered in this study Thus, for complexes between $Eu(DPM)_3$ and 1,8-naphyridine or 1,10, phenanthroline, the europium atom must, on a time-averaged basis, be located equidistant from the two nitrogen atoms. Such a position involves large values for the angle χ for many protons and hence small values for $\Delta\nu$. One very interesting point emerged from this work. Complex formation appeared to be more rapid and observed shifts were larger for pyrimidine than for pyridine. Pyrimidine is a much weaker base than pyridine, so the basicity of these ligands (as measured by their pK_a values) is not a satisfactory criterion for their ability to combine with the shift reagent. The results obtained in this study are summarized in Table 9.1.

TABLE 9.1 Eu(DPM)$_3$-INDUCED DOWNFIELD SHIFTS (Δ PPM) IN THE PMR SPECTRA OF π-DEFICIENT HETEROCYCLES IN CCl_4. CONCENTRATION— 4×10^{-4} *M* + 0.3 *M* EQUIV OF Eu(DPM)$_3$ (10)

Compound	Solvent	Δ1	Δ2	Δ3	Δ4	Δ5	Δ6	Δ7	Δ8	Δ9	Δ10
Pyridine	CCl_4		9.1	3.2	2.9						
	$CDCl_3$		7.9	2.7	2.4						
2-Me	$CDCl_3$		5.2	2.1	2.1	1.7	4.6				
3-Me	$CDCl_3$		7.5	1.5	2.3	2.6	7.4				
2,3-Di-Me	$CDCl_3$		4.9	0.8	1.6	2.2	4.5				
2,5-Di-Me	$CDCl_3$		4.0	1.6	0.8	1.2	4.0				
Pyridazine	CCl_4			4.4	2.4						
Pyrimidine	CCl_4		12.2		7.6	4.4					
Quinoline	CCl_4		6.9	2.2	2.2	1.4	1.1	1.0	5.8		
	$CDCl_3$		6.1	1.9	1.9	1.2	0.9	0.9	5.1		
Isoquinoline	CCl_4	8.6		9.6	3.3	2.1	1.1	1.1	2.1		
	$CDCl_3$	7.3		7.8	2.8	1.7	0.9	0.9	1.7		
Naphthyridines											
1,6-	CCl_4		3.2	2.0	3.0	10.5		10.8	5.0		
1,7-	CCl_4		2.2	1.6	2.5	3.6	10.5		9.1		
	$CDCl_3$		1.6	1.0	1.7	2.3	6.9		6.7		
1,8-	$CDCl_3$		0.5	1.5	2.0						
Cinnoline	CCl_4			4.6	2.2	1.4	0.8	0.8	1.8		
Phthalazine	$CDCl_3$	3.7				1.4	0.9				
Quinazoline	CCl_4		12.9		11.1	3.2	1.6	1.6	3.2		
	$CDCl_3$		9.5		8.3	2.4	1.2	1.2	2.4		
Triazanaphthalenes											
1,4,5-	$CDCl_3$		1.5	1.0			1.0	1.4	2.1		
1,4,6-	$CDCl_3$		2.2	2.8		8.7		9.0	4.0		
Pteridine	$CDCl_3$		9.2		8.6		2.5	1.8			
Acridine	CCl_4	1.1	0.5	0.3	4.7						
Phenanthridine	CCl_4	1.7	1.0	[a]	6.4		6.0	1.2	[a]	[a]	1.7
Phenanthroline	$CDCl_3$		0.2	0.4	0.1	0.1	0.1	0.1	0.4	0.2	

[a] Peaks from these three protons form a broad envelope.

REFERENCES

1. Schwarzhaus, *Angew. Chem., Int. Ed.*, **9,** 946 (1970).
2. De Boer and van Willigen, *Progr. Nucl. Magnetic Resonance Spectrosc.* **2,** 111 (1967).
3. McConnell and Chesnut, *J. Chem. Phys.*, **28,** 107 (1958); McConnell and Holm, *ibid.*, **27,** 314 (1957).
4. McConnell and Robertson, *J. Chem. Phys.*, **29,** 1361 (1958).
5. Hinckley, *J. Am. Chem. Soc.*, **91,** 5160 (1969).
6. Sanders and Williams, *Chem. Commun.*, **1970,** 422.
7. Sanders and Williams, *J. Am. Chem. Soc.*, **93,** 641 (1971).
8. Beaute, Wolkowski, and Thoai, *Tetrahedron Lett.*, **1971,** 817.
9. Mazzocchi, Tamburin, and Miller, *Tetrahedron Lett.*, **1971,** 1819.
10. Armarego, Batterham, and Kershaw, *Org. Magnetic Resonance*, **3,** 575 (1971).

10 CORRELATIONS AND ADDITIVITY RULES

The information presented in this chapter is intended to bring together isolated data for simple "aromatic" systems from many places within this book and to suggest trends which might be of use to the chemist in his day-to-day work. Assignments in spectra where strictly comparable data are not available can sometimes be made successfully by the application of general principles which to date have not been formulated on a broad basis. The following discussion contains no references but is aimed at correlation of the large amount of data found in the tables in previous chapters. Emphasis is given to the limitations of each "rule" and to the pitfalls to be avoided in its routine use.

I. THE AZINE SYSTEMS IN NONPOLAR SOLVENTS

It was stated in the Introduction that chemical shifts, as reported by most chemists, were subject to at least a ± 0.1 ppm uncertainty, and for this reason the differential shifts discussed in this section are confined mainly to those above 0.2 ppm. Throughout the rest of this chapter these are referred to as "large" shifts. Figure 10.1 shows the chemical shifts for a number of mono- and bicyclic azines in CCl_4 or $CDCl_3$.

Introduction of one nitrogen atom into the benzene ring produces characteristic shifts of the α-, β-, and γ-protons (see Chapter 2). Thus, the α-protons which are adjacent to the local anisotropic influences of the nitrogen atom and its associated lone pair are strongly deshielded. Protons in the γ-position are deshielded by about 0.3 ppm while the β-proton actually is shifted slightly upfield, leading to a difference between the resonance positions of these two protons of about 0.4 ppm. This difference is also evident in the spectra of quinoline and isoquinoline where H-4 of quinoline (γ to N-1) absorbs about 0.5 ppm, downfield from H-4 of isoquinoline (β to N-2). This behavior is found in all bicyclic systems containing a quinoline or an isoquinoline-type ring.

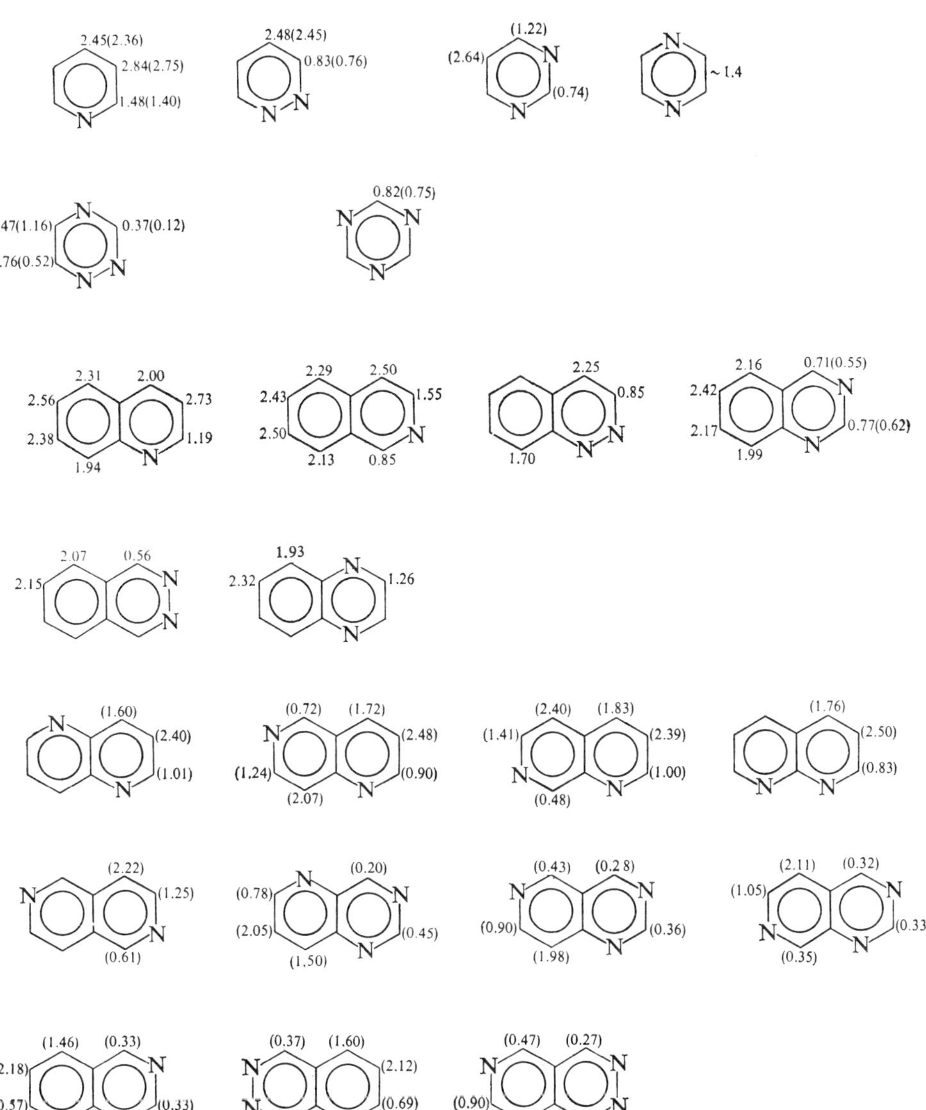

Fig. 10.1 Chemical shifts (τ) of six-membered nitrogen heterocycles. Solvents—CCl_4 ($CDCl_3$).

The introduction of a second nitrogen atom β or γ to the first produces a strong deshielding of protons α to it but has little influence on more distant protons. It is important to note that when the second nitrogen atom is β to the first that the deshielding of the proton is much less than expected, and certainly any additivity rule does not hold in this case. However, the reduction of deshielding is predictable and an allowance can be made for it. The lack of effect at the β- and γ-positions is even more noticeable for the addition of a third nitrogen atom, for example, H-2 of quinazoline and *s*-triazine absorb at essentially the same position. Many more examples of this phenomenon can be found by examination of Fig. 10.1.

When a nitrogen atom is introduced into a position α to another nitrogen atom, as in pyridazine, anomalous shifts are observed. The presence of the two adjacent nitrogen atoms causes an extra deshielding of about 0.6 ppm of protons α to each nitrogen atom, over and above that expected for the local anisotropic effect of the additional nitrogen atom. It is difficult to say why this added downfield shift occurs but, because the anomalous effects are restricted to the α-protons, it can be assumed that they are due to variations induced mutually by the adjacent nitrogen atoms in either the nitrogen anisotropy, the nature of the nitrogen lone pair, or both. The relatively high basicity of pyridazine (pK_a 2.24) compared with other diazines (pyrimidine, pK_a 1.23; pyrazine, pK_a 0.65) suggests that the lone pairs of pyridazine are more readily accessible to protons than those of pyrimidine or pyrazine, and provides support for modification of these lone pairs. Also, the methyl group of 2-methylpyridazine absorbs at lower field than that of α-picoline (by about 0.2 ppm) and H-8 of cinnoline absorbs about 0.2 ppm downfield from H-8 of quinoline. These observations again support the hypothesis that considerable modification of the lone pairs has occurred, in such a way as to increase their general deshielding effect on adjacent protons.

The introduction of a nitrogen atom into one ring of a naphthalene system also affects the chemical shifts of ring protons in the other ring. A nitrogen atom in an α-position causes a downfield shift of the peri proton of about 0.2 ppm, and this can be attributed directly to "through-space" effects of the lone pair on the nitrogen atom. Shifts of other protons can be explained in terms of favorable resonance contributions from various canonical forms. Typical shifts of the protons of an azine or an azanaphthalene on insertion of another ring nitrogen atom are collected in Table 10.1, and since these are additive they can be used, in most cases, to calculate the expected chemical shifts of protons in polyaza systems to within 0.1 ppm of the observed values. The one case where this approach gives a considerable error is in the calculation of the shifts of H-1 and H-4 of phthalazine by insertion of a nitrogen atom into isoquinoline. However, insertions of other nitrogen atoms into the phthalazine ring system obey the additivity rules set out in Table 10.1.

TABLE 10.1 ADDITIVITY RULES FOR REPLACEMENT OF CARBON BY NITROGEN IN AZANAPHTHALENE SYSTEMS IN NONPOLAR SOLVENTS. ALL INSERTION SHIFTS IN PPM ±0.1 ppm

A. ONE NITROGEN ATOM PER RING

I. INSERTION OF N AT POSITION 1

Shifts of	H-2	H-3	H-4	H-5	H-6	H-7	H-8
	−1.5	+ Small	−0.5	−0.1	−0.1	−0.2	−0.2

II. INSERTION OF N AT POSITION 2

Shifts of	H-1	H-3	H-4	H-5	H-6	H-7	H-8
	−1.6	−1.3	+0.2	−0.2	−0.35	−0.3	−0.3

B. TWO OR MORE NITROGEN ATOMS PER RING[a]

I. INSERTION OF N INTO QUINOLINE-TYPE SYSTEMS—FIRST N AT POSITION 1

	N into 2		N into 3		N into 4	
Shifts of	H-3	H-4	H-2	H-4	H-2	H-3
	−1.0	+0.2	−0.5	−1.4	+0.1	−1.5

II. INSERTION OF N INTO ISOQUINOLINE-TYPE SYSTEMS—FIRST N AT POSITION 2

	N into 1		N into 3		N into 4	
Shifts of	H-3	H-4	H-1	H-4	H-1	H-3
	−0.7	−0.25	−0.3	−1.9	−0.1	−0.8

[a] Note "double proximity effects" where two nitrogen atoms are α or β to each other. Example: Pteridine—calculated from pyrazine and pyrimidine by insertion of nitrogen atoms in the benzenoid ring:

	$\tau 2$	$\tau 4$	$\tau 6$	$\tau 7$
Calculated	0.25	0.33	0.80	0.65
Observed	0.20	0.35	0.85	0.67

These shifts can also be used successfully with simple substituted derivatives of the "aromatic" ring systems.

Little can be said about the values of coupling constants observed for azines. Coupling constants between protons β and γ to the ring nitrogen atom, for example, J_{34} of pyridine, have values typical of a benzene ring (7.6 Hz). Those between protons α and β to the ring nitrogen, for example, J_{23} of pyridine, J_{34} of pyridazine, J_{45} of pyrimidine, have values about 2.5 Hz less than the β–γ coupling (i.e., ~5.0 Hz), and those α,β to two ring nitrogen atoms, as in pyrazine and 1,2,4-triazine, are reduced by a further increment of 2.5 Hz (to ~2.6 Hz). Protonation effects on coupling constants are not always consistent and must be used with great caution if details of molecular structure are to be derived from them. Additivity relationships for direct ^{13}C–H coupling constants in polyaza systems are discussed in Chapter 2.

II. SUBSTITUENT EFFECTS

Substituent effects on chemical shifts within a series of compounds have often been used as an aid in structural determinations and in assigning peaks in the spectra to specific protons. Unfortunately, the method has sometimes led to erroneous results, and from the discussion below this is not unexpected. In this section shifts induced by most common substituents for solutions in CCl_4, $CDCl_3$, or DMSO are discussed. It must be emphasized that use of polar solvents such as DMSO causes considerable variation in some of the substituent shifts found for less polar solvents, but because the only comparative study of substituted pyridines used DMSO as solvent, most shifts for the pyridine system will involve this solvent. In many instances additivity rules have been proposed but a consideration of the data listed in this book shows that these rules must be regarded to hold only for closely similar compounds with essentially the same substitution patterns.

A. The Methyl Group

The methyl group is probably the most common substituent encountered in studies of the NMR spectra of "aromatic" heterocyclic systems. Its effects on the various systems have been discussed many times and methyl derivatives have often been used as models in attempts to unravel the spectra of parent heterocyclic systems. Substituent shifts for a number of methyl-substituted heterocycles in CCl_4 or $CDCl_3$ are given in Fig. 10.2.

From Fig. 10.2 it can be seen that the introduction of a methyl group into pyridine causes an upfield shift of all the remaining ring protons, with those ortho or para to the methyl group moving about 0.23 ppm, and those meta to it about 0.18 ppm. Unfortunately, these shifts cannot be applied to the

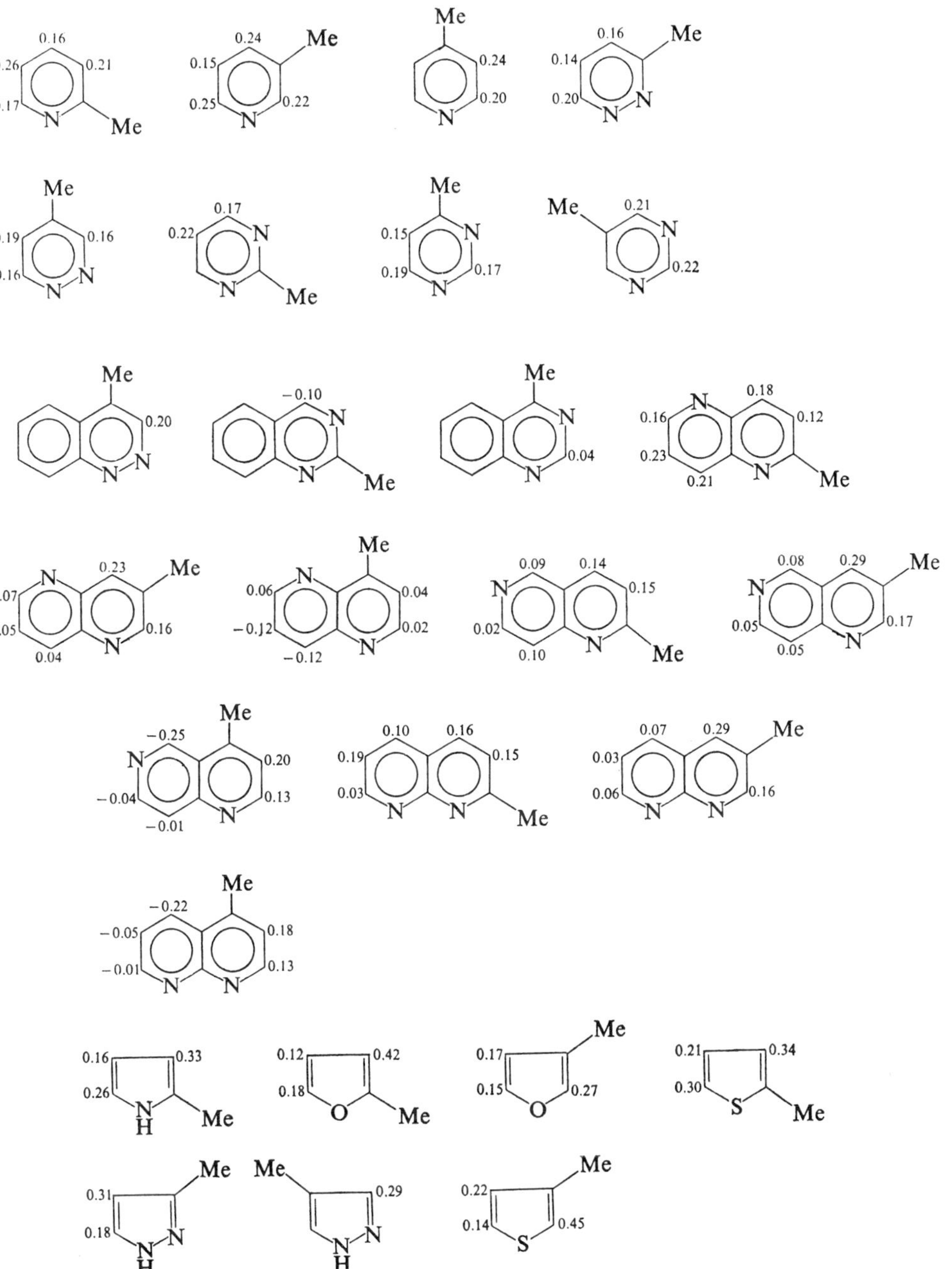

Fig. 10.2 Substituent shifts in ppm to high (+) or low (−) field for the methyl group.

pyridazine or pyrimidine systems where the total effect is usually less and where ortho protons are usually not shielded as much as would be expected. It would be most difficult to use methyl derivatives as models in these systems. In the azanaphthalene series, the situation is complicated by the influence of the second aromatic ring, and in some cases the methyl substituent even causes downfield shifts of the ring protons. A 4-methyl group in cinnoline deshields H-3 by roughly the same amount as in 4-methylpyridazine, but 2- or 4-methyl substituents in quinazoline exert considerably different effects than those observed for the corresponding pyrimidines. Within the 1,5-, 1,6-, and 1,8-naphthyridines a number of trends are apparent. The most noticeable of these is the fact that a 4-methyl substituent induces mainly very small or negative shifts in the protons of the other ring. This is an unusual effect not found in any other bicyclic systems known to the author. In many cases the shifts in these systems are consistent enough for the methyl derivatives to be used as satisfactory model compounds.

In the five-membered ring systems, pyrrole and pyrazole, shifts induced by *C*-methyl substitution are much more consistent. Protons on carbon atoms adjacent to the substitution site are shielded by about 0.3 ppm while those meta to the methyl group move only about 0.17 ppm. Methyl substitution in the furan system produces results which depend very much on the α or β orientation of the substituent while those for 2-methylthiophen are similar but slightly larger than similar shifts in 2-methylpyrrole.

It is interesting to note that shifts observed for the introduction of a methyl group into the pyridine system are much smaller when DMSO is used as solvent than those discussed above for CCl_4 or $CDCl_3$ solutions, and the random nature of the reported results is probably a direct consequence of their small size and hence is of no significance.

B. The Cyano Group

Acceptable data on shifts induced by this group are only available for pyridines in DMSO and for thiophens and 2-cyanofuran in CCl_4. The electron-withdrawing nature of the group is obvious from the general downfield shift of all protons with the largest effect at the position α to the substituent, 0.5–0.9 ppm for the compounds above, with lower shifts, 0.0–0.6 ppm at more distant protons. Although values obtained for the cyanopyridines, the cyanothiophens, and 2-cyanofuran all fit into these ranges, the difference in solvents suggests that this may be more a coincidence than a correlation. The observed shifts are given in Fig. 10.3.

C. Carbonyl-Containing Groups

The aldehyde, acetyl, methoxycarbonyl, carboxy, and acetamido groups all act in a very similar manner and for this discussion may be grouped together.

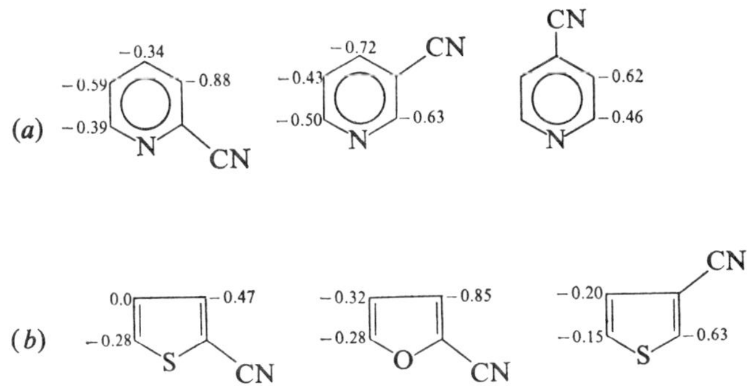

(a) [2-cyanopyridine: −0.34, −0.59, −0.88, −0.39; 3-cyanopyridine: −0.72, −0.43, −0.50, −0.63; 4-cyanopyridine: −0.62, −0.46]

(b) [2-cyanothiophen: 0.0, −0.47, −0.28; 2-cyanofuran: −0.32, −0.85, −0.28; 3-cyanothiophen: −0.20, −0.15, −0.63]

Fig. 10.3 Substituent shifts in ppm to high (+) or low (−) field for the cyano group. Solvents—(*a*) DMSO; (*b*) CCl_4.

The observed shifts are shown diagrammatically in Fig. 10.4. Again, the general electron-withdrawing nature of these groups causes deshielding of all ring protons. The largest effect occurs at positions adjacent to the substituent and can be ascribed to the anisotropy of the carbonyl bond. When the substituent is in position 2 of these ring systems, whether they be pyridines, pyrroles, furans, or thiophens, the effect on the adjacent proton (H-3) is greater than for substitution in other positions, and this is caused by hindered rotation about the N–C–C=O bond, which is known to favor the cis N–O conformation. Where the substituent is β or γ to the heteroatom, rotamer populations are more equal and the two adjacent protons are less deshielded. It is obvious from Fig. 10.4 that satisfactory additivity rules could be formulated for these substituents with greatest consistency obtainable for the shifts of protons adjacent to the carbonyl group. In the use of such rules care must be taken to check that other substituents on the ring do not interact sterically with the carbonyl substituent and upset the rotational isomerism of the model.

D. The Amino Group

The only simple six-membered amino compound for which comparative data are available is 3-aminopyridine in DMSO. Electron donation by the amino group shows up as a general upfield shift of all ring protons, with most effect in the mesomerically activated 4- and 6-positions. Similar effects are observed for 2- and 3-aminothiophens.

E. The Nitro Group

Spectra of a few nitro compounds are included in tables in earlier chapters and enough data are available for trends to be established (see Fig. 10.5).

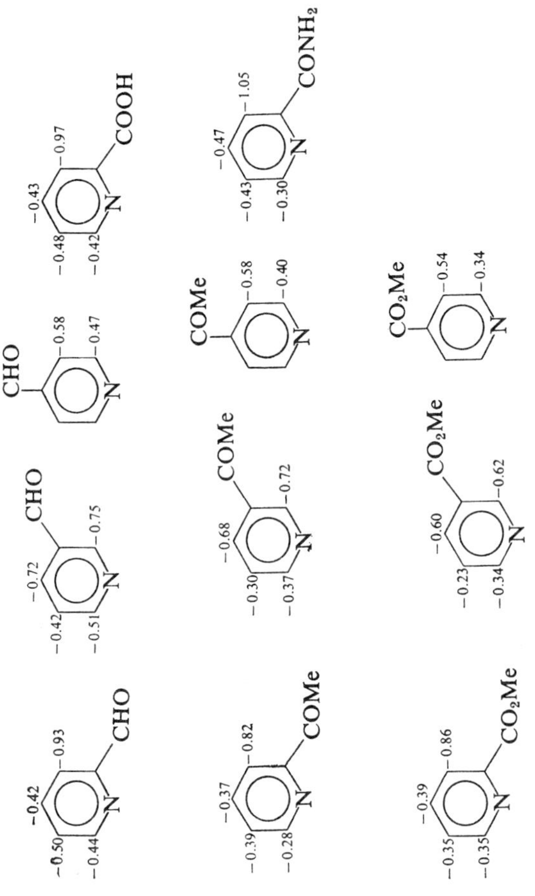

−0.42
−0.50 | 0.93
−0.44 | N | CHO

−0.72 | CHO
−0.42
−0.51 | N | −0.75

CHO
−0.58
N | −0.47

−0.43
−0.48 | −0.97
−0.42 | N | COOH

−0.37
−0.39 | −0.82
−0.28 | N | COMe

−0.68 | COMe
−0.30
−0.37 | N | −0.72

COMe
−0.58
N | −0.40

−0.47
−0.43 | −1.05
−0.30 | N | $CONH_2$

−0.39
−0.35 | −0.86
−0.35 | N | CO_2Me

−0.60 | CO_2Me
−0.23
−0.34 | N | −0.62

CO_2Me
−0.54
N | −0.34

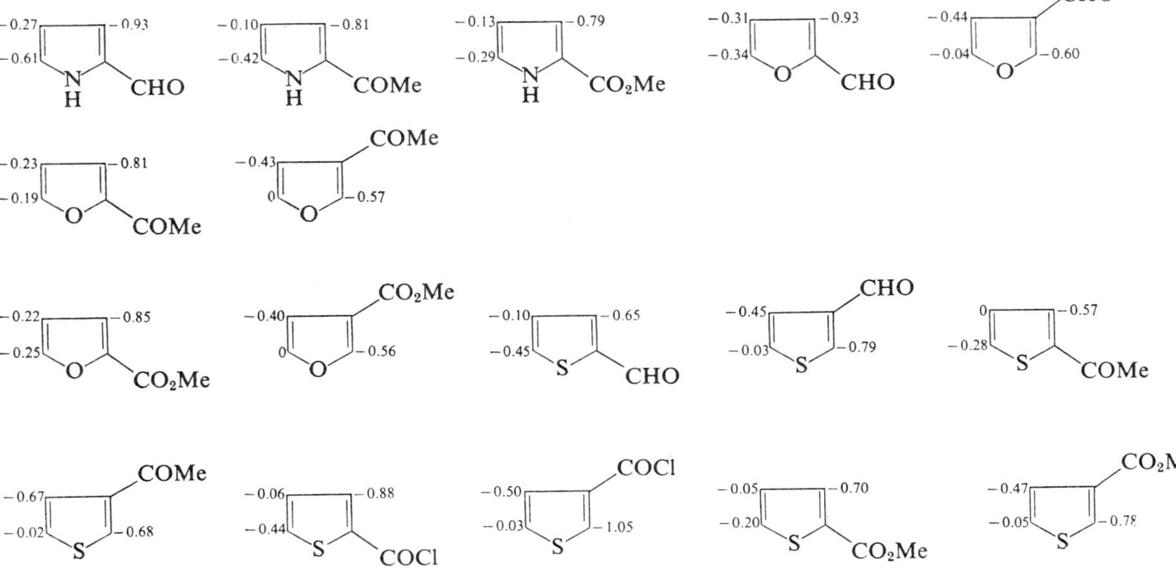

−0.27 −0.93 −0.61 N H CHO
−0.10 −0.81 −0.42 N H COMe
−0.13 −0.79 −0.29 N H CO₂Me
−0.31 −0.93 −0.34 O CHO
−0.44 CHO −0.04 O −0.60
−0.23 −0.81 −0.19 O COMe
−0.43 COMe 0 O −0.57
−0.22 −0.85 −0.25 O CO₂Me
−0.40 CO₂Me 0 O −0.56
−0.10 −0.65 −0.45 S CHO
−0.45 CHO −0.03 S −0.79
0 −0.57 −0.28 S COMe
−0.67 COMe −0.02 S −0.68
−0.06 −0.88 −0.44 S COCl
−0.50 COCl −0.03 S −1.05
−0.05 −0.70 −0.20 S CO₂Me
−0.47 CO₂Me −0.05 S

Fig. 10.4

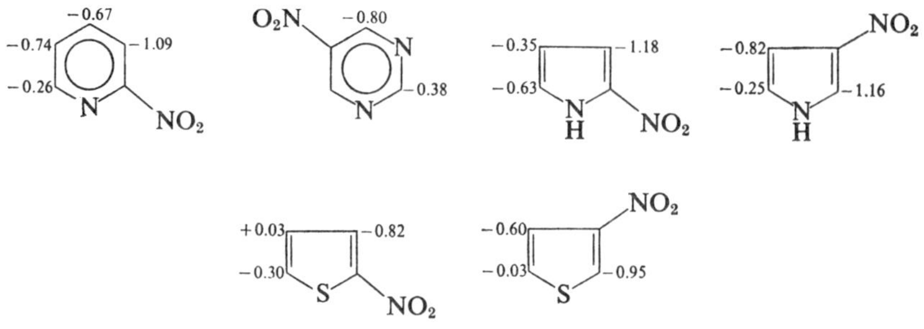

Fig 10.5 Substituent shifts in ppm to high (+) or low (−) field for the nitro group. Solvents—various.

The nitro group is probably the strongest electron-withdrawing group for which substituent shifts are available, and this reflects in the very strong deshielding of all protons. Signals from protons adjacent to the substituent move downfield by about 1 ppm because of the combined effect of reduced electron density at the carbon atom to which the proton is attached and the "through-space" anisotropic deshielding of the nitro group.

F. The Hydroxyl and Methoxyl Groups

The only comparative data available for compounds with these substituents are for a number of pyridines, pyridazines, and pyrimidines. Substituent effects, summarized in Fig. 10.6, are almost entirely confined to protons ortho or para to the substituent, in agreement with the mesomeric electron-donating influence of these two groups. If the values for 4-methoxypyridine are neglected, a shift of an ortho proton by about 0.5 ppm and of a para proton by about 0.4 ppm by a methoxyl substituent can be postulated.

(*a*)

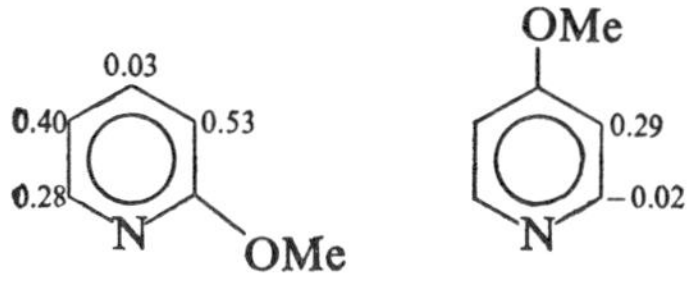

(*b*)

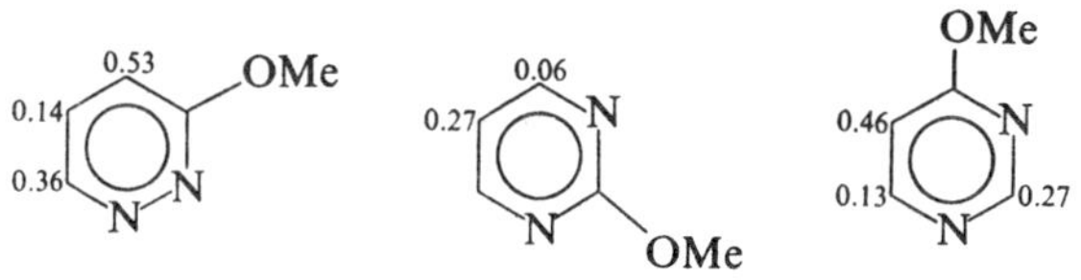

Fig. 10.6 Substituent shifts in ppm to high (+) or low (−) field for the methoxyl group. Solvents—(*a*) DMSO; (*b*) $CDCl_3$.

However, 4-methoxypyrimidine poses a problem because the substituent is both α and γ to nitrogen atoms and could be looked upon as an analog of 4-methoxypyridine, for which the rule does not hold. Thus, caution is needed in applying shielding corrections for this substituent.

G. Halogen Groups

The discussion of halogen substituent shifts does not include information on the fluoro group, the NMR properties of which are covered in the various sections dealing with fluorine substitution in specific heterocyclic systems. Also, data for iodo compounds are extremely scarce so this group is omitted as well. However, spectra of chloro- and bromosubstituted compounds are plentiful and a number of generalizations can be made from the information collected in Fig. 10.7. It has often been said that substitution by chlorine

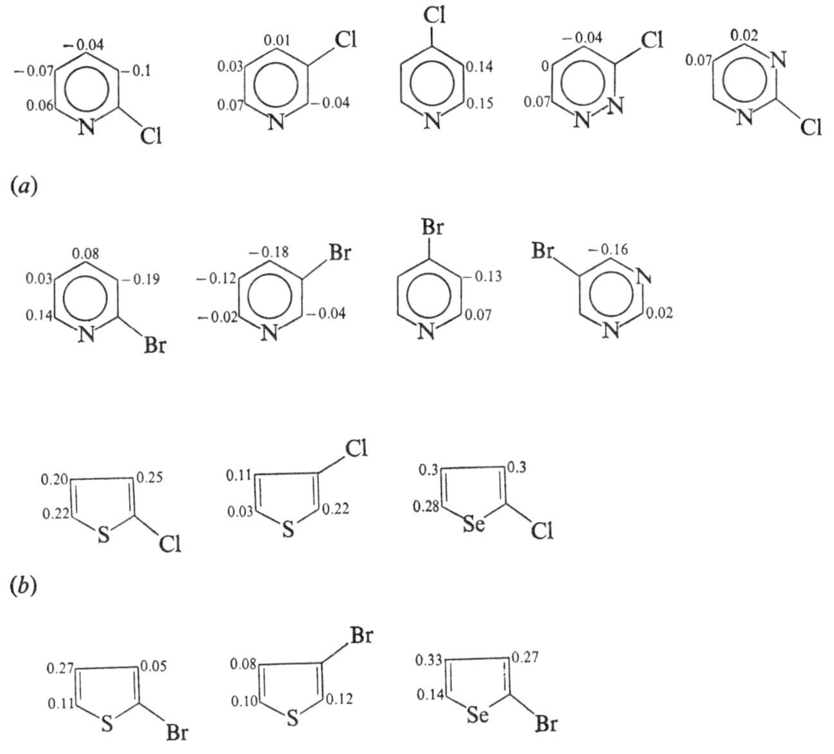

Fig. 10.7 Substituent shifts in ppm to high (+) or low (−) field for halogen substituents. Solvents—(*a*) DMSO; (*b*) CCl_4.

causes very small shifts in the remaining ring protons and hence that chloro derivatives are good models for the respective unsubstituted compounds. For nitrogen heterocycles and for furans this is essentially true, but for thiophens and selenophen moderate upfield shifts are observed for all of the remaining ring protons. A similar pattern emerges for bromo substitution, although the effects for π-deficient heterocycles are a little larger than those observed for chloro substitution. Halogen shifts have been shown in some cases to be additive, but in others they definitely are not, and care must be exercised in the use of such rules.

III. SOLVENT EFFECTS

Solvent effects are discussed at length in sections dealing with specific ring systems, and particularly for pyridines in Chapter 2. However, a number of general statements can be made which may be useful in the assignment of peaks in the spectrum of an unknown compound.

The spectra of neat liquids change markedly as the molecular aggregates are modified by the addition of a solvent, and extrapolation of chemical shifts to infinite dilution has often been used to obtain data free of the influences of solute–solute interaction. Nuclear magnetic resonance studies of ring systems are carried out normally in solvents as nonpolar as possible, with CCl_4 and $CDCl_3$ probably being the most widely used. In almost all cases, a small downfield shift of all protons occurs when the solvent is changed from CCl_4 to $CDCl_3$ (approximately -0.1 ppm) and if this is taken into account, meaningful comparison can usually be made between spectra of compounds in these two solvents.

As solvents of higher polarity are used, specific solvation effects become more and more important. The order and magnitude of shieldings within a ring system can often be varied enough by change of solvent so that peaks overlapping in the first solvent are clearly resolved in the second. Such techniques have obvious uses in the interpretation of complex spectra. Thus, a change of solvent from CCl_4 to methanol, acetone, DMSO, or D_2O causes a general but unequal downfield shift of all ring protons in heteroaromatic systems. These polar solvents strongly solvate the dipolar lone pairs of these compounds, and the downfield shifts of protons α to the lone pairs in π-deficient nitrogen heterocycles are usually less than those exhibited by other protons in the ring. This effect can be ascribed to a modification of field from the lone-pair dipole in such a manner as to partially cancel the solvent shift of the α-protons. However, in the case of indole, change of solvent from CCl_4 to DMSO causes a large shift of the α-proton and a small shifts of the β-proton, and this must be tied up with the tendency for this and other π-excessive systems to protonate on carbon.

By far the most widely studied solvent shifts are those induced by benzene. For most heterocyclic systems the observed shifts are upfield and the shifts for protons in different positions in a ring system are different enough and constant enough to be of diagnostic use in structural determinations. The specific nature of these shifts is discussed or referred to in the relevant sections of the book.

It is very interesting to note that a change of solvent can greatly modify substitution shifts. For instance, a 4-methyl group induces a shift of about +0.15 ppm in the α-resonance of pyridine when pentane is used as solvent, but of −0.01 ppm when DMSO is used. Many examples of this type of behavior can be found in the tables in other chapters. Care must be exercised if substituent shifts obtained for compounds in one solvent are to be applied to problems where another solvent is used.

INDEX

In the main, this index lists the heterocyclic systems rather than individual compounds mentioned in the text or tables. Naturally, all entries refer to some aspect of nuclear magnetic resonance. The letter *f* after a page number indicates that the subject is treated on that and the following page or pages.